Handbook of FABRICATION PROCESSES

Handbook of FABRICATION PROCESSES

O. D. Lascoe
Professor Emeritus, Industrial Engineering
Director, Manufacturing and Engineering Laboratory
Purdue University

ASM INTERNATIONAL
Metals Park, Ohio 44073

First printing, April 1988
Second printing, September 1989
Third printing, February 1994

Library of Congress Catalog Card Number: 87-71947
ISBN: 0-87170-302-5
SAN: 204-7586

Editorial and production coordination by
Carnes Publication Services, Inc.

PRINTED IN THE UNITED STATES OF AMERICA

In memory of

O. D. Lascoe

(1912–1988)

Acknowledgments

The author has expressed his appreciation to the following individuals and companies for their contributions to this handbook: Harry Conn, formerly Chairman of the Board, and Ted Brolund, President, W. A. Whitney Corp.; Ford Motor Co.; Ron Fowler, Director of Metal Fabricating Institute; participants in *Plastics for Tooling* seminars held at Purdue University; Irv Poston, General Motors Corp.; Niagara Machine & Tool Works; The U.S. Baird Corp.; Wysong & Miles Co.; Extrude Hone Corp.; Danly Machine Corp.; Verson Allsteel Press Co.; Welty-Way Products, Inc.; E. W. Bliss Co.; E. F. Houghton & Co.; J. A. Richards Co.; Durable Punch & Die Co.; Richard Nordlof, Mechanical Tool & Engineering Co.; Federico Strasser, Bell & Howell Co.; Uni-Punch Co.; L. E. Winter; C. M. Head, Manufacturing Research; Harold P. Schick, General Tire & Rubber Co.; Robert H. Goodale and Edward Ferrari, General Electric Co.; Richard S. Vanker, Pioneer Engineering & Manufacturing Co.; John Delmonte and R. J. Hough, Furane Plastics; B. L. Harrison, Hamilton Standard Division of United Aircraft; J. J. Janush, Wheelabrator Corp.; and to his graduate students who conducted many applied research projects on materials processing.

Preface

Sheet-metal fabrication processes are receiving greater attention and are more widely applied by the metalworking industries because of cost and material savings. This book compiles proven theories and operations tested in industrial applications.

Many texts and reference books have been written about chip-producing machine tools and the processes for producing preworked shapes. This volume focuses on the non-chip-producing machine tools that shape metals by shearing, pressing and forming. New materials and advances in tooling are discussed, as well as the need for applied science in optimizing the operations for sheet-metal fabrication processes. Examples of each forming process are given. The text also describes the mechanics of each process so that a logical decision can be made concerning the best operation for a specific result.

Every so often, the happy coincidence occurs that the right book is written by the right author; O. D. Lascoe, Professor Emeritus, Purdue University, is certainly considered by many to be among the foremost authorities on fabrication processes. His career spanned more than four decades. Until he was incapacitated by an accident in June 1987, Professor Lascoe was an active consultant and author, and an international expert in metal cutting, metal forming, EDM and fine measurements. [*Editor's note*: Professor Lascoe died in February 1988.]

O. D. (Orville Dwayne) Lascoe was born in Sibu, Romania, in 1912. His family immigrated to the United States when he was 12 years old and settled in Gary, Indiana. While his father worked in the steel mills, O. D. did odd jobs and went to school. After finishing high school and working as a machinist and diesetter in Indiana and Kentucky, O. D. enrolled in Western Kentucky University, where he obtained a Bachelor's Degree in 1938. He also studied at Purdue University, where he obtained a Master's Degree in 1945.

With his work experience and academic degree he joined the staff of the machine shop in the department of General Engineering at Purdue University in 1942. In 1955 the machine shop was attached to the newly formed Department of Industrial Engineering and renamed the Machine Tool Laboratory. A few years later, the Department of Industrial Engineering became the School of Industrial Engineering, and O. D. became the Director of the Manufacturing Engineering Laboratories, a title he held until his retirement in June 1978.

When Professor Lascoe took charge of the machine shop at the end of World War II, the machines in the shop were still run by overhead line shafts and leather belts and the primary illumination came from skylights and windows. O. D. redesigned the entire laboratory. He traveled about 40,000 miles in search of suitable equipment for the laboratory and was very selective about picking machines available from government war surplus. It has been estimated that he secured about $2,500,000 worth of tools and equipment. When the remodeling of the laboratory was completed in 1946, Purdue had the finest machine-tool laboratory of any university in the United States. In addition to the machine-tool laboratory, a precision-measurement laboratory (Naval Gage Laboratory) had been installed and equipped with instruments provided by the U.S. Navy.

When Professor Lascoe retired in 1978, the Manufacturing Engineering Laboratories were equipped with 190 pieces of major manufacturing equipment ranging from conventional machine tools and presses to CNC machines, EDM, ECM, Electroshape, and testing and measuring equipment.

During his tenure at Purdue, Professor Lascoe was an undergraduate teacher and researcher. Thousands of engineers credit him for a considerable part of their formal education, including Neil Armstrong and Gus Grissom. He supervised 13 Ph.D. researches and 45 Master's theses.

He designed and directed 94 conferences, seminars and training programs for industry (Chrysler Corp., TRW, Shell Oil, Gulf Oil) and for technical and trade societies such as the Society of Manufacturing Engineers (SME), Society of Tool and Carbide Engineers (STCE), Society of the Plastic Industries (SPI), Numerical Control Society (NCS), and National Machine Tool Builders Associations (NMTBA), to name a few. It is estimated that 17,000 people outside the university have benefited from his conferences.

During his years at Purdue O. D. received numerous awards and citations:

— SME National Education Award, 1963
— SPI Man of the Year Award, 1972
— Automation, Citation of Honor, 1962
— Citation from President Yen-Chia-Kan, Republic of China (Taiwan), 1975
— Society of Carbide Engineers (Service Award)
— Best Teacher Award, Schools of Engineering at Purdue

On his retirement, two scholarships were set up in his name: The O. D. Lascoe Award for Excellence in Manufacturing Engineering (for engineering seniors), sponsored by the W. A. Whitney Corp.; and the James Danly–O. D. Lascoe Award for the outstanding junior student majoring in Manufacturing Engineering, sponsored by the Danly Company.

Professor Lascoe was a member of a number of professional societies and held membership in the following honorary societies: Sigma Xi (Research); Tau Beta Pi (Engineering); Pi Tau Sigma; Iota Lambda Sigma (Industrial Education); and Alpha Pi Mu (Industrial Engineering).

Some of his contributions to the advancement of knowledge in his field were as follows:

— Represented U.S. Department of Commerce as Machine Tool Consultant, Pozan, Poland, 1960
— Machine Tool Consultant, National Science Aid Program, Alahabad, India, 1970
— Represented U.S. Department of Commerce as Machine Tool Consultant in Korea, Hong Kong and Taiwan, 1972
— Industry and University Survey for National Science Council, Taipei, Taiwan, 1974
— Consultant to National Machine Tool Builders Association in training programs and product-liability lawsuits, since 1949
— Consultant in Manufacturing Engineering for General Electric, Emerson Electric, Chrysler Corp., General Motors, TRW, etc.

Besides publishing numerous technical articles in academic and trade magazines, Professor Lascoe authored a textbook on machine-shop operations and setups for the American Technical Society. He also was a contributor to the *Die Design Handbook*, published by ASTME; the *Production Handbook*, published by Ronald Press; and the *Grey Iron Handbook*, issued by the Grey Iron Foundry Association.

One of Professor Lascoe's best traits was that he was very much in tune with the needs of his students in and out of the classroom. The barbecue parties at his home for his students in general, and the local SME students chapter in particular, were yearly events. He was always viewed by his students as a father figure as well as an excellent teacher. This book is, in so many ways, a tangible record of his legacy.

Joseph ElGomayel
Purdue University
West Lafayette, IN
January 1988

Note: On his retirement in 1978, Professor Lascoe entrusted the management of the Manufacturing Engineering Laboratories to his former student and assistant, Dr. Joseph ElGomayel, currently Professor in charge of the Laboratories. The editorial staff of ASM International wishes to express their gratitude to Dr. ElGomayel for his thorough review of the galley proofs after Professor Lascoe's devastating accident. We would also like to thank Professor Lascoe's daughter, Donna Maltony, who in a period of great difficulty for herself and her family, took the time to answer questions, find missing artwork, and locate an excellent reviewer.

Sunniva Refsnes
ASM International

Contents

Section 1: Fabricating Processes and Equipment

Section 2: Stamping and Forming Operations

Section 3: Plastics for Tooling

Section 4: Structural Shapes

Section 5: Nontraditional Machining

Section 6: Definitions: Design Considerations

Handbook of FABRICATION PROCESSES

Section 1
Fabricating Presses and Equipment

1A: Press Construction and Application

Although the modern press (Figure 1A-1) does not have the glamour and sophistication of the numerically controlled machine tool, it is a fine precision machine tool bearing only superficial resemblance to its big, heavy, cumbersome predecessor (Figure 1A-2). Today's press, the backbone of mass production, is built with meticulous attention to accuracy, clearances, parallelism, and deflections, and sometimes even has sophisticated electrical controls.

The biggest reason for this change has been a dramatic increase in the degree of sophistication of today's tooling. The intricate dies that run in modern presses require infinitely better guiding and deflection characteristics than those available only a few years ago. Such dies have been made possible through the use of precision diemaking machines working with better-grade die steels, carbide inserts, etc., which give them long life and extremely close clearances. Dies of this nature are expensive and demand a precision press for satisfactory operation. One example of today's tooling is the progressive die, which combines many operations in one press to produce a finished part with each stroke. This, coupled with automatic feeding and ejection equipment, leads to greater press speeds for more production at less cost per piece.

To select the proper press for a particular job, one must have an intimate knowledge of the work to be performed and be capable of selecting a press of compatible design. This requires the ability to analyze various types of press construction, because there are many specialized types of presses on the

Figure 1A-2. Outdated press

Figure 1A-1. Modern press

market today, each with relative advantages for certain applications.

To enumerate, the conventional open-back inclinable (OBI) gap press is available with a mechanical clutch (Figure 1A-3), a friction clutch (Figure 1A-4), or a positive pneumatic clutch (Figure 1A-5), and with either one or two connections. Standard-speed and high-speed OBI's can be used in conjunction with automatic feeds for high-production work. Horn, adjustable-bed, and deep-throat presses are also available for

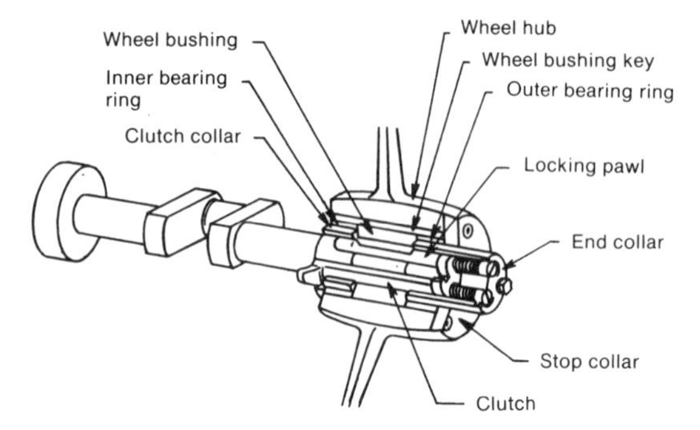

When press is tripped, clutch rolls into engagement with wheel, locking shaft and wheel together.

Figure 1A-3. Crankshaft equipped with mechanically actuated positive clutch

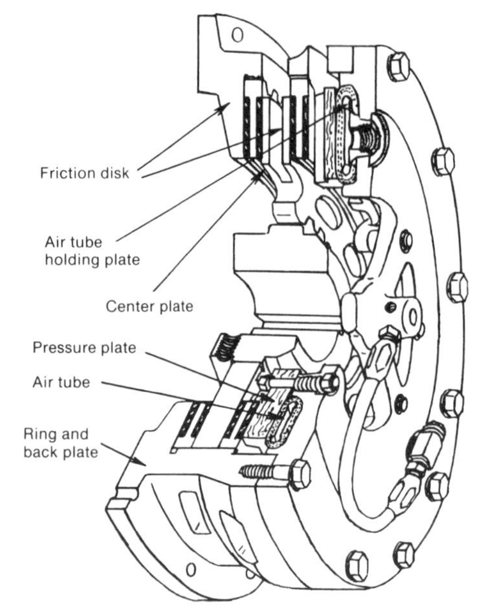

Axial pressure exerted by air tube forces pressure plate against friction disk to engage clutch. When air is released, clutch disengages.

Figure 1A-4. Disk-type friction clutch

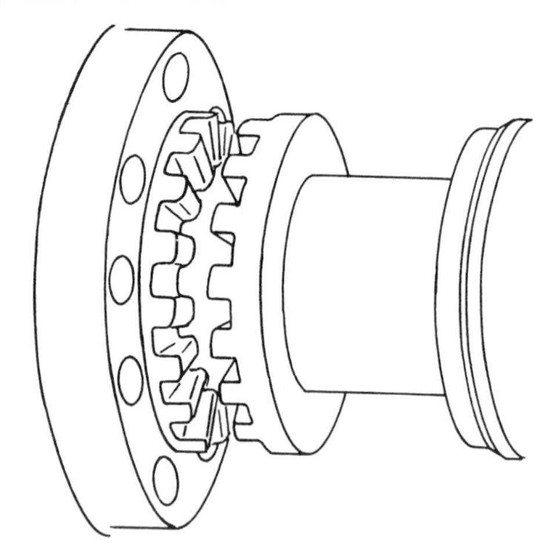

Pressure exerted by pneumatic chamber on sliding sleeve engages jaws. Disengagement is done by springs.

Figure 1A-5. Widely used type of pneumatically actuated positive clutch

specialized requirements. Drives for straight-side presses can be broken down into three classifications: crankshaft, eccentric gear, and eccentric-shaft (Figure 1A-6). Major builders of presses today have specialized in that they are building straight-side machines for particular applications. In other words, some are designed and built primarily for certain types of jobs and can fall into the high-production category (or heavy-duty blanking). It is important to have an intimate knowledge of the type of press for the work to be performed so that the proper selection can be made. There are so many details which are important when selecting a press, such as the type of electrical controls, auxiliary devices, feeds, etc., that sometimes the basic machine is overlooked.

There are three basic components in a press — namely, the frame, the drive, and the gibbing. If these components are properly designed for the required application, chances are that the press selected will be correct for the job. It is extremely important to remember that these areas must be analyzed thoroughly before an intelligent selection can be made.

PRESS FRAMES

There are two major types of press frames: the straight-side frame and the gap (OBI)

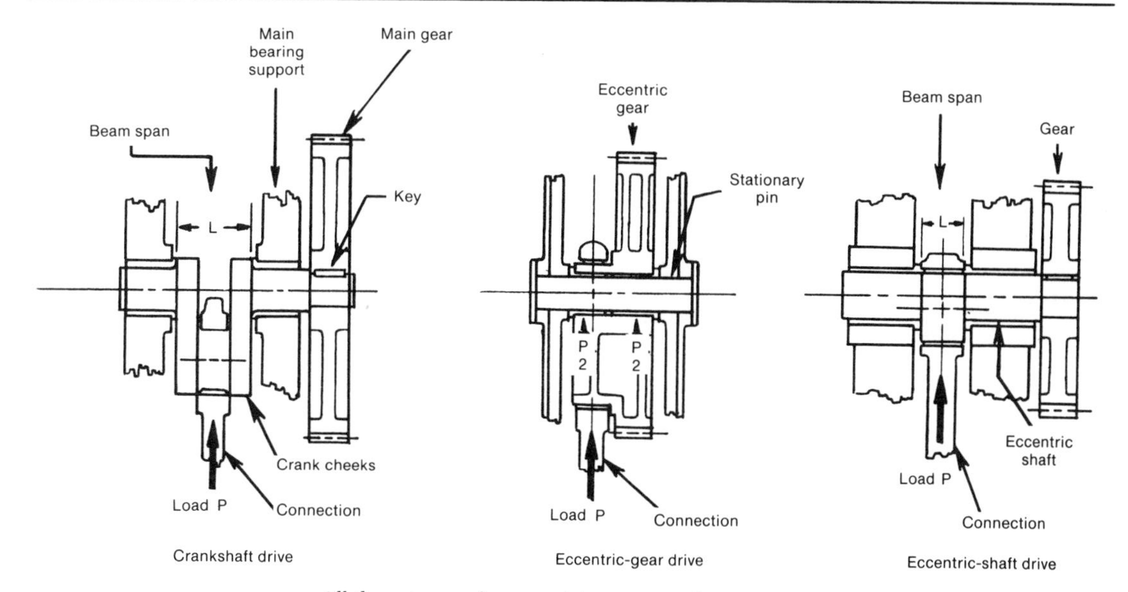

All three types of eccentricity convert the rotary motion of the shaft into a reciprocating motion of the slide. Each type has its own useful characteristics.

Figure 1A-6. Cross sections of crankshaft, eccentric-gear, and eccentric-shaft press drives

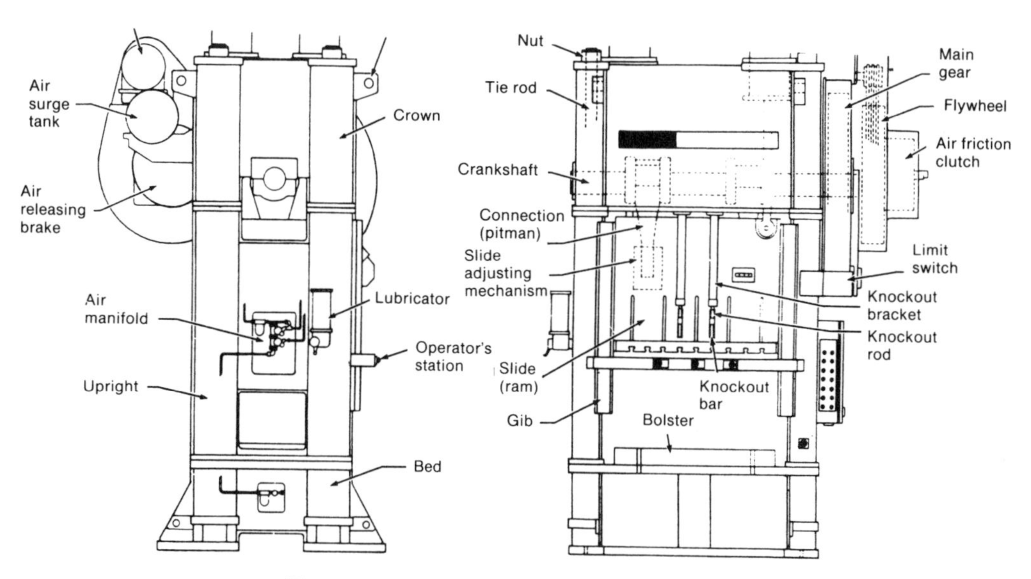

Figure 1A-7. Typical straight-side press frame

frame (Figures 1A-7 and 1A-8). The basic job of the frame of the press is to contain the load and control the deflection. It follows directly that the bed and slide members of the press, onto which the die shoes are fastened, must exhibit a similar degree of rigidity to back up the die shoes when under load. If the press members deflect excessively under full load, the die shoes will also deflect, following to some degree the deflection of the press members. When this occurs, both upper and lower die elements are distorted. As a result, the finished stamping may not come up to full expectations because the center of the

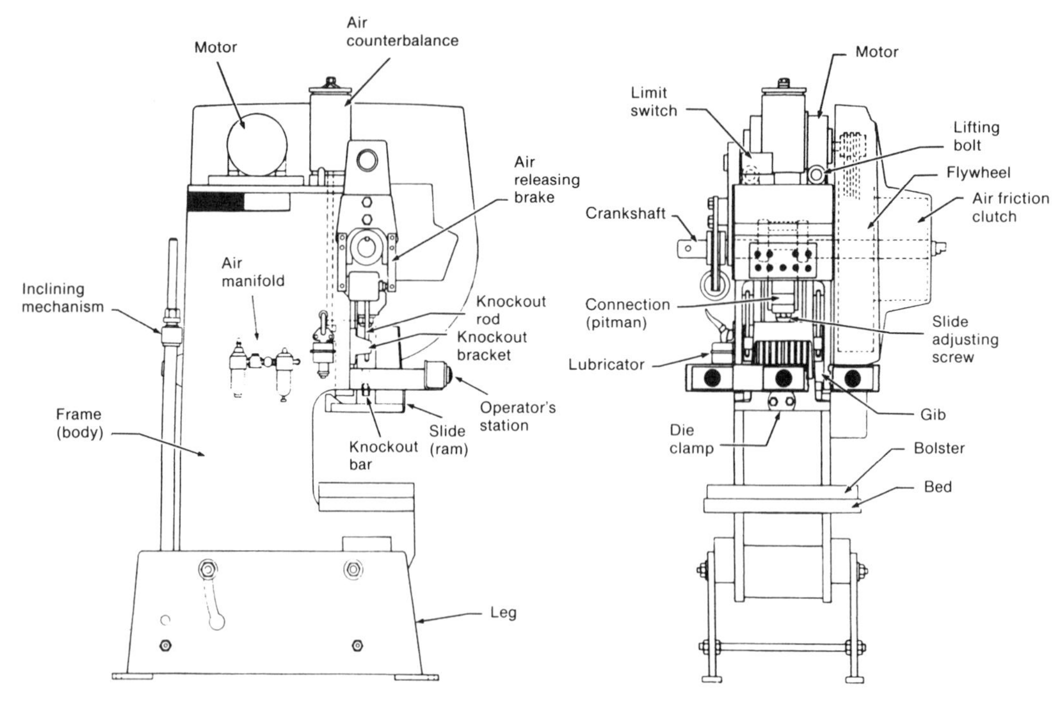

Figure 1A-8. Typical OBI (gap) press frame

die has been deflected away from its normal position.

Since all materials and structures follow the laws of elastic deformation, both builders and users have recognized the impossibility of building a press which has zero deflection (nor would such a press be desirable).

There are two basic types of materials used in the construction of press frames — namely, steels and cast irons. You will note in Figure 1A-9 a curve for cast iron superimposed on the curve for steel. The stress design factor used by reputable builders is relatively low, and consequently there does not appear to be much difference in the deflection of the component parts, whether they are made of steel or of cast iron. However, one of the big advantages of steel is the straight-line relationship to the yield point. This provides a big safety factor should the machine be overloaded. Another advantage of steel is that it is basically easier to repair than cast iron. The steel plate used in presses comes as a homogeneous rolled plate, which eliminates the possible hazards of blowholes and slag inclusions sometimes found in cast frames. Steel fabrications also lend themselves to the utilization of uniform cross sections, which cut down stress raisers that might lead to trouble later on. Basic engineering sections can be used in the design of fabricated presses, which allow the use of standard formulae with empirical relations that closely resemble the actual deflection characteristics of the frame. This, of course,

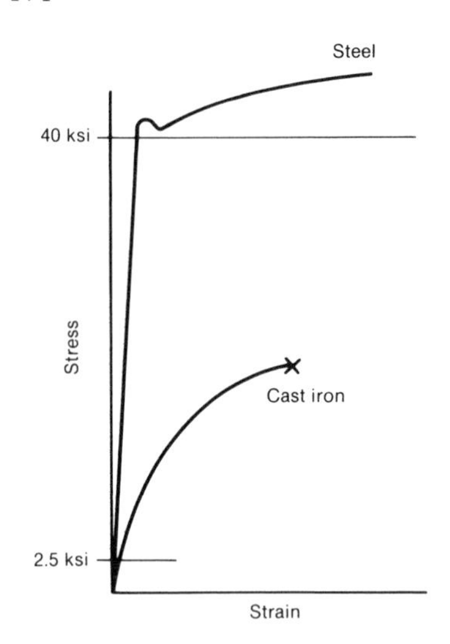

Figure 1A-9. Stress-strain curves for steel and cast iron

is an important point when designing a press for the very-low-deflection characteristics required in many cases with carbide-type tooling. Generally speaking, however, fabricated steel machines do not weigh as much as cast iron machines, since with fabricated steel we can take advantage of standard structural shapes to provide rigidity where it is required. It is not necessary to have the excessive mass found in a cast iron structure. Fabrication has become more and more common in recent years, and most press builders will quote on fabricated machines. It is, however, extremely important that the big weldments which form the basic component of the frame be thoroughly stress relieved. Also important in conjunction with stress relieving is to have the frame grit blasted to remove objectionable scale and provide a good base for primer paint.

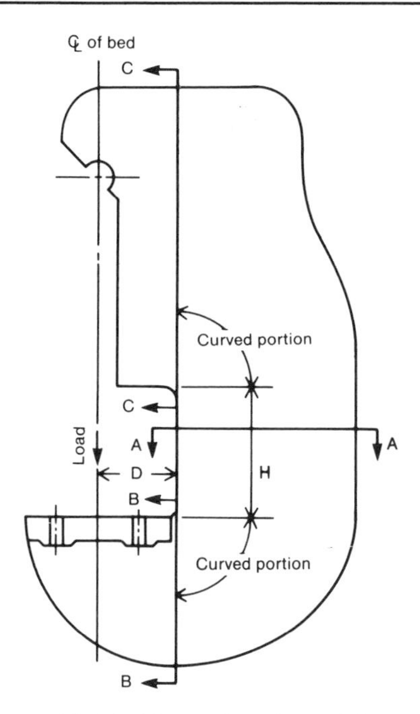

Figure 1A-10. Typical right-to-left shaft-inclinable press frame

Gap Frame

The gap-frame press in reality is like a C clamp. As the load is imposed, the mouth of the C will have a tendency to "alligator," or open up. Gap-frame presses feature versatility of application and material handling. Their most distinguishing feature is unobstructed die space accessible from all sides except the rear. In some cases, even the rear is accessible.

Because of their extreme versatility, gap-frame presses are used for a great variety of operations such as piercing, blanking, and drawing.

Standard gap-frame presses are built with one-piece frames. Designing these presses with the proper rigidity is, therefore, considerably more complex than for presses with straight-side frames.

Figure 1A-10, which shows a typical right-to-left shaft-inclinable press frame, indicates some significant structural points. Combined bending and direct stress loads are encountered in both the upper and lower cantilever beams (C-C and B-B). These loads extend forward the distance D to support, respectively, the main bearing structure above and the press bed below. It is noteworthy that the upper cantilever beam has a section C-C which is very deep in relation to the throat, D. Therefore, deflection in this upper cantilever beam is so small that it can be neglected in calculations.

The total deflection of the gap area is made up of both vertical and angular deflection and is a function of gap height and depth. The *angular deflection* is the more important of the two as far as the press user is concerned, because it may result in misalignment between punch and die.

When a gap-frame press is fabricated from steel plate, the side plates which form the uprights of the frame have a uniform cross-sectional width and thickness. The strength of the frame might simply be described as the depth of this plate times the thickness at the throat, which gives us an areal measure. The thicker and deeper the plate, the stiffer the frame can be. Sometimes on larger gap-frame presses a reinforcing rib is welded around the throat to increase the resistance to deflection.

In designing a line of OBI machines from the smallest to the largest it is important to have a frame properly designed for each machine tonnage. This is called "controlled deflection," which means that, at maximum tonnage, if a perpendicular is dropped from the slide and from the bed, these lines will intersect at a point approximately at the stock-line height. This then helps to eliminate tooling wear due to the angularity or alligator effect in the machine. It is also interesting to note that properly designed OBI's have their bearings in the frame set on a 45-degree angle. This, then, transfers the shaft load directly to the frame and eliminates loading the bolts that attach the bearing caps.

Press builders advocate selection of gap-frame presses of sufficient capacity to perform operations without having to resort to tie rods for reinforcement. This does not imply that a gap-frame press cannot be used safely up to its rated capacity. However, where angular deflection may be a determining factor for good die life, such as in the use of large-area blanking dies, conservatism is suggested in selecting a press.

Straight-Side Frame

The straight-side frame has the advantage over the gap frame in that it eliminates angular deflection. Most straight-side frames are four-piece frames, consisting of the crown, two uprights, and the bed. As with OBI's, straight-side frames can be made from cast iron or steel.

Basically the straight-side frame has to resist "oil-canning" and twisting. With progressive dies, the load can be, and quite often is, on one side or the other of the centerline. This then imposes a heavier load on one side of the press. Such off-center loads can create unusual conditions in the frame which must be resisted. In a cross-sectional view of the frame can be seen a central plate which runs the full length of the uprights as well as the length of the crown and bed. To this plate, the gibs are attached. The instant that the press load is applied, this load has to be transferred from the tooling to the slide and through the gibs to the frame. At this instant, the gibs are what wire is to electricity — a means of conveying the load. It is now the job of the frame to do something with the load. To eliminate twisting, a triple-box construction is used (Figure 1A-11). Continuous welds running the full length of the press provide high resistance to twisting under load.

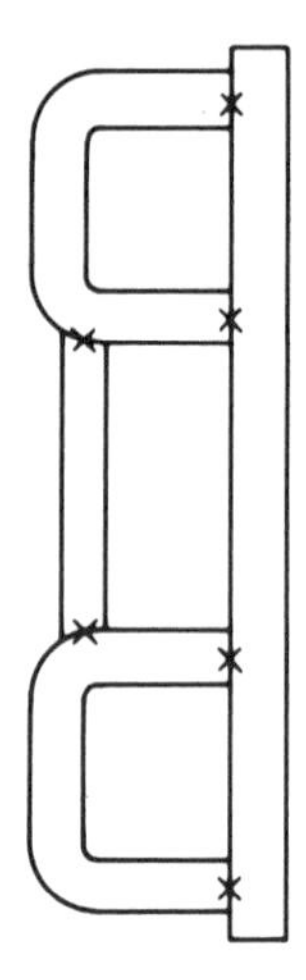

Figure 1A-11. Triple-box straight-side press frame

The type of work has a great bearing on how stiff the press bed should be. For instance, when carbide tooling is used with close clearances, it is generally felt that it is important to have a stiffer bed to support the lower die shoe. This in turn extends the life of these very expensive dies.

Any discussion of straight-side presses would be remiss if it did not cover the function of the tie rods, since straight-side four-piece frames are held together with tie rods.

The relationship among tie-rod stress, frame stress, and press load in a 200-ton press is shown in Figure 1A-12. Line AB shows the increase in tie-rod stress under increasing press loads. Line AD shows the decrease in frame stress under increasing press loads. Point A indicates the stress in the tie rods and frames as being equal when there is no work load in the press. Point B indicates the stress in the tie rods when there is no compression left in the frame. (This is the "lift-off" point. The press work load equals the tie-rod load.) Point D indicates zero compressive stress in the frame at the "lift-off" point. Point C indicates the tensile stress in the tie rods at the maximum rated tonnage of the press. The distance CF indicates the press load at 100% capacity. Point F indicates the compressive stress in the frame at the maximum rated tonnage of the press. Line EB shows what the tie-rod loads would be if the tie rods were not shrunk. BH shows the rate at which tie rods will yield beyond the "lift-off" point.

Let us follow what happens to the tie rods and press frame as we shrink the rods. With the rods assembled through the frame parts, the tie-rod nuts are brought into firm contact with the crown and bed surfaces. The rod is then heated by any one of several accepted methods. As the temperature of the rod increases, its length increases and the lower tie-rod nut moves away from the lower bed surface. When the rod has been sufficiently lengthened by the application of heat, the lower nut is tightened a predetermined amount. As the rod starts to cool, it tends to shrink to its original length, but is prevented from doing so by the semirigid frame parts between the tie-rod nut faces. If the frame parts were absolutely rigid — i.e., if they would not shorten (compress) under the load — the tie rod, when it had cooled off completely, would be stressed in tension to 145 MPa (21 ksi). The frame, of course, would be stressed in compression to a load equal to the tie-rod load.

But since the frame parts are not absolutely rigid, and will shorten somewhat under load, the final load equilibrium point between the tie rods and frame will be somewhere between zero and the "lift-off" point. The exact point at which this equilibrium will occur is dependent upon the relative stiffness (area and modulus of elasticity) of the tie rods and the frame parts. In general-purpose press design, this point is targeted to fall somewhere between 90 and 110 MPa (13 and 16 ksi) in the tie rods. This is, therefore, the residual stress in the tie rods with a proportionate stress in the frame parts under a no-load press condition. The tie rods are in tension and the frame parts are in compression. As a work load is developed in the press, the stress in the tie rods increases while the stress in the frame parts decreases. This change varies directly as the press work load. Figure 1A-12 shows a simple graphical representation of the relationship among tie-rod stress, frame stress, and press load. The distance GA on the diagram is the stress adjustment in the tie rods, resulting from the shortening of the frame structure under the load developed in shrinking the tie rods.

Under full-load press conditions, the tie rods will exhibit a tensile stress of between 117 and 131 MPa (17 and 19 ksi). This stress is

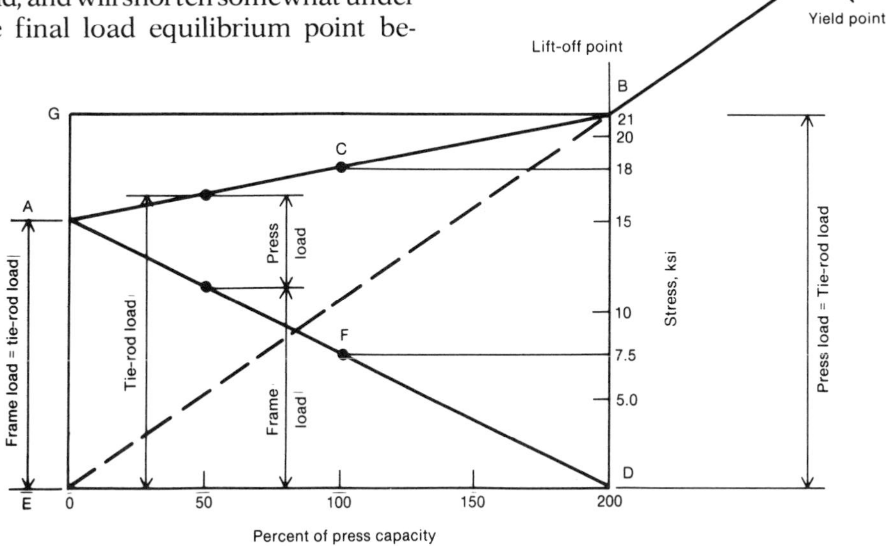

Figure 1A-12. Relationship among tie-rod stress, frame stress, and press load in a 200-ton press

made up of two components: a stress between 45 and 59 MPa (6.5 and 8.5 ksi) holding the frame parts in compression, plus a stress of approximately 72 MPa (10.5 ksi) from the press work load. (Refer to Points C and F in Figure 1A-12).

In order for the crown to lift off the uprights, the frame parts must decompress completely and return to their original height. For this to occur, it would be necessary to stretch the tie rods to the length they had attained while they were hot prior to their shrinking. It is, therefore, apparent that at the "lift-off" point, the load in the tie rods must equal the work load in the press. This commonly is set by design at 150 to 200% of press capacity. Or, stated in other terms, it would require a 50 to 100% press overload to lift the crown from the top of the uprights. Note also in Figure 1A-12 that the tie-rod stress fluctuation is on the order of 21 MPa (3 ksi) — from 103 MPa (15 ksi) at no load to 124 MPa (18 ksi) at full load. Such a stress fluctuation is well within the endurance limits of the types of steels from which tie rods are made, and long tie-rod life may be anticipated provided the rods are properly shrunk.

While the frame members will not separate at their joint surfaces under normal loads, the distance between the bed and the crown will nevertheless increase slightly from no-load to full-load conditions.

It might, at this point, be in order to point out the fallacy of overdesigning press frames for rigidity. It is not generally realized that a too-rigid press structure can be as harmful and troublesome as a too-flexible structure. There are, as a matter of fact, a few cases in which frame structures had to be weakened before satisfactory productivity, press life, and die life were possible. The design of a frame is not wholly dependent on the press tonnage. The type of machine, the class of work for which it is intended, the type and construction of the dies which are contemplated, and even the speed of operation all have some bearing on frame-structure design. A frame that is too unyielding can be the cause of very severe damage to dies or to other parts of the press solely because of the commercial tolerance of the stock being run through the press. The tie rods on all presses serve as some protection against faulty die setting, doubles, and other inaccuracies, but they do not provide safety against breakage of press or die components.

PRESS DRIVES

The drive on a press is made up of the motor/flywheel combination, clutch and brake, and gearing, if used (Figures 1A-13 and 1A-14). Since an understanding of the function of a drive is a prerequisite for an understanding of the press itself, four basic terms used in connection with the drive will now be defined:

Clutch: A means of efficiently transmitting and controlling the energy made available mechanically (rotating flywheel).

Torque: A twisting effect that produces rotation. The clutch has to create the twisting effort to set the drive in motion.

Tonnage: The rated capacity of the press. The pressure the ram will safely exert at or near the bottom of the stroke. Depends on the physical size of the press and its components.

Energy: The ability of the flywheel/motor combination to produce, store, and deliver power. The motor power is stored in the form of flywheel energy by the heavy flywheel rotating at high speed. This energy is then used to move the rated tonnage through a distance. The product of this tonnage and distance is the energy load on the press (inch-tons). The main function of the motor is to bring the flywheel back up to speed when energy is absorbed by work being done in the machine.

Types of Clutches

The function of the clutch is to deliver and control the surge of force that is required for the working of metal. When the press is running continuously, the function of the clutch is to transmit torque from the flywheel to the crankshaft or eccentric. If the

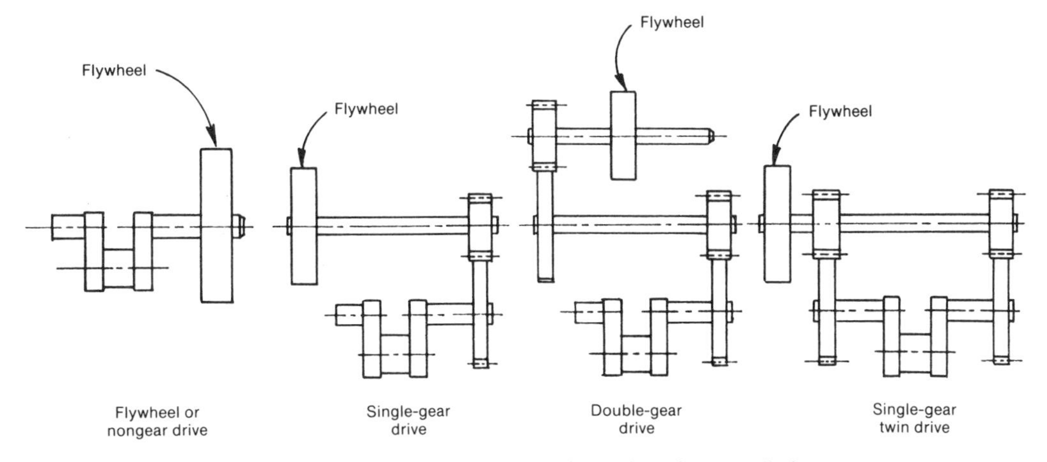

The first three press drives, from the left, are called "single-end drives" because the crankshaft is driven from one end only. When the crankshaft is driven from both ends, the drive is termed a "twin drive" (far right).

Figure 1A-13. Types of press drives

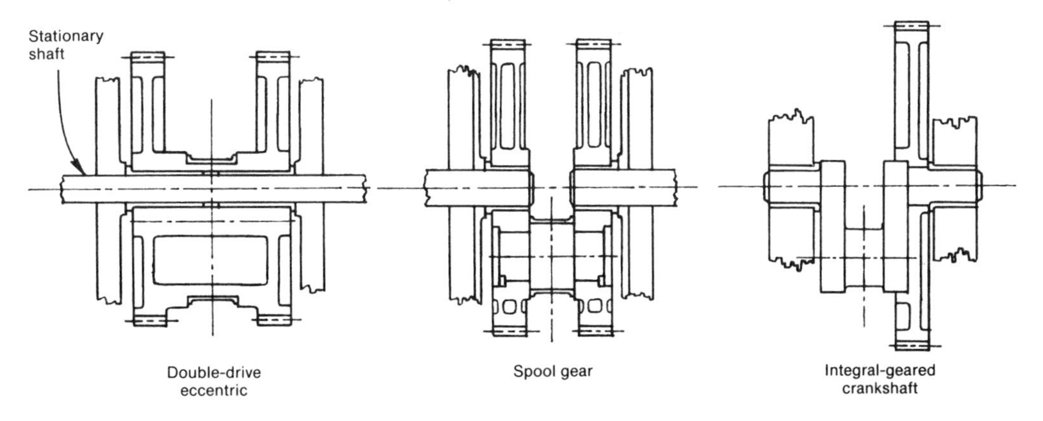

Variations of eccentric-gear drives include the double-drive eccentric, the spool gear, and (at far right) the crankshaft with sides of cheeks keyed directly to the main gear.

Figure 1A-14. Eccentric-gear drives

press is single-stroked, the clutch must accelerate the gears and other rotating members from zero to their full operating speed, as well as transmit the required torque. These rotating members must be accelerated for each stroke of the press. The brake must decelerate these rotating parts in order that the slide may be stopped. Thus, the selection of the proper clutch is dependent upon two completely different factors:

1. The torque required to deliver adequate force to the slide
2. The acceleration of parts that must be started at each stroke of the press.

The selection of the brake is dependent upon the deceleration of the moving parts that must be stopped at each stroke of the press. The brake must be of sufficient size to stop the slide and dies for emergency or inching operations.

There are two basic types of clutches:

1. Positive
2. Friction.

Positive Clutches

Positive clutches are used on many small and medium-size presses, particularly on gap-frame types. Positive-engagement clutches are either mechanically or pneumatically actuated. Mechanical actuation is ordinarily done by means of a foot treadle, although the treadle can be replaced by an air cylinder.

All positive clutches engage keys, pins, or multiple jaws. Mechanically operated units can be engaged and disengaged only once for each stroke. They use some type of throwout cam or surface to disengage the keys, pins, or jaws near the top of the stroke. The foot treadle releases the throwout cam and springs engage the driving members to start the press. An auxiliary cam may be used to release the foot-treadle or air-cylinder linkage at a given point of the stroke. Then the press will single stroke even if the treadle linkage is kept actuated. This single-stroking mechanism can be disconnected for continuous operation. The clutch will then remain in engagement as long as the treadle is depressed. These clutches use a drag friction brake which is usually on and not released during press operation.

Mechanically operated clutches are compact in size and are usually characterized by simplicity of operation.

By comparison to friction clutches, which accelerate the press-drive members by slippage of friction surfaces, the acceleration of the drive members by a positive clutch is very rapid. However, it must be remembered that these clutches are always mounted on the crank or eccentric shaft. Therefore, the inertia of the parts that must be started on each stroke of the press is kept relatively low.

Friction Clutches and Brakes

Most friction clutches use air pressure to exert force to clamp friction surfaces together, and the brakes use springs to exert this force. Brakes are air-released. The brakes are spring-operated rather than air-operated so that the press will stop in event of a power or air failure. This type of clutch provides relatively shock-free acceleration and deceleration of the press-drive parts. Friction clutches may be mounted on either the crank or eccentric shaft, the intermediate gear shaft, or the driveshaft. Such factors as press size, press speed, how fast the press is to be single-stroked, type of clutch-brake, and proper attention to the inertia of the press drive determine the location of the clutch.

Two basic types of friction clutches are manufactured: the combination clutch-brake unit in which all parts are assembled on one common sleeve mounted on the shaft; and clutch and brake units separately mounted on the shaft with a single valve for synchronization. As the name denotes, the combination clutch-brake is designed as one unit. A common air chamber is used to operate the clutch and release the brake. This means that the clutch and brake are mechanically synchronized. This design inherently means that most of the clutch-brake parts must be started and stopped each time the press is single-stroked, and is a high-inertia design.

With the advent of automation and the much faster single-stroking rates of presses, it became necessary to reduce the stored energy of the parts to be started and stopped by using low-inertia clutches and brakes which are separate units.

Torque Control

One of the big plus factors of the modern friction clutch is that it controls torque through air pressure. The torque limitations of a press are determined by the size of the drive shaft, clutch, gears, and crankshaft. A torque load can be imposed when a load is encountered on the upstroke. The press drive has to transmit flywheel energy in the form of torque to the slide. Too great a torque load will cause failures of the torque-carrying members, such as shearing of keys, breaking of gear teeth, and excessive wear of clutch members. The clutch is one link in this torque-transmission chain which can be controlled and which limits torque overload on the rest of the system. If more torque is required than the clutch can deliver at a rated air pressure, it will slip. This is an important feature of the modern-day press and should be understood in order to take advantage of it in the field.

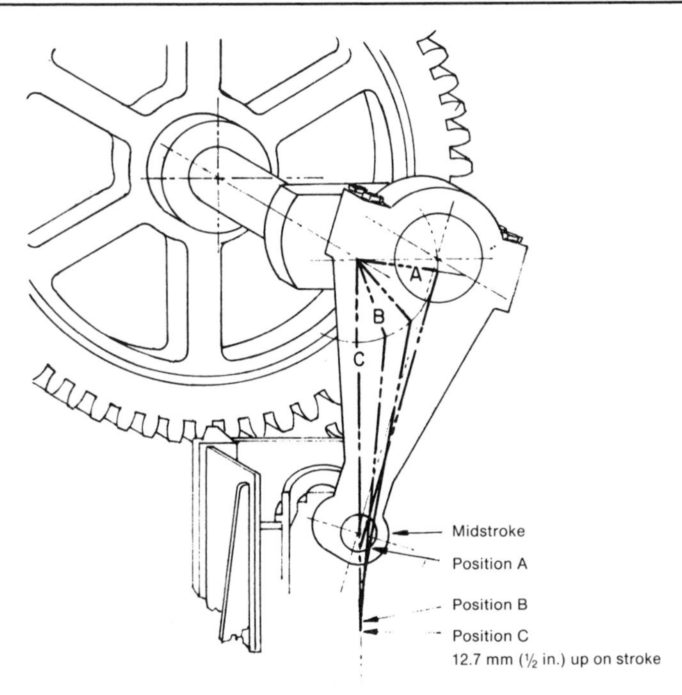

The torque capacity of an eccentric press varies with the length of the moment arm, C. This arm becomes longer and longer as the press moves higher on the stroke, as shown by arms B and A.

Figure 1A-15. Tonnage-stroke relationship

Tonnage

Tonnage is the force in tons exerted by the press against a workpiece. The amount of force or tonnage that the press can exert is a function of the torque it develops and the strength of its component parts (Figure 1A-15). The tonnage of a press is a function of the size of its physical parts. They must be adequate for the loads applied. A properly designed press has a controlled deflection since all parts will deflect under load.

Energy

Energy is probably one of the most misunderstood facets of a press. We are so used to referring to tonnage that, in many cases, we either neglect or ignore the energy requirements of the machine. Presses should be rated for energy as well as tonnage. All jobs require tonnage, and this tonnage has to be pushed through some distance. Multiplication of tonnage by distance gives the energy, in inch-tons, required to do the work. This energy comes from the motor/flywheel combination. When the press does work, energy is taken out of the flywheel, causing it to slow down. It is the job of the motor to bring the wheel back up to speed before the next stroke is begun. Blanking jobs usually require very little energy since the tonnage is actually being pushed through a very short distance. As opposed to this, drawing operations require a great deal of energy because the metal is probably drawn through a sizeable distance. High-speed presses are, by nature, generally flywheel-type machines and are used in blanking operations. In some cases the energy is taken out of the flywheel at such a high rate that the flywheel is actually of limited value, and it is necessary to use a motor large enough to directly supply the energy required for the work. In these cases, the flywheel becomes a sheave and not a storehouse of energy. On the other hand, the drawing press is a slower-operating machine, is usually geared, and allows a greater length of time between each stroke for the flywheel to come back up to speed; consequently, relatively smaller motors can be used.

1B: Mechanical and Hydraulic Presses: Advantages and Disadvantages

MECHANICAL VERSUS HYDRAULIC PRESSES

The gap between the advantages of selecting a mechanical press over a hydraulic press and vice versa has been narrowed somewhat in recent years by improvements in design and the addition of features to each type that duplicate the desired actions of the other. The choice today becomes more a question of production requirements, economics, and versatility.

The selection should be based on the type of work to be performed within the press. A hydraulic press was once the only press considered for deep draw work, and a mechanical press for blanking, forming, and shallow draws. With the addition of varidraw drives on mechanical presses and larger pumps and improved controls on hydraulic presses, this no longer need be the case (Figure 1B-1).

In order to fairly equate one to the other, we will compare the advantages and disadvantages of each type of press and then compare them again with respect to job application. Note: The information presented below should be considered to be general and by no means to establish limitations of either type of press.

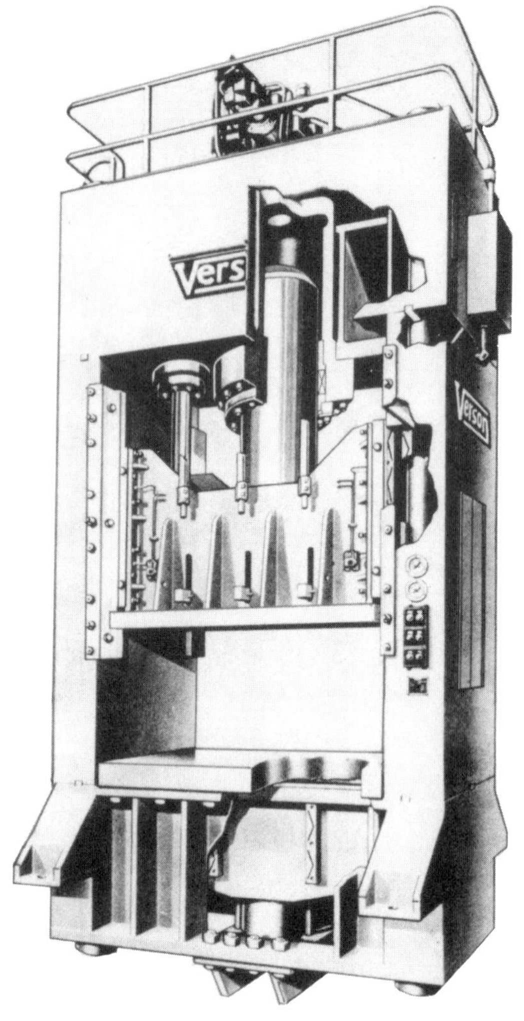

Figure 1B-1. Cutaway view of a hydraulic press

Advantages of the Single-Action Hydraulic Press (Figures 1B-2 and 1B-3)

1. The first and most important advantage is that a hydraulic press cannot be overloaded because the system is protected with two separately adjusted relief valves.

2. Full tonnage can be developed through the entire stroke, whereas on a mechanical press the tonnage capacity of the press diminishes the higher up on the stroke the work is performed.

3. The tonnage of the press is readily adjustable up to the maximum of its rating, allowing for low-tonnage operation with fragile dies.

4. The stroke is adjustable for the work being done. For example, a press with a 30-in. (760-mm) stroke can be set to operate with a 3- or 4-in. (76- or 102-mm) stroke on a shallow draw.

5. Die setting is easier because it is not necessary to adjust for material thicknesses or variations in stock.

6. The drawing speed remains constant throughout the stroke.

7. In general, for long-stroke presses, a

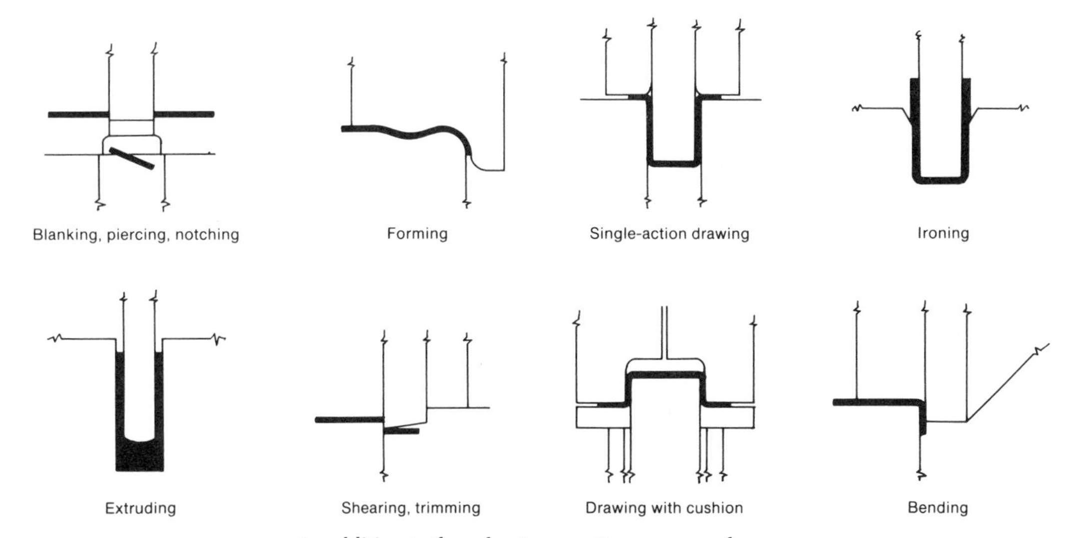

In addition to these basic operations, many others can be performed in single-action presses with special adaptions and designs.

Figure 1B-2. Typical operations of single-action hydraulic presses

hydraulic press is less expensive than a mechanical press.

Disadvantages of the Single-Action Hydraulic Press

1. A much larger motor is required as compared with a mechanical press because there is no flywheel in which energy can be stored. However, the hydraulic press does not require more electrical current than the mechanical press, because the motor operates at approximately full load only through the time of the press cycle.

2. Not generally applicable to blanking operations since the shock of breakthrough is detrimental to piping, gaskets, and press connections. However, this disadvantage has been lessened somewhat in recent years through the use of better welding techniques, manifolding, and flexible joints.

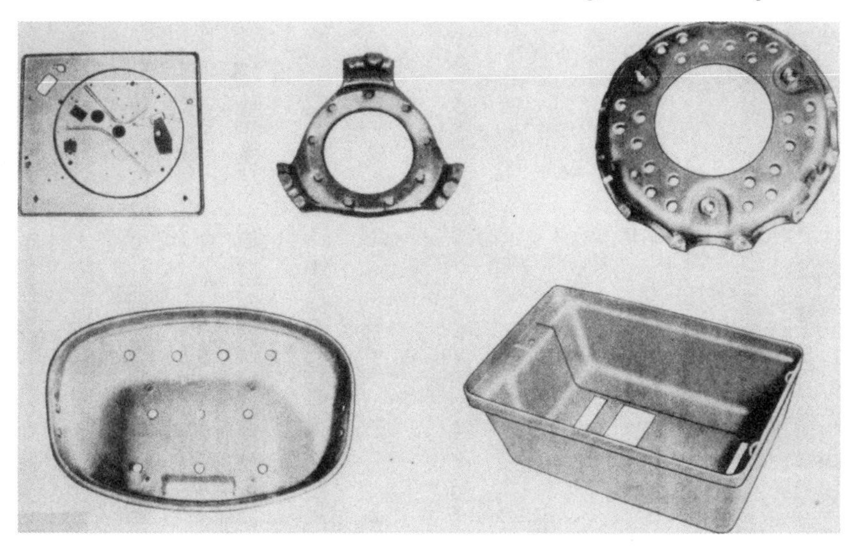

Figure 1B-3. Typical parts produced in single-action hydraulic presses

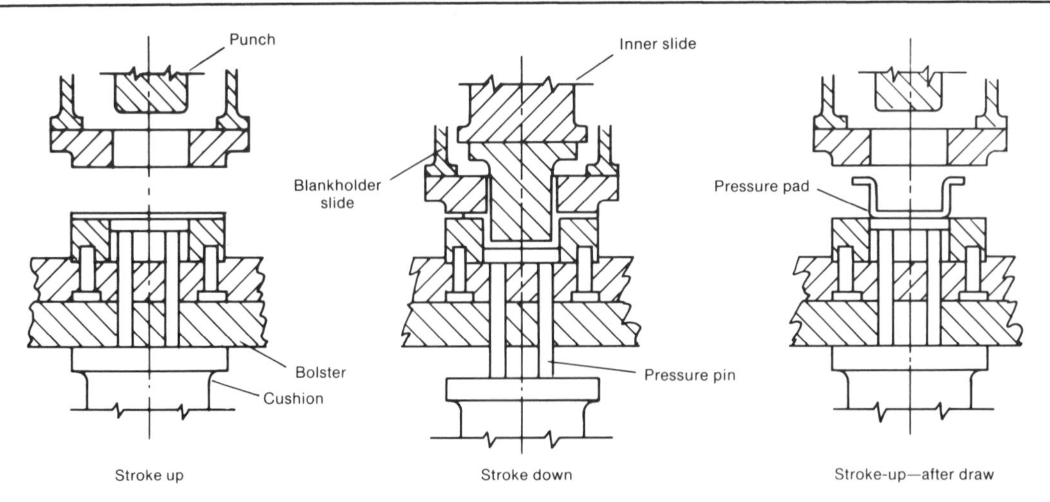

Figure 1B-4. Typical operation (deep drawing) performed in double-action hydraulic press, shown in three positions of stroke

3. Generally considered more difficult to maintain than mechanical presses, chiefly because breakdowns of mechanical presses are of the visual type and easily detected, whereas tracing of a hydraulic breakdown requires a thorough understanding of the circuit because the source of trouble is seldom visual.

4. Nested or multiple-die setups are not advisable unless pressures are reasonably well balanced. This is seldom possible.

Advantages of the Double-Action Hydraulic Press (Figures 1B-4 and 1B-5)

1. The double-action hydraulic press has four corner blankholder adjustments at the operator's fingertips. The tonnage set on each corner can readily be seen on the gauge above the control.

2. As with the single-action press, neither the blankholder nor the punchholder can be overloaded.

3. Full tonnage is available on both punchholder and blankholder throughout the stroke.

4. Tonnage is adjustable on both the punchholder and the blankholder.

5. Full adjustment of stroke is available on both punchholder and blankholder.

6. Die setting is an easier task since no adjustment need be made for material thicknesses or variations in stock.

7. Drawing speeds are constant throughout the stroke.

8. The double-action hydraulic press can be used to perform single-action operations by tying the blankholder and punchholder together to act as one. This is not possible on a double-action mechanical press.

Disadvantages of the Double-Action Hydraulic Press

The disadvantages of the double-action hydraulic press are basically the same as those of the single-action hydraulic press.

Advantages of the Single-Action Mechanical Press

1. The single-action mechanical press is faster than the conventional hydraulic press.

2. The mechanical press is by far the most suitable for blanking operations because the breakthrough shock is not detrimental to the machine.

3. Does not require as large a motor as the hydraulic presses, because it can store energy in the flywheel and then dissipate the energy throughout the press stroke.

4. Can be easily adapted for roll and transfer feeds and for progressive dies.

5. In general, mechanical presses with short strokes are more economical than hydraulic presses.

Disadvantages of the Single-Action Mechanical Press

1. The single-action mechanical press with cushion has a rated tonnage which is full tonnage a certain distance up — usually 1/2 in. (13 mm) above the bottom of — the stroke. Less power is available for work encountered at points above the bottom of the stroke. For instance, a 300-ton (270-tonne) mechanical press with a stroke of 30 in. (760 mm) entering a draw 4 in. (102 mm) above the bottom of the stroke would have an available tonnage of only 114 tons (103 tonnes). At higher points on the stroke, the tonnage factor would be even lower.

2. The mechanical press will not adjust itself for stock variations as does the hydraulic press, and thus requires extreme care when setting dies and making allowances for material-thickness variations. If the punch is adjusted too low, the press may stick on dead center. This requires the tedious task of unsticking the press.

3. A single-action mechanical press, when used on a draw which is equal to almost half of the stroke, enters the work at the high-midstroke velocity, the drawing speed being reduced as the punch continues downward. This is in direct contrast to the hydraulic press, which slows down before entering the work, then proceeds to draw at a constant rate, giving the metal a chance to flow to its best advantage.

Advantages of the Double-Action Mechanical Press

1. Double-action mechanical presses are generally faster on long strokes than are double-action hydraulic presses.

2. As in the case of the single-action mechanical press, a smaller motor is required.

3. Although maintenance of the double-action mechanical press is much more complicated than that of the single-action mechanical press, it is much less technical than maintenance of the hydraulic presses, particularly the double-action hydraulic press. This advantage is variable, depending on the number of hydraulic presses in the plant and the relative familiarity of the maintenance men with each type.

Disadvantages of the Double-Action Mechanical Press

1. In comparison with the double-action hydraulic press, much more time and care is required in the adjustment of the double-action mechanical press (as with the single-

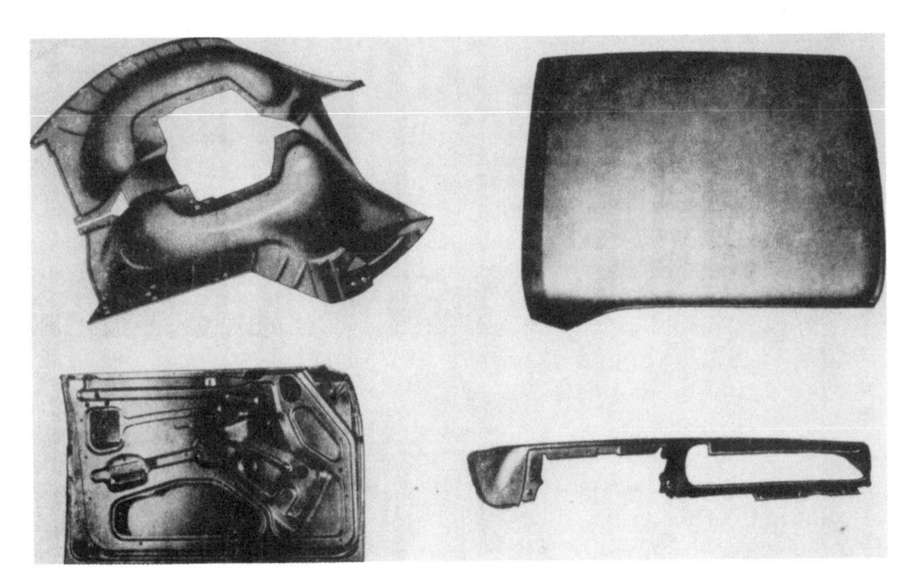

Figure 1B-5. Typical parts produced in double-action hydraulic presses

action mechanical press), due to variations in stock thickness and other factors.

2. The same variations in drawing speed occur with the double-action mechanical press as with the single-action mechanical press, where the tonnage diminishes the higher up on the stroke the work is performed.

Closing the Gap Between Hydraulic and Mechanical Presses

Having listed the basic differences between the two types of presses, we will now review what has been done to close the gap between them. Considering first the question of differences in speed, the hydraulic press can be speeded up by the addition of larger pumping equipment; the mechanical press, by the addition of a varidraw drive. The speed of the press is determined by the speed at which the material can be drawn. This speed cannot be exceeded or the material may tear. In the case of the mechanical press, this then determines the speed of the press through its entire cycle, which is generally slower than a mechanical press is capable of running. By installing a varidraw drive, we can maintain the recommended drawing speeds through that portion of the stroke where the draw occurs and then increase the speed of the press through the balance of the cycle. This can as much as double the production of the press.

To date one of the largest advantages of hydraulic presses over mechanical presses has been the fact that a hydraulic press cannot be overloaded. This advantage has now been eliminated by the introduction of overload units that can be installed in mechanical presses. These units basically consist of a hydraulic cylinder, installed within the connections of the press, that maintains a constant and adjustable pressure that, when exceeded, causes the cylinder to collapse and stop the press, preventing any damage to the press proper. After the obstruction (the cause of the overload) has been removed and the overload reset, the unit will automatically build up to the predetermined pressure and be ready to resume its function.

Press Selection for Specific Operations

Having discussed the advantages and disadvantages of each type of press, let us now review them with respect to the type of operation to be performed.

Deep drawing: For draws up to 4 in. (102 mm) deep, a mechanical press would be recommended for reasons of production capacity alone. For drawing work exceeding 4 in., the choice is a question of economics and production requirements.

Blanking and trimming: The choice for these types of operations should be a mechanical press, which is better able to sustain the shock of breakthrough than is a hydraulic press.

Forming: A mechanical press is the better choice for forming, because the stroke requirement for this type of work is generally short.

Coining: A mechanical press should definitely be used for coining, because generally in coining work the material must be moved. The requirement here is for a hammering type of force rather than the squeezing action performed by hydraulic presses.

Straightening: A hydraulic press is definitely the choice in this case, because the tonnage and stroke are adjustable and the possibility of overloading the press is eliminated.

Die spotting and die tryout: Again, a hydraulic press is recommended because of the versatility of being able to adjust both the stroke and the pressure.

FORMING BY HYDRAULIC AND CRANK PRESSES

Drawing Operations

Deep drawing of sheet metal will cause a compression or shrinkage at right angles to the direction of draw, and this compression must be absorbed by either a thickening of the metal or the formation of wrinkles. Aluminum alloy sheet will stretch readily, but has a high resistance to shrinking, and as a result wrinkles will form during drawing unless the dies are specially designed to prevent them. The time element also enters largely into drawing processes because the material will rupture if drawn too rapidly, but can be stretched if the drawing process is carried out slowly or in easy stages. The dies used for drawing purposes must be designed to allow for any shrinkage that can be foreseen, and the design of the part to be formed should be such that any resulting wrinkles will be in a portion of the blank that can be trimmed away after forming.

Controlling Factors in Drawing

1. **Size of blank vs. size of finished piece:** A cup may be successfully drawn in one operation if the inside diameter of the finished piece is not less than 0.6 times the diameter of the blank; if the operation is carried out in three steps using successively smaller dies, the ratio may be reduced to 0.4 times the blank diameter.

2. **Pressure with which edges of blank are held during drawing:** This pressure should be sufficient to permit some slip between the die and the pressure plate, but must be sufficient to cause actual drawing of the metal under the punch. Experience will determine the pressure to use in each case.

3. **Hardness, or temper, of sheet:** The greater the hardness, the greater the tendency to rupture and the greater the force required for drawing.

4. **Exact shape of punch and die:** When the radius of curvature of the punch is too large, the blank will tend to be perforated rather than stretched. If the radius is too small, the blank will rupture in the vicinity of the radius because of its inability to stretch over a sharp corner. The radius may be replaced by a 45-degree bevel with suitable radii at the corners, but as the hardness of the metal increases, the bevel becomes less effective.

5. **Condition of die face:** Cleanness is highly important. Any dirt on the face of a die or punch will have a tendency to scratch the surface of the sheet, and unless this scratch is rubbed or polished out it may serve as the origin of a crack. If the sheet has been anodized and the scratch penetrates the coating, it may serve as a point from which oxidation can start.

6. **Lubrication of sheet:** The importance of proper lubrication cannot be overemphasized.

7. **Temperature:** Ordinarily, drawing may be carried out with the work metal at room temperature; but if failure occurs under these conditions, satisfactory results may be sometimes obtained by heating the sheet to a temperature between 100 and 180 °C (212 and 356 °F).

Characteristics of Hydraulic Presses

The load pressure of a hydraulic press can be set at any required maximum so that if for any reason this maximum is reached, the press will automatically release itself before any damage occurs. This characteristic is important in drawing aluminum alloys, which have a tendency to work harden. As the metal hardens, the pressure required to draw it increases. Spoilage may be avoided by setting the release pressure low enough to avoid rupture. Should excessive work hardening occur, the press will automatically release and the partly formed piece can be removed and annealed. Fluid pressure permits a rapid approach of the punch to the work and the pressure and speed at the time of contact, thus avoiding a blow that might harden the metal. Once contact has been made, the speed at which the draw takes place can be controlled, which makes it possible to make a fairly deep draw in one stroke at low operating speed.

The low speed of the hydraulic press permits the use of low-cost punches and dies, because these parts receive little or no shock at the time of contact and comparatively little strain during drawing. Also, the stroke is held (dwells) at the bottom position, which gives the metal time to set and prevents springback to a considerable extent.

When the number of pieces to be produced justifies the cost, hardened steel or mild steel dies and punches may be used, but they must be carefully applied, because they are apt to foul because of the adhesion of particles of the sheet to the steel. Satisfactory results can generally be obtained by using zinc alloy dies and punches that can readily be replaced at low cost when they become worn.

Single-Action Presses

Single-action presses, either overdrive or underdrive, are the most commonly used because they are well suited for most press operations. Single-action presses are the simpler of the two types, having only one slide, and are generally more versatile. Although the single-action drive has only one slide motion, cushions can be provided to enable single-action presses to perform blankholding, stripping, knock-out, and lift-out operations.

Double-Action Presses

With two separate actions provided by two slides — the blankholder for holding the part and the inner slide for drawing — the double-action press is used primarily for deep drawing operations. These are operations which reshape the metal by forcing it to flow plastically, and which would normally cause the part to wrinkle if not held under pressure while being reshaped.

Although the double-action drawing operation can sometimes be accomplished in a single-action press using a cushion as a blankholder, this is generally less practical than using a double-action press. A single-action press requires considerably more energy to depress the cushion than is required by the blankholder in a double-action press. Therefore, a double-action press can perform a deep drawing operation with a smaller flywheel and motor.

Double-action presses provide more positive blankholder control and more equal blankholder pressure than are possible with single-action presses.

1C: Fundamentals of Press-Brake Operation

PRESS BRAKES: HOW THEY WORK

A press brake is one of the most common machine tools found in the metalworking industry, and yet it is one of the least understood (Figure 1C-1).

A press brake basically consists of four slabs of steel and a crosspiece (Figure 1C-2). Two side housings are held together at the top by the crosspiece and at the bottom by the bed.

The ram slides up and down in front of the side housings and does all its work by pressing against the bed. A gibbing and slide arrangement (Figure 1C-3) keeps the ram from falling off the side housings, and two power sources are mounted on both side housings, above the two ends of the ram to alternately push it down and pull it up during the stroke cycle.

Both the ram and the bed are fitted to accept a wide range of dies for many different operations: bending, flanging, drawing, cutoff, parting, blanking, hemming, curling, staking, notching, coining, piercing, ribbing, lancing, corrugating, beading, seaming, pipe forming, channel forming, embossing, bulging, trimming, perforating, slotting, shearing offsets, seminotching, slitting, louvering, coping, crimping, riveting, and tube forming.

MECHANICAL AND HYDRAULIC PRESS BRAKES

There are two major types of press brakes, classified by the nature of their drive systems: mechanical and hydraulic.

The mechanical press brake was the first on the scene, and for years dominated the industry.

Mechanical drive systems range from

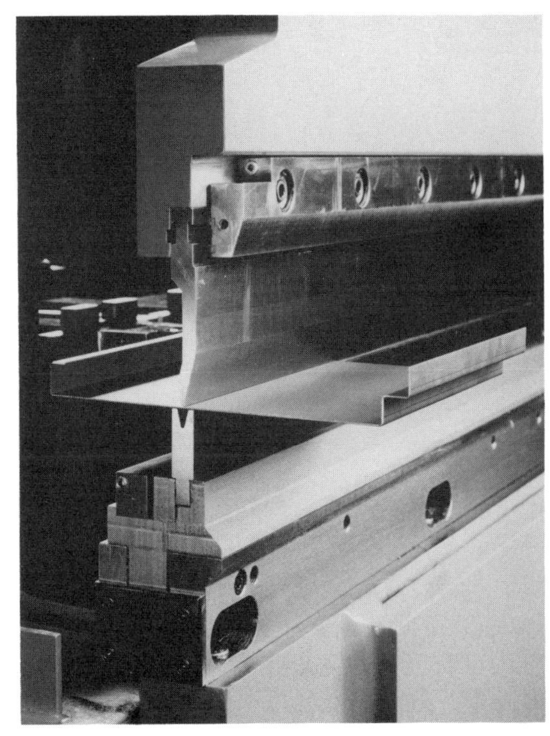

Figure 1C-1. Press brakes

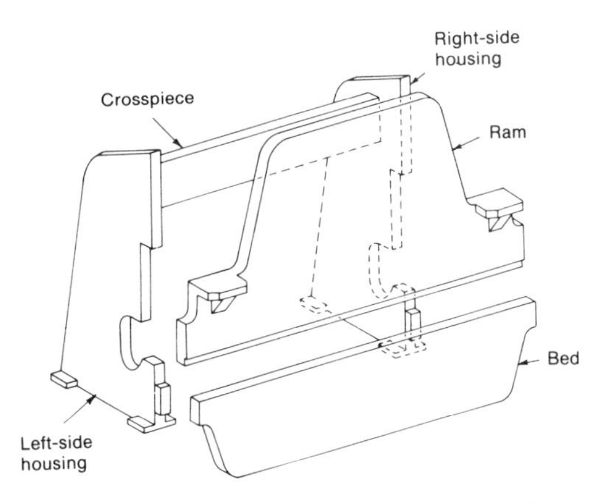

Figure 1C-2. Schematic illustration of a press brake

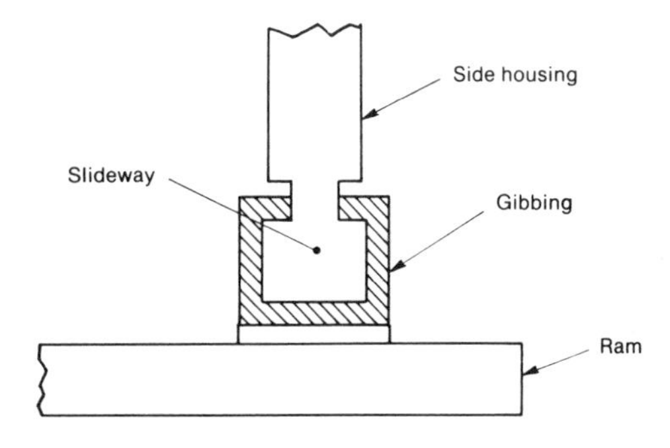

Figure 1C-3. Cross-sectional view of gibbing and slideway on one end of a press-brake ram

simple, nongeared belt drives to single- and double-geared drives, with mechanical linkages from motor to flywheel to clutch to gears to crankshaft to crank arm to ram (Figure 1C-4). They all have one thing in common: a crankshaft action that converts rotary motion into straight, reciprocating motion (Figure 1C-5).

During a stroke cycle, the crank arm drives the ram down to the bottom of the stroke and then back up to the top.

In continuous operation, the ram can be cycled very rapidly for high-speed strokes, due to the mechanical advantage of the crank arm and momentum in the flywheel. The work capability of mechanical drive systems is defined in terms of available tonnage throughout the stroke. Mechanical manufacturers generally cite two reference points for available tonnage: midstroke and bottom-of-stroke (Figure 1C-6). While the bottom-of-stroke tonnage is the maximum stated ton-

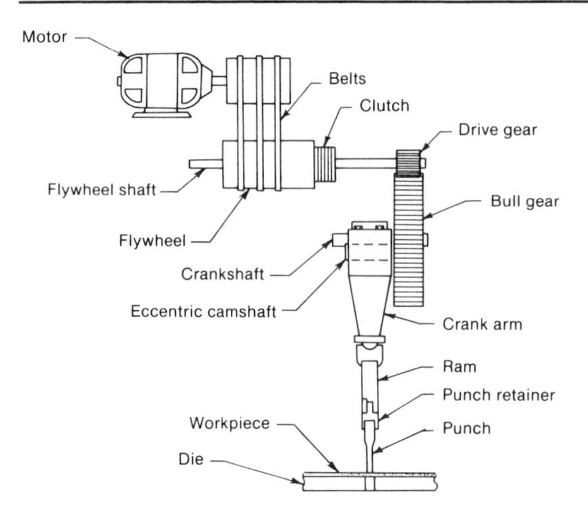

Figure 1C-4. Drive mechanism of a mechanical press brake

nage rating for a given press brake, it is not the same as the midstroke, and this difference is due to the nature of the drive system. At midstroke, the crank arm is not in its optimum work position, and the available tonnage is normally about two-thirds of the full maximum bottom-of-stroke rating.

As the crank arm continues down from midstroke, the available tonnage rises slowly until near the bottom of the stroke, where the available tonnage rises very sharply. This is because the full mechanical advantage of the crank arm appears near and at the bottom of the stroke. A toggle effect is achieved, and the available tonnage can sometimes exceed three times the maximum published rating for the press brake. This is very useful for certain types of operations that are performed within the last few fractions of an inch from the bottom of the stroke. The full rated tonnage is achieved between ⅛ and ½ in. (3.2 and 13 mm) from the bottom, depending on the complexity and quality of the drive system. JIC standards establish minimum distance versus tonnage ratings, and, generally, the more complex and expensive drive systems achieve maximum tonnage ratings higher up the stroke. These criteria apply across the board to mechanical presses as well as press brakes. One major press manufacturer states that "When press capacity is mentioned in terms of tons, a logical question of any purchaser should be, 'At what

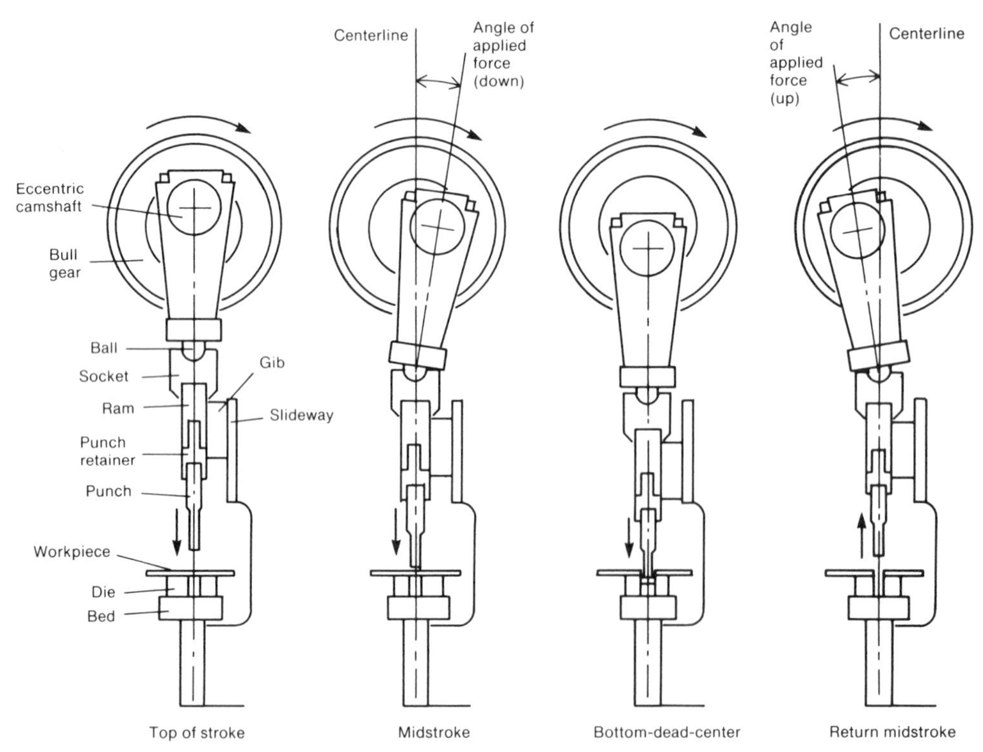

Figure 1C-5. Stroke cycle of a mechanical press brake

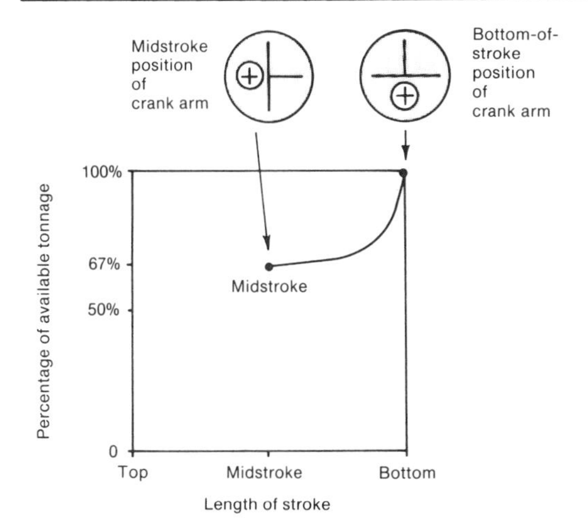

Figure 1C-6. Available tonnage curve for a typical mechanical press brake

point above bottom stroke is this capacity reached?' " This is a vital point, because above it the available tonnage drops off rapidly.

The reason this is important is because of the danger of overloading the brake. If the operation being performed on the brake requires tonnage at or near the maximum rating of the brake, it should be performed within this area of maximum available rated tonnage at the bottom of the stroke. If it is performed higher up the stroke, the machine will be in danger of being overloaded because of the design limitation of the drive system. Overload can be very serious, and can occur when one is least expecting it, even in routine jobs. Commercially available metal varies as much as 50% in hardness and as much as 10% in thickness from specification among both pieces and sources. Foreign metal sometimes varies even more. These common variations increase tonnage requirements when they occur. In addition, a simple mistake, such as underestimating the tonnage requirement for a job or inserting two thicknesses of metal, can also increase the tonnage requirement.

Manufacturers of mechanical press brakes advise their customers not to use their machines under overload conditions, but not because the machine is incapable of delivering extra tonnage. At the bottom of the stroke, a mechanical press brake can deliver as much as three times its maximum rating. The caution against overloading is given because the drive system and even the structure of the press brake can be damaged by the effect of the extra stresses placed on the mechanical linkages, the slide system, and the frame.

When overload occurs at midstroke, it is sometimes called "torque overload" because the stresses affect the rotating members of the drive system. When it occurs at the bottom of the stroke, it is termed "bottoming overload" and is a function of the increased mechanical advantage of the toggling action of the crank arm. Between these extremes, an overload demonstrates characteristics of both.

There are special clutch options that permit successive engagement-disengagement of the clutch during the downstroke to allow inching, and other options permit reversal of the motor to withdraw the ram, but generally the crank arm and ram must bottom before the ram returns upward.

The speed of a mechanical press brake is cited in terms of strokes per minute and is usually stated without a load for continuous or intermittent operation (Figure 1C-7).

The difference lies in the loss of flywheel energy during operation. Intermittent, or stop-start, operation usually involves just one stroke — that is, the ram cycles down and then back up to the top of the stroke and

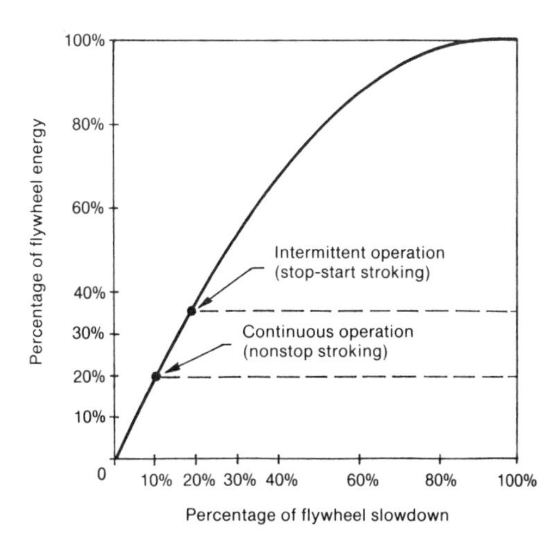

Figure 1C-7. Flywheel slowdown and loss of energy during operation

then stops. This action drains up to 35% of the flywheel energy and causes up to a 30% slowdown of the flywheel. If there is a slight pause before the next stroke cycle, the motor will have time to restore the flywheel's full energy.

In contrast, under continuous operation, the flywheel loses only about 20% of its energy and suffers only about a 10% slowdown. This lower energy loss is due to the momentum of the flywheel in continuous cycling, which reduces the energy loss and makes energy restoration faster and easier. There are formulas available that can aid in selecting a drive system and motor that will result in the least energy loss under load for a given job.

Stroke control is accomplished through operation of a clutch/flywheel engagement. Hand levers, foot treadles, and air-electric controls are available. Different clutch options are available, from mechanical clutches to air-disk assemblies, and even automatic clutch speed-shifting. Generally, the more complex and precise the control, the more expensive the press brake will be.

Hydraulic press-brake drive systems are relatively less complex, and comprise a motor, a pump, a valving system, and a hydraulic cylinder with the piston connected to the ram.

The cylinder is usually double-acting — that is, oil is pumped under pressure into the top, above the piston, to make the ram move down; to make it move up, oil is pumped into the bottom, under the piston. With oil under pressure both above and beneath the piston, the piston is locked in place anywhere in the stroke cycle. The pressure of the oil under the piston is generally held between 10 and 15% of the pressure of the oil above the piston. This is done to keep the ram from dropping by gravity, and also to help control the ram throughout the stroke (Figure 1C-8). The downward movement of the ram during a stroke cycle means that the pressure of the oil entering the cylinder above the piston is much greater than the pressure of the oil underneath the piston. The upward movement of the ram is caused by a reversal of this action. The valving system ensures the desired control and direction of flow either into or out of the cylinder.

There is no mechanical linkage to trans-

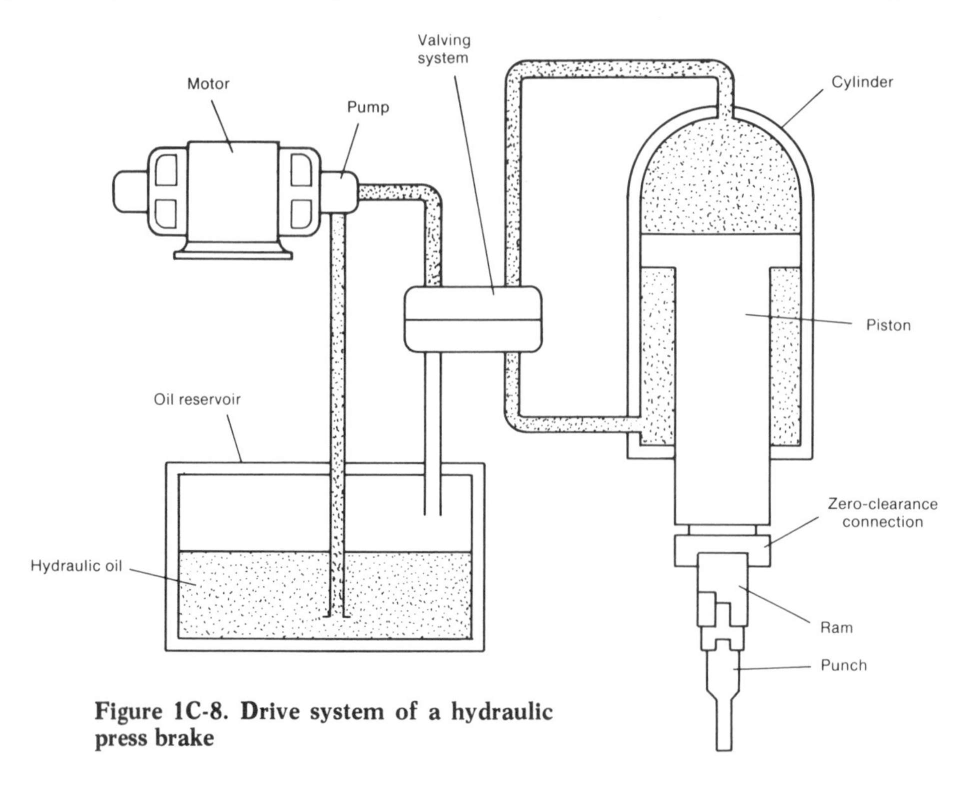

Figure 1C-8. Drive system of a hydraulic press brake

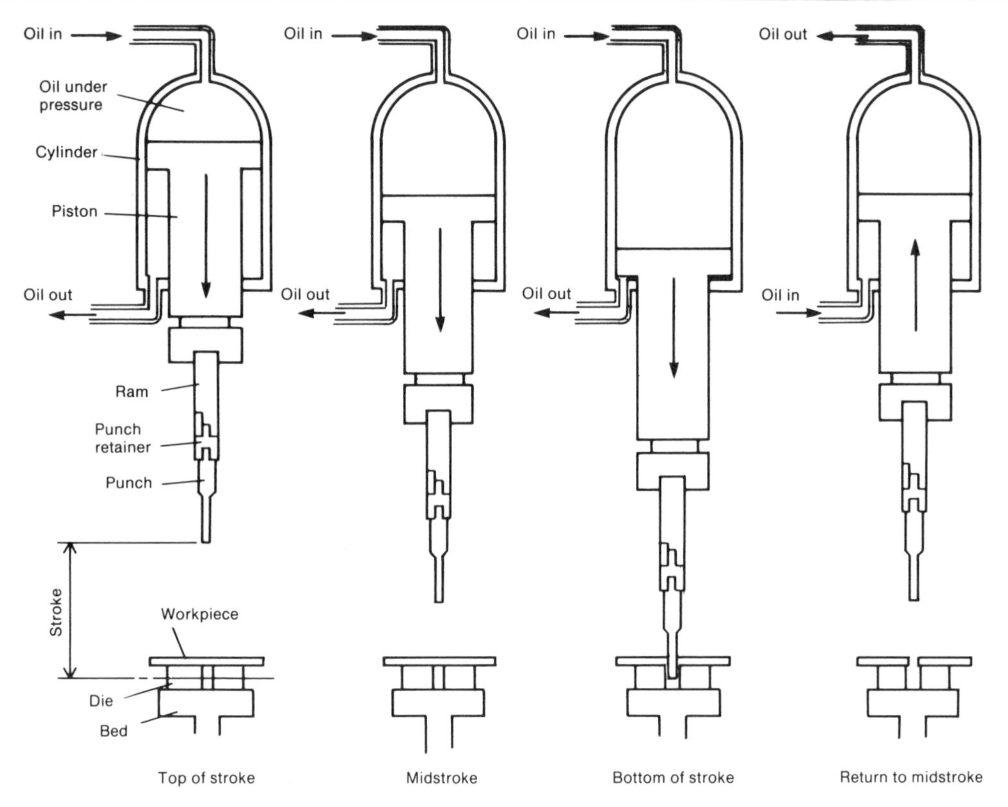

Figure 1C-9. Stroke cycle of a hydraulic press brake

late rotary action into straight, reciprocating action, as in a mechanical press brake. The rotary action of the motor in a hydraulic brake serves to drive a pump, which then forces the oil throughout the system.

Because the pressure is a steady, controlled flow, the available tonnage of a hydraulic press brake is the same at all points during the downstroke (Figure 1C-9). The ram can also be instantly stopped, anywhere during the downstrike, by simply releasing the foot pedal. Stroke control is much more precise than in a mechanical brake, and involves much more complex action.

A hydraulic press brake also permits much longer strokes, for deep drawing, multiple punch-stepping, or large, complex, multiple-action dies. Extra-long cylinders can be installed for extremely long strokes, as long as 5 ft (1.5 m). This is not practical on a mechanical press brake, because the crankshaft eccentric limits the length of the stroke. Standard strokes on hydraulic brakes are normally at least three times as long as those on mechanical brakes of equal rating.

The length of the hydraulic stroke can also be varied to suit the job that is being done (Figure 1C-10). Strokes as short as ½ in. (13 mm) can be set anywhere throughout the stroke range with limit switches that usually are very easy to set and change. The shorter the stroke, the faster the stroke cy-

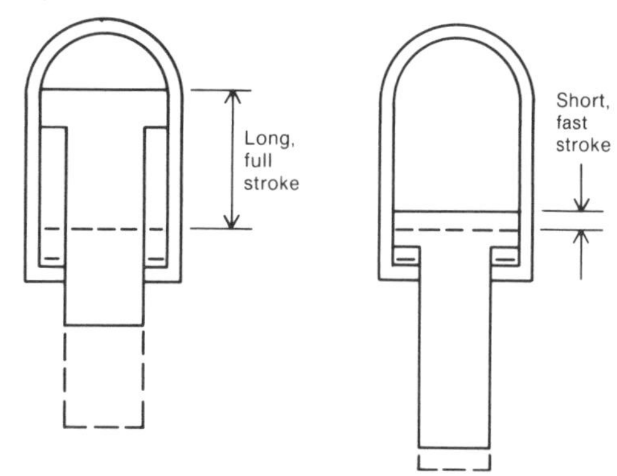

Figure 1C-10. Cylinder in a hydraulic press brake, showing adjustable piston travel

cle. For operations requiring very short strokes, such as punching and blanking, a hydraulic press brake can achieve speeds that compare favorably with those of mechanical press brakes, especially when the material is hand fed. Overloading is not the problem with hydraulic brakes that it is with mechanical brakes. The oil pressure into the cylinders can be set to deliver a given tonnage. Any job requiring more tonnage will cause the ram to stall, but there will be no damage to the press-brake drive system or its structure, because it will not deliver any more than its rated tonnage.

Now, it is true that increasing the pressure of the oil will also increase the tonnage available on the ram, and hydraulic manufacturers normally build in about 10 to 15% reserve as an overload safety feature. However, they do not promote this capability. If they did, human nature being what it is, we would soon find operators using press brakes in this reserve range continually, and overloading in this area could overstress the hydraulic system and the structure. The frame and other components are naturally designed around the rated tonnage, with something to spare. Increasing the load capability of these components would increase the cost and result in what is already available: the next larger size rating. The bending capacity of a hydraulic press brake far exceeds that of a mechanical because of its long, full power stroke; it is possible to bend much thicker plate on a hydraulic press brake than on a mechanical brake of the same size. On a hydraulic brake, the chief limitation on bending capacity is the length of the piece to be bent, because of the tonnage requirement. On a mechanical press brake, the limitations include length and thickness — or, in other words, tonnage and stroke. (Figure 1C-11). For example, if you recall the available-tonnage curve for a mechanical press brake, you will remember that at midstroke the available tonnage rating is only two-thirds the bottom rating. Above midstroke, it is even less.

Now, relate that to the length of power stroke required to make a 90° air-bend of a given section of mild steel plate, in a standard V-die (Figure 1C-12). The start of the bend requires about 80 to 90% of the over-all bending tonnage requirement. When the male die achieves a bend close to 20°, most of the work is complete, because the yield point of the metal has been reached. After that, the rest of the bend requires very little increase in tonnage.

The tonnage required to start the bend should be compared with the available-tonnage curve for the mechanical press brake to see if it can deliver the tonnage at that point in the stroke without going into over-

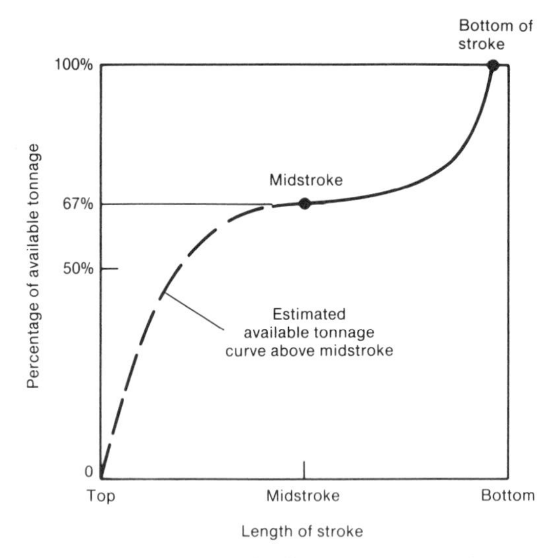

Figure 1C-11. Available tonnage curve for a mechanical press brake

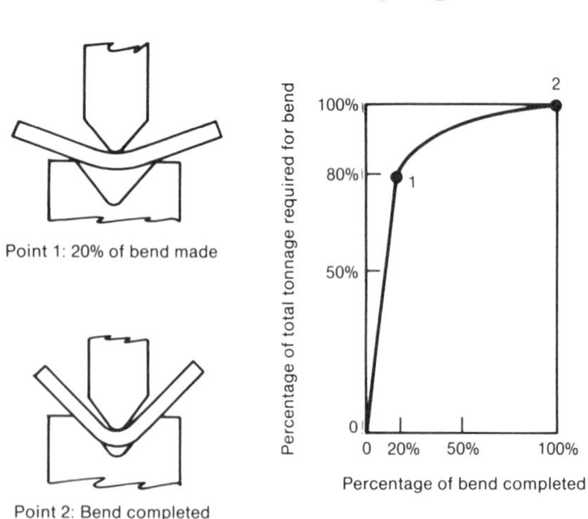

Figure 1C-12. Position of male die, and percentage of bend completed vs. percentage of total required tonnage, at two selected points in cycle

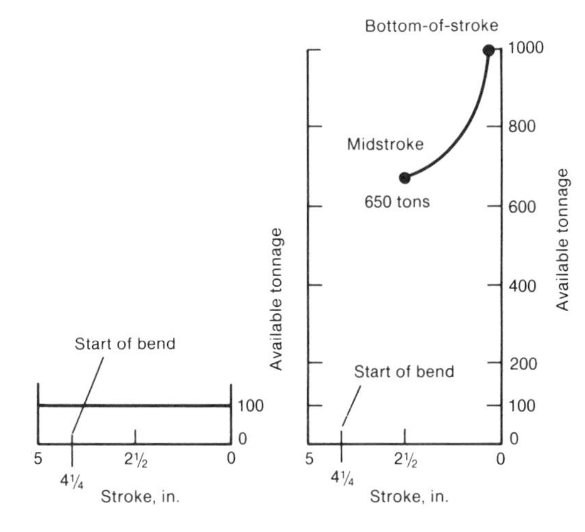

Work metal, mild steel 1 in. (25 mm) thick by 1½ ft (0.46 m) long. Die opening, 10 in. (255 mm) wide. Tonnage requirement, 81 tons. Required bend stroke, 4¼ in. (108 mm).

Figure 1C-13. Comparison of stroke capabilities of two standard press brakes: a 100-ton hydraulic brake (left) and a 1000-ton mechanical brake (right)

load. If the beginning of the bend is above midstroke, remember that the available tonnage in this area is even less. For heavy plate, from ½ in. (13 mm) thick and up, the recommended width of the required die is ten times the thickness of the metal. This puts the beginning of the bend for heavy plate up in the stroke area of a mechanical press brake where it is at a tonnage disadvantage, compared with a hydraulic.

Thus, for example, it is possible to bend short lengths of mild steel plate 1 in. (25 mm) thick on a hydraulic press brake as small as 100 tons, over a V-die 10 in. (255 mm) wide, requiring a 4¼-in. (108 mm) power bending stroke, whereas the smallest mechanical press brake that could perform the same bend in one hit without going into overload would have a 1000-ton rating and a standard 5-in. (127 mm) stroke (Figure 1C-13).

At the other end of the scale, with light-gauge sheet metal and plate up to ¼ in. (6.4 mm) thick, mechanical press brakes perform very well. Their cycling speeds, especially with automated feeding and unloading equipment, are extremely high. They are particularly suited to long runs of the same bending or forming operation. Small hydraulic brakes, however, have several advantages that compare favorably with small mechanical brakes: an inherent versatility in doing the wide range of operations normally found in metal fabricating shops; precise operating control of the stroke; speed in set-up and changeover; flexibility in handling both deep-stroke heavy plate bending and short-stroke sheet metal work; and quiet, shockless operation.

In examining both types of press brakes, tonnage-distribution characteristics must be considered. (Figure 1C-14). The construction we have been discussing involves a long ram with two sources of power — one at each end. The power distribution in this design is the full tonnage of both power sources along the entire length of the ram as long as the load is distributed along the entire length of the ram.

If a short job is performed, and the load is concentrated in the center of the ram, the full tonnage of both power sources will be available for the load. If the load is concen-

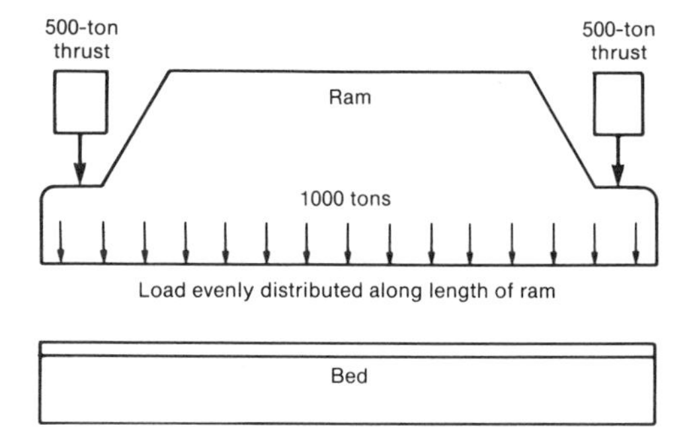

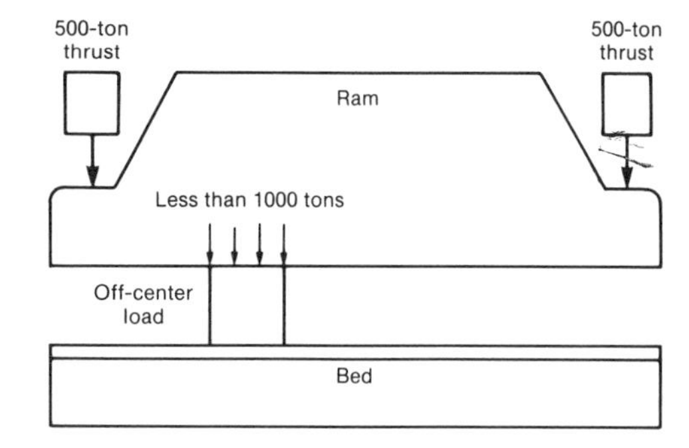

Figure 1C-14. Effects of load distribution in a 1000-ton press brake

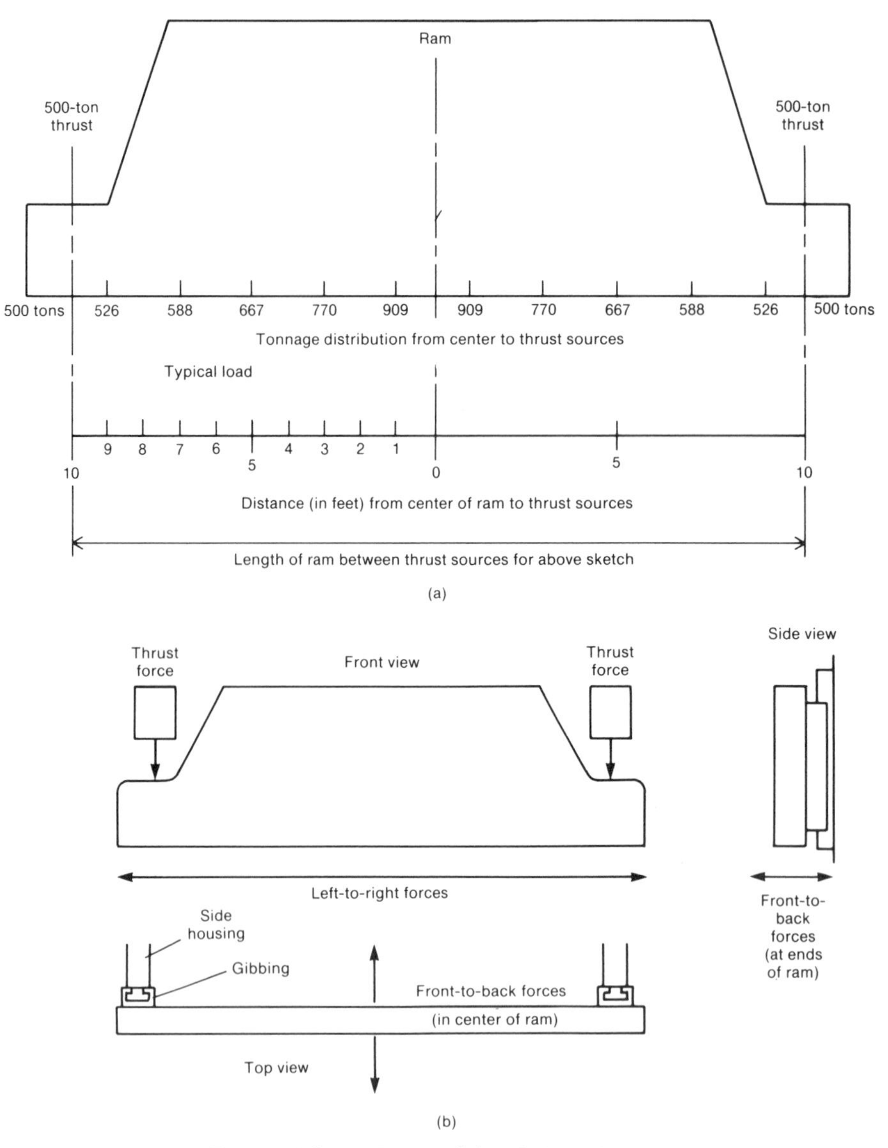

Figure 1C-15. Typical load distribution from center to thrust forces (a), and forces that act on the ram (b)

trated under only one of the power sources, only half the rated tonnage will be available to perform the job (Figure 1C-15). Moving the load toward the center of the ram permits the other power source to participate in the operation, and so the available tonnage rises as the load is moved until the short load is exactly centered. At that point, the full tonnage of both power sources is concentrated on the load.

Thus, an operator should not count on the full rated tonnage of the press brake if the tonnage on a short job is concentrated at either end of the machine. When horns are used, the available tonnage drops even more, depending on the horn length and the distance of the load concentration from the power source.

The shape of the ram is designed to resist deflection best in the vertical, left-to-right direction. The depth of the center of the ram adds to its stiffness and rigidity as a beam. This characteristic aids in leveling the ram from right to left. From front to back, however, the ram is relatively weak, and is dependent on the gibbing and slideways, which are mounted only on the rear of the ram, at both ends. Front-to-back deflection can be a problem even with a stiffener along the rear length of the ram.

Operators should therefore center the load in the dies from front-to-back, especially on press brakes fitted with wide platens. Loads off-center from front to back can cause great stress, if not damage to gibbing and slideways, because this is their weak direction.

The shape and mounting of a press-brake ram also leads to a unique advantage in permitting press brakes to be connected in tandem (Figure 1C-16). Both mechanical and hydraulic press brakes can be so connected to accommodate extra-long work. The controls can be designed to permit either machine to be operated independently, or both together. Tandems are increasing in popularity because designers are specifying longer beams and structurals to minimize welding. It may seem like science-fiction, but there is no reason why more than two press brakes cannot be joined to provide even longer bending capacities. Three or more machines can be connected in this way.

SELECTION CRITERIA

With so many features, options, accessories, and whatnot, it seems difficult for a prospective purchaser to separate what is needed from what is not. Too often, the easiest course is to have the press brake salesperson make the recommendation. Relying on the salesperson's integrity is hardly businesslike, even if it does provide a scapegoat. Also, it happens that the most important factors are unintentionally sidelined. With a general background of the inherent differences between mechanical and hydraulic press brakes, a prospective purchaser is in a better position to evaluate the many factors, especially if the following selection criteria are used.

The two primary factors that should be used in establishing selection criteria are: (1) the job to be performed and (2) the human operators who will perform the job. All other factors are subordinate to, and in many cases dependent on, these two.

Job-Related Selection Criteria

The requirements of the job or jobs, when carefully analyzed, will help indicate the size, type, and often the accessories that are needed.

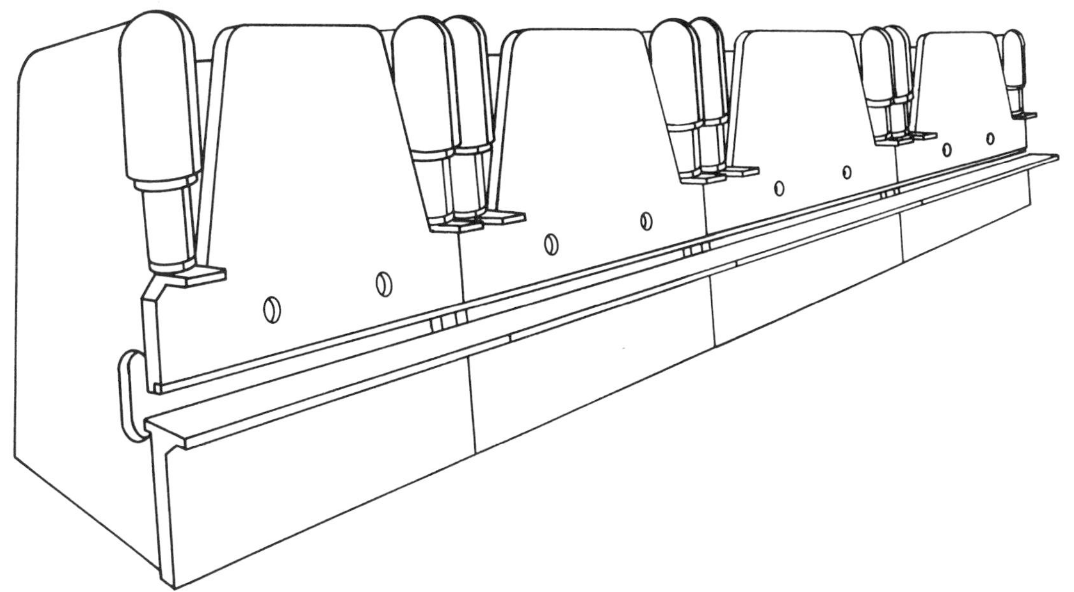

Figure 1C-16. Extension of the tandem concept

1. **What is the nature of the job?** There are over 40 types of forming, cutting, and drawing operations that can be performed on a press brake. Some can be done singly, some in combination; some with narrow dies, some with wide platens; some take a short stroke, others need very deep strokes; some can be automated, and for others automation is neither functionally practical nor economically feasible. The time-honored custom of the "trial bend" to see whether the machine can "take it" is not very wise when the tonnage requirement is approaching the maximum rating of the machine, nor is it sensible when performing jobs in combination. Building complex dies without considering the characteristics of the press brake can be needlessly expensive.

2. **What kind of tooling will be required?** Fortunately, there are many standard dies that are easily used with or adaptable to commercially available press brakes. These dies can be used for many of the operations common to the industry. Most press-brake manufacturers offer a selection of standard dies and can either make special dies on order or recommend a reliable source. Tooling will be covered more thoroughly elsewhere.

3. **What type of metal will be used?** General industry ratings of press-brake capabilities are based on mild steel with a tensile strength of about 60 ksi (415 MPa). Metals with higher tensile strengths will require proportionately more tonnage and will also have different working characteristics. For example, some high-strength metals are best worked by preheating before bending to prevent fracture because they are so much more brittle than mild steel.

4. **What are the tolerances of the raw stock?** Piece-to-piece mill variations in native metal may run as high as 50% in hardness and 10% in thickness, and foreign metals may vary even more. This requires planning for reserve tonnage to prevent overloads or stalling of the machine.

5. **How thick is the metal to be formed?** Specifications for dies vary with the thickness of the metal to be formed. Generally, sheet metal is defined by thickness up to about 3/16 in. (4.8 mm), and thicker metal is considered to be plate. Sheet-metal dies can be ruined by working them with plate; on the other hand, the precision and sharp bends usually associated with bending of sheet metal can rarely be achieved with plate dies. As mentioned earlier, heavy thicknesses of plate require longer strokes, because wider dies are needed.

6. **How long is the metal to be bent or formed?** The length of the bend, or the circumference of the hole or blank, affects the amount of tonnage required to complete the operation. The longer the bend or the larger the hole, the more tonnage is required. Performing multiple operations on the same stroke increase the tonnage requirement. Press-brake manufacturers publish tonnage charts that make tonnage calculation simple; in some cases, such charts are affixed to the machines for quick reference.

7. **What is the volume of the job?** The expected volume and rate of output will help determine the type of press brake required. Will the machine be performing one job continuously at high production speeds, or will it be necessary to adapt the brake quickly to many different kinds of operations? If the former, a mechanical brake would be better suited, whereas a hydraulic brake would be better for the general-purpose application.

8. **What about material handling?** This is one of the most neglected areas. Feed and takeoff considerations should properly depend on the job to be performed, but it is often the reverse. Feed and takeoff can become so complex and unwieldy that they affect the operating efficiency of the press brake. Conveyor systems can be adapted to the rear of the press brake for feedthrough, but there is still the problem of returning the workpieces to the front for any required additional work. Automation is one answer, and is adaptable to both hydraulic and mechanical brakes, but it is not justified except for long runs. The great majority of jobs coming into metal fabrication shops are not of this nature. Thus, material handling is, and will continue to be, a major problem.

Human Factors

Selection criteria based on human factors are as follows:

1. **Do you need experienced operators?** Most mechanical press brakes require a high degree of operator skill in double-clutching if they are to be operated productively. Learning to operate them is similar to learning how to operate the clutch on a manual-shift automobile. Some mechanical brake manufacturers offer brakes so easy to operate that inexperienced personnel can learn quickly. A press brake requires a trained operator and, in some cases, a helper who actually performs the job.

2. **How easy is the press brake to operate?** The machine should be simple to start and run; instantly responsive to the operator's control; precise in performing the desired operation; and repeatable in doing it over and over with the same initial accuracy.

3. **How long does it take for setup and adjustment?** The time and effort required to set up the press brake and the changeover from one set of dies or job to another, should be minimal, to avoid unnecessary downtime. Adjustments should be easy to make and, once made, should not require constant checking.

4. **How safe is the press brake to operate?** Of course, the operator must be constantly alert, but must complex procedures be followed and precautionary equipment used to prevent accidents? Will the operator be able to stop the machine instantly in case of an accident or the immediate likelihood of an accident?

5. **How much noise and disturbance does the machine make?** Will operation of the press brake shake the floor and affect other nearby workers? Will the noise level be acceptable to the operator and nearby personnel? Some shops already have a high noise level, from operation of saws, grinders, drop hammers, shears, etc., and so this may not be a problem. Other shops do not generate the same level of noise, and the personnel may not be accustomed to noisy machines.

6. **How much fatigue will the operator experience?** Can the press brake be operated at a speed comfortable to the operator to minimize fatigue and the accidents due to loss in alertness caused by fatigue?

7. **Again, what about material handling?** Feed and takeoff should be arranged to permit the operator easy access to pallets and containers and yet freedom to operate the machine safely and efficiently.

Additional Considerations

The physiological and psychological aspects of operating the machine affect the operator's state of mind and physical well-being, and thus greatly influence the rate of productivity and efficiency with which the press brake is used. The operator is also likely to have a positive attitude if aware that the machine purchase decision considered these aspects.

Cost is often the first consideration in purchasing a press brake, but it should not be. Buying a press brake is not just a purchase of capital equipment, it is an investment that can return its cost within a few months, depending on work load.

The costs of hydraulic and mechanical press brakes are roughly similar. Pricing changes constantly, if only because of inflationary pressures, but prices of competitive makes are usually within 10 to 15% of each other, and the differences can sometimes be justified on the basis of standard or special features included in the prices.

Other factors that may have a bearing on the decision process include repair and maintenance probabilities, and service availability. It may also be appropriate to examine the specifics of procurement and financing, such as foundation requirements, delivery capabilities, shipping and handling problems, setup and erection requirements, discounts, and terms of payment and financing.

The analysis of the criteria developed in considering job and operator requirements will naturally lead to a determination of what accessories and controls will be required. Some of these will be standard on some machines, others will be optional. They include such items as ram tilt or slope adjustments, clutch options, automatic lubrication tonnage indicators and controls, multiple speeds, die cushions, wide platens, manual and/or powered back gauges, ram-leveling controls, oil filters and coolers, operating control options, T-slots, and many others. There are so many of these individual features that every press brake made is an individual in its own right. This proliferation of features offers a customer a great deal of freedom in tailoring the machine exactly to specific applications, but it also generates confusion and uncertainty. When do you stop adding accessories? Again, remember the two basic factors that are used to establish the criteria for selection: the job, and the operator. In terms of these two factors, the characteristics and limitations of both mechanical and hydraulic press brakes as described in this article will help determine which type is best suited for your application.

BEND MECHANICS AND TOOLING

Bend Mechanics

Bending and forming involve complex phenomena, but a general idea of what happens during these operations may be useful in understanding what a press brake does. The mechanics of bending a section of mild steel plate ½ in. (13 mm) thick and 1 ft (0.3 m) wide into a 90° bend will serve as an example.

In a cross-sectional view of a male and female die set about to bend the section of mild steel, the metal can be seen to be stressed at three points, like a beam: downward in the center by the male die, and upward at the two shoulders of the female V-die (Figure 1C-17). As the male die presses down, the metal section resists until the male die achieves enough tonnage to overcome the elastic limit of the metal (Figure 1C-18). Inside the metal, the molecules at the upper surface are being crowded toward each other in compression, while at the lower surface the molecules are being stretched away from each other under tension. Somewhere in the center, the two opposing forces cancel each other out to create a zone or plane of zero stress that extends throughout the metal. When a cross section is taken through the stressed area, this plane of zero stress is found to be approximately in the center between the two surfaces, and is termed the neutral axis. Prior to the bend, the metal section was 1 ft (0.3 m) wide. After the bend, the neutral axis across the width is still exactly 1 ft, but the outside surface that was under tension is now slightly longer than 1 ft, having been stretched, and the inside surface that was under compression is slightly shorter than 1 ft, having been compressed (Figures 1C-19 and 1C-20).

If the metal were bent on a very tight radius — for example, over a vee die opening eight times as wide as the metal thickness or less — the neutral axis will be distorted closer to the inside compressed radius than to the outside surface under tension. Widen-

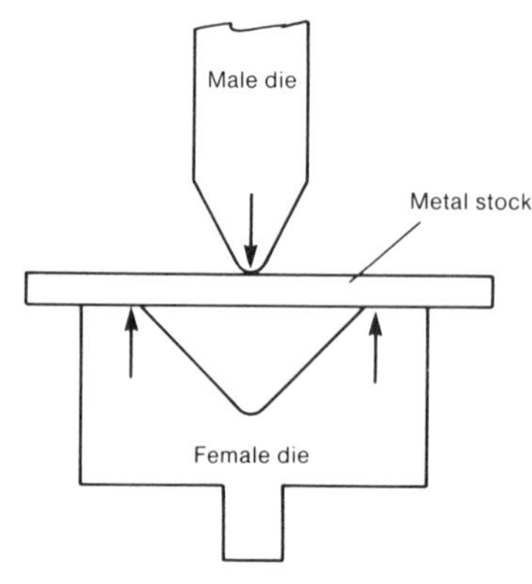

Figure 1C-17. Three points of stress on metal stock during bending.

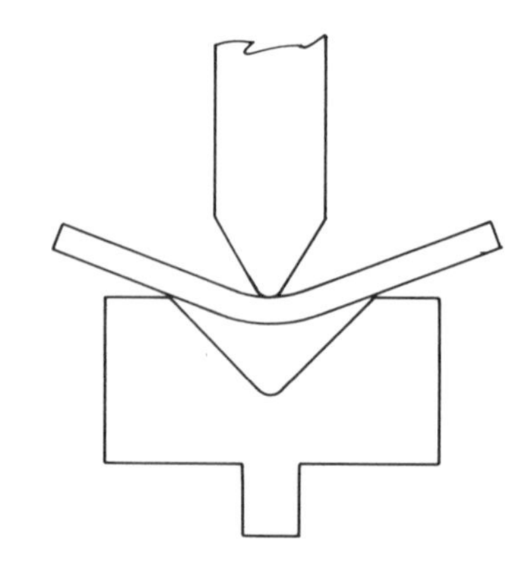

Figure 1C-18. Metal stock being pressed beyond elastic limit

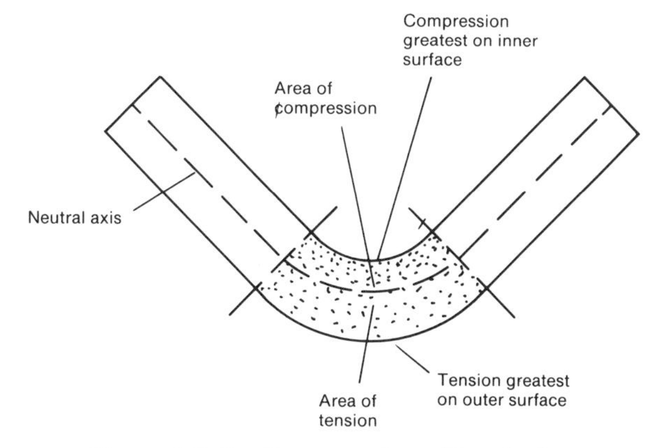

Figure 1C-19. Cross section of bend, showing compression and tension areas

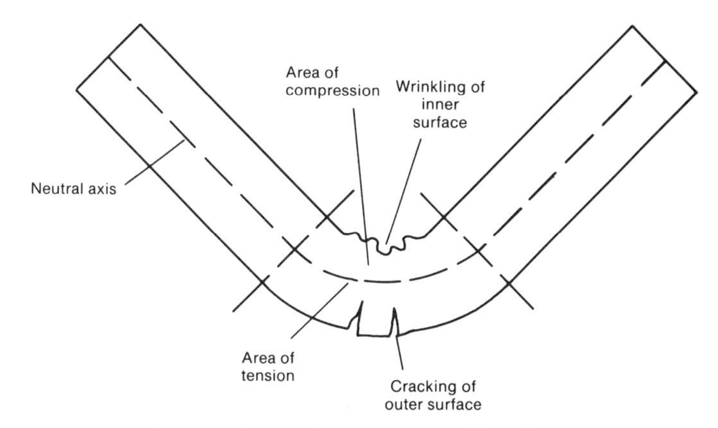

Figure 1C-20. Cross section of bend, showing wrinkling and cracking

ing the die opening to ten times metal thickness or more makes the radius of the bend more gradual, and thus the neutral axis will be closer to the center between the two surfaces.

As the bend is made, the metal on both sides of the neutral axis flows to accommodate the two opposing stresses. If the bend radius is too small, the inner surface may wrinkle or buckle as the molecules are crowded into the small area on the inner side of the neutral axis. On the outer side of the neutral axis, the area under tension is relatively much greater, and so the molecules may be stretched beyond their cohesive bonds. Cracks, fractures, and splitting will then occur. The slower the bend, the more time is allowed for the metal to flow. The thicker the metal, the more tendency to crack over narrow V-dies. Generally, sheet metal is bent over V-dies with openings up to eight times metal thickness, and plate over ½ in. (13 mm) thick is bent over V-die openings ten times metal thickness; these recommendations apply to mild steel with a tensile strength of approximately 60 ksi (415 MPa). Harder metals may require wider V-die openings, and in some cases may be too brittle to be formed at all at ambient temperatures.

The compressive and tensile stresses create opposite forces in the metal, which have an effect known as springback. This is why male dies are normally 88° in nose radius in order to overbend the metal. The normal springback in mild steel will force the metal back after bending by about 2° to achieve the desired 90° bend. Overbending is a common practice in air bending; its success depends on the skill and accuracy of the diemaker.

Another method of countering springback is to bottom the dies by slamming the male die into the metal and forcing it down into the V-die with great pressure. High compressive stresses thus permanently deform the metal into the desired bend, with very little or no springback. Some dies are designed expressly for bottoming, with a small bead on the tip of the male die to help distribute the compressive stresses on the inner radius of the bend and actually make the inner radius larger. A drawback of this type of die is that the thickness of the metal at the bend is much less than that of the rest of the metal, and thus is proportionately weaker.

Another name for bottoming is "coining," and while it can be performed on both mechanical and hydraulic press brakes, the tonnage requirement should be carefully determined to fall within the rated maximum of the press brake.

This is especially critical for mechanical brakes, because of their ability to supply more than the rated maximum tonnage at the bottom of the stroke. Many mechanicals are used for bottoming operations because of the well-known mechanical advantage of the toggle-action crank arm. Unfortunately, many of these mechanical brakes are also being overloaded, making them susceptible to unexpected downtime for repairs. Any resultant dissatisfaction with the machine is truly not the fault of the press brake.

Bottoming can be done on a hydraulic press brake, contrary to popular opinion, and in fact it is much safer, because the hydraulic brake will not deliver more than its maximum rated tonnage. Required bottoming pressures are normally 20 to 80% greater than the tensile strength of the metal, depending on its hardness.

A third method of minimizing springback is stretch-forming. The metal is gripped between two opposing sets of clamps and then stretched beyond the elastic limit as the clamps pull it over the die configuration. Very little springback occurs because the metal is formed as it is stressed beyond the elastic limit. At present, this process is limited to relatively shallow drawing of sheet metal.

A fourth method of minimizing springback is forming by the Guerin process, using hard rubber or urethane. The female die is eliminated; the hard rubber pad serves as a universal female die, for it can accommodate most common male dies. The male die presses the metal into the rubber pad, displacing the rubber, which then tends to flow up and around the male die under pressure, wrapping and forming the metal to the exact contour of the male die. The rubber is held in a chase designed to guide the flow of the rubber up around the male die. The metal is not formed between hard male and female die surfaces, but rather between one hard surface, the male die, and the relatively soft surface of the rubber. The metal is thus allowed to flow more easily into the desired shape. Extremely sharp bends can therefore be achieved, and such bends are superior in strength in the radius area to bends formed by the bottoming method.

Press-Brake Tooling

The two major types of bending methods are air-bending and bottoming. Because bottoming dies use pressure to cancel out springback, they are machined to the exact radius desired in the bend; for example, a 90° bottoming die produces a 90° bend. In contrast, air-bending dies are machined to 88° to produce a 90° bend.

Air-bending is commonly used for forming heavy sheets and plates over 10 gauge, and bottoming is used only for light-gauge forming (about 14 gauge and lighter).

Gooseneck dies permit forming of two bends close to each other. The die is designed so that the second bend may be made without the die interfering with the return flange. This design, however, makes it mandatory that the die not be overloaded, and thus it should not be bottomed. Gooseneck dies can also be provided with a radius on the tip for other than sharp corner bends.

Hemming and seaming dies can consist of a sharp knife die set and a flat die set. The operation is performed in two strokes, with the knife die making the first bend and the flat die closing up the hem or seam. The flattening stroke requires the most tonnage, and there is often no need to bottom the knife die to produce the required bend.

Bead-forming dies are available for forming two main types of beads: the "central" bead and the "tangential" bead.

To obtain a good bead in thick material, or a small bead in thin material, a wire mandrel should be used during the final closing operation. The wire may be left inside the finished bead or be withdrawn, depending on the final use of the product and the tightness of the bead. It is important when forming the bead to keep the burred edges of the metal to the inside of the bend to avoid scoring of the die surface.

Bottoming methods are not recommended because of the die shapes involved. Generally, the pressure required to form beads without bottoming is about four times that of a standard 90° bend.

Pipe-forming dies are not merely scaled-up versions of bead-forming dies. Small tubes or pipes are best formed as with three strokes required. Hydraulic press brakes are especially useful in forming pipe because they can be set to hold the pipe closed while the pipe seam is being sealed by tack welding or banding.

Medium-size pipe can be bumped progressively on larger-radius dies. Each of the edge quadrants should be formed first and the two center quadrants last. Very large pipes may be formed by this method, using larger-radius dies and forming a half-circle at a time. Two half-circles are then welded together to make a complete pipe section. Large-diameter pipe will probably be too large to fit in the press-brake throat, but it can be accommodated between the side housings or on a horn, depending on its length. An alternative method is to preform the edges in a press brake and form the remainder of the pipe in a roll-former. This method has been successfully used for extremely thick metal that is difficult to get started in the roll-former because of the length of the pipe.

Channel-forming dies can consist of a standard V-die set or a gooseneck die, depending on the size of the channel. Two strokes are needed, and each bend should be accurately gauged. This method reduces die cost, minimizes tonnage, and ensures that the bottom and sides of the channel remain flat.

Channels can also be formed in a single stroke, which is not only faster but also more accurate. The drawback is that channel-forming dies are single-purpose dies, are relatively expensive, and require at least four times the tonnage of 90° air bends and thus a larger press brake. Production quantities therefore must justify the added expense.

Deep channels are possible with standard V-dies. A reverse bend is made in the center of the metal to open up the channel so that the straight knife die will fit to form the bottom corners. After both bottom corners have been formed, the bottom is straightened to finish the deep channel. Hydraulic press brakes are especially well adapted to this operation, because a long stroke is required to open the dies far enough to remove the part.

Radius-Forming Dies. A standard knife die and V-die set can be used for small-radius bends on sheet metal. The knife die should have a radius no smaller than the required bend radius. As the metal thickness increases, greater allowance must be made for springback because this is an air-bending operation. Larger radii may require a male die with a machined radius. If this radius is approximately ¾ in. (19 mm) or less, the metal will follow the male die in air-bending over a channel type die or a simple V-die. If the radius is larger and a full 90° bend is required, the metal will leave the male die as the die moves down. In this case, a matched bottom radius die is needed to force the metal back against the male die contour. This becomes a bottoming operation with higher tonnage requirements, and, of course, added expense for the matched die set.

Progressive bumping can also be used to form radii in short increments. For best results, the knife die should have a well-rounded nose. To avoid flat sections under the nose of the male die, reducing the V-die opening is recommended. The advantages of this method are low die cost and the ability to compensate for differences in springback. Any desired radius can be formed by adjusting the length of the stroke.

U-Section Dies. U-sections can be formed with simple radius dies by the bottoming method or with springloaded pad dies. In both cases, a second stroke with flat dies may be used to compensate for springback.

Offset dies can perform two close bends in one stroke, increasing production up to 500% compared with single bends, but the tonnage requirement can range from 3 to 15 times that of a simple 90° bend. Pressures are highest for Z-bends with sharp right-angle corners. Offset dies are also more expensive.

Offset bends can also be made in two strokes with standard V-dies including a narrow bottom V-die to allow clearance for the sheet to fold down. The sheet must be turned over between strokes, but the tonnage requirements are low, and a single standard die size can be used to form many different shapes.

The illustrations at the end of this article show a variety of press-brake operations and dies.

Hard rubber or urethane pad dies are becoming more widely known and used. The pad is enclosed in a chase that forces the rubber to flow up and around the male die as the latter presses down into the die. This method is advantageous for forming highly polished or plated sheets that must not be scratched during forming. The metal is formed to the exact contour of the male die, and therefore the shape of the male die is critical.

The pad material has a limited life, but simple precautions can be followed to minimize the frequency of replacement. For example, sharp corners such as those found around holes or slots should not be pressed into the material, to protect against cuts; the material should be free to expand a little sideways; the length of the material should be the same as that of the part being formed to prevent "digging in" at the ends; and in general, the die should be designed so that the male die penetration is as shallow as possible. Deep penetration can cause extrusion and splitting of the material.

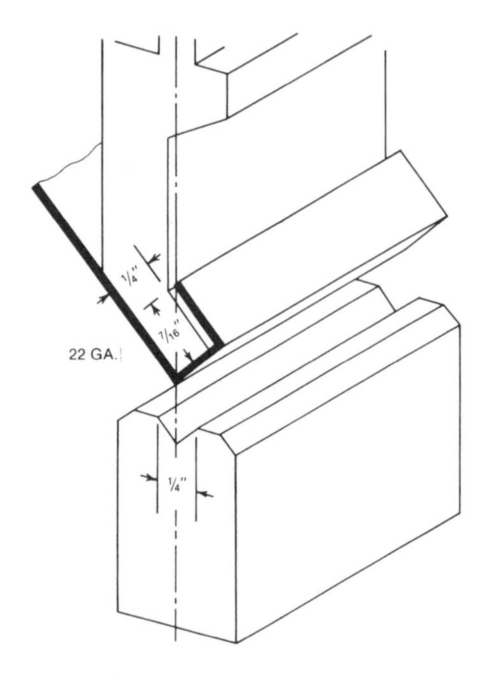

90° forming gooseneck punches

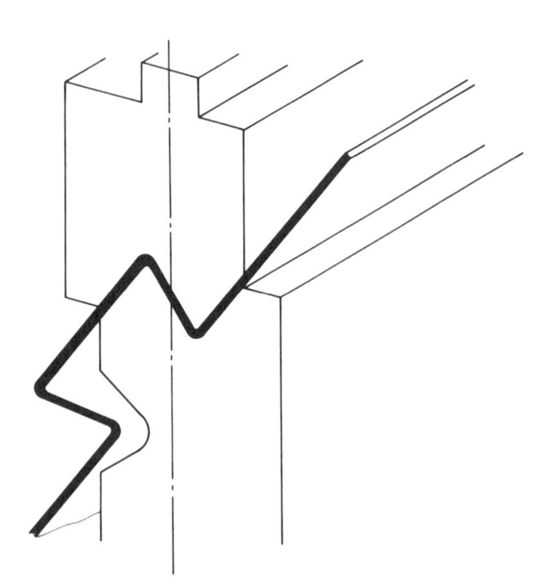

Acute-angle offset dies

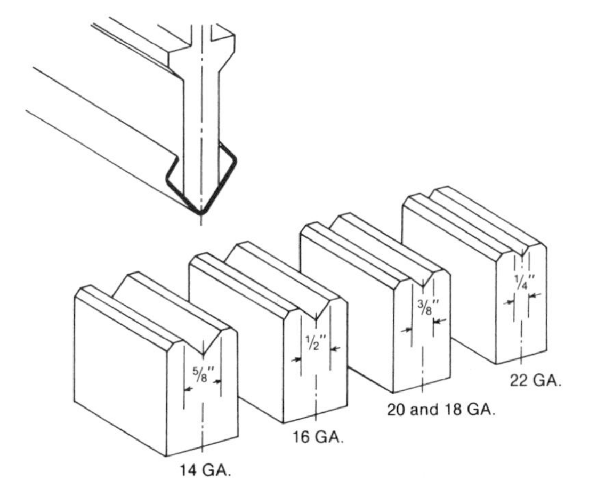

90° forming punches and dies

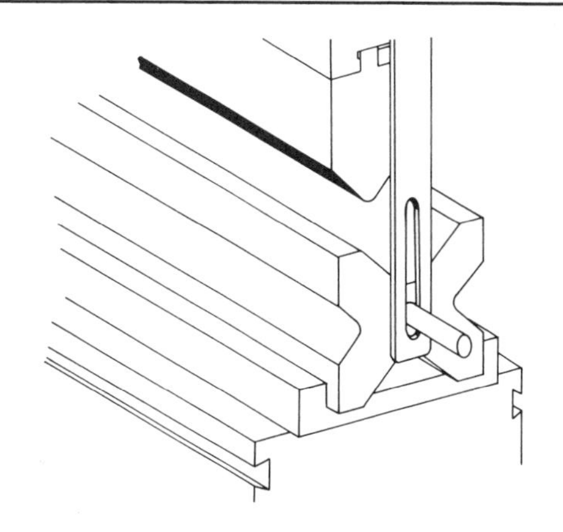

Four-way dies

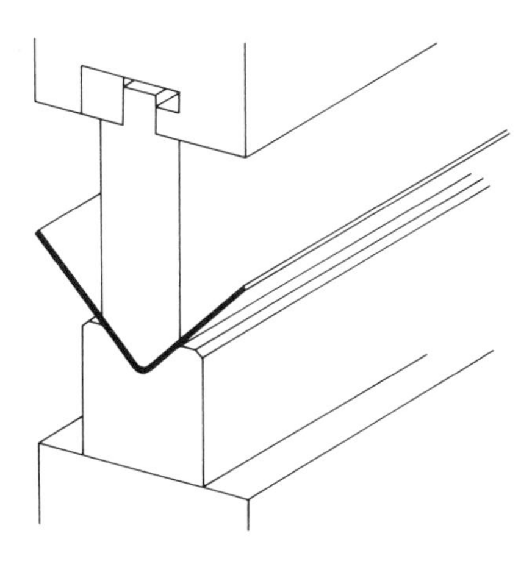

Principle of bending on a press brake

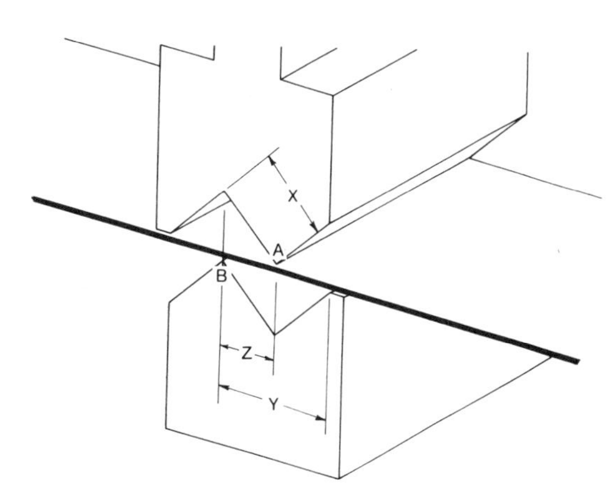

Offset dies

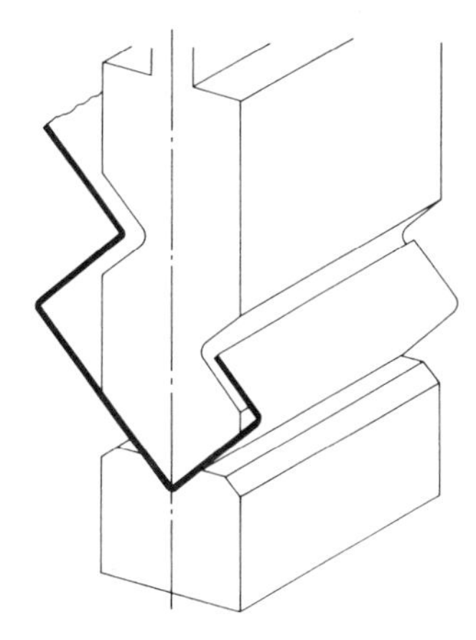

Typical modification of Chicago standard-type press-brake dies

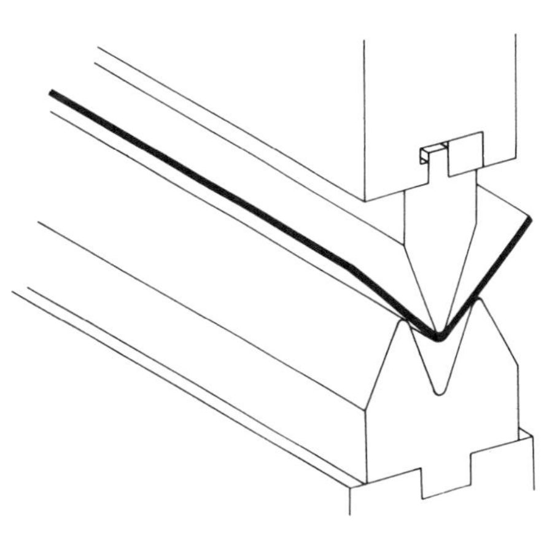

Principle of forming on a press brake

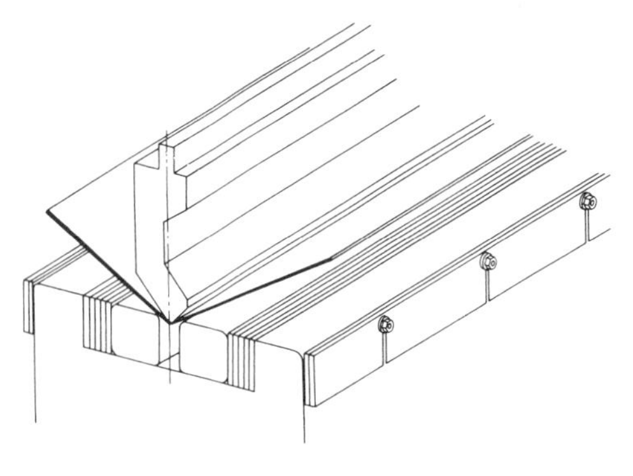

Adjustable female dies

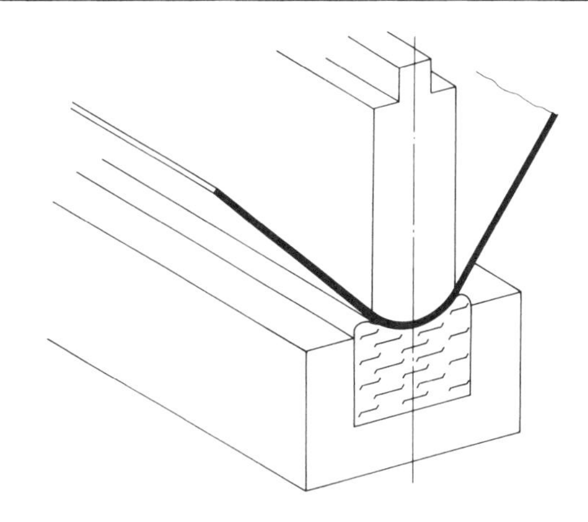

Radius-forming dies

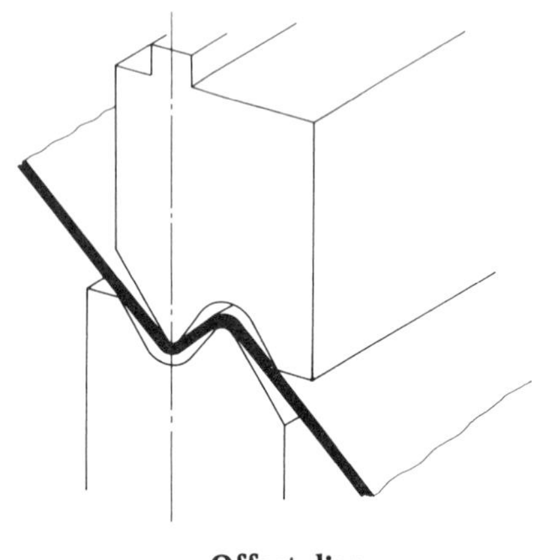

Offset dies

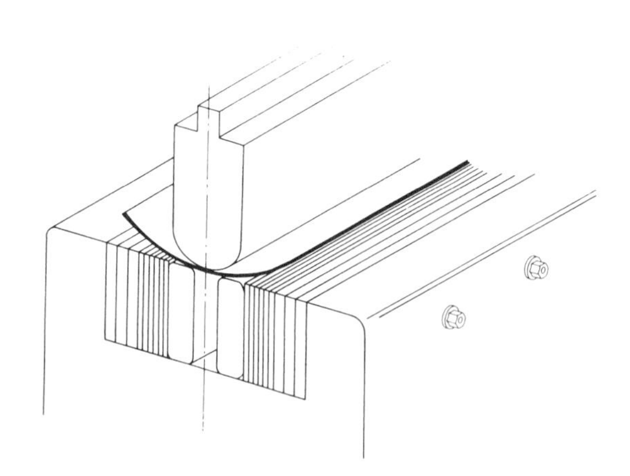

Adjustable female dies

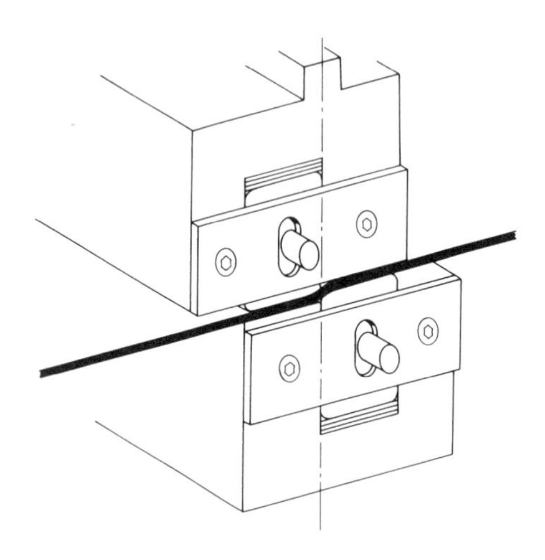

Adjustable offset dies

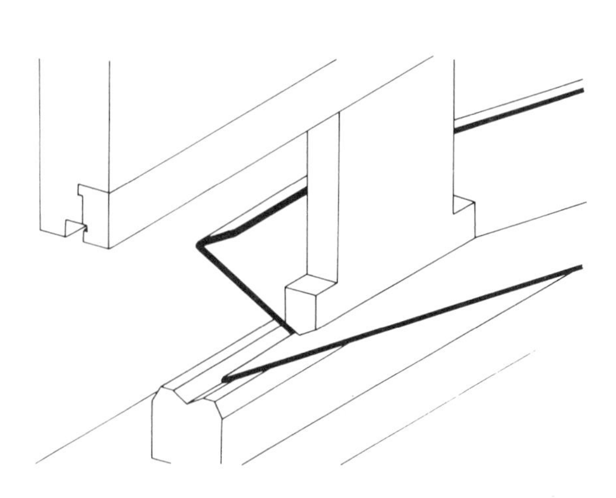

Box-forming dies

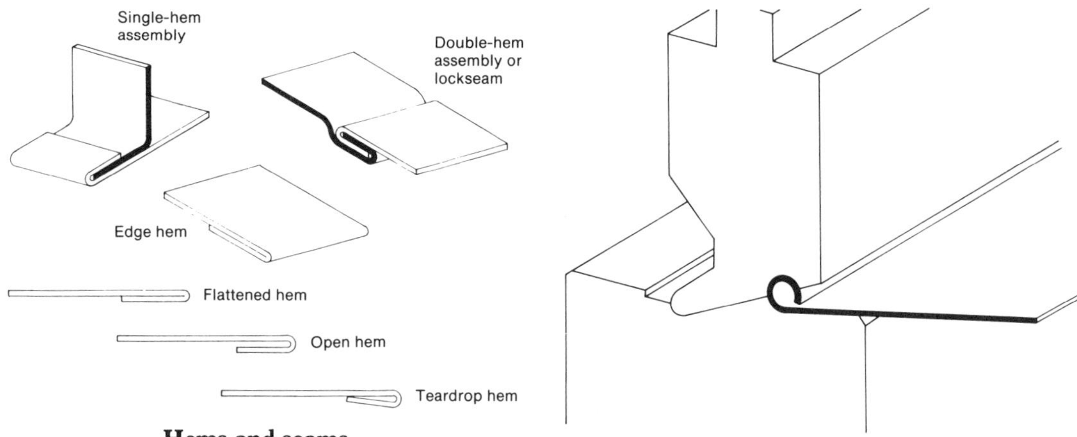

Hems and seams

Curling dies

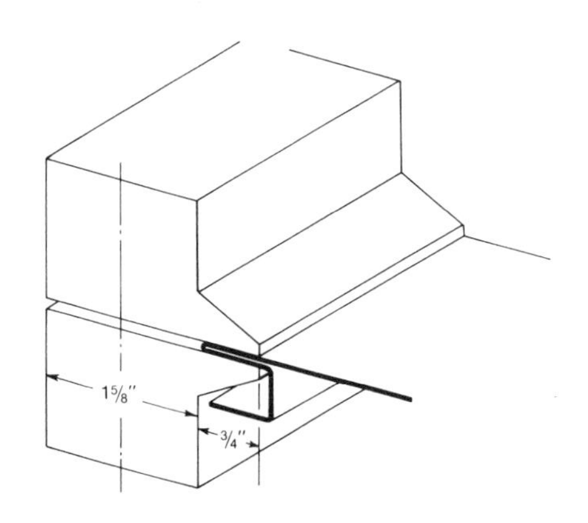

Hemming die set

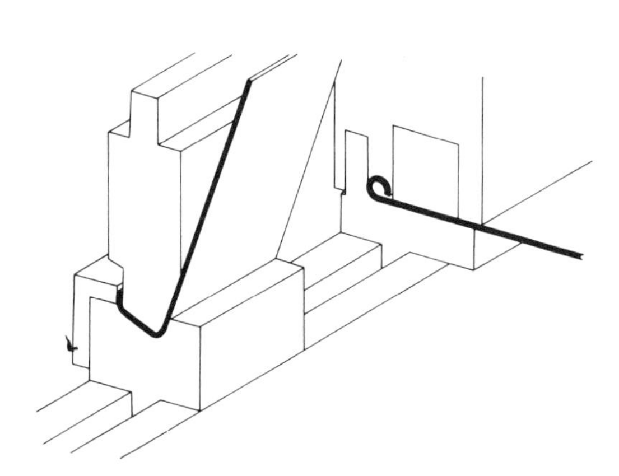

Curling dies

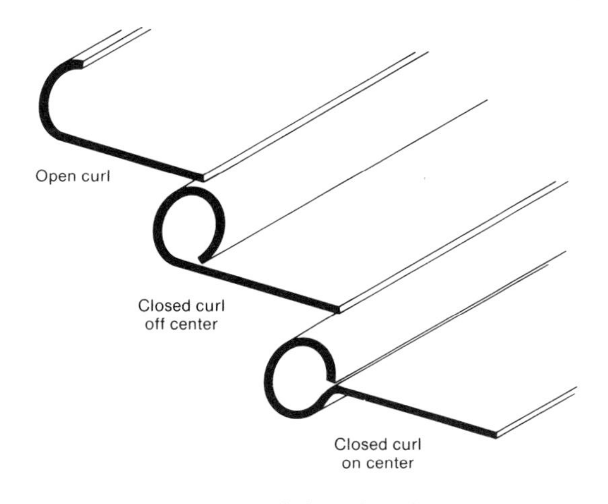

Open and closed curls

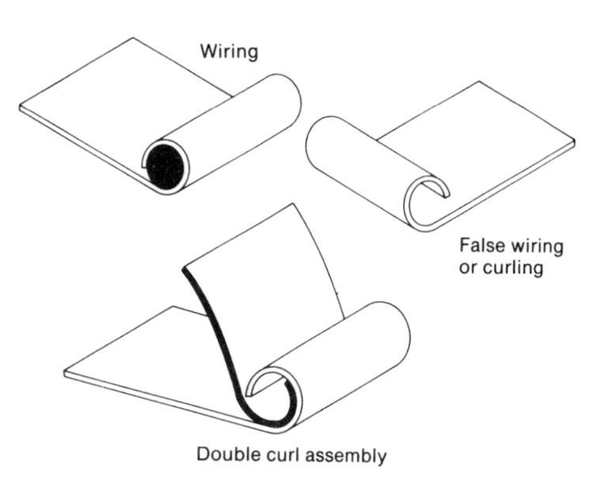

Wiring and curling

1D: Power Squaring Shears: Design and Application

Power squaring shears, once considered to be almost a necessary evil, are coming into their own. In keeping with the increasing demands of the industry for greater production and precision, power squaring shears have been improved to the point where they are now considered to be extremely accurate production machine tools. Most of these changes have taken place in recent years as builders have realized that industry is demanding more return on investments in equipment.

It is well known that one power squaring shear can produce blanks for any number of succeeding operations and that in many cases only one shear is required for the average shop. This makes the power squaring shear an extremely valuable and important tool with so many other machines and operations depending on it. Therefore, it is vital that the proper power shear be selected based on factors which will be discussed in detail in this article.

First and foremost in the selection of a new power shear is the job which it will be expected to do. Some of the questions to be considered are:

1. High or low production output
2. Nominal or extremely accurate off-cut blanks
3. Thickness of material
4. Type of material
5. Maximum sheet size to be sheared
6. Maximum off-cut length
7. Skill of the operator.

All of these factors must be taken into consideration in choosing a machine on the basis of the three major variables, which are:

1. Iron vs. steel construction
2. Mechanical vs. hydraulic power
3. Overdrive vs. underdrive.

IRON VS. STEEL CONSTRUCTION

Many arguments have been advanced in favor of both types of construction. Properly designed machines of steel construction with underdrive, either welded or bolted, generally produce excellent shearing results (Figure 1D-1). Shock-vibration transmission, a

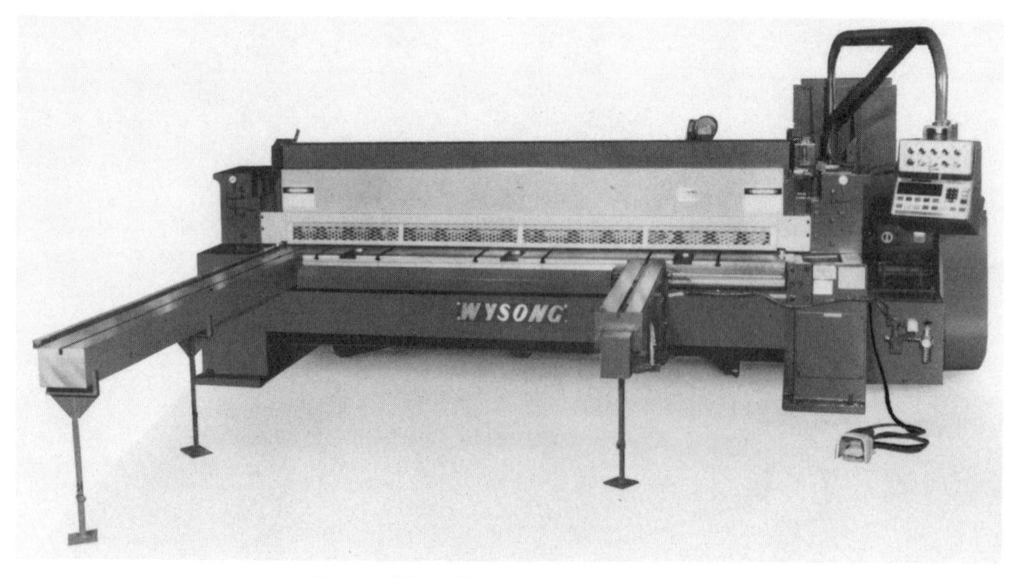

Figure 1D-1. Power squaring shear

characteristic of steel, may make it necessary to provide a special foundation for steel shears. Also, extreme caution should be exercised to avoid accidental overload, which can seriously damage these machines.

Nodular cast iron machines are constructed from high-tensile-strength nodular or ductile iron castings. The tensile strength of nodular iron is about 40% greater than that of steel while the damping qualities are the same as those of high-grade gray iron castings. Special foundations are not required for nodular cast iron machines because of the shock-absorbing qualities of the iron castings. Also, it is generally found that nodular cast iron machines are somewhat heavier than steel-fabricated shears, which means that the cross-sectional area of the major frame components of nodular iron machines is greater.

Aside from the shock-damping characteristics of nodular iron castings, shear material, as used by major builders, is not a factor in shear accuracy provided the structural design for the material is sound. It must be remembered that the major load on a shear is a shifting or traveling load and that torsional stresses as well as horizontal and vertical stresses are developed during the cut. To ensure accurate shearing, the crosshead and bed must be rigid enough to resist deflection in both the horizontal and vertical planes and also twisting from torsional stresses.

Since steel weldments and iron castings have their own individual advantages and characteristics, it has been determined that the most advantageous approach is a combination of the two designs. The bed and ram are the blade-carrying members, and it is ideal that they be fabricated from steel with design characteristics that provide the greatest possible degree of rigidity or stiffness. The end-frame members supporting the bed and ram are ideally made from ductile iron castings to absorb the shock of shearing and prevent its transmission to the foundation. This design approach in effect combines the advantages of both concepts in a unit that is designed to provide the ultimate in accurate, production shearing capabilities.

MECHANICAL VS. HYDRAULIC

Other factors being equal, mechanical shears are faster than hydraulic shears. Generally, a mechanical shear will operate approximately twice as fast as a hydraulic model of comparable capacity (Figure 1D-2). The hydraulic machine is designed to operate at 30 strokes per minute. The compar-

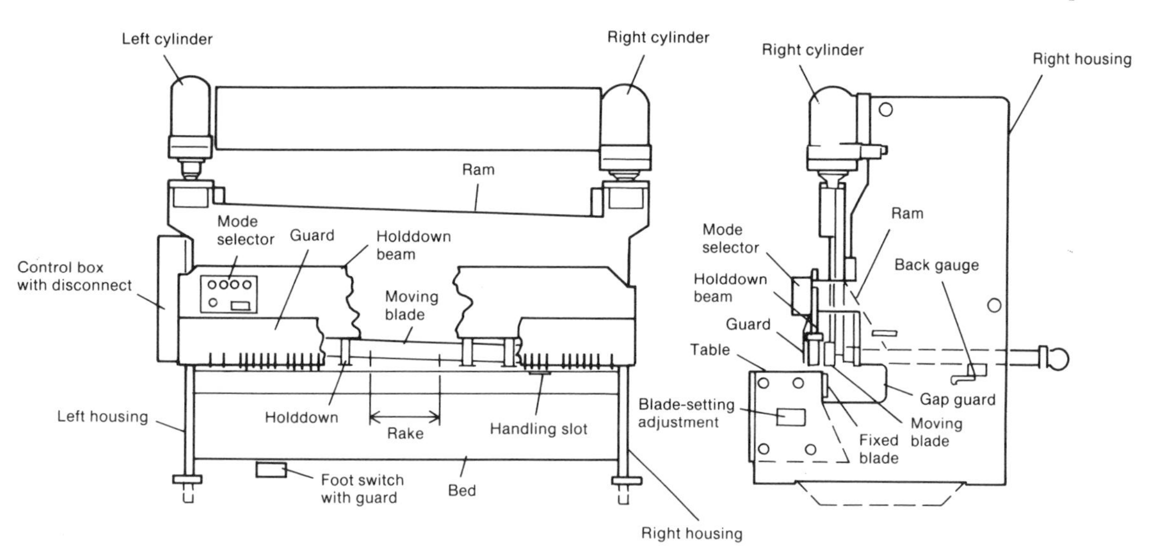

Figure 1D-2. Typical hydraulic power shear of the squaring, overdrive type

able-size mechanical shear (Figure 1D-3) will operate at 60 strokes per minute.

Hydraulic shears are usually associated with heavy plate fabrication where operating speed is not critical. The major advantage of the hydraulic design is the provision for rake adjustment. By increasing the rake angle, heavier metal thickness can be handled and the capacity of the shear extended. It should be remembered that when this is done with most makes, the knives must overlap at the low end so that the available shearing length is reduced. Of course, the higher rake angle will have a decided effect on the camber, curl, and bow of the off-cut.

In the shearing range from light gauge stock through ½-in. (13-mm) plate, the trend is toward higher, more accurate production, which favors the mechanical shear. In this capacity range, another factor is initial cost, which is generally lower for the mechanical model.

The choice between hydraulic and mechanical equipment will be strongly influenced by the nature of the work for which it is purchased. In most cases, the decision will be fairly clear cut.

UNDERDRIVE VS. OVERDRIVE

Many years ago, the overdrive gap-frame shear was very popular because the operator could make an adjustment in the position of the knifebar and notch or slit with the aid of the gap. With this design, slitting is accomplished by moving the sheet through the shear and making multiple cuts without overlapping the blades at the left-hand side. In order to use the slitting unit, it is necessary to adjust the position of the knifebar to prevent overlap of the blades. After the slitting operation is completed, another adjustment is required to lower the knifebar to its normal position. The time required to make these adjustments oftentimes will cost more than the savings realized in slitting long sheets. Due to the fact that the drive mechanism in a gap shear is mounted over the gap, the twisting load created during the shearing cycle tends to open the gap in a "breathing" action. In order to reduce deflection at the end frames, it is necessary that extremely heavy fabricated plates be used along with additional ribbing in the form of cross ties between the end frames and behind the knifebar. The breathing action will have a tendency to change the blade setting during the shearing cycle, but this condition is not objectionable if the blades are not set for extremely accurate shearing or for thin-gauge stock.

The underdrive, gapless-frame machine is a low silhouette unit which allows the operator full visibility around the machine. The fully enclosed end frames add additional stability to the frame components, and thus it is easier to maintain an accurate blade set-

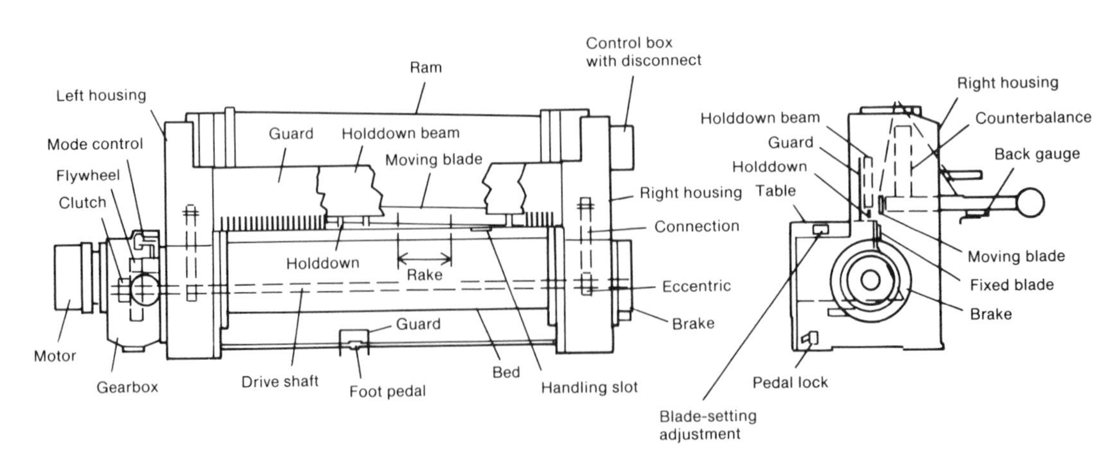

Figure 1D-3. Typical mechanical power shear of the squaring, underdrive type

ting throughout the shearing operation. This is extremely important when working with very close blade settings.

OTHER FACTORS IN SELECTION

Also essential to accurate production shearing are adequate bearing areas and gibbing to guide and confine the ram. As an aid in maintaining accurate blade settings, either the knifebar gibway or the guide gibs should have a nonmetallic liner. Phenolic Grade C is normally used because it absorbs a slight amount of lubricant and develops an extremely smooth operating surface. With nonmetallic liners, it is not necessary to provide running clearance in the gibways so that the knifebar cannot move during the cutting cycle. A variation of ±0.001 in. (±0.025 mm) is not unusual for the blade setting, and with hand scraped blade seats, a setting of 0.002 in. (0.05 mm) or less may be maintained. These very close blade settings are required for shearing of thin, high-tensile-strength sheets and some types of plastics and paper. The nonmetallic liners eliminate knifebar "float" and guarantee that precise blade settings are maintained throughout the cutting cycle. To lengthen blade life, it is recommended that the blades be set to the widest possible clearance to produce good shearing results on the thinnest material to be sheared. The same blade setting may then be used to shear all materials up to rated capacity.

RAKE, RAM ANGLE, AND CLEARANCE

Rake is the slope of the knife from right to left. The amount of rake is determined by shear design with definite consideration given to shearing results. Too-high or too-low rake will have a definite effect on camber, twist, and bow of the sheared material. The load required to shear the material depends on the thickness and type of material and the rake of the upper blade. In a normal shearing situation, a portion of the material is sheared and the rest is broken through due to the shearing action. For example, in the case of mild steel up to ¼ in. (6.4 mm) thick, we assume that one-half of the material will be sheared and the balance will break. Therefore, in calculating the required shear load, only one-half of the shear strength is used, and the formula is as follows:

$$\text{Shearing load} = \frac{T^2S}{2 \tan \text{slope}}$$

where T is metal thickness, S is shear load of the material allowing for breakthrough, and tan slope is equal to slope (in inches) divided by 12.

The shearing load increases rapidly as the metal thickness increases. For example, ⅜-in. (9.5-mm) mild steel is only 50% thicker than ¼-in. (6.4-mm) mild steel, but the shear load increases 225%. For this reason, it is not recommended that a machine be used for cutting heavier than rated capacity even in narrow pieces.

Design engineers generally agree that the knifebar should move in a plane slightly away from vertical, as shown in Figure 1D-4. This will cause the upper blade to come in contact with the stock immediately above the edge of the lower blade, resulting in a square sheared edge. As the blade moves downward at a slight angle from vertical, the

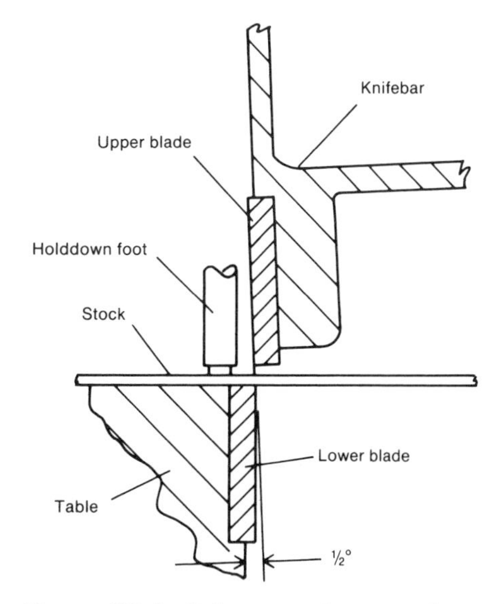

Figure 1D-4. Schematic showing plane of movement of knifebar in shearing

blade clearance at the point of edge contact will come into play and the clearance will increase as the blades overlap. This additional clearance is required to prevent damage to four-edge blades. The clearance at overlap of the blades is also an aid in shedding narrow strips as they are cut. Actual tests have indicated that an angle from ½ to 1° is optimum and produces the best-quality sheared edge. A greater or lesser degree of angle creates greater deformation in the stock during the shearing operation, thereby affecting squareness of cut and edge condition.

If the operator is certain that the shear is rigid and stable enough to resist deflection, the blades may then be set for shearing a wide range of thicknesses from light-gauge material through capacity. It should be remembered, however, that best shearing results are obtained when the blades are accurately set for the thickness to be sheared. Therefore, the edge conditions will vary between the lighter and heavier gauges if only one setting is used for all gauges. Normally, however, the edge differences are slight and quite acceptable as compared with the time required for adjusting the blades for each thickness.

Use of a fixed minimum setting eliminates the possibility of shearing thin stock with too much clearance. When this happens, the shear piece can wedge between the knives, causing severe jamming and possible damage to the machine. Correct initial setting and accurate grinding of the knives are essential to good shear performance.

The shearing process is extremely complex, involving such factors as crystal structure, slippage planes, brittle fracture, and anisotropy. These are of interest primarily to the metallurgist. At the same time, it isn't enough for the shear user to say, "The knife goes about one-half of the way through and then the piece breaks off," even though this is essentially true and can be confirmed by examining the edge of the off-cut.

It is known that both tensile and compressive stresses are involved. As the knife contacts the metal being sheared, the top surface is momentarily under tension while the bottom, supported by the lower knife, is under compression. As the elastic limit of the metal is exceeded, it is stressed in shear until its ultimate strength is exceeded, at which time it fractures, breaking completely away from the parent metal. If knives are sharp and the clearance is correct, the sheared edge will be clean with little burr and will be very close to the perpendicular.

TWIST, CAMBER, AND BOW

The internal stresses that are present during shearing help to explain the three most common bugaboos in shearing: twist, camber, and bow.

Twist is perhaps the most common condition. This is the tendency of the off-cut to curl into a spiral or corkscrew. Once twist is present in the off-cut, it is difficult to remove. Twist is a function of internal stresses but also relates to the rake or slope of the top knife. Too much or too little rake will have a definite effect on the amount of twist. It has been generally determined that the range from 5/16 to 7/16 in./ft (26 to 36 mm/m) is nominal for metal thicknesses up to ½ in. (13 mm). A degree of rake much higher will increase the amount of twist, and a much lower rake will create a twisting action due to excessive forging of the edge. Rake is also affected by stresses developed within the material during the shearing action, and so it follows that the wider the off-cut piece, the less the tendency to twist. Heavy stock will twist more than thin stock and soft stock will twist more than hard material. Dull blades have an effect on twist due to forging of the edge, and so it is recommended that blades be kept as sharp as possible. Excessive blade clearance also affects the degree of twist.

Camber is the tendency of a sheared strip to take on an arc while lying flat. The rake of the shear blade has little effect on camber, which is usually caused almost entirely by stresses within the material. Knife adjustment, to a small degree, affects camber, but inferior material with excessive internal stresses is the greatest offender. If camber is severe after leveling and knife adjustment have been checked, the only answer is to overshear and edge trim. As with twisting, camber is most severe on narrow strips. In shearing of material into strips, fine-quality material should be used. In any operation where camber is objectionable, it is recommended that stress-relieved stock be used.

Bow is the tendency of a piece of material to hump in the center. Again, bow is almost always due to deficiencies of stresses within the material. If there is considerable variation in thickness across the sheet, bowing will result, and very little corrective action can be made at the shear. If bow is a serious problem, it is recommended that high-quality, stress-relieved stock be considered.

DETERMINATION OF SHEARING METHOD

In a well-organized shop, the shear department foreman usually lays out the work, determining which gauging method should be used and what the cutting procedure will be. On fairly large orders, this is by no means a simple problem. It is so complex that it actually has a name — "The Two-Dimensional Cutting Stock Problem." It is defined as "finding the least wasteful and most economical way of cutting large sheets of stock into smaller rectangles to meet incoming orders." Given a large number of orders, there are literally millions of possible combinations. Very recently, IBM Corporation developed a program for the IBM 7094 computer aimed solely at solving this problem.

The average shear used will never require a computer for laying out shearing jobs, but the problem is still a difficult one. For example, there are more than 240 ways of shearing 3-by-6-ft (0.9-by-1.8-m) sheet into 3-by-6-in. (76-by-152-mm) blanks. More than half of them are wrong in that they waste stock. More than three-quarters of them are wrong because they require extra operations. Here are recommended methods for two very common shearing applications. Note that many other shearing methods are possible.

1. **Shearing small blanks from rectangular sheet.** Assuming a 3-by-6-ft sheet is to be sheared into small rectangular blanks, the best procedure will usually be to trim one long edge of the sheet by gauging the end against the squaring arm. Judgment now enters the picture. Whichever dimension of the small rectangle divides most evenly into the width of the sheet should be set on the back gauge. The sheet should then be sheared into strips using the back gauge. The sheared strips are now trimmed using the squaring arm. They are fed through and gauged against the back gauge. (The value of power-operated back gauges for this kind of work should be readily apparent.) The last strip will have to be turned around and gauged against the front stops. A skilled operator, feeding the stock firmly against gauges, can work within 0.002 to 0.004 in. (0.05 to 0.10 mm) and better with this method.

2. **Shearing large blanks from thin stock.** This common situation creates a problem because thin sheet will sag if it is not supported, which makes the use of back gauges impractical. Usually, the best procedure is to trim one long edge of the sheet by gauging against disappearing two-position stops in the support arms. Now one short end can be trimmed square by locating the trimmed long edge against the squaring arm and gauging against a preset stop therein. The sheet is then reversed, end for end, and the opposite short end is edge trimmed, gauging the same trimmed long edge against the squaring arm. The sheet is then turned 90° and the fourth trim cut is made by gauging the short end against the stop in the squaring arm. We now have a square sheet which can be cut into blanks of the required width by gauging against swinging stops in the squaring arm. This job requires more, slower, and less-accurate operations than back gauging. If the quantity of thin material being processed is large, magnetic sheet supports should be specified. With them, even thin, ferrous material can be cut more rapidly and more accurately.

MAINTENANCE AND SAFETY

No hard and fast rules can be set down for shear maintenance. The common-sense practices that apply to any precision machine tool should be followed. All shear manufacturers supply maintenance manuals with their products. Many of them are quite complete and contain a great deal of useful information. A maintenance schedule in keeping with the manual should be set up and adhered to.

Good practice requires that the shear be located centrally to reduce material handling. This is a much-neglected practice. In many shops, it sometimes appears that the shear is placed wherever there happens to be an open space. If the shear is serving a battery of presses, the traffic in and out of the shear area will be very heavy. Imaginative engineering can do wonders here. By relocating the shear, it may be possible to set up simple conveyor systems that will substantially reduce materials-handling cost.

Properly used, a shear is a safe tool. Statistics indicate that there are fewer accidents involving shears than involving power presses or press brakes. Most of the accidents that do occur are caused by carelessness and poor housekeeping, in that order. Carelessness cannot be eradicated, but it can be reduced by an adequate safety program. Poor housekeeping is inexcusable. The shear that is installed off in a corner is apt to be a dirty shear — perhaps because the supervisor doesn't pass by it every hour or so. Trim is allowed to accumulate. Dirt also accumulates, sometimes in sufficient quantity to affect the performance of the valves and the electrical controls that are the operator's chief safeguard. The dirty shear, badly maintained, is an inefficient mechanism. The well-maintained shear, properly adjusted to do its job and used as any precision tool should be used, is one of the most productive and accurate machines in the plant.

HIGH-SPEED AND SPECIAL SHEARS

As shearing production requirements steadily increase, the need for higher-speed blanking shears is also increasing. Shears are now available with a cyclic rate of 200 or 300 strokes per minute. The 200-stroke machines are generally used in a cut-to-length line or for high-speed blanking operations following a coil line.

There are relatively standard machines which utilize an air clutch and air brake rather than the slower jaw-type clutch. With dc controls, the time cycle for engaging is constant, and the shear timing cycle is used for gauging. In a coil line, the edge of the sheet contacts a position stop and microswitch arrangement actuating the shear, which makes a quick cycle while the coil continues to move, with a hump forming in front of the machine. As soon as the hold-down releases the sheet, the hump drops and the cycle is repeated. Forty or fifty continuous cuts per minute on a single-stroke basis are possible with this type of equipment.

For high-speed blanking, the 200-stroke machine is equipped with an automatic probe gauge utilizing electronic sensors in the back-gauge angle. As the edge of the sheet comes into contact with the activated sensors, the machine is automatically actuated through a remote-control clutch-tripping device.

Three or four sensors are mounted in the back gauge, depending on shear length. Any one or a combination of two or three sensors may be activated, and a pilot light indicates when contact with the sensor is made. Should the blank edge not become seated firmly against the back-gauge angle, the sensor contact will not be completed and the pilot light will indicate which sensor is involved. With this arrangement, the operator is responsible only for feeding the material into the machine and making sure that it is seated firmly against the back-gauge angle. A selector switch allows the operator to actuate the machine automatically with the probe gauge or manually with the foot switch.

The 300-stroke-per-minute high-speed shears are designed especially for coil-line operation. These machines have the reversed knifebar design with all knifebar ribbing toward the front of the machine. There are no overhanging projections behind the machine, allowing the conveying equipment to be moved very close to the edges of the blades.

The top blade is of high-carbon, high-chromium bowtie design. The bowtie top blade shears flat with little camber or curl in the sheared blanks. With the knifebar ribbing on the front, the standard holddown is eliminated and a special pneumatic holddown is normally used. The 300-stroke high-speed shears are used in extremely high-production coil lines and will make from 60 to 70 cuts per minute.

Industry is placing ever-increasing demands on shear manufacturers for increased production and efficiency. With high-speed machines, the actual shear time is only 3% of the total time required for the complete shearing operation. The balance of time is required for handing the material into the machine in the front or away from the shear in the back. A new system is now available for supporting, conveying, and stacking sheared blanks on the back of the machine. This system includes sheet lifters to hold the stock level with the bed line, thereby eliminating inaccuracies which might be caused by drooping of the stock. After the blank has been sheared, it drops onto the conveyor and is carried to a stacking device at the rear of the machine. A built-in scrap separator is actuated by the operator when a scrap or trim cut is made. The scrap separator closes automatically to pass production blanks to the stacker. The scrap box is mounted on rollers for easy scrap removal. When used with an automatic probe-gauge shear, the handling equipment becomes a semiautomatic shearing system designed to provide the ultimate in high-production, accurate shearing operations.

In some shearing operations, back-gauge accuracy of 0.002 or 0.003 in. (0.05 or 0.075 mm) is just not good enough. For example, assume that a large sheet-metal blank is prepunched and it is required that the blank be sheared into smaller pieces with a tolerance of ±0.001 in. (±0.025 mm) between the punched holes. The accumulative error of 0.002 in. (0.05 mm) in the back-gauging operation would amount to 0.020 in. (0.5 mm) or more after ten cuts. To eliminate this problem, the back gauge is eliminated, and gauging is done by means of locator pins in the shear bed and the prepunched holes in the blank. The bed of the shear is modified to accommodate a pin-gauging unit with a direct reading system graduated to 0.002 in. (0.05 mm). The steel bushings in the bed portion of the unit are adjustable to a tolerance of ±0.001 in. (±0.025 mm) from the cutting edges of the blades. Just above the pin-gauge unit is mounted an air-operated locator pin which is plunged into the locator bushing. The back-gauge and side-gauge units are used only for rough locating of the prepunched hole over the locator bushing within approximately 0.015 in. (0.38 mm). The air-operated pin is then actuated to accurately position the blank for shearing. Only a few minutes are required for setup, and the machine is then actuated in a normal manner for continuous shearing using the pin-gauge unit. An accuracy of 0.001 in. (0.025 mm) is obtainable provided the holes are accurately punched in the blank.

High-speed shears with a probe gauge and handling equipment and the automatic pin-gauge arrangement constitute just one step in the direction of automatic shearing systems, which may someday be controlled both in the front and the rear by tape-controlled or punch-card systems. The dreams of yesterday's metal fabricators are beginning to come true.

1E: Power Bending Rolls for Metal Cylinders

Bending of steel sheet and plate into plain cylinders, cones, ovals, multiple-ribbed or beaded cylinders, and other circular shapes can be handled readily on modern bending rolls. Such rolls are widely used from the smallest shop to the largest fabricating plant. Rolls vary in size from simple hand units mounted on the worker's bench to machines weighing 500,000 lb (227,000 kg) or more. Bending capacities range from the thinnest-gauge stock to the thickest plate. Regardless of machine size or capacity, the bending operation is similar — namely, production of a cylindrical shape or shell on horizontal rotating rolls.

There are two classes of bending rolls: the initial pinch type and the pyramid type. The roll arrangement determines the classification, as indicated in Figures 1E-1 and 1E-2.

Of the two machines, the initial-pinch-type roll is the most popular and widely used at the present time. In this machine, the three roll forgings are of the same diameter, with one front roll mounted almost directly below the top roll. During the bending operation, the upper roll is in a fixed position between the drive end of the machine and the swing-type end support. When the end support is opened the upper roll automatically pivots upward on its own trunnion. This permits easy removal of the formed section.

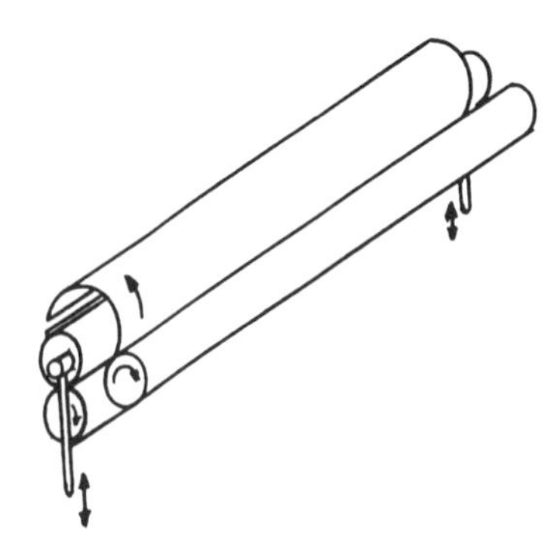

Figure 1E-1. Pinch-type rolls

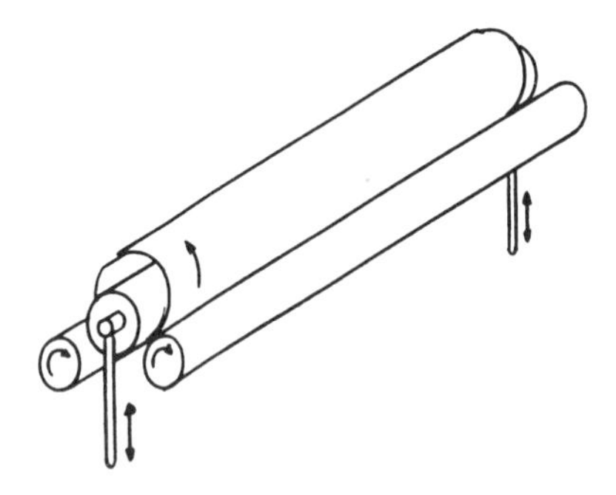

Figure 1E-2. Pyramid-type rolls

Adjustment of the lower front roll (up and down) compensates for various sheet and plate thicknesses. Then, too, this vertical movement provides the pressure needed for the pinching action and friction that carry the stock through the rolls. The rear roll is also adjustable. The diameter of the cylinder or shell to be rolled is determined in part by raising or lowering the rear roll. Also, the diameter of the upper pinch roll must be considered when determining cylinder size. Usually all three rolls are geared and operate at the same peripheral speed. Thus, with positive rotation of the rolls, it is possible to form light-gauge sheet with an effectiveness equal to that of heavy plate. This is possible because the sheet is pulled through the rolls and is not forced to rotate an idler roll, which is common on pyramid-type equipment.

Pinch-type rolls offer several distinct advantages in forming steel sheet and plate. Arrangement of the rolls with positive adjustment and feed makes it possible to form nearly true cylinders with each pass-through of the plate. The flat on the leading edge is determined by the tangent point of contact between the top roll and the point where the material first touches the rear or radius roll, as shown in Figure 1E-3. In order that flats may be eliminated on the leading edges of the material being formed, such edges are preformed by reversing the rotation of the rolls and running the stock from the rear of the rolls. Only a small portion of the sheet or plate is so formed. Ordinarily, if a large number of cylinders or parts are to be made,

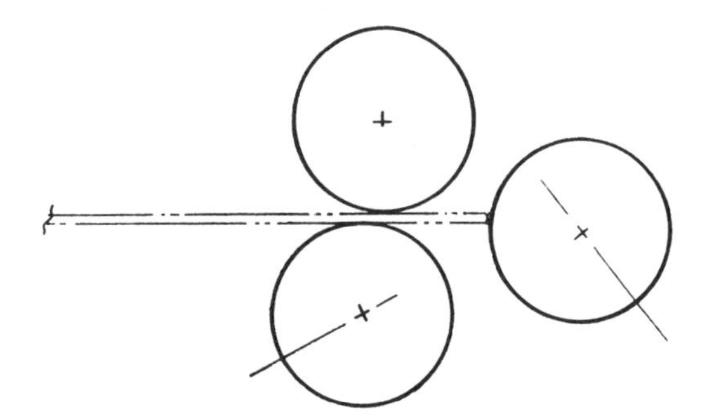

Figure 1E-3. Illustration of how distance from tangent point on upper roll to right-hand roll determines extent of flat on part

the entire run is preformed in this manner to save time. After this initial bending, the material is then run through the front of the rolls in the normal manner. Flats cannot be removed so simply in pyramid rolls. It is necessary to preform the edges on a press brake. This method increases handling time and therefore expense.

At the present time, many pinch-type rolls are being used to form cylindrical shapes to be made into pipe. Ordinarily, such cylinders can be formed having an inside diameter within 1 to 2 in. (25 to 50 mm) of the diameter of the top roll. Assuming that the leading edge is preformed, the best results in bending sheet or plate for pipe purposes are accomplished through a single pass of the stock through the rolls. No attempt should be made to completely close the pipe on the rolls. The cylindrical shape should be formed oversize and then removed for welding. Closure for butting the ends together can be effected in a fixture or by using wire, rope, chain, or any other means which, on being tightened, will close the gap between the edges. Figure 1E-4 is a schematic illustration of a cylinder ready to be closed into a pipe section.

The pyramid-type bending roll consists of three rolls. The lower two, of like diameter,

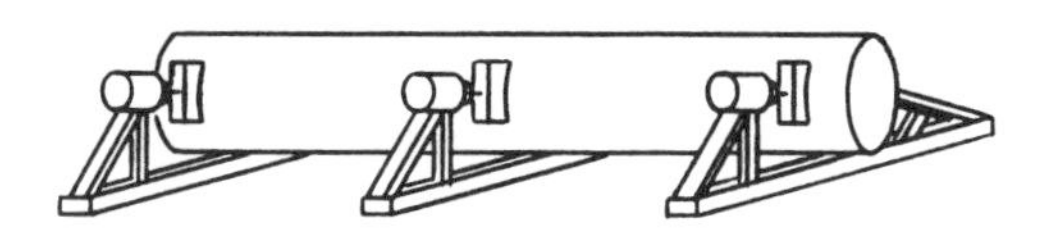

Figure 1E-4. Clamping fixture with hydraulic cylinders

are mounted side by side on the machine frame. The third, or idler, roll, usually larger in diameter, is centered in pyramid fashion about the lower rolls. One end of the idler roll is mounted on a trunnion in the machine drive end. The other end of this roll rotates in a hinged or swing-type end support. In addition, this roll can be raised or lowered, and, through this adjustment, it is possible to vary the diameter of the cylindrical shape produced. Ordinarily, the inside diameter of the cylinder will be 4 to 6 in. (100 to 150 mm) greater than the diameter of the idler roll.

The hinged end support performs the same function on the pyramid machine that it does on the pinch-type roll. It is opened to permit withdrawal of the work (see Figure 1E-5).

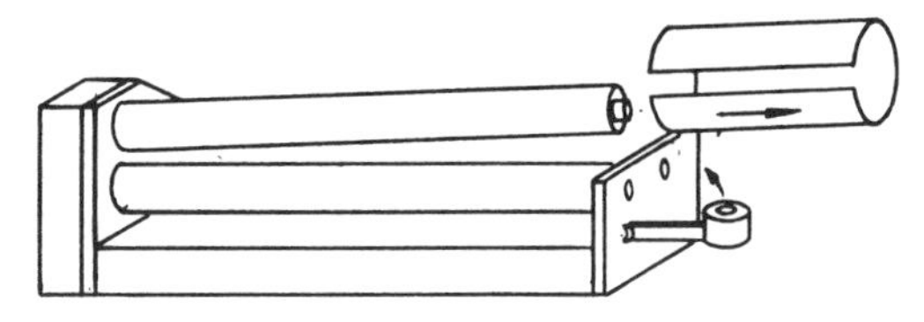

Figure 1E-5. Removal of work from pinch-type rolls

The idler roll on pyramid machines is raised or lowered through power elevating screws on large machines and with hand adjustment screws on smaller machines. Adjustment screws on both ends of the rolls can be synchronized, or, if desired, only one end of the roll need be raised. With this latter adjustment, the roll bender is then set to form conical shapes of the desired size. Some machines are equipped with dial indicators to show the amount of vertical movement. Indicators of this type simplify roll settings for different cylinders and for resetting when running interrupted jobs.

The majority of bending rolls used today range from 4¼ to 8 in. (108 to 203 mm) in diameter and from 3 to 10 ft (0.9 to 3 m) in length. At least one manufacturer is realizing that there is a market for modern, up-to-date bending-roll design, and features are now being offered which were unheard of a few years ago. The better-quality bending rolls are of all-steel, welded construction and are so designed to have torque-resistant frames that minimize deflection.

Some of the features offered by modern bending rolls are:

- **Safety treadle.** This treadle gives the operator full-range accessibility to stop the machine. Many operators also use the treadle as an operational switch.
- **Front operation of rear or radius roll.** The rear roll actually is the radius roll, and the position of this roll determines the diameter of the finished cylinder. The operator controls the position of this roll from the front of the machine by a single hand wheel.
- **Automatic drop-end latch.** By unlatching the drop end and lowering it, the top roll is raised, thereby allowing removal of the cylinder. When the drop end is raised, the top roll repositions itself and is automatically locked in proper alignment. On the larger 6″, 7″ and 8″ diameter rolls, the drop end is operated by air power.

The questions arising from rolling operations are numerous. In order to answer these questions, the principle of rolling must be understood. Rolling is a procedure of controlled deflection of the fibers of the material being rolled until they are permanently deformed. The yield strength of the material must be exceeded and, in some materials, springback must be considered. If the position of the three rolls and the diameter of the cylinder are known, the size of the roll required to do a certain job may be calculated. The first step is to determine the position of the rolls by simple line layout (Figure 1E-6) illustrating relative roll positions with material in place. The load on the rear roll is determined by the following formula:

$$\text{Load on rear roll} = \frac{L(T^2) \times \text{yield strength of material}}{6 \times \text{offset distance}}$$

where L is the length of the roll, T is the thickness of the material, and the offset distance is the distance from the tangent point of contact of the material with the top roll to the point where the material touches the rear roll.

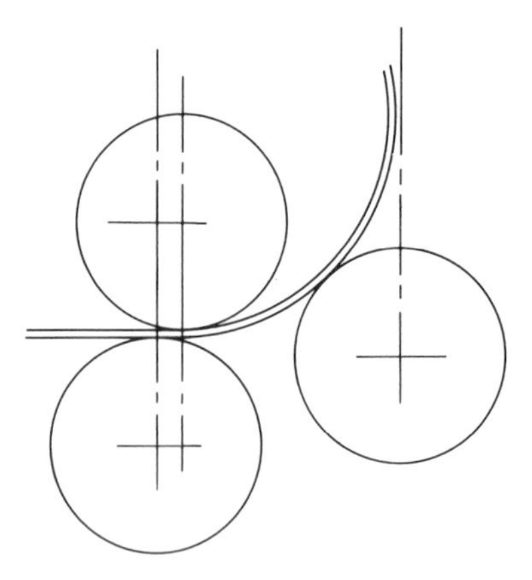

Figure 1E-6. Roll-position layout

After the rear-roll load is calculated, the loads on the top and bottom rolls may be determined as follows:

$$\text{Load on top roll} = (x)\,\frac{c}{a}$$

$$\text{Load on bottom roll} = (x)\,\frac{b}{a}$$

where x is the load on the rear roll, c is the total tangent distance at the point of contact of the material with the bottom and rear rolls, a is the tangent-point-of-contact distance between the front and top rolls, and b is the tangent-point-of-contact distance between the top roll and the point where the material touches the rear roll.

Deflection is always present in any rolling operation and must be considered in choosing a roll for a given job. Deflection of the rolls varies as the cube of the length of the rolls. Therefore, a cylinder formed on a short roll could not be formed on a longer roll due to excessive deflection. As indicated, the greatest load will occur on the top roll and the smallest load on the rear or radius roll. To compensate for this deflection, the rolls are crowned for the heavier loads.

The question "Why not use an alloy steel or tool steel for the rolls to reduce deflection?" is frequently raised. This question is best answered by a bulletin released by the Bethlehem Steel Company:

"WHY USE ALLOY STEEL?
Where deflection is the only limiting factor there is obviously no advantage in using high strength steel. Increasing size, changing shape or shortening the span are the only ways of increasing stiffness of the steel part."

The reason for this is that the modulus of elasticity is practically the same for all types of steel. Where continued deflection is to be considered, alloy steel rolls will provide greater fatigue life than standard carbon steel rolls, but the deflection factor will be the same. To calculate the amount of deflection, the following formula is used:

$$\text{Deflection} = \frac{5}{384} \times \frac{WL^3}{EI}$$

where W is the load exerted on the roll, L is the length of the roll (in inches), E is modulus of elasticity, and I is moment of inertia.

This formula emphasizes that deflection varies as the cube of the length and is little affected by the type of steel used.

Although bending rolls are offered as standard units, every job should be considered special. Each machine should be engineered according to the customer's exact rolling requirements in relation to the type, length, and thickness of the material to be formed. Other factors to be considered are the diameter of the cylinders to be formed and whether or not special grooving is required. It is not enough simply to order a machine with rolls 6 in. in diameter and 96 in. long. It might be necessary to provide special grooving in the rolls, and the particular job involved might require that the rolls be crowned to compensate for deflection. Without proper crowning, the finished cylinder would be larger in diameter at the center than at the ends, and so it is vitally important that specific working requirements be outlined.

Grooves are often required in rolls, ranging from a single groove to several grooves in each roll. The purpose of these grooves is to accept preformed flanges on the blanks to be rolled. If a cylinder is to be formed with one or more flanges toward the inside, grooves must be cut into the top roll only. If, however, the flanges are on the outside of the cylinder, the grooves are required in the bottom and rear rolls. These grooves are usually about 15 to 20% wider than the thickness of the material to be rolled. Cylinders with flanges toward the outside are much easier to roll than cylinders with inside flanges. This is due to the fact that the material is in tension during rolling of outside flanges (the material stretches).

The material has to be compressed in rolling of inside flanges. This is a more difficult operation because the loads are greater and the grooves must have little or no clearance for the material thickness in order to prevent distortion in the flanges. A cylinder with symmetrical flanging (flanges on both ends) is much easier to control than one with a single flange. Where a cylinder has only one flange, the resistance to rolling is greater on that side and a "feed-in" table is required to maintain alignment of the blank with the groove. This is not true where flanges of the same depth are located on opposite ends of the cylinder, because of the reactive resistance of the opposing flanges balancing the load.

Rolls are often grooved for rolling small angle shapes, as in Figure 1E-7. This is a difficult operation where a single part is to be rolled, because of the eccentric load resulting from rolling an angular shape. The

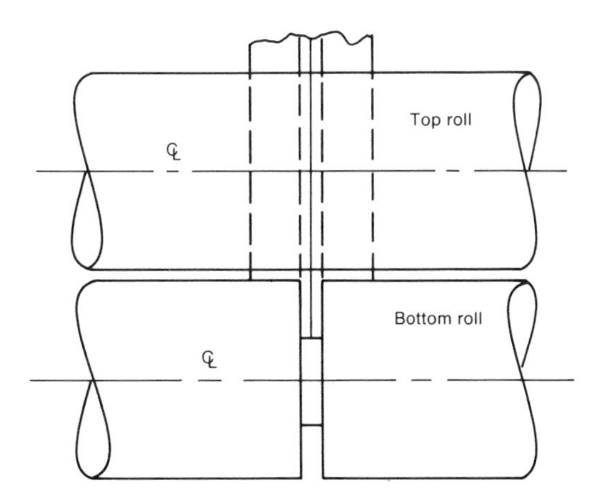

Figure 1E-7. Grooved rolls, which allow forming of small flanged parts

flanged section extending into a groove in the roll offers more resistance to being rolled than the flat area between the rolls. This causes the part being formed to take on a helical shape similar to that of a coil spring.

This problem exists only when rolling a single angle shape. By simply having grooves in the rolls wide enough to accommodate a double thickness of the angles, the parts are rolled back to back. The reactive force of one angle counteracts that of the other angle, and two perfectly formed rings result.

Cones may be formed on bending rolls through the use of a cone-rolling attachment. This is a device attached to the right-hand end of the machine adjacent to the top and bottom rolls. This attachment makes cone rolling possible by retarding the speed at which the minor radius is fed through the rolls while the larger or major radius is fed through the rolls at the peripheral speed of the rolls. To accomplish this, the left end of the bottom roll must be adjusted to pinch the cone blank while the right end (adjacent to the attachment) must be adjusted loose to allow the material to slip. Where cones with steep angles are to be rolled, it is sometimes necessary to misalign the rear roll to give additional angularity in the rolled part. We might add that forming conical shapes on a roll using an attachment such as we have described is not recommended where quantity production is required.

For greater efficiency and higher production, a true-circle (TC) roll has been developed and has been highly acclaimed by the industry.

TC rolls are unique in that they produce cylinders with no flat on the leading edge and with virtually no flat on the trailing edge. Whereas the flat produced on a standard machine may amount to as much as 2 in. (50 mm) or more, TC rolls reduce the flat to less than 1/2 in. (13 mm). It is not necessary to preform blanks, which is a required operation in rolling with conventional machines. The small amount of flat produced by TC rolls is so close to the welded seam that it becomes practically indiscernable. Some manufacturers with conventional bending rolls perform the preforming operation on a press brake so that the TC roll will actually do the work of two machines.

TC rolls utilize three standard rolls as on conventional machines, but with the addition of a forming shoe just ahead of the top and bottom rolls, as shown in Figure 1E-8. The forming shoe acts as a third roll between the top and front rolls. The rear roll is used as a fine adjustment to form the cylinder to the final dimension.

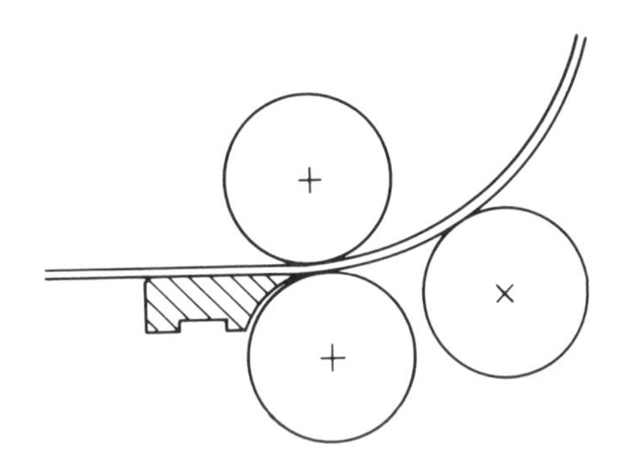

Figure 1E-8. Forming shoe acting as a third roll between top and front rolls

TC rolls are available with an automatic cycling feature, which automatically feeds the blanks into the rolls and then opens the drop end after the cylinder has been formed. An air-operated ejection cylinder pushes the cylinder from the rolls, and the drop end then returns to normal position for the next blank.

Special forming rolls have been built for forming of culvert pipes, grain bins, and prefabricated buildings from corrugated blanks. Modified machines have been built with special grooving for flat blanks with flanges or beaded portions and also for forming of flat bar stock or angles.

1F: Rubber-Pad Forming and Hydroforming Processes

Rubber-pad forming employs a rubber pad on the ram of the press and a form block on the platen. The form block usually is similar to the punch in a conventional die, but it also can be the die cavity. The rubber acts somewhat like hydraulic fluid in exerting nearly equal pressure on all workpiece surfaces as it is pressed around the form block. The advantages of the rubber-pad forming processes are:

1. One rubber pad takes the place of many different die shapes, returning to its original shape when the pressure is released.
2. Tools have fewer components, and are made of easier-to-machine materials, than conventional tools.
3. The forming radius decreases progressively during the forming stroke, unlike the fixed radius of conventional dies.
4. Thinning of the work metal (such as occurs in conventional deep drawing) is almost completely eliminated.

The disadvantages are:

1. Rubber wears out quickly, or tears on sharp projections. The average life of a rubber pad is about 20,000 pieces.
2. Without special or accessory equipment, rubber pads exert less pressure, than conventional dies, and definition may be less sharp.
3. Shrink flanges may have wrinkles that need hand work.

GUERIN PROCESS

The Guerin process is the oldest and most basic of the production rubber-pad forming processes.

Presses. Almost any hydraulic press can be used in the Guerin process. For maximum forming capability, the press tonnage and area of the rubber pad must be suitable for the operation under consideration. Generally the rubber pad is about the same size as the press ram; however, it can be smaller.

Tools. The main tools are the rubber pad and the form block (Figure 1F-1). The rubber pad is fairly soft and is usually three times as deep as the part to be formed. The pad can consist of a solid block of rubber, or of laminated slabs cemented together and held in a retainer as shown in Figure 1F-1. The slabs can also be held loose in a flanged retainer. The retainer is generally made of steel or cast iron, and is about 1 in. deeper than the rubber pad. It is also strong enough to withstand the forming pressures that are generated (up to 20 ksi, or 138 MPa, in some applications, although an upper limit of 2 ksi, or 14 MPa, is more common).

The minimum pad thickness is $1^1/_3$ times the height of the form block, as shown in Figure 1F-1. However, pad thicknesses generally vary from 6 to 12 in. (152 to 305 mm), and the most commonly used thickness is 8 or 9 in. (203 or 229 mm).

Form blocks (punches) are made of wood, plastic, masonite, cast iron, steel, or alloys of aluminum, magnesium, zinc, bismuth, or kirksite.

VERSON-WHEELON PROCESS

The Verson-Wheelon process was developed from the Guerin process. It uses higher pressure and is designed primarily for forming of shallow parts, using a rubber pad as either the die or the punch. A flexible hydraulic fluid cell forces an auxiliary rubber pad to follow the contour of the form block and exert a nearly uniform pressure at all points.

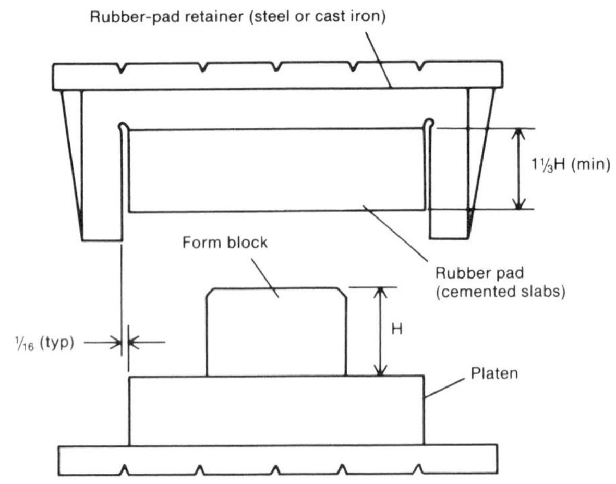

Figure 1F-1. Tooling and setup for rubber-pad forming by the Guerin process

The distribution of pressure on the sides of the form block permits forming of wider flanges than can be formed by the Guerin process. Also, shrink flanges, joggles, and beads and ribs in flanges and web surfaces can be formed in one operation to rather sharp detail in aluminum, low-carbon steel, stainless steel, heat-resisting alloys, and titanium.

Presses. The Verson-Wheelon press has a horizontal cylindrical steel housing the roof of which contains a hydraulic fluid cell (Figure 1F-2). Hydraulic fluid is pumped into the cell, causing it to inflate or expand. This expansion creates the force needed to flow the rubber of the work pad downward, over and around the form block and the metal to be formed.

Below the chamber containing the rubber pad and the hydraulic fluid cell is a passage, extending the length of the press, that is wide

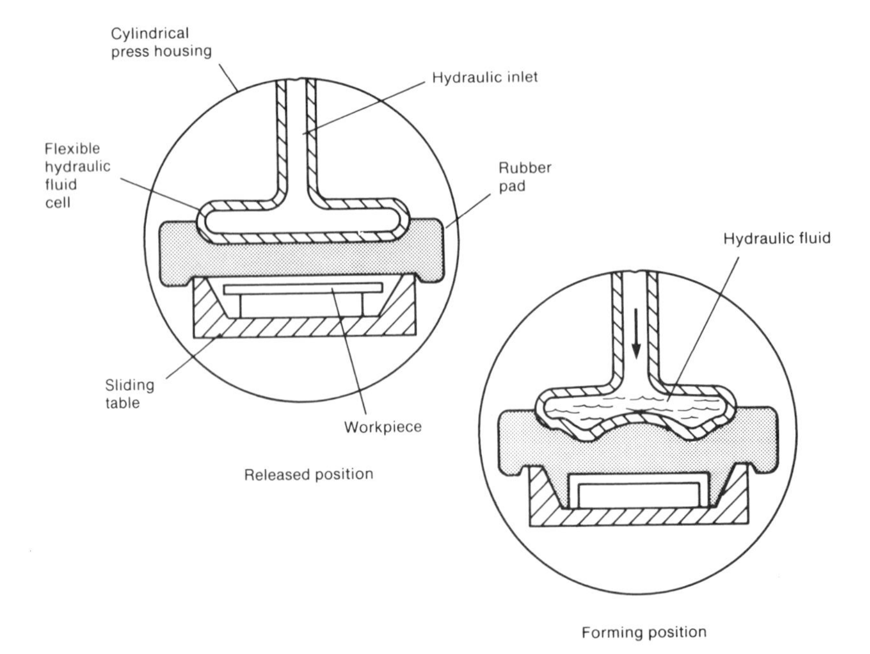

Figure 1F-2. Principles of the Verson-Wheelon process

and high enough to accommodate a sliding table containing form blocks. At each end of the passage is a sliding table that alternately can be moved into position for forming.

The rubber pad used in the Verson-Wheelon process has a hardness of about Durometer A35. It is usually protected from sharp corners on the form block and blank by a throw sheet or work pad that is harder and tougher than the pad itself. The throw sheet is much less costly to replace than the rubber pad below the fluid cell.

Machines are available with maximum-pressure ratings of 5, 7.5, and 10 ksi (34, 52, and 69 MPa). Sliding tables range in size from 20 by 50 in. (51 by 127 cm) to 50 by 64 in. (127 by 163 cm). The larger machine can form parts having flange widths up to 9⅜ in. (238 mm).

Heating elements can be used with the sliding tables for producing parts made of magnesium. The maximum temperature is 800 °F (425 °C), and special heat-resisting throw pads are used to protect the hydraulic fluid cell.

Tools. The form blocks are made in much the same way as are the form blocks for the Guerin process.

MARFORM PROCESS

The Marform process was developed to apply the inexpensive tooling of the Guerin and Verson-Wheelon processes to deep drawing and forming of wrinkle-free shrink flanges (Figure 1F-3).

A blankholder plate and a hydraulic cylinder with a pressure-regulating valve are used with a thick rubber pad and a form block similar to those used in the Guerin process. The blank is gripped between the blankholder and the rubber pad. The pressure-regulating valve controls the pressure applied to the blank while it is being drawn over the form block.

In forming of a soft aluminum alloy blank, the diameter normally can be reduced 57%, and reductions as high as 72% have been obtained. A shell depth equal to the shell diameter is normal when the minimum stock thickness is 1% of the cup diameter. Depths up to three times shell diameter have been reached with multiple-operation forming. Minimum cup diameter is 1½ in. (38 mm).

Foil as thin as 0.0015 in. (0.038 mm) can be formed by placing the blank between two aluminum blanks about 0.030 in. (0.76 mm) thick and forming the three pieces as a unit, the inner and outer shells being discarded.

Presses. The Marform process is best suited to a single-action hydraulic press where pressure and speed of operation can be varied and controlled. A marforming unit can be installed within the shut height of the press (distance from the top of the bed to the

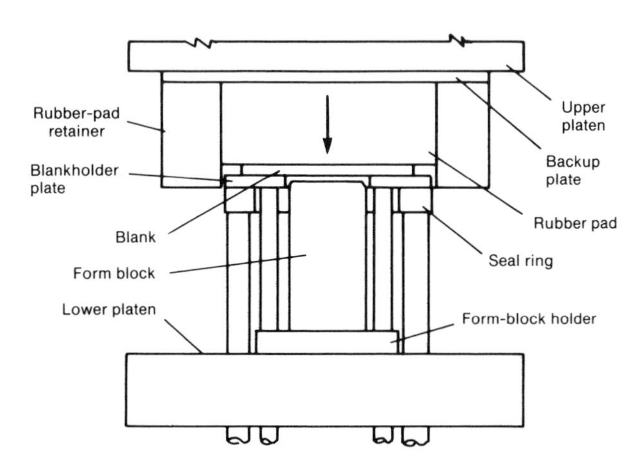

Figure 1F-3. Tooling and setup for rubber-pad forming by the Marform process

bottom of the slide). However, a press that incorporates a hydraulic cushion system in its bed has been designed specifically for Marforming.

Rubber pressures used depend on the press tonnage and surface area of the rubber pad. Present installations range from 5 to 10 ksi (34 to 69 MPa).

Tools. The rubber pad is similar to that used in the Guerin process. It normally is 1½ to 2 times as thick as the total depth of the part, including trim allowance. The rubber pad can be protected from scoring by the use of a throw sheet, which is either cemented to the pad or thrown over the blank.

RUBBER-DIAPHRAGM AND HYDROFORMING PROCESS

This process differs from those previously described in that the die cavity is not completely filled with rubber, but with hydraulic fluid retained by a 2½-in.- (64-mm-) thick cup-shape rubber diaphragm. This cavity is called the pressure dome (Figure 1F-4). A replaceable wear sheet is cemented to the lower surface of the diaphragm (Figure 1F-4).

More severe draws can be made by this method than in conventional drawing dies because the oil pressure against the diaphragm causes the metal to be held tightly against the sides as well as the top of the punch.

First-draw reductions in blank diameter of 60 to 70% are common. When redrawing is necessary, reductions can reach 40%. Low-carbon steel, stainless steel, and aluminum in thicknesses from 0.010 to 0.065 in. (0.25 to 1.65 mm) are commonly formed. Parts made of heat-resisting alloys and copper alloys are also formed by this process.

Presses. A special press, called a Hydroform press, is used for this process. A lower hydraulic ram drives the punch upward; the upper ram is basically a positioning device. A hydraulic pump delivers fluid under pressure to the pressure dome. The blankholder is supported by a bolster, and does not move during the operation.

The largest press of this type can draw a blank 25 in. (635 mm) in diameter to a maximum depth of 12 in. (305 mm). Punch diameter can range up to 19 in. (485 mm). Maximum dome pressure is 15 ksi (103 MPa). Maximum rating is 1500 cycles per hour. The practical production rate in cycles per hour is usually about two-thirds the machine rating. However, the operation often takes the place of two or three conventional press operations.

In a variation of this process, the punch is stationary and the blankholder is actuated by the die cushion of a single-action hydraulic press.

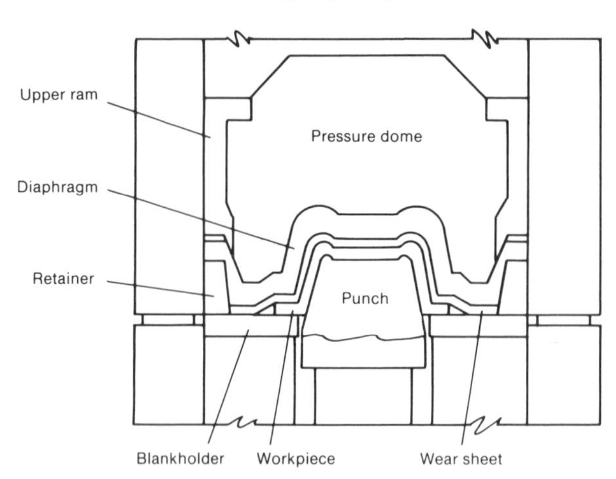

Figure 1F-4. Rubber-diaphragm forming in a hydroforming press

Tools. Punches can be made of tool steel, cold rolled steel, cast iron, zinc alloys, plastic, brass, aluminum, or hardwood. The choice of material depends largely on the work metal to be formed, the number of parts to be made, the shape of the part, and the severity of the draw.

Blankholders are usually made of cast iron or steel, and are hardened if necessary. Clearance between punch and blankholder is not critical — it may be 50% or more of the thickness of the metal being drawn.

For short runs, an auxiliary blankholding plate can be placed on a blankholder that is already in place. The auxiliary blankholder plate should not overhang in the punch clearance more than its thickness, and it should not be larger than the blankholder.

Rubber strips are placed on the blank to break the vacuum caused by dome action during drawing. Blankholders can be contoured to match the shape of a preformed blank, or to preform a blank as an aid in forming (Figure 1F-5 illustrates the forming of an automotive tail-lamp housing.)

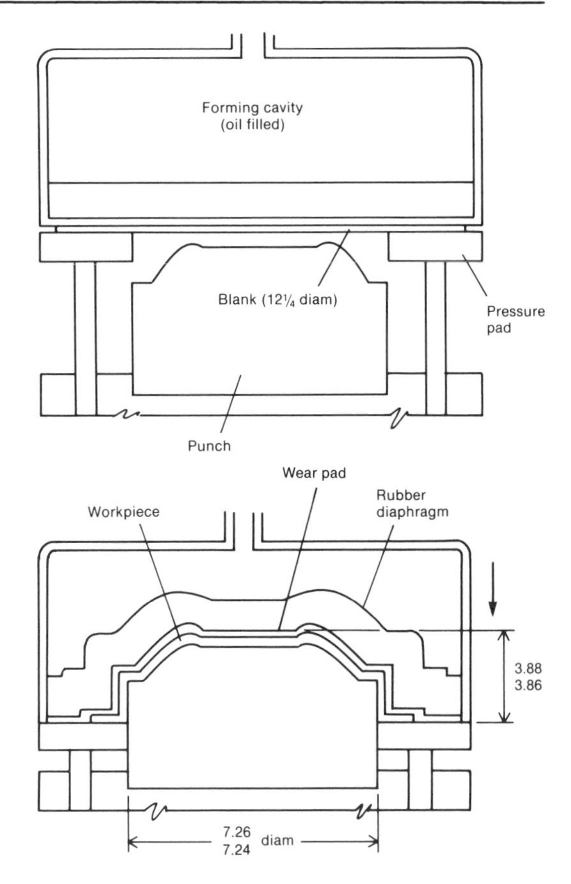

Figure 1F-5. Forming of an automobile tail-lamp housing in one draw in a rubber-diaphragm press (dimensions are in inches)

PRINCIPLES OF DIFFERENT PROCESSES

Use of hydraulic action in the forming of thin metal parts has been experimented with since the beginning of the century. However, the techniques have only recently been developed industrially.

This type of process calls for special machines such as diaphragm-forming hydrobuckling machines, or traditional presses such as gasket-draw and pressure-controlled machines, or more simple machines which are without gasket or control. Let us now look at the principles of these different processes.

Hydraulic Forming With Diaphragm

During the 1950's, this process was proposed on the market of special machines where oil pressure is applied by a synthetic rubber diaphragm on the metal to be formed (Figure 1F-6). Contrary to rubber stamping, where the elastic bulk is crushed by greatly increased local pressures (to the detriment of the part's sunken areas), the pressure here is uniformly distributed, and very compact concave rounded forms result.

Example: Round-off/thickness ratio of soft sheet metal: 3 to 4 for a pressure of 1000 bars; about 2 for a pressure of 2000 bars

Decrease in sheet/tool friction permits draw-value ratios superior to those obtained by the traditional process. The draw ratio is the ratio between the blank diameter (D) and the punch diameter (d), while the LDR (Limited Drawing Ratio) corresponds to the highest possible D/d ratio before fracturing occurs.

Example of a part: Headlamp parabola; traditional drawing (transfer press); 5 shifts hydraulic draw with diaphragm; 1 blow

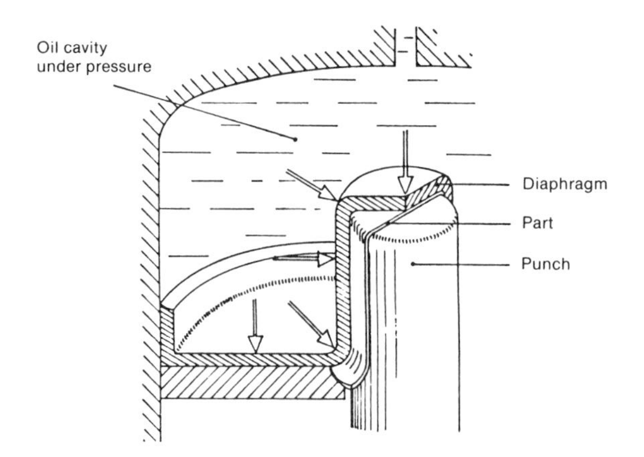

Figure 1F-6. Hydraulic forming with diaphragm

Hydrobuckling

The hydrobuckling process, patented by P. Cuq, can be included here because it incorporates metal-deformation control when there is minimum friction and an absence of thickness reduction.

The part is formed by two combined actions, from a blank obtained by traditional drawing (Figure 1F-7):

1. Hydraulic pressure (P) created by a pump inside the part, which flattens the metal against an exterior die or mold.
2. Mechanical collapsible load action (F).

The pressure P is adjusted in ratio to the upstroke of the piston to fit the form so that the metal areas which flatten the sunken parts of the mold are not obstructed by friction and therefore undergo no expansion — no thickness reduction. The word "buckling" is not quite suitable as it implies a brutal, more or less controlled deformation.

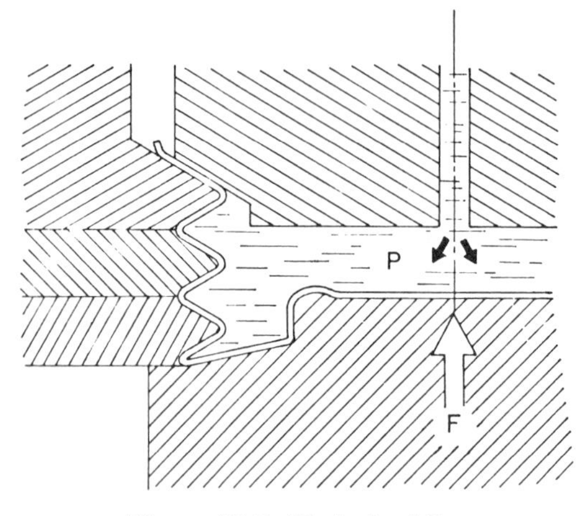

Figure 1F-7. Hydrobuckling

This technique has been used for several years in the production of pulleys for automobile trapezoidal belts. In relation to the traditional process (roller), the support facings of the belts show not even the slightest bulge.

Hydraulic Forming With Gasket and Pressure Control

During the past decade, a technique has been developed which requires, in principle, nothing but a traditional machine — a double-effect hydraulic press (Figure 1F-8). Impermeability is ensured by means of a gasket between the die plate and the blank (or the flange), against which the blank grip leans. Pressure starts up when the punch goes into action, and its value is controlled during the entire operation in the leakproof chamber of the die cavity.

This process is most useful in making conical or vertical-wall draws, if there is enough play between the punch and the die to allow sufficient pressure to form metal beads around the punch at the start of the operation. This pressure should be high: 300 to 500 bars (30 to 50 MPa) for soft steel, 400 to 800 bars (40 to 80 MPa) for stainless steel. Here the path of deformation (which we mentioned earlier) becomes most important. We know that a sheet has important defor-

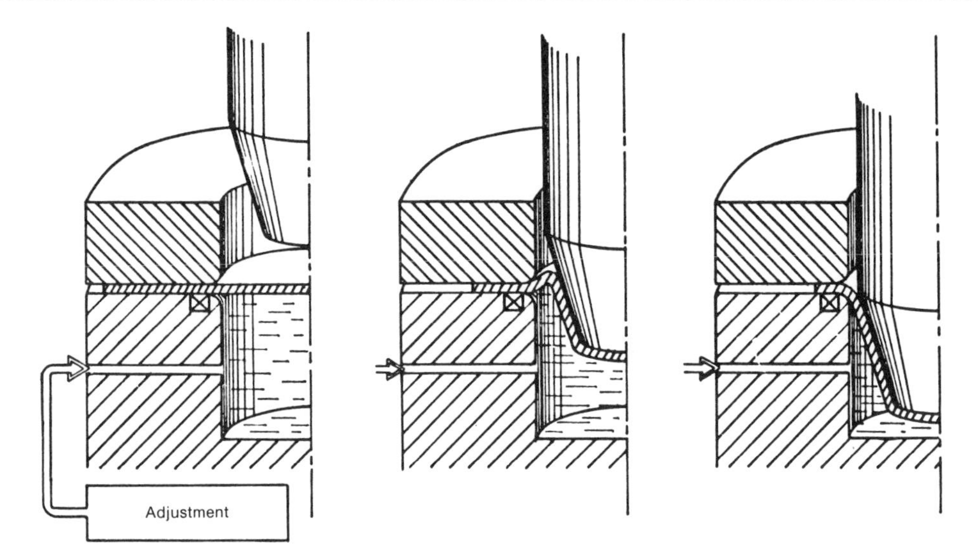

Figure 1F-8. Hydraulic drawing with joints and pressure control

mation possibilities if it has previously been subjected to a restraining effort, but these possibilities are reduced if it has been subjected to expansion. The formation of beads around the punch can be compared with a beginning of return, or, again, to a first-draw blow corresponding to a less-severe reduction ratio. The metal starts deforming in the direction of restraint.

In practice, it has been found that it is not always easy to define pressure control, or the height of the beads. To enable us to follow the deformation process in order to obtain an optimim path compatible with the part's form, markers such as the Limitation Firming Curve and control of deformation paths are valuable tools in the development of these techniques.

ANALYSIS OF THE "AQUADRAW" HYDRAULIC PROCESS

The process described by Granzow (4) and known as the ARMCO Steel Corporation's "Aquadraw" process has an extremely simple principle: equipment does not need to be adjusted and a double-effect hydraulic press is perfectly suitable.

A fluid (water, or a "solvent oil" emulsion) fills the die cavity over which the blank-grip presses the blank down with a force of F_{sf} (Figure 1F-9). When the punch penetrates the die, fluid pressure rises. The fluid then escapes and creates a film between the blank and the die plate, preventing all contact between the sheet and the die.

Surface Aspect

A first result is found in the aspect of the formed part:

- Absence of scratches, particularly on

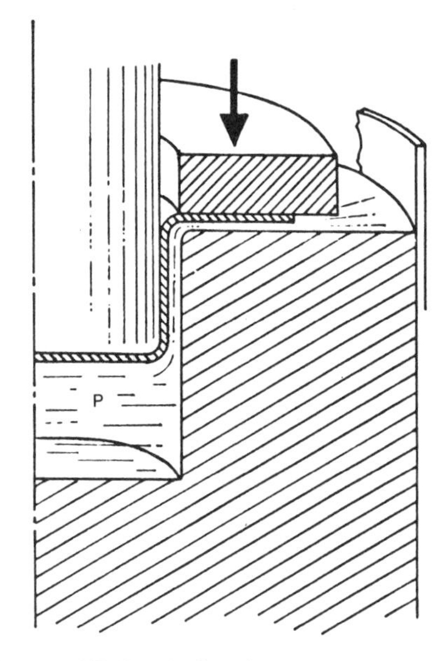

Figure 1F-9. Hydraulic drawing by the Aquadraw process

soft steel (although Granzow found some marks on stainless steel)

- A matte surface, although we noted an increase in roughness after hydraulic drawing of TC sheet (blank roughness, $R_a = 1.12$; after traditional drawing, $R_a = 0.80$; after hydraulic drawing, $R_a = 1.62$).

Performance

The elimination of blank/die friction permits best use of the plastic possibilities of metal. We have shown that, for soft steel, it is possible to increase the maximum draw ratio (D/d) to 25%:

Soft steel	TC	Qualities XE	XES
Traditional drawing	2.0	2.0	2.15
Hydraulic drawing	2.20	2.50	2.60

For example, Granzow obtained increases in D/d of 2 to 2.5 for austenitic stainless steel and 1.87 to 2.21 for ferritic steel.

Thinning

We have noticed that thinning in the formed part is distributed fairly regularly. Measurements have shown in particular that the thinning peak found in traditional drawing in the round-off of the punch disappears here (Figure 1F-10).

The reason for this is simple. High pressures (400 to 700 bars, or 40 to 70 MPa energetically flatten the sheet against the punch as soon as it enters the die cavity. Granzow was even able to note the possibility of printing punch details on annealed brass, which decreases sheet/punch displacements and, at the same time, sheet deformation in the vertical wall and punch round-off.

Again, for the same reason, the *punch radius* (round-off) has little importance in the successful production of parts by this process. Conical or hemispherical forms, for example, can be produced without folds. Granzow's tests on annealed aluminum have yielded a maximum D/d ratio of 1.75 for traditional drawing with a 3.2-mm (0.126-in.) round-off punch, and this value increases to 2.0 for larger round-offs (19 to 50 mm, or 0.75 to 1.97 in.), whereas the hydraulic process gives a constant ratio of 2.5 for all round-offs.

Industrial Limitations of the Process: Forces

In general, the necessary forces are considerable — about four times greater than in traditional drawing.

Instrument apparatus with which we have equipped our double-effect hydraulic press (100 tons of plunger jack, 50 tons of blank grip) permits us to analyze the development of the operation (Fig. 1F-11) by registering displacement, the speed of the press slide, and forces supported by the slide on which the punch is fixed. We have also included a pickup apparatus which indicates the pressure in the die cavity.

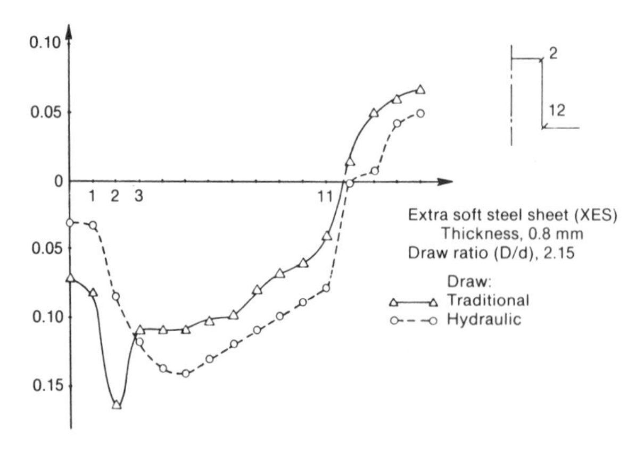

Figure 1F-10. Thickness variations along the length of a generator

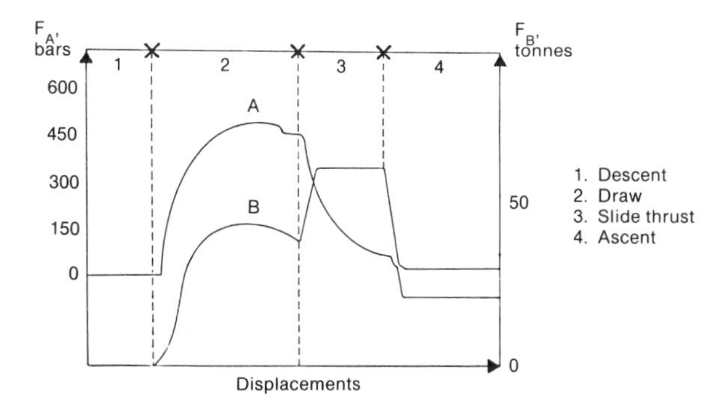

Figure 1F-11. Register of pressure P in the die cavity (curve A) and total force F (curve B)

Example: Comparative forces, in tons, for the two processes. Speed of the slide: 60 to 75 mm/s.

Soft steel sheet	Draw Traditional	Draw Hydraulic
Thickness, 0.8 mm; D/d ratio, 2.17	11 tons	48 tons
Thickness, 0.5 mm; D/d ratio, 2.2	8 tons	34 tons

Knowledge of the pressure (P) of water in the die during the operation will enable us to evaluate the necessary force, and consequently to choose the appropriate press.

An experimental study gave us the following results:

- Pressure P in the die cavity varies similarly to the D/d draw ratio (Figure 1F-12).
- When the blank-grip force F_{sf} varies between its two extreme values (folds; fracturing) it has little influence on the value of P.

Draw ratio, D/d	2.3		2.3			
Blank-grip force (tons)	6	8	9	12	16	19
Pressure, P (bars)	645	660	585	585	585	585

- In speed ranges permitted by our press (15 to 75 mm/s), draw speed has no influence on pressure.

Fluid pressure in the round-off die zone is sufficient to prevent sheet/tool contact (absence of friction observed experimentally). It therefore creates the horizontal forces necessary for stamping (Figure 1F-13).

We think therefore that we have found an order of magnitude for pressure P1 at the rounded-off die by deducing that P1 creates,

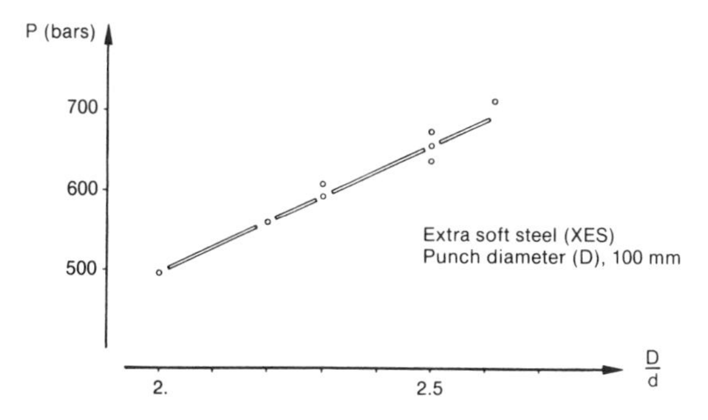

Figure 1F-12. Pressure in the die cavity

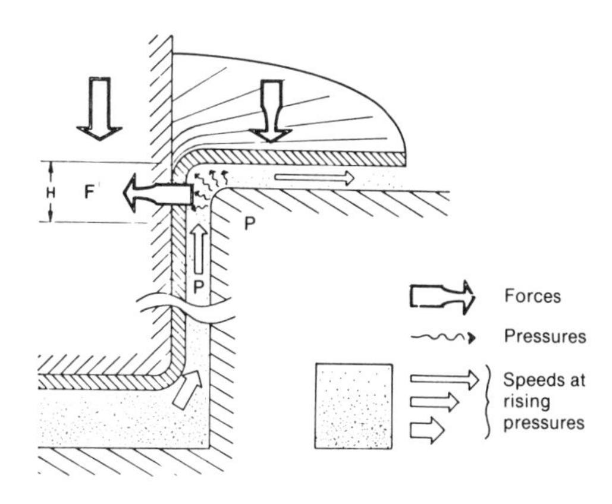

Figure 1F-13. Distribution of pressures and speeds

around the punch, and at height h, where the sheet is not yet flattened against the punch (h being close to the value of zero round-off of the die), a force F equal to the force of traditional drawing from which the portion due to friction has been removed.

Example: ES-quality soft steel; thickness (e), 0.8 mm
Resistance to tension (R), 33 kgf/mm^2
D/d ratio, 2.17
Traditional draw force: $F = d \cdot eRK(B)$
Hence pressure $P1 = R \cdot \frac{e}{P} \cdot K(B)$
(The K coefficient for a B ratio of 2.17 is close to 1.)
The calculated value of P1 is
$33 \cdot \frac{0.8}{5} = 5.28$ bars (corresponds to the measured value).

This evaluation cannot be precise, because:

- P1 differs from P in the die due to load losses, and the speed will therefore increase forces.
- The h-height zone where force F is created is not strictly equal to 0 (die round-off) and depends on play (note that even the slightest uncertainty regarding h creates a big error in P1).
- In the traditional process, the portion due to friction can be 20% (for example) of the real force.

Conclusion

The simplicity of the hydraulic drawing process is an asset in its development. Equipment is no more complex than for the conventional process, and is placed in a press of traditional design — a double-effect hydraulic press sufficiently powerful to supply extra forces.

Cadence is similar to that of traditional drawing presses handled by an operator, provided there is automatic fluid feed in the die cavity and an outlet system (such as a spring at the bottom of the die). The process is therefore perfectly competitive for a workshop equipped with a double-effect hydraulic press, and for small or average sets.

This process could, however, be developed more rapidly by adapting it on a mechanical press after having adjusted the influences of speed and blank-grip forces.

The hydraulic process increases draw performance, which in certain cases permits a reduction in the number of blows or the use of a lower-quality metal.

There is no need for lubrication, and hence lubricant costs, application, and elimination are precluded. To be watched is any anticorrosion component (such as very highly diluted solvent oil) that has been added.

Because of the absence of friction, the parts produced show no scratches and have a matte aspect, so there is a possibility of forming coated sheets with this process.

1G: Multiple-Slide Machines and Tooling

The multiple-slide machine, sometimes called a four-way, four-slide, or multislide machine, is a somewhat specialized item of stamping equipment although it is very versatile within a limited area of stamping applications. From the design standpoint, a multiple-slide machine may be described as a horizontal stamping press using a system of cam-controlled tools for producing small parts from coil stock.

Numerous small parts are produced by these machines. In addition to stampings from sheet metal, these machines are also used to fabricate parts from wire. Brackets, clips, electrical connectors, and clamps are typical of the types of stampings made on such machines.

MACHINE DESIGN AND OPERATION

The design of a multiple-slide machine is a radical departure from the normal concept of pressworking mechanisms. Nevertheless, the basic principle used in shaping and cutting of stampings is the same as that of any conventional press. Moving slides, fitted with suitable tools, strike the metal while it is supported on tools equivalent to dies, forcing it to assume the desired shape.

Machine Construction

The base of the multiple-slide machine is essentially a large steel table. A hole is provided near the center of the surface through which completed parts are discharged (Figure 1G-1).

The driving mechanism of the machine consists of an electric motor that is linked by gears or belts to the shafts which operate the tools. The shafting arrangement is the most unusual characteristic of the machine. It consists of four shafts which completely surround the bed of the machine to form a rectangle. The shafts are connected at all four corners of this rectangle by gears. Power is applied at one point of the shaft arrangement and, due to the gearing, gives motion to all four shafts. The slides of the press are connected to cams mounted on the shafts. Split-type cams are often used for this purpose to facilitate conversion to various jobs.

Through this shafting arrangement, slides can be introduced to the die area from all

Figure 1G-1. Multiple-slide machine

four sides of the press. In addition, several slides may be attached to each shaft, increasing the flexibility of the machine and the number of operations that may be performed. It should be noted that the motion of the slides is in the horizontal plane, another departure from the conventional method of pressworking.

Another important component of the multiple-slide machine is the piercing head. Die sets for piercing, cutting, trimming, blanking, or embossing are mounted in this head. The die components are mounted on edge on a solid backup plate. The punch plate retains the punch components and is mounted on a slide driven by cams on the front shaft. Stock is fed from the left side of the press. Sometimes two or more piercing heads may be mounted on a single press. The slides for these heads may be mounted on either the front or rear shaft, providing flexibility in tooling. For instance, a burr resulting from the piercing operation may be located on the desired side of the part by locating the tooling on the proper ram. The piercing head is actuated by solid-type cams for reasons of strength. In addition, the cams used in this application are not changed as frequently as are other cams.

After being formed, the part is stripped from the forming tools or the post and dropped from the press.

ADVANTAGES OF MULTIPLE-SLIDE MACHINES

Multiple-slide machines perform in a manner similar to presses equipped for progressive or transfer operations — that is, a number of separate operations, usually all the operations necessary to complete the part, are accomplished in a single machine. In fact, due to its method of operation, this machine may perform the full sequence of operations on a part that could not be completed in a single conventional press even though transfer or progressive dies were used. This minimizes the number of presses needed to produce a stamping and provides the added advantages of reduced handling costs and conservation of floor space.

Tool Cost

Advantages with respect to tooling cost may also be realized through use of multiple-slide machines. The tooling used is generally simpler in construction than the average comparable die sets used in conventional presses. Piercing, shearing, and forming units are of simple design and are relatively inexpensive to build.

Die details are mostly standard parts and can often be made at the plant where the machine is located. Forming units are fastened to the cam-driven rams by simple standard methods. The maintenance cost of this type of tooling is relatively low.

Operating Cost

The per-piece cost of operating a multiple-slide machine is comparatively low. This is a direct result of a high rate of production. As many as 20,000 to 70,000 stampings can be made in a 16-hour shift on the fastest models, depending on the length of the stamping and the number of stampings produced per stroke.

FACTORS IN SELECTING MULTIPLE-SLIDE MACHINES

Multiple-slide machines can be applied to the production of numerous types of small parts. However, there are certain general limitations to observe in assigning work to this type of machine. These limitations are related to part size, design, and production volume.

Size Limitations

Dimensional specifications for stampings produced on multiple-slide machines must be held within the capabilities of the machines themselves. As indicated previously, these machines are applied to the production of relatively small stampings. The largest machines are capable of handling stock up to a maximum of 3 in. (76 mm) in width. The longest feed length possible is 12½ in. (318 mm). Stock thicknesses up to 3/32 in. (2.4 mm) can be handled. This range makes it possible for the largest machines to handle practically any thickness of stock used in automotive body or chassis stampings.

Smaller machines are limited in application to even smaller parts since the opening in the piercing head, which varies proportionately with press size, governs the width of material that may be processed. The smallest machine can accommodate a maximum stock width of 3/4 in. (19 mm) and a feed length of 3 in. (76 mm). Maximum stock thickness on smaller machines is 1/32 in. (0.8 mm).

The stroke of the piercing-head ram varies from 3/8 in. (9.5 mm) on the smallest machine to 3/4 in. (19 mm) on the largest. The throw of the forming cams varies from 7/8 to 2 in. (22 to 51 mm).

Design Limitations

The range of stampings that may be made on multiple-slide machines is somewhat limited in respect to part design. Basically, the use of these machines is limited to parts whose shapes can be achieved by shearing operations and pure bending operations. This confines the use of multiple-slide machines to types of operations such as piercing, notching, slitting, cutting and straight flanging, bending, and sizing. Extrusions are generally impractical because a solid stripper is used in the piercing head. An extrusion usually would prevent or interfere with the movement of the stamping to succeeding stations. Any forming in the piercing head must be done in a manner that will not hinder movement of the part.

Drawing or stretching operations are considered unfeasible both from the standpoint of tonnage requirements and the operational limits of the die setups that may be used. The press and tooling designs do not permit the use of air cushions or blankholding mechanisms required for these operations.

Volume Requirement

The volume requirement for a part is a further consideration in contemplating the use of a multiple-slide machine. Such a machine is a high-volume piece of equipment and as such can be used most profitably for high-volume operations. To avoid excessive downtime and loss of production in changeovers, parts not qualifying as high-production items should generally be produced by some other method.

TOOL PLANNING AND DESIGN FOR MULTIPLE-SLIDE MACHINES

The principles followed in planning and designing tools for multiple-slide machines are similar to those used in layout of progressive dies for conventional presses. A strip development is made in which the operational elements are divided and set up in the most effective sequence. This indicates the die stations that must be built and their relative locations (Figure 1G-2).

Sectional Construction

In the construction of the dies, it is general practice to employ sectional construction. This is done to minimize maintenance problems and reduce downtime. Since a number of slides can be put into operation, the work load can be spread over a considerable distance. When numerous piercing or cutting operations are involved, several piercing heads may be employed. This permits spacing of the work to avoid crowding delicate punches together. It also reduces the necessity of building a long, heavy, and complicated progressive die. Further, versatility and changeability of the tooling setup is increased where multiple piercing heads are used. Making changes in any stage of the die is easier, because it is unnecessary to remove and disassemble a large, complex progressive die.

Component Strength

Provisions should be made in designing multiple-slide dies for adequate strength and wearing qualities in all components. Punches, die inserts, and forming tools should be made of materials capable of withstanding rapid repetitive shock. Special hardened inserts should be supplied in all areas where tools are subject to exceptional wearing stresses. Rigid tool mountings are necessary to prevent deflection under load.

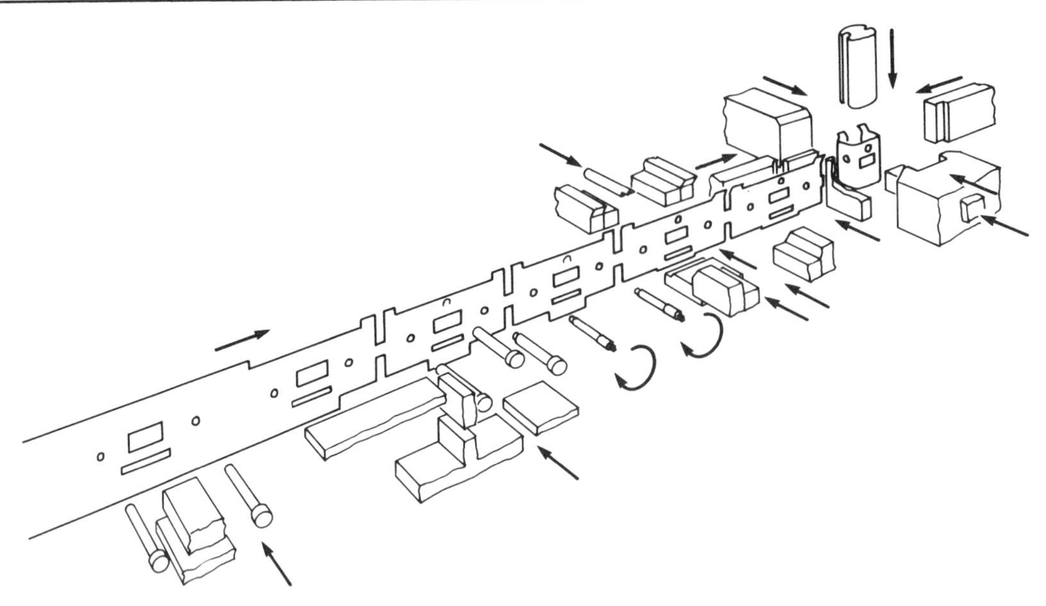

Figure 1G-2. Hypothetical multislide application showing wide variety of functions that can be performed in production of complete stampings directly from coil stock

DESIGNING FOUR-SLIDE PARTS

Forming of parts from metal strip on four-slide machines (Figure 1G-3) is a fast, low-cost production method. Often parts made in multiple operations in power presses could be produced at much lower cost in four-slides. These machines automatically produce metal parts from wire or strip by the combined action of a power-press section for stamping and cam-actuated, sliding tools for forming. With a change in either the tools or the machine adjustments, a wide variety of forms and parts can be produced.

Similarities to Presses

Four-slide machines are similar to power presses in many ways. Both have feeding mechanisms which receive strip or wire material from a continuous coil. The material is fed into a die area in which press operations are performed in sequence at one or more stations. There the similarities end.

Conventional power presses have tool motions in only one direction, so that cam systems must be devised to form most complex parts. These systems are limited in what they can accomplish, and parts must be transferred from one machine or area to another for different operations. On the other hand, standard four-slides provide six or more forming slide motions within the forming area. These include four or more motions in the forming plane and two or more motions in a plane perpendicular to the forming plane.

Press Operations

A four-slide machine can perform most press operations within its strip-size capacity. Basic metal-processing operations that can be performed in a four-slide press section are: piercing, notching, coining, swaging, extruding, lancing, embossing, forming, beading, dimpling, stamping, and stippling. These press operations are completed before the forming operation.

Other Operations

Besides performing press operations prior to work in the four-slide area, the machine can do other work before and after the forming steps — for instance, small brackets with

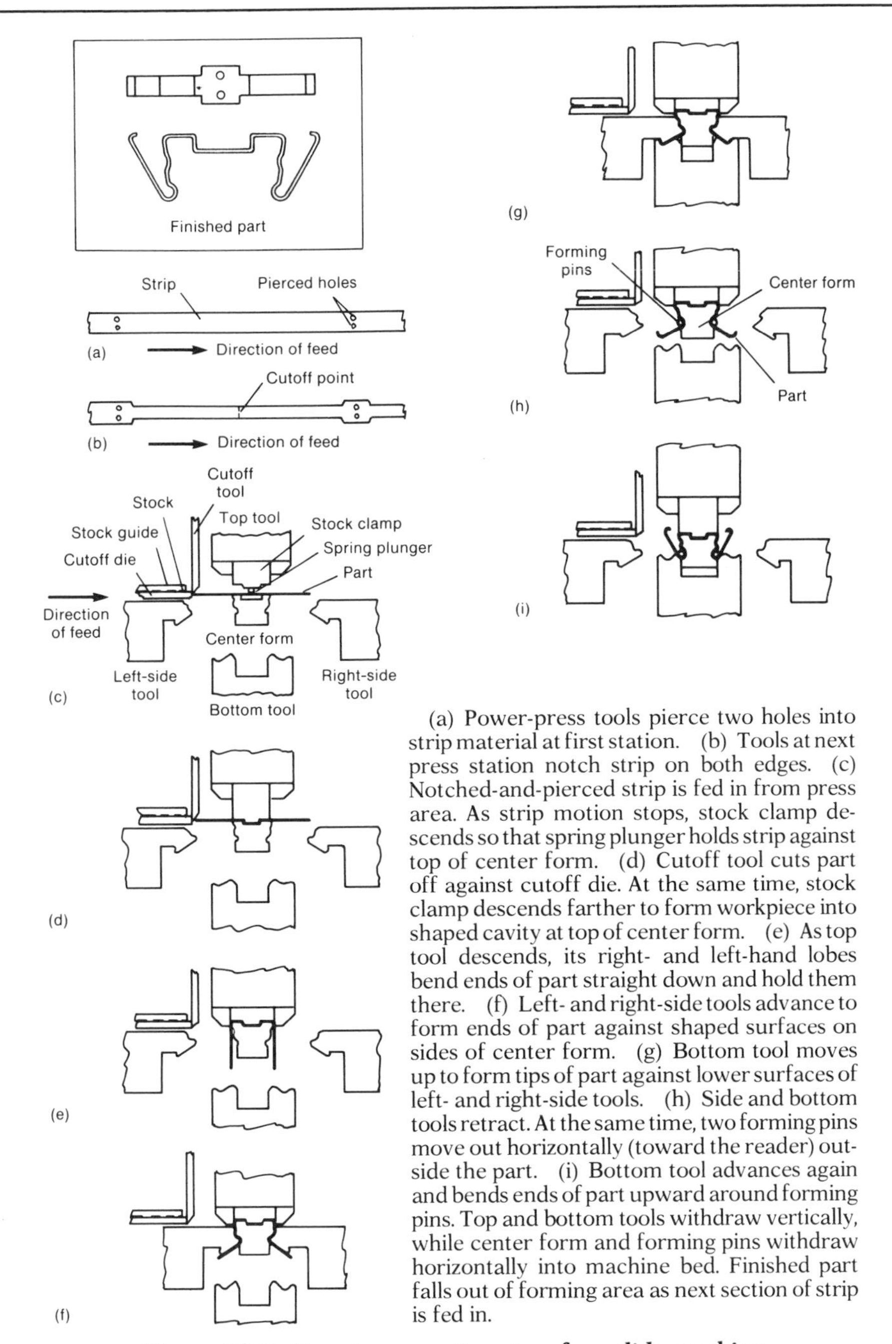

(a) Power-press tools pierce two holes into strip material at first station. (b) Tools at next press station notch strip on both edges. (c) Notched-and-pierced strip is fed in from press area. As strip motion stops, stock clamp descends so that spring plunger holds strip against top of center form. (d) Cutoff tool cuts part off against cutoff die. At the same time, stock clamp descends farther to form workpiece into shaped cavity at top of center form. (e) As top tool descends, its right- and left-hand lobes bend ends of part straight down and hold them there. (f) Left- and right-side tools advance to form ends of part against shaped surfaces on sides of center form. (g) Bottom tool moves up to form tips of part against lower surfaces of left- and right-side tools. (h) Side and bottom tools retract. At the same time, two forming pins move out horizontally (toward the reader) outside the part. (i) Bottom tool advances again and bends ends of part upward around forming pins. Top and bottom tools withdraw vertically, while center form and forming pins withdraw horizontally into machine bed. Finished part falls out of forming area as next section of strip is fed in.

Figure 1G-3. Forming operations in a four-slide machine

three holes, one of which is threaded. All three holes are pierced, and one of them extruded, in the first die station in the press area of the four-slide. The extruded hole is then tapped in the press area.

Hose clamps can be formed from strip material which is first stamped to size in the press area. It is then cut off and formed around the center form so that the ends butt at the lugs. The part is pushed farther out on the center form into a welding station where the ends are pressed together and butt welded. By doing the welding independently of forming — so that forming and welding are performed concurrently — the production-cycle time is only that of the slower operation and not the total of the two.

In fabricating a small electrical part, a flat contact strip is dimpled and a hole pierced in the press area. After the strip has been formed on the center form, a threaded stud is fed from a magazine and dropped into the pierced hole. A staking head, located in the position of the right-side tool, finishes the cycle by staking the stud in place.

Materials

The limitations on making parts in four-slide machines are usually related to material properties, not machine capabilities. The mechanical properties of common strip materials which affect their suitability for four-slide production are tensile strength, yield strength, hardness, and ductility.

Obviously, tensile strength and hardness must be taken into account when planning for production by almost any mechanical method. However, yield strength and ductility, by themselves, do not directly determine four-slide producibility. Instead, "relative formability," in terms of bending characteristics and forming radii, is used.

Five families of strip materials commonly used in four-slide production are:

1. Low-carbon, cold rolled strip steels
2. Spheroidize-annealed, cold rolled spring steels
3. Type 300 and 400 stainless spring steels
4. Copper alloys
5. Beryllium-copper alloys.

High-formability steels: Low-carbon, cold rolled strip steel and spheroidize-annealed spring steel are the most common high-formability materials used in four-slide production. Cold rolled strip steel (available in five tempers) is the basic ferrous material for four-slide parts other than springs. AISI 1075 spheroidize-annealed spring steel is the main material for spring parts, while AISI 1095 is used for spring parts in thinner sections. These two materials are hardened and tempered after forming to attain desired spring properties.

Tempered spring steels: The behavior of tempered spring steels in forming is difficult to predict. To control the mechanical properties, the strip mill should be required to fulfill two of three critical specifications: hardness, tensile strength, and minimum inside-bend radius without fracture. In addition, the allowable limits of reproducibility of mechanical properties should be established with the mill.

Stainless steels: Type 302 is the most common stainless steel specified for four-slide parts. Other stainless steels that may be used (austenitic types, such as the 200 and 300 series) have mechanical properties similar to those of type 302. Series 400 stainless steels have properties similar to those of AISI 1075 spheroidize-annealed cold rolled spring steel.

Nonferrous strip materials: The nonferrous alloys provide a wide selection of mechanical properties. However, there may be large differences in the hardness and tensile strengths of various alloys that have roughly equivalent formability. Beryllium-copper alloys are available in four heat treatable conditions or in up to five mill-hardened conditions.

The primary advantage of four-slide machines is their versatility, meaningful because it can result in reduced unit part cost. Rapid changes can be made in the combination of press and forming operations, and production can be shifted quickly from one part to another.

Forming operations are more flexible than similar operations on power stamping presses because the stroke, shape, and timing of each forming tool can be independently adjusted. For example, changes in the design of a strip part can often be accomplished by single tool adjustments. In a stamping press, an entirely new die may be needed.

Families of parts, instead of requiring a different die for each size, sometimes need only one or two additional tools; the rest is accomplished with tool and feed adjustments.

Despite these differences, four-slide machines and high-speed stamping presses have some common production characteristics:

1. High volume
2. Low-cost materials (standard mill-product strip and wire stock)
3. Predictable material-strength characteristics
4. Uniformity of parts
5. Excellent surface finish (surface appearance of original stock is retained).

As a basis for considering which process to use, four-slide operations can be compared with press operations in some other ways:

1. **Production speed:** Four-slide machines have a wide range of output rates in parts per hour. The production rate depends on the material and the forming operations. Presses are capable of output rates as high as those of four-slides on simple parts, but parts which require complex forming can be produced at higher speeds on four-slides.

2. **Tooling cost:** Because the tool motions in a four-slide are already built into the machine, toolmaking is confined to shaping of the working surfaces of cams, center forms, and slide tools. In progressive-die forming, the entire cam-motion system must be designed and produced.

3. **Tool adjustments:** Because the forming motions for a part are built into a fixed set of dies and cams, the initial setup time for equivalent parts is usually greater for a four-slide. However, the time required for tool adjustment and maintenance during production runs is less on four-slides.

4. **Combined operations:** Four-slides perform tapping, welding, and other secondary operations that must be done separately in press work. Four-slides can also interlock sections of a single part and assemble two or more parts.

5. **Heavy press operations:** Four-slides cannot do heavy coining or swaging, or cannot blank parts with a large number of die stations or high-tonnage requirements.

6. **Deep drawing:** Draws exceeding 1/8 in. (3.2 mm) are difficult to do on four-slides because of limitations on the lifter motions needed to clear the draws from the die.

In general, press parts require relatively little forming. The parts are quite simple and do not necessarily have a strip shape. Four-slide parts are more complex, involve a higher degree of forming, and usually have the basic strip shape integrated in their design.

Tolerances

Thickness. If close dimensional tolerances are expected, parts should be designed with mill dimensional standards in mind. Standard thickness tolerances for tempered and untempered cold rolled carbon spring steel are the AISI mill standards. These standards cover approximately three-quarters of the four-slide parts that are made. Closer thickness tolerances on cold rolled carbon spring steel are now available from specialty mills.

Edges and Width. Strip edges specified for four-slide strip parts are by far the most common in four-slide parts:

1. A prepared edge of a specified contour (round, square, or beveled), which is produced when a very accurate width is required, or when the finish of the edge suitable for electroplating is required, or both
2. An approximately square edge produced by slitting
3. An approximately square edge produced by rolling or filing for the purpose of eliminating the slitting burr.

Since most four-slide strip parts use the full width of the strip, AISI width tolerances on the raw strip material are also significant.

Camber. "Deviation of a side edge from a straight line" is the definition of camber. Standard AISI camber tolerances are 1/2 in. in 8 ft (12.7 mm in 2.44 m) for widths between 1/2 and 1½ in. (12.7 and 38.1 mm), and

1/4 in. in 8 ft (6.35 mm in 2.44 m) for widths over 1½ in. (38.1 mm).

Finishes

While many types of finishes can be produced in cold rolled carbon steel strip, three are most commonly used:

1. **Dull finish:** A finish without lustre, produced by rolling on rolls roughened by mechanical or chemical means. This finish is especially suitable for lacquer or paint adhesion, and is beneficial in aiding drawing operations by reducing the contact friction between the die and the strip.
2. **Regular bright finish:** A finish produced by rolling on rolls having a moderately smooth finish. It is suitable for many requirements, but not generally applicable to plating.
3. **Best bright finish:** Generally of high lustre produced by selective rolling practices, including the use of specially prepared rolls. This is the highest-quality finish produced and is particularly well-suited for electroplating.

Common finishes on hardened-and-tempered cold rolled carbon spring steel include black, scaleless, bright, polished, and polished-and-colored (blue or straw) finishes.

Design Hints

Design recommendations for producing metal parts in four-slides and power presses are similar to the extent that they are determined by material characteristics or by part shape. However, there are also special differences.

The following design recommendations will give optimum part quality at maximum production speed.

External Contours. Design recommendations for external contours are as follows (see Figure 1G-4):

- Lugs or ears (Figure 1G-4a) should be formed with their bend lines at an angle equal to or greater than 45° to the rolling direction.
- Components requiring rounded ends (Figure 1G-4b) should have a radius equal to or greater than ¾W, except when otherwise indicated.
- A rounded end with a radius equal to ½W may be used if a relief angle of 10° or greater at the point of tangency with the part edge is also used (Figure 1G-4c).
- Corners along the stock edge should be as near square as practical (Figure 1G-4d).
- The side that is to be free from burrs should be specified (Figure 1G-4e).
- All notches (Figure 1G-4f) should extend inside the stock edge at least 1½T, but not less than 0.020 in. (0.51 mm).
- Tapers should be recessed at least 1T from the edge of the part (Figure 1G-4g).
- Parts should have straight edges on the flat blanks wherever possible (Figure 1G-4h).
- To form a square corner, the minimum bend allowance should be 1½T or R + ½T if the corner is on a tapered end, and 2T or R + T if on a square end (Figure 1G-4i).
- Relief slots for tabs, and short flanges whose edges are flush with the external blank outline, should have a depth of at least T plus the bend radius (Figure 1G-4j).
- When flanges extend over a portion of the part, a notch or circular hole should be used to eliminate tearing. Notch depth should be T plus bend radius (Figure 1G-4k). A hole should have a diameter of 3T.

Internal Contours. Design recommendations for internal contours are as follows (see Figure 1G-5):

- A radius of at least ½T should be used at the intersections of all edges which do not lie along the stock edge, and should be specified as maximum if radius is not critical (Figure 1G-5a).
- Punched (pierced) holes should have a diameter of at least 1T (Figure 1G-5b), but not less than 0.030 in. (0.76 mm).
- The minimum distance from the edge of any round, punched hole to the blank edge should be at least 1T (Figure 1G-5c), but not less than 0.030 in. (0.76 mm).
- Edges of adjacent punched holes should

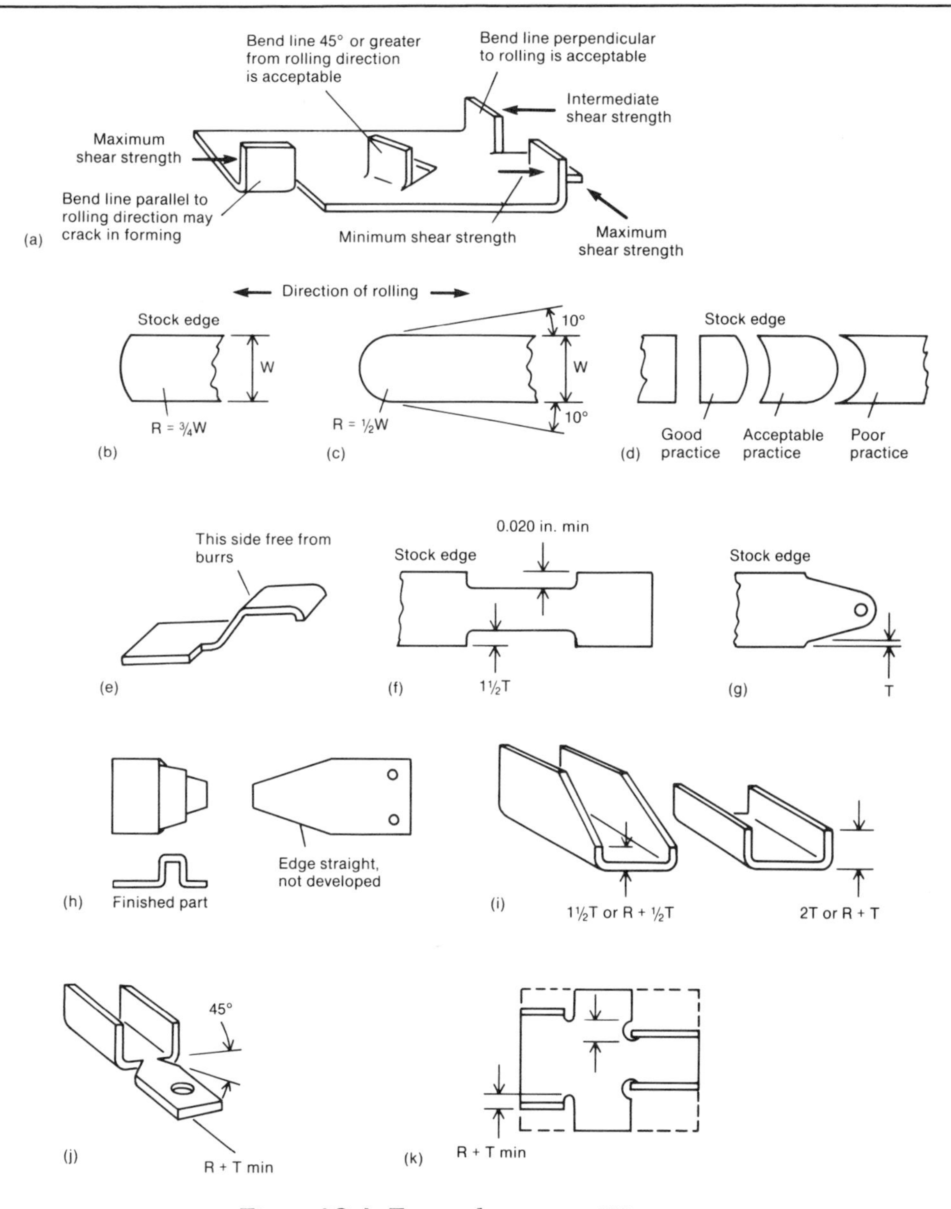

Figure 1G-4. External contours (W = part **width; T = stock thickness. See text.)**

be a minimum of 2T apart (Figure 1G-5d), but not less than 0.060 in. (1.52 mm).

• Extruded holes should have a minimum spacing of 6T between their edges, and should be at least 4T from the blank edge. Depth of the extrusion should be a maximum of 30% of its outside diameter (Figure 1G-5e).

• Holes should be spaced a minimum of 1½T from bend tangent lines (Figure 1G-5f).

• Threaded screw or bolt holes should be at least 1½ times the screw diameter from the centerline of the hole to the edge of the part (Figure 1G-5g).

• Slots that are parallel to the bend should be a minimum of 4T from bend tangent lines (Figure 1G-5h).

• Beads should be a maximum of 2T high and have a minimum inner radius of 1T (Figure 1G-5i).

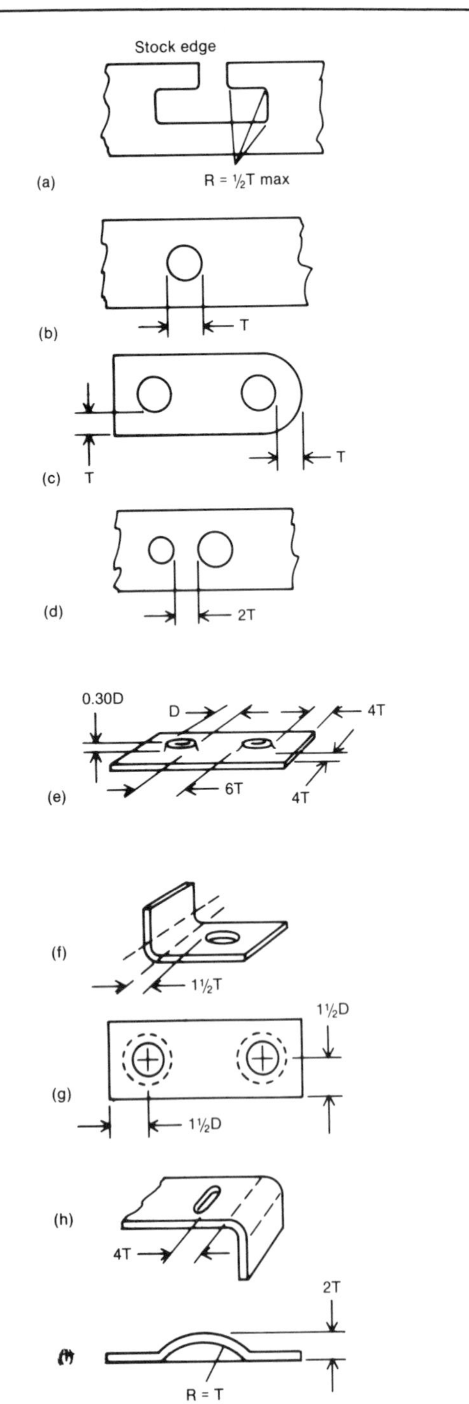

Figure 1G-5. Internal contours (T = stock thickness. See text.)

Tabs and Slots. The following recommendations are made for design of tabs and slots (see Figure 1G-6):

- Widths of tabs and slots should be a

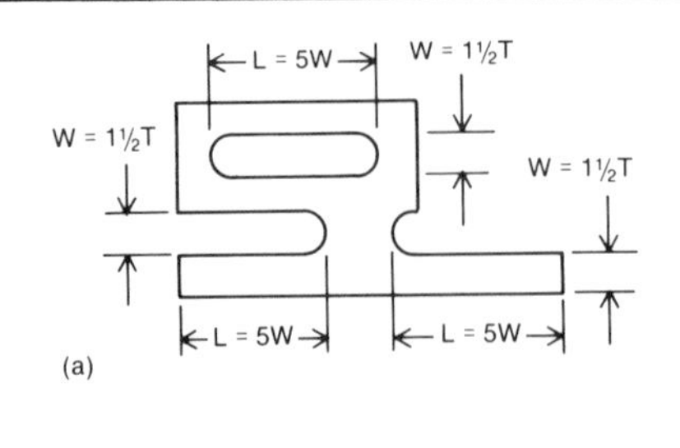

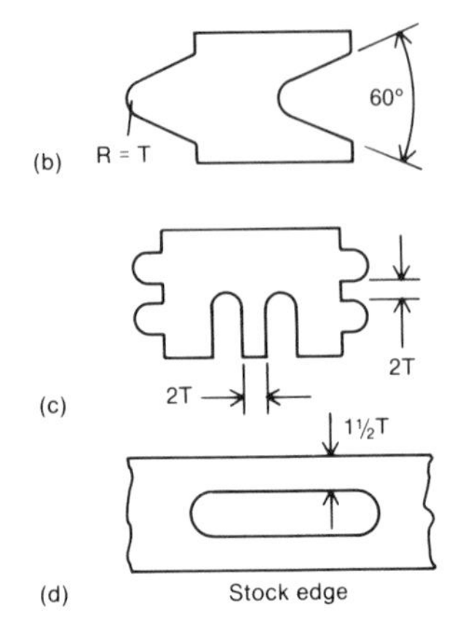

Figure 1G-6. Tabs and slots (T = stock thickness. See text.)

minimum of 1½T or 0.020 in. (0.51 mm). Their length should be a maximum of 5 times their width (Figure 1G-6a).

- Triangular tabs or slots should have minimum end radii of 1T and should form included angles of 60° or more (Figure 1G-6b).
- Adjacent tabs or slots should be spaced at least 2T or 0.030 in. (0.76 mm) apart (Figure 1G-6c).
- Internal slots should be at least 1½T or 0.030 in. (0.76 mm) from the edge of the stock (Figure 1G-6d).

Lugs, Bridges, and Curls. Design recommendations for lugs, bridges, and curls are as follows (see Figure 1G-7):

- Lanced lugs should have tapered sides (Figure 1G-7a), except when otherwise indicated.
- A taper is not required when the lug margin is pierced with a clearance prior to forming (Figure 1G-7b).
- Bridges should have a maximum ratio of lanced length to bridge height of 4:1 (Figure 1G-7c).

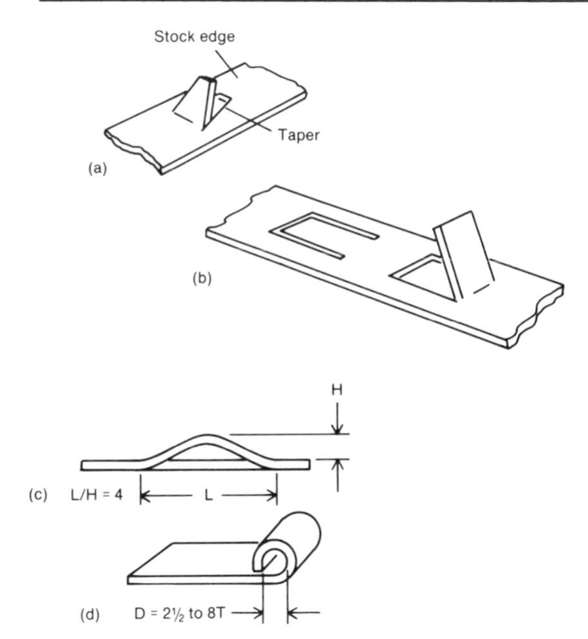

Figure 1G-7. Lugs, bridges, and curls (T = stock thickness. See text.)

- Curls should have inside diameters of 2½T to 8T (Figure 1G-7d).

CUTOFF METHODS FOR FOUR-SLIDE PARTS

The four-slide machine is properly credited with built-in versatility in both its press and forming operations. The four-slide's ability to produce small metal parts — complex or simple, to precise or commercial tolerances — in one machine cycle at high speeds is due also to the flexibility of the cutoff operation. Although the cutoff's function is simply to separate continuous strip or wire into individual parts, the ends of the parts can also be shaped and even formed, as needed, at the same time.

The various four-slide cutoff methods are specifically illustrated in Figures 1G-8 through 1G-16 as performed on a vertical-slide (Verti-Slide) machine. With the exception of the rear-motion cutoff in Figure 1G-13, however, all the methods described here apply to any type of four-slide machine. In addition to those shown, there are many other techniques which are now being applied to cutting off material in four-slides — and more will undoubtedly be developed in the future by ingenious tool engineers.

The Standard Strip Cutoff Method. A standard cutoff head with a standard cam is usually used with conventional tooling. A straight cutoff (Figure 1G-8) is made with a standard cutoff punch and standard die block.

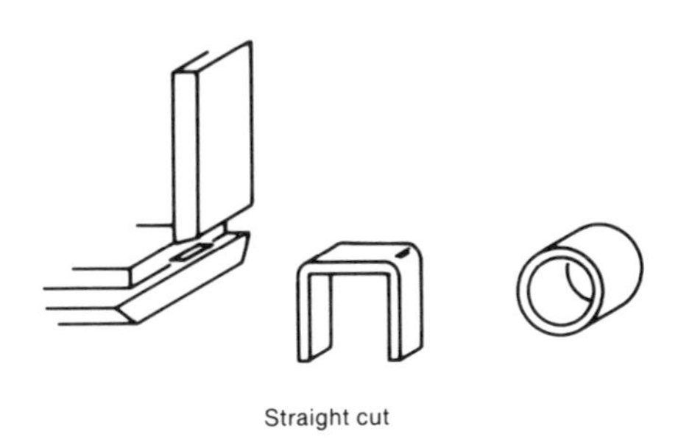

Figure 1G-8. Four-slide cutoff methods: standard strip method

With Limited Space. A straight cutoff can also be made with the standard cutoff die block, but with the cutoff punch or blade mounted on the side of the top forming tool (Figure 1G-9). This method is often used when the distance between cutoff die and center form is small.

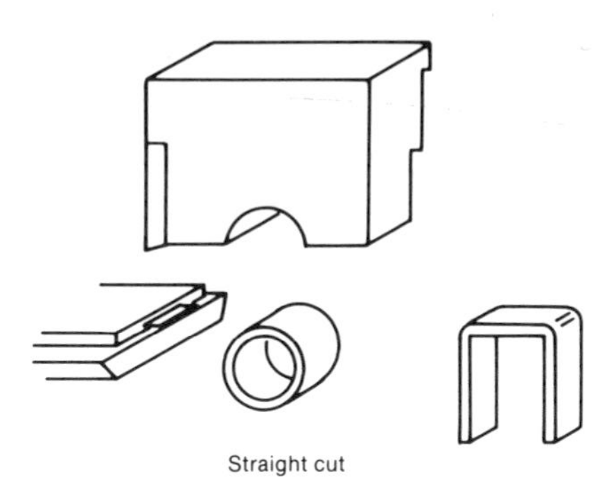

Figure 1G-9. Four-slide cutoff methods: limited space

With Stationary Forming Arbor. The standard cutoff attachment can be combined with a stationary forming arbor (Figure 1G-10). The left-hand tail or leg remains in place until forming by the other forming slides has been completed.

For Mating or Matching Cutoffs. Using a standard cutoff attachment, the cutoff

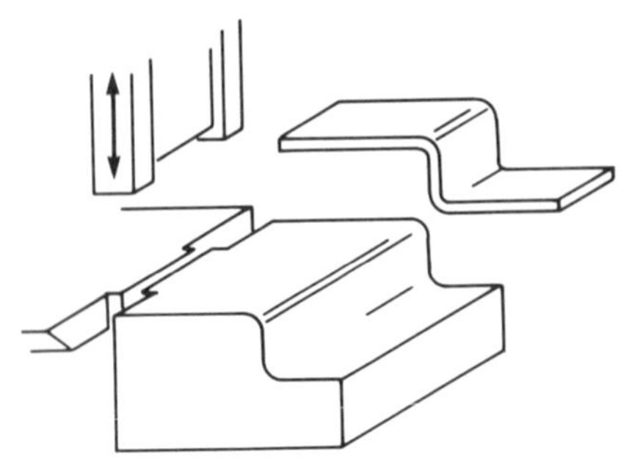

Figure 1G-10. Four-slide cutoff methods: stationary forming arbor

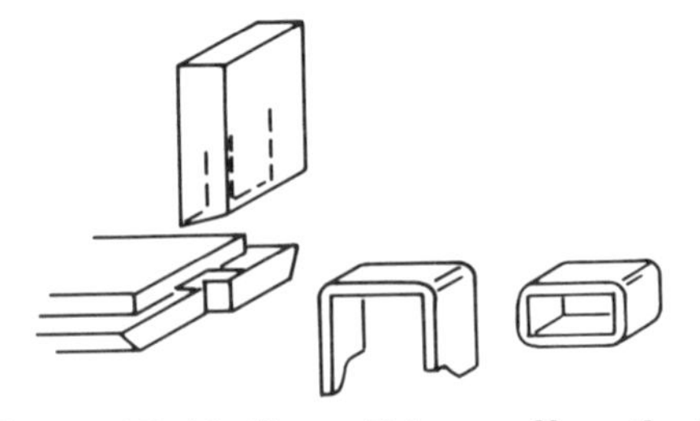

Figure 1G-11. Four-slide cutoff methods: mating or matching cutoff

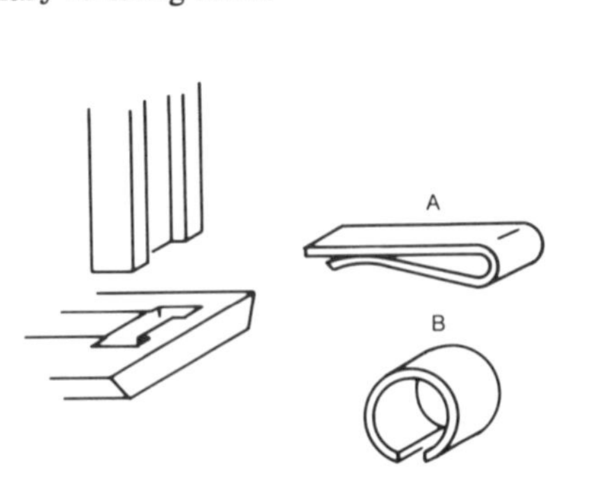

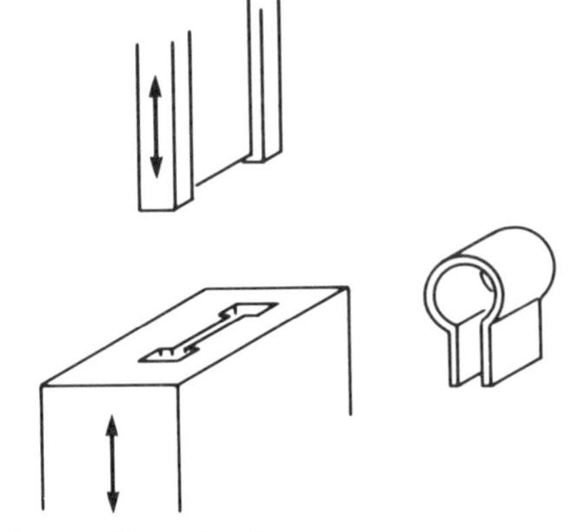

Figure 1G-12. Four-slide cutoff methods: double cutoffs

die block and cutoff blade can be designed to produce any desired mating or matching cutoff shapes (Figure 1G-11).

Double Cutoffs. Unmatched cutoff shapes can be formed with a double cutoff die (Figure 1G-12, left).

Another type of double cutoff (Figure 1G-12, right) employs the standard cutoff attachment together with the bottom slide. This design is usually used only when the bottom tool is not required for forming operations. If the bottom tool is also needed for forming, special cams can be used.

With Long Legs. Wherever long legs must be formed down past a double cutoff die, a sliding die may be mounted in the standard cutoff attachment (Figure 1G-13). This sliding die can be actuated by the left slide cam or by use of a specially designed cam motion. Again, note that independent end shapes are produced.

Rear-Motion Cutoff. The patented rear-motion cutoff (Figure 1G-14) allows complete freedom of the Verti-Slide forming slides for more intricate forming operations.

After the strip has been fed all the way in over the center form, the rear-motion slide containing the cutoff die moves in under the strip. The cutoff punch moves down to cut the strip and then returns to its retracted position. The die then withdraws into the tooling plate, allowing the end of the cutoff piece to be formed downward.

Wire Cutoffs. A typical wire cutoff (Figure 1G-15) uses standard cutoff components with regrinding for the particular wire diameter. A cutoff blade on the side of the top tool may be used. A quill-type cutoff is also available (Figure 1G-16).

MULTISLIDE AUXILIARY EQUIPMENT

Four-slide forming machines accept wire or coil stock directly from a reel, straighten, feed, and cut off the required length. Figures 1G-17, 1G-18, and 1G-19 illustrate auxiliary equipment used in typical operations performed by these machines.

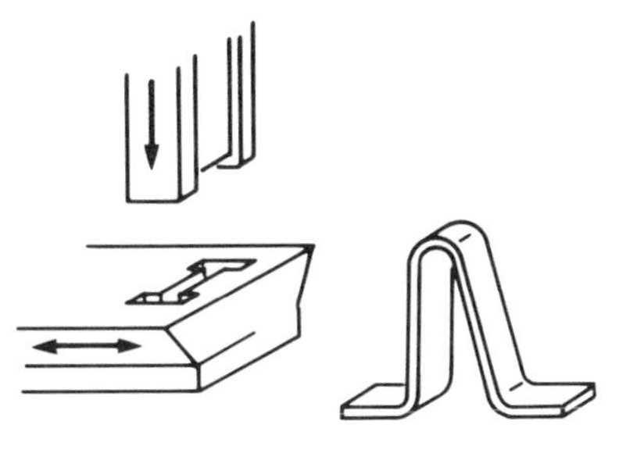

Figure 1G-13. Four-slide cutoff methods: long legs

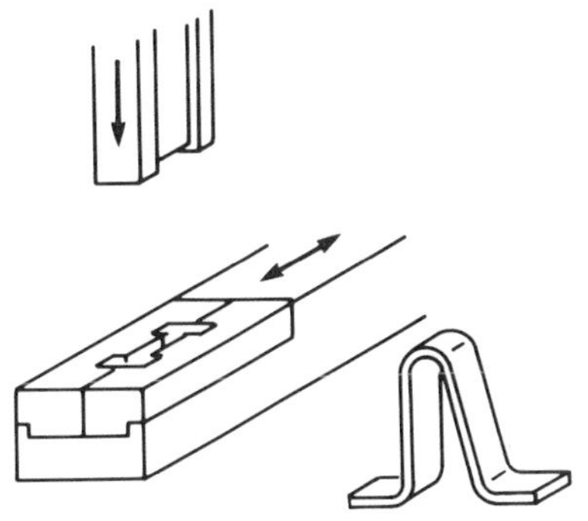

Figure 1G-14. Four-slide cutoff methods: rear-motion cutoff

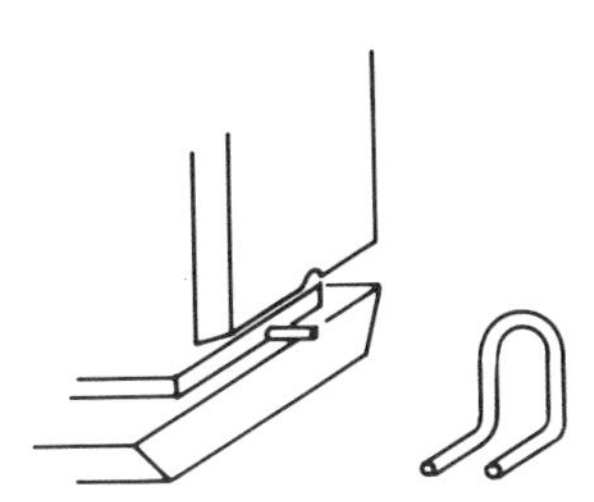

Figure 1G-15. Four-slide cutoff methods: wire cutoff

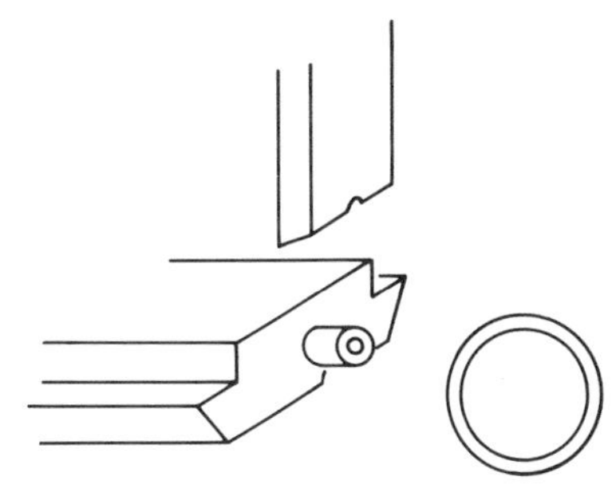

Figure 1G-16. Four-slide cutoff methods: quill-type cutoff

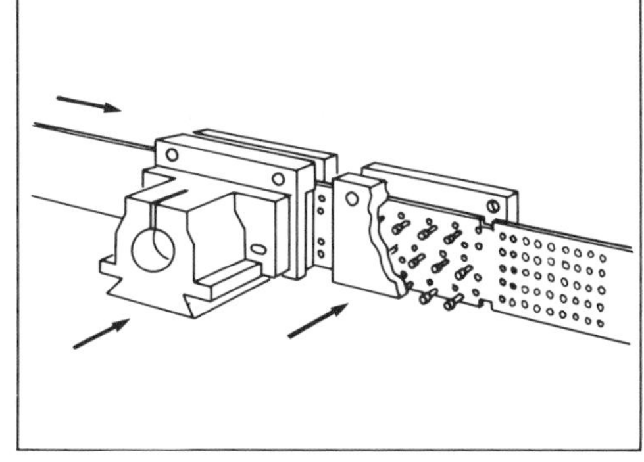

(a)

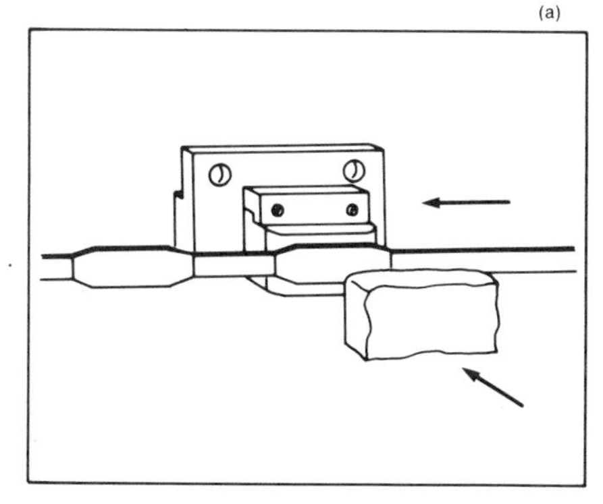

(b)

(a) Heavy-duty die heads. (b) Toggle press, which provides high tonnage at the bottom of the stroke for heavy swaging, coining, or embossing.

Figure 1G-17. Multislide auxiliary equipment: heavy-duty die heads and toggle press

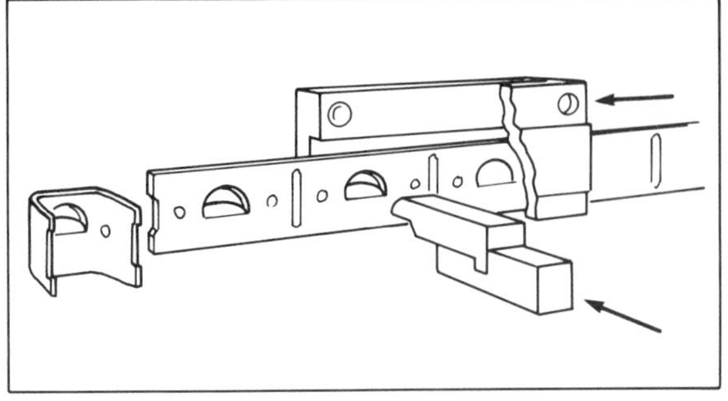

(a)

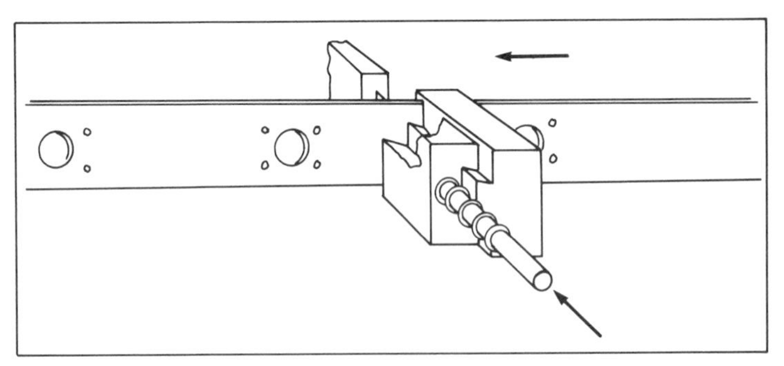

(b)

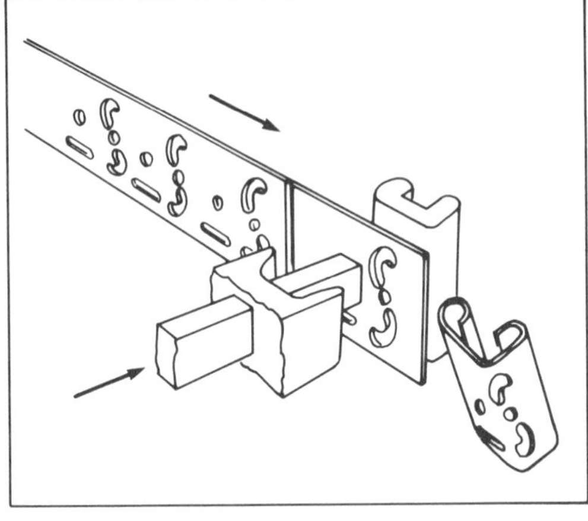

(c)

(a) Rear auxiliary slide, for operations requiring motion in direction opposite die head. During die-head dwell, this unit allows a variety of operations, including shearing, forming, extruding, preforming, and lancing. (b) Rear-position knockout, employed to provide positive cam-actuated control, in place of die spring pads, for return of material flush with die surface. (c) Positive blankholder, used in the front-forming-tool position to prevent slippage or bulging during cutoff and forming, and providing the ability to pilot and position the blank against the form.

Figure 1G-18. Multislide auxiliary equipment: rear auxiliary slide, rear-position knockout, and positive blankholder

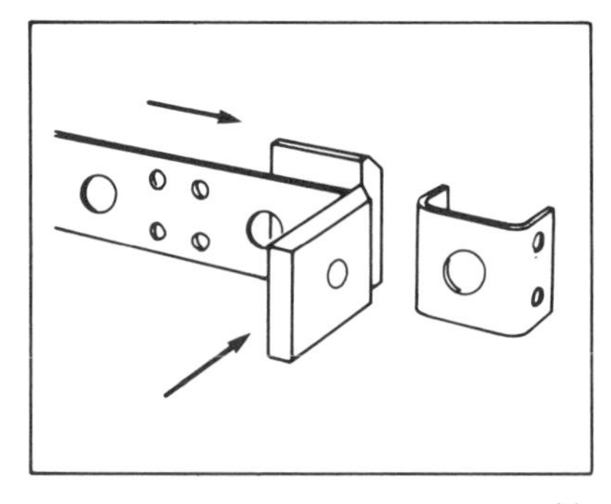

(a)

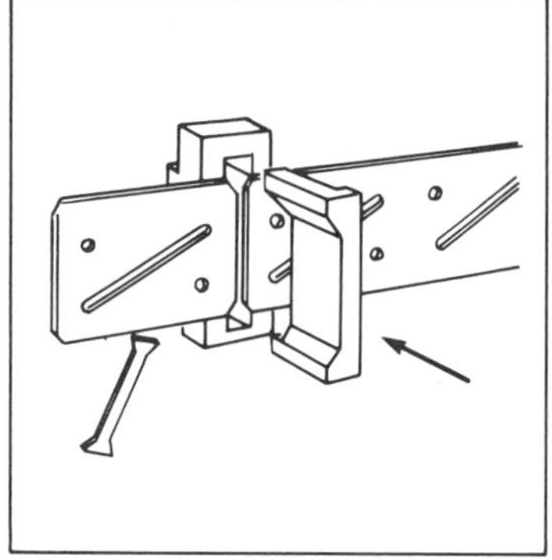

(b)

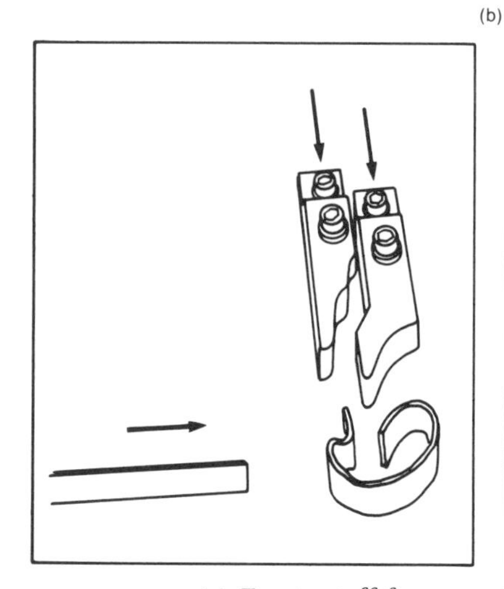

(a) Front cutoff, for separating a blank from the strip prior to forming when scrap is not removed between blanks. Mounted between dies and forming position. Individual timing of cutoff permits separation at instant blank is held by forming tools or positive blankholder. (b) Rear cutoff, for separating blanks by removal of slugs of straight or shaped configurations. Small backplate is furnished for cutoff-die mounting, allowing slug-removal-type parting when left-hand forming slide is used. (c) Split slide. A split front, right-hand, or rear slide can be furnished to operate in the standard slide opening, permitting tooling double movements for versatility in forming, locating, and assembly operations.

Figure 1G-19. Multislide auxiliary equipment: front cutoff, rear cutoff, and split slide

1H: Flexible Manufacturing Systems

One of the latest concepts in the metal fabricating industry is the introduction of the Flexible Manufacturing System (FMS) — a system designed for highly efficient sheet and plate fabrication (Figures 1H-1, 1H-2, 1H-3, and 1H-4).

This system incorporates a series of machines and work stations linked by a common control, providing automatic production of a family of parts for continuous processing.

The computer-integrated manufacturing concept facilitates and improves the manufacturing process. Computer control of the entire system allows constant monitoring of inventory levels, cycle times, tool usage, scrap levels, and other data.

Properly designed material handling can incorporate vertical storage for efficient use of floor space by employing robots in a variety of positions for automatic loading/unloading systems.

There are four levels of flexible manufacturing systems:

Level 1: A single processing machine that is equipped to perform at least two operations in an automatic press with programmable workpiece positioners for x-axis and y-axis movements.

Level 2: Incorporates a fabricating cell and consists of one or more processing machines linked with appropriate material-handling equipment.

Level 3: Consists of two or more processing machines and augments them with other equipment and controls.

Level 4: Would incorporate the future automatic factory. The system would include certain types of fabricating centers that change tools and load and unload automatically.

TOOL MANAGEMENT

Fabrication of sheet metal or plate requires a variety of tools. So the problem of having the right tool on the machine at the right time in a flexible fabricating system is a complex one — especially when one machine will be used to produce a wide variety of parts, each requiring several different tools.

The first step in managing the tool requirements in an FMS is to integrate automatic parts-nesting technology with tool

This system includes laser press with automatic tool changer and tool storage, bidirectional pallet transfer for vacuum-lift feeder, vacuum feeder, carousel for unloading small parts through trap door in table, automatic vacuum unloader for large parts, outfeed pallet transfer for large parts, General Electric 2000P control, and software by W. A. Whitney.

Figure 1H-1. Flexible Manufacturing System (FMS) for sheet-metal parts

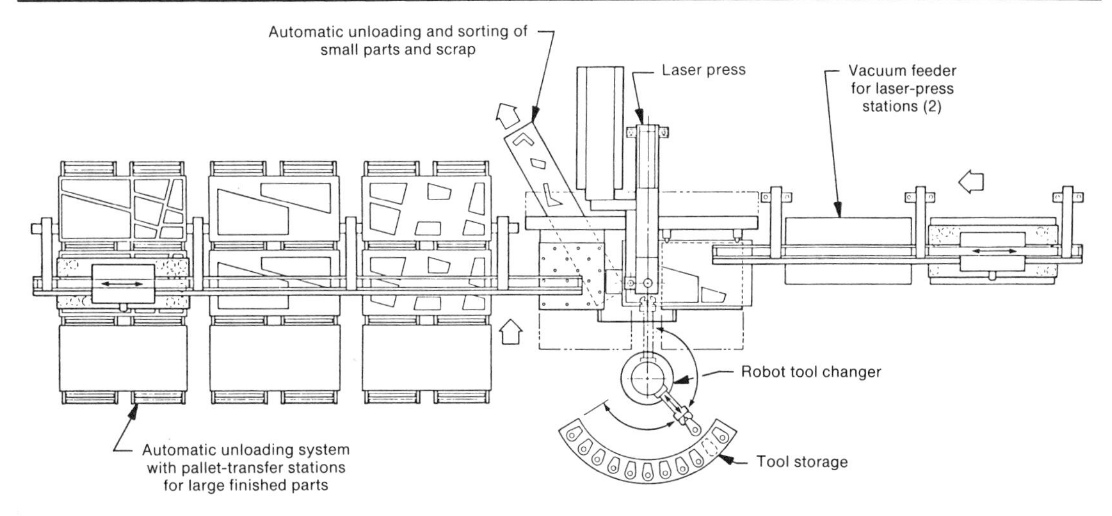

Figure 1H-2. Flexible Fabricating System (FFS) for sheet-metal parts

management so that, as parts are nested, the appropriate tool is selected from the tool library file. During production, an automatic tool changer selects each tool sequentially from the tool-storage racks and is ready to change tools at the proper time, resulting in maximum machine utilization. The tool-storage racks for retrieval and replacement of tools are modularized to allow for resharpening or replacement of tools as they are used, and are monitored by the computer without stopping production.

These tool racks can take a variety of shapes depending on the number of tools the machines will need in order to fabricate the parts in production. Some systems can incorporate up to 300 tools, while other "cell-type" systems require as few as 20 to 80 tool stations.

W. A. Whitney has developed the applica-

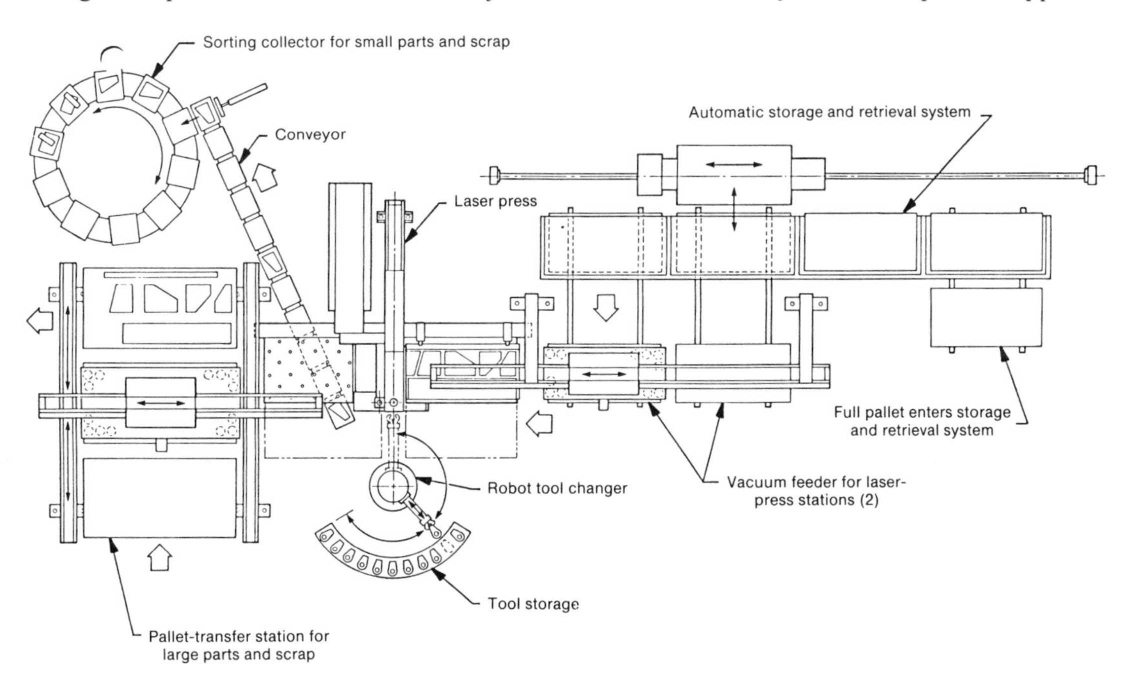

Figure 1H-3. Flexible Fabricating System (FFS) for sheet-metal parts

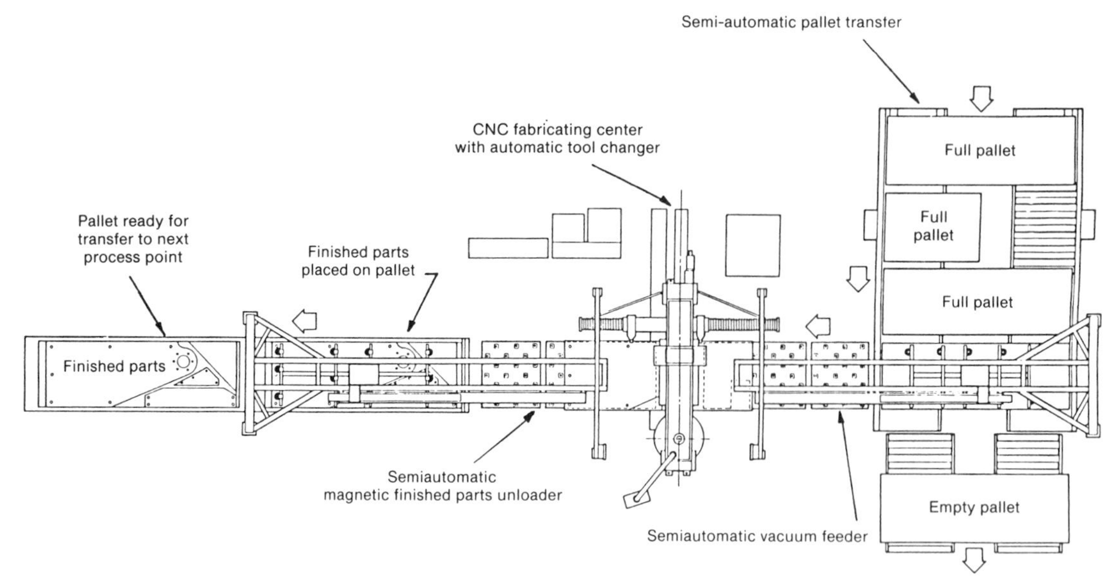

Figure 1H-4. Partial Flexible Fabricating System for sheet-metal and plate parts

Tool changer has storage capacity for 180 sets of tooling, tool racks on rollers to facilitate exchange of complete racks of tools, and light curtain for stopping tool changer when light beam is interrupted.

Figure 1H-5. Automatic tool changer with tool-storage rack

A set of tooling includes a punch with punchholder, a stripper, a die adapter with die, and an aluminum cartridge tool holder.

Figure 1H-6. Close-up view of punch tooling on storage rack

tion of cartridge tool changers which make robotic tool changing reliable and cost-efficient (Figures 1H-5 and 1H-6).

CUTTING SYSTEMS

The W. A. Whitney Flexible Manufacturing System employs the latest cutting concepts — gas-laser cutting and air-plasma cutting. Each manufacturing application has a specific set of requirements which may be satisfied by either of these cutting technologies.

Principles of CO_2 Gas-Laser Cutting

Carbon dioxide (CO_2) lasers convert energy from an electrical discharge into a highly collimated, monochromatic, invisible beam of light of extremely high energy. This light (photon) energy, when directed to a material surface, causes vaporization and melting. The vaporization is achieved rapidly enough, with little heat being conducted into the surrounding material, which results in an extremely narrow heat-affected zone (HAZ), no physical distortion, straight-sided cuts, and deposit of minimal dross on most materials.

CO_2 lasers require three gases: carbon dioxide, nitrogen, and helium. The basic gas which supplies the molecular action required for photon generation is CO_2. Nitrogen acts to sustain and reinforce the molecular action. Helium is added as a cooling agent. The generated photons are directed by mirrors in the resonator cavity in such a way as to sustain the "lasing action."

For maximum productivity, oxygen-assisted cutting (Figure 1H-7), which causes an exothermic reaction, significantly increases cutting rates. The gas flow directs vaporized metal from the cutting area, ensuring full impingement of the laser beam on the material. The gas jet blows away the molten slag, producing a relatively dross-free (slag-free) cut. The cutting rate is inversely proportional to the thickness of the material and is directly proportional to the laser output power above a certain threshold.

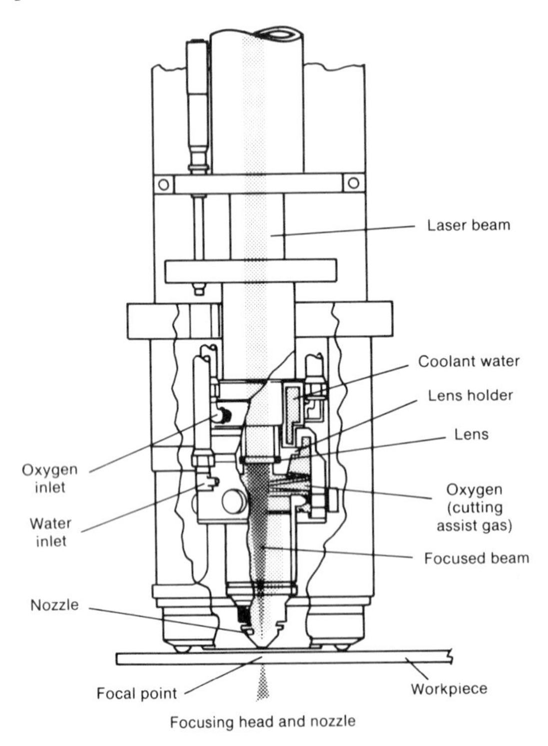

Figure 1H-7. Gas-laser cutting system

Principles of Air-Plasma Cutting

The air-plasma arc is the most efficient type of plasma arc. It simply combines compressed air with electricity to form the plasma. Air-plasma cutting achieves a significantly cleaner cut than other cutting methods and results in vritual elimination of dross in mild steel.

Operating costs of the air-plasma system are about one-half those of dual gas and water injection systems. The Whitney system (Figure 1H-8) is self-contained in the machine, and no external gas or water supplies or environmental devices are required.

A waterfall slag-collector system, in concert with vacuuming above and below the work area and a torch-cooling system, permit 100% duty cycle.

Dual gas and water injection plasma systems are available on customer request.

Advantages of air-plasma cutting are:

- Cutting is three to five times faster than conventional gas cutting methods.
- Applicable on a wide variety of metals and alloys, such as stainless steel, chromium-nickel alloy steel, aluminum, copper, etc., which are not workable with conventional methods. The only limitation is that the material must be an electrical conductor.
- Workable on a wide range of material thicknesses. Best between 0.030 and 1.00 in. (0.8 and 25 mm).
- Air plasma is more efficient than other types of gas plasma.

The air-plasma process is the most efficient and productive cutting process on the market today. Use of compressed air as the cutting medium eliminates expensive and cumbersome gas setups. Cutting speeds up to 350 ipm (150 mm/s) are obtained with long-life consumables for up to four hours of continuous cutting.

AUTOMATIC PARTS NESTING

The "just-in-time" concept of production involves many different factors. One of the most vital techniques of manufacturing processes is the use of automatic parts-nesting systems.

By analyzing daily fabrication requirements, including manageable cost variables such as machine production rates, inventory control, labor costs, and scrap rates, the automatic nesting system produces parts when they are needed, in the quantities required. The system will lay out different parts in any multiples nested at random, based on schedule requirements rather than only on efficiency of material utilization.

Incorporated into the automatic nesting

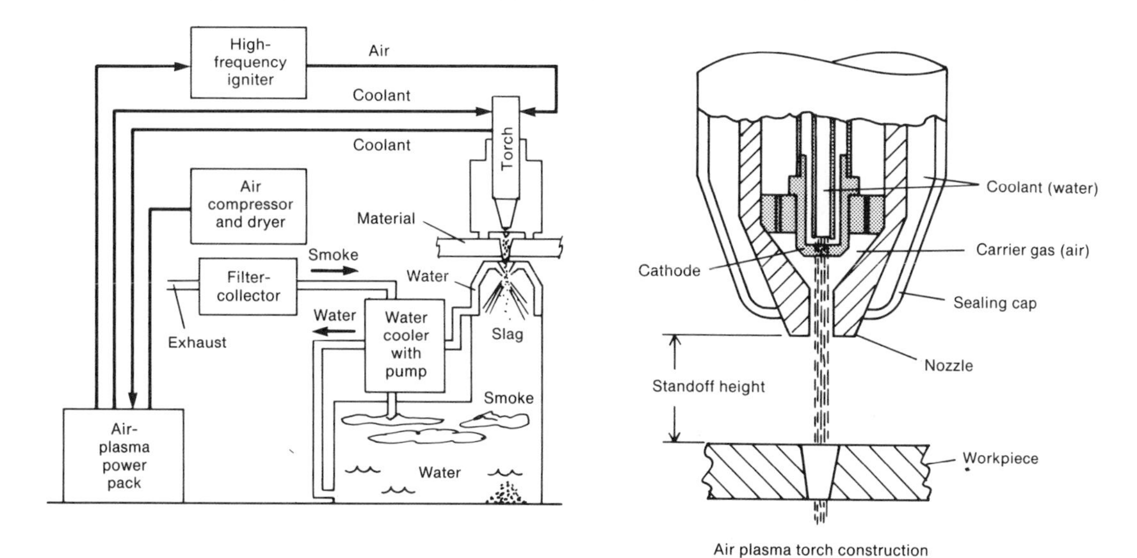

Figure 1H-8. Air-plasma cutting system

system are subfiles for raw-material, machine, tool, and cost management. Additionally, all incoming orders are entered into the active file to constantly update the system on the types of parts needed.

Parts nesting is not necessarily optimized to minimize scrap, but instead to generate precisely, by part thickness and material type, those parts required for the day's production (Figures 1H-9 and 1H-10).

Large parts are stacked on a pallet at the unloading station. At the end of a production run, the pallet is indexed to the station where stacks of parts are removed and sent to the next manufacturing operation.

Figure 1H-9. Automatic large-parts unloader with outfeed pallet-transfer system

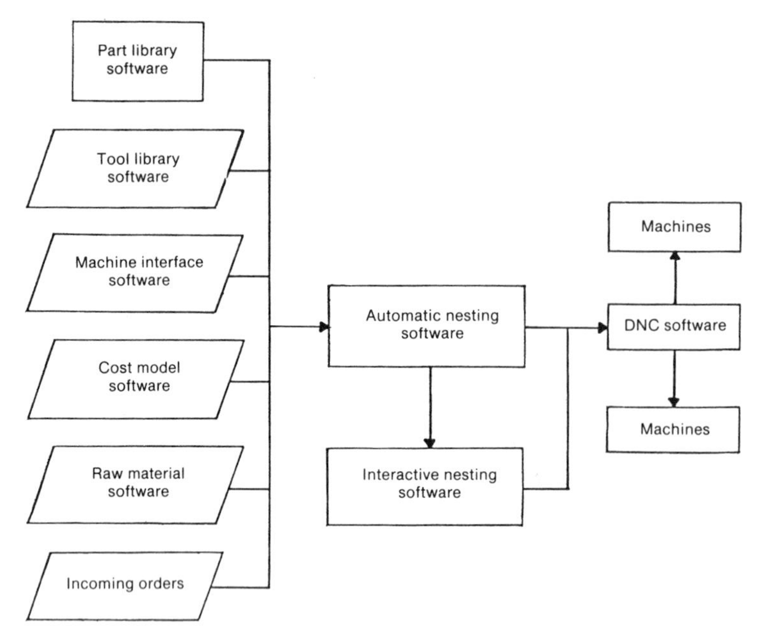

This system provides increased machine productivity and reduced setup time; capacity to produce parts in "just-in-time" environment; reduced in-process and raw-material inventories; increased materials utilization through intelligent nesting; reduced labor burden in many fabricating processes; optimized raw-material inventory; improved information reporting for informal management decisions; increased machine flexibility through better management of direct numerical control of machinery.

Figure 1H-10. Automatic nesting system

AUTOMATIC MATERIAL LOADING

To maximize the efficiencies of an FMS, automatic material loading is a vital aspect. Depending on the specific needs of the manufacturer, automatic storage and retrieval systems can be developed for 4-by-8-in. (100-by-200-mm) sheet or plate sections or for 5-by-10-in. (125-by-250-mm) blanks.

Storage of raw material is provided in a number of configurations from single stacks of blanks on index pallets to elaborate libraries of pallets that are automatically moved to provide an uninterrupted flow of materials.

Automatic loading of the fabricating center is provided by either a pneumatic or a magnetic feeder that selects one sheet for delivery to the work stations (Figure 1H-11).

AUTOMATIC SORTING AND STACKING

For an FMS to effectively produce random parts in random quantities, two factors are needed: automatic parts nesting, and automatic sorting and stacking. These interconnected processes allow an FMS to generate parts at random with outstanding efficiency.

Automatic sorting is needed especially for small parts that are difficult to remove mechanically or pneumatically from the work station. Consequently, a computer-driven trap door is used to sort each part into a designated bin on a carousel below the work table. The carousel is indexed automatically by the computer so that each part ends up in the selected bin (Figure 1H-12).

Includes powered roller conveyor, pallet transfer, automatic vacuum-lift feeder, automatic sheet-thickness gauging, and air-knife sheet separator

Figure 1H-11. Bidirectional pallet transfer and vacuum-lift sheet feeder for FMS for sheet-metal parts

Larger parts are removed from the work table by a programmable platen having a matrix of cushion cups or electromagnets. Parts are stacked on pallets and then automatically indexed to an unloading area for further processing.

Automatic tool changing and storage are performed by a changer robot with indexable, two-position, tool-insert transport.

The tool-storage system has a handling capacity of 180 tool cartridges and provision for changing tooling in the tool room or at the system location.

FUNCTIONAL DESCRIPTION OF W. A. WHITNEY'S CIM NEST SYSTEM

The term "material nesting" is generally conceived of as referring strictly to a method in which the prime goal is to reduce the scrap rate of the raw material being processed. This simplistic approach to the problem is not a good solution in the vast majority of cases. Additionally, two inseparable components of the problem must be recognized and addressed if the true lowest-cost solution is to be obtained: the labor rate/machine burden costs of the manufacturing process involved; and the cost of not adhering to the schedule generated by the plant's MRP (material-requirement planning) or other production-scheduling system. A nesting program sent to the shop floor is of little value if the parts placed on the machine by the nesting system were needed several days ago or if they require excessive setup time by the machine operator. In most nesting situations, the three categories of the problem (material utilization, labor rate/machine burden costs, and schedule adherence) must be balanced by making intelligent trade-offs to obtain the most cost-effective solution.

The carousel is automatically indexed for sorting of parts. Collector pans are removed from the carousel at the end of a production run and are replaced with empty ones.

Figure 1H-12. Carousel for sorting of small parts unloaded through a trap door in the table of the laser press

Functional Specifications

Graphical Part Programming. This is a multifunctional program which enables the user to construct an NC part program by entering data via a simple menu-driven graphical format. This program minimizes the degree of mathematical ability required and allows the user to see the results of each step before proceeding. It will provide immediate output without cutter-location files or postprocessing. Each target machine can be specified in order that unique machine output will be achieved without the expense of writing a postprocessor.

Cost-Model File. An integral part of W. A. Whitney's CIM NEST system is the cost-model file. This file contains information, entered by the user of the system, which describes how much weight will be given to each of the three categories mentioned above as an automatic nest is generated. The file contains information regarding the costs of owning and operating the Plasma-Punch, the rewards and penalties given to part orders which are either not yet due or past due, part-priority assignments, and the importance of order cohesiveness. This data, combined with the cost of the raw material being nested, is used to generate the nest solution with the lowest total cost in dollars and cents. The cost model allows the user to tailor the nesting algorithm to the best output for each manufacturing environment rather than forcing the factory to adapt to the output of a rigid nesting algorithm. In other words, it allows the entire system to be truly flexible, providing its users the opportunity to implement the "just-in-time" concept of production.

Part-Program Interface. The CIM NEST system is capable of accepting either NC (tape image — i.e., G01, M04, etc.) code or APT Source code as input describing the actual part geometry to be produced. This feature provides the user the option of interfacing with the wide variety of part-programming systems available or with a CAD/CAM systems APT Source generator. The system is capable of accepting part-program data in either English or metric (SI) units and will output nests in either system. This internal conversion feature adds greatly to the flexibility of the system and eases the task of companies that are converting to metric.

Machine Interface. The system includes a universal postprocessor file which is configured to properly drive the Plasma-Punch machine. Included in this file is the information required to build the nest: information such as repositioning speeds, tool offsets, clamp interference zones, and punching speeds is all contained here. An added benefit of this design approach is the possibility of creating part programs which are generic in nature so as to allow them to be used on future machines which may not be identical to the one first used.

Raw-Material File. Every raw material to be processed by the system has its own file containing details of dimensions, plasma cutting speeds, plasma kerf, part-spacing allowances, number of sheets on hand, and the cost of the material.

Scheduling (MRP) Interface. The nesting system is order-driven; an order consists of a part number, the required quantity, a due date, and an optional priority number. W. A. Whitney can configure their software to allow direct downloading from a user's business computer. As parts are produced, the system automatically updates the status of the orders.

Automatic Nesting. The automatic nesting portion of the system draws information from the cost-model file, the machine file, the scheduling file, and the raw-material file to create the lowest-total-cost nest. Because every nest is built for a specific production date, and because typical nesting time is a matter of few minutes, a powerful tool is provided for the just-in-time method of manufacturing. During the nesting process, factors such as the punching tools required and the locations of punch hits within individual parts are considered so as to make the proper trade-offs among machine setup time, schedule adherence, and materials efficiency. This concept of optimization gives this nesting system the ability to provide the core of a fully automated flexible manufacturing cell controlling not only the operation of the Plasma-Punch, but also material-handling equipment and unloading robots. Other features of the nesting routine include the ability to automatically nest smaller parts within cutouts of larger parts, to recognize parts that have grain constraints, and to efficiently nest parts which are irregularly shaped.

Interactive Graphics. If desired, every nest may be viewed by the operator on the system's CRT. At this time the operator is given the opportunity to check material efficiencies, production time, tool requirements, and order status. Additionally the operator may interactively adjust the displayed nest through commands such as Move, Rotate, etc. From this level of the software, reports can be generated detailing material utilization and time required for part production and tool wear, as well as shop-floor documents including plots of nests.

1I: Die Sets: Applications and Functional Requirements

A die assembly is a complete punch-press tool that is used to produce large numbers of interchangeable stamped parts. It consists of mating pairs of punches and dies, their retention plates, a stripping device, and a subassembly called a die set. When the press is actuated, the mating components are forcibly brought together to perform the required operations.

The upper die assembly is fastened to the moveable press ram and the lower die assembly is secured to the stationary bolster plate or press bed. The upper member *must* travel in a precisely controlled path to maintain the alignment and orientation of mating components during the working portion of the press stroke.

Every punch press must permit relative movement between the ram and the press frame in order to satisfy its primary function of delivering energy. The "running" clearance provided, however small, may have an adverse effect on the operational alignment of mating punch-and-die combinations. For example, even when press parts are closely fitted and precisely aligned, the running clearance may actually exceed the required punch-to-die clearances within the die assembly.

The impact nature of the punching process further complicates the alignment problem. It generates deflecting forces that must be effectively opposed so that alignment can be maintained.

Obviously, the alignment system of the press requires that the die assembly provide additional guidance and support at the point of impact and throughout the working cycle. It is in this area that a good die set demonstrates but one facet of its tremendous value.

Caution: Although the die set can be relied upon for the ultimate alignment, it cannot be expected to compensate for a press in poor condition or to operate properly if subjected to heavy deflecting loads.

The guide post/guide bushing combinations, properly fitted, ensure precise control of the moveable member at every critical point of the stroke to maintain the relative position of each pair of mating components.

FUNCTIONAL REQUIREMENTS

A good die set is not just a subassembly. It provides the foundation for a basic working system upon which a press tool can be assembled, aligned, inspected, and put into operation. It enables the die builder to assemble many separate components into a single integral unit that provides many cost-saving benefits:

1. **Ease and accuracy of setup:** The die assembly is installed in the press as a single unit which minimizes setup time and ensures proper alignment of mating components.
2. **Improved and consistent stamped-part quality:** The ensured accuracy of each setup maintains the original degree of punch-to-die clearance uniformity and thereby enhances piece-part quality.
3. **Increased die life:** Correct register of punch-to-die reduces the rate of wear on both components and minimizes the risk of component breakage.
4. **Ease of die maintenance:** Cutting components can be sharpened in assembly, as units, without removing them from the die set — a distinct advantage over removing the components and sharpening them individually.
5. **Simplified die-repair procedures:** Broken or worn components can be removed and replaced without disturbing their relative positions. Alignment and orientation of mating components is virtually ensured.

DIE-SET NOMENCLATURE

The die set is a modular unit consisting of a stationary lower plate (*die holder*), a moveable upper plate (*punch holder*), and at least two sets of precisely fitted *guide posts* and *guide bushings*. Smaller die sets are equipped with a projecting stem (*shank*) which extends from the top surface of the punch holder (Figure 1I-1).

Generally, the guide bushings are assembled to the punch holder and the guide posts to the die holder. This arrangement is normally reversed when ball-bearing bushing assemblies are utilized.

THE SHANK: FUNCTIONS AND VARIATIONS

The shank is used to center the die assembly in the press and to secure the upper die member to the press ram. A large die set may also be equipped with a shank, but it should be used only for centering purposes — not to secure the upper die assembly to the ram.

The shank may be an integral part of the punch holder or it may be a separate unit that is secured by means of a threaded end or individual fasteners. Integrally welded shanks are structurally stronger, cannot loosen in operation, and provide the least interference with knockouts or with adjacent fasteners and components.

Floating shanks are designed to compensate for a ram face that is not perpendicular to the cavity for the die set shank. The top of the punch holder is machined to receive a swivel adapter which contains a separate

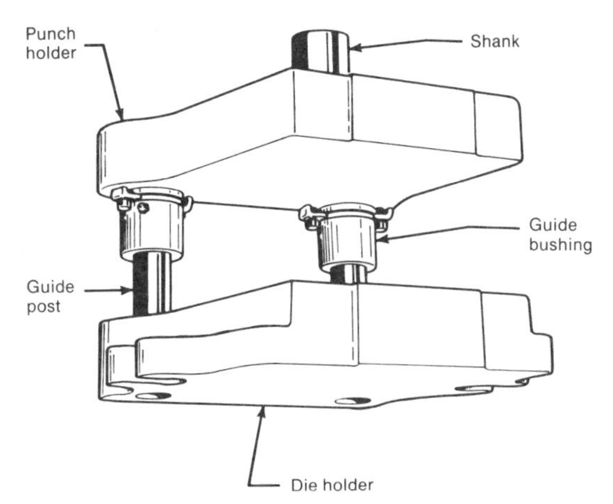

Figure 1I-1. Die-set nomenclature

shank. The shank is permitted to move until it conforms with the ram face. Parallelism between the punch holder and the die holder is not affected.

ALIGNMENT CONSIDERATIONS

The die set is expected to maintain the alignment of all mating components during the die-building stage and throughout the productive life of the die assembly. In essence, a good die set must withstand the rigors of use without prematurely losing its ability to control alignment.

Many factors must be considered and controlled in order to satisfy this objective:

1. Dimensional specifications and limits
2. Allowances for cylindrical fits
3. Material types related to the various die-set components
4. Geometrical deviations that may be anticipated
5. Permissible variations in the alignment of guide posts and guide bushings
6. Available styles of die sets and components.

The United States Standard for Die Sets, USAS B5.25-1968, provides detailed data for each of the foregoing factors. (Copies of this standard are available at a modest fee from The American Society of Mechanical Engineers, United Engineering Center, 345 East 47th Street, New York, New York 10017).

Functionally, standards data provide product definition by means of material specifications, dimensions, and permissible tolerances. In this sense, they tend to establish an area or range from which selections can be made. They fail, however, to provide the basis on which to make a selection.

APPLICATION REQUIREMENTS

The ultimate alignment of mating punches and dies can not be better than that provided in the die set. Alignment must be established during the building stage when practically no deflecting loads are applied. Alignment must be maintained when the tool is in operation and producing stamped parts.

Under operating conditions, the entire die set and the individual components are continuously subjected to high impact loads and deflection forces. Consequently, all selections should be predicated on the basis of resistance to shock, deflection, and wear.

PUNCH AND DIE HOLDERS

These plates must satisfactorily oppose rapidly applied compression loads. The plate material must have high compressive strength and toughness. Because the die holder may be mounted on parallels or span large bolster-plate openings, it must also exhibit a high degree of stiffness or resistance to bending. The ability to resist these destructive forces is largely dependent on the material selected and the physical proportions of the plates. In general, rigidity increases with plate thickness.

Because of the need for many fastener holes and extensive machining operations, the plate should be free from internal defects such as porosity, hard spots, or shrinkage.

When a die set requires milled pockets, large counterbores, cutouts, or burnouts, the plate should be stress relieved. It is recommended that the die-set manufacturer perform the machining before the plates are bored for guide posts and guide bushings.

Based on these considerations, the most widely used plate materials are:

1. Hot rolled steel containing 0.15 to 0.30% carbon and 0.30 to 0.90% manganese
2. Gray or cast iron and specialized high-test irons such as Meehanite
3. A combination wherein one die-set member is made of steel and the other is cast
4. Tool steels (hardened or partially hardened), aluminum, magnesium, and other special alloys.

The steel plate and the high-test irons are the most popular materials. Of the two, the advantages of all steel die sets are clearly established:

1. Greatest resistance to impact loads and to loads that are not uniformly distributed
2. More uniform density
3. Ability to be flame-cut and welded.

PLATE FLATNESS AND PARALLELISM

The opposite surfaces of the individual plates must be both flat and parallel to avoid an adverse effect on the ultimate alignment of mating components (Figure 1I-2).

Components mounted on a surface that is not flat may cause the line of action to be canted away from the perpendicular and result in sheared or broken cutting edges. Should a component be assembled over a small bump in the surface, a rocking action may be imparted by the punching load, causing component breakage or premature wear.

Punch-to-die alignment will be impaired if the opposite surfaces of a plate are not parallel even though both may be flat (Figure 1I-3). The components will be tipped away from the perpendicular and the lines of action of the upper and lower die assemblies cannot coincide.

Parallelism parameters are applicable to the assembled die set as well as the individual plates (Figure 1I-4).

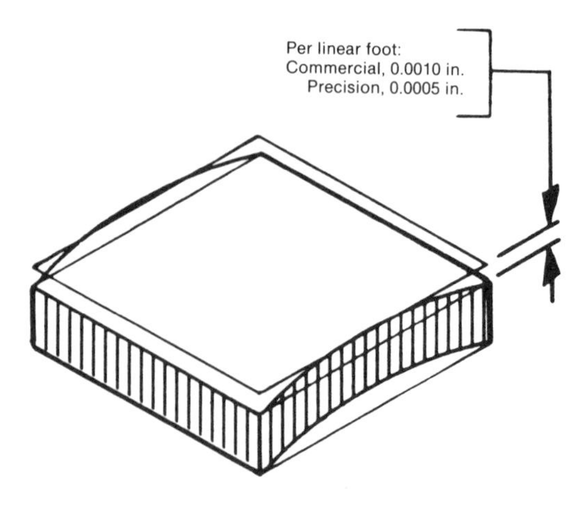

Figure 1I-2. Plate flatness

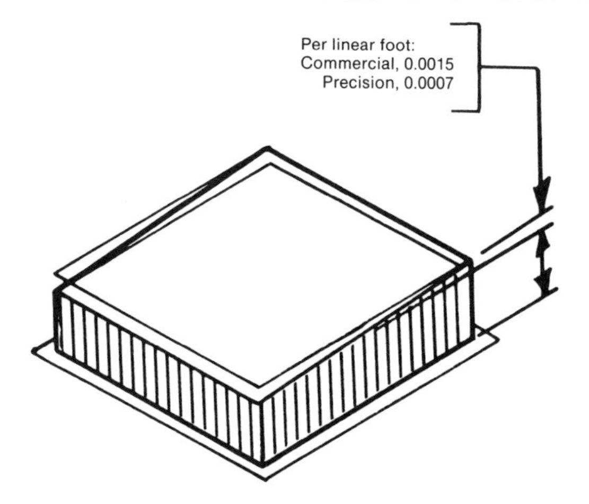

Figure 1I-3. Plate parallelism

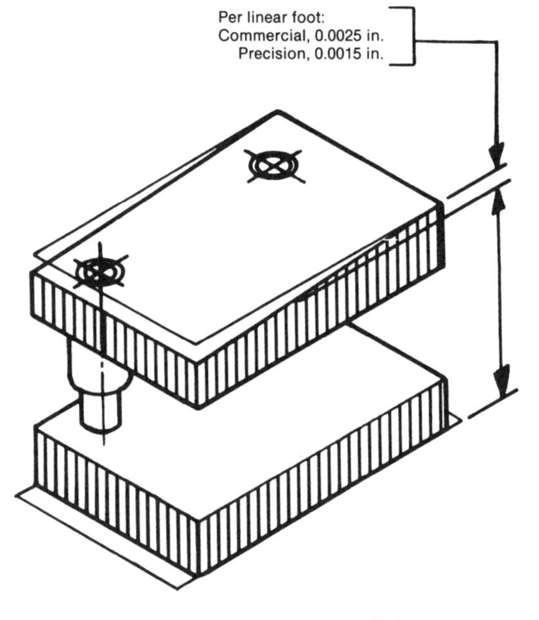

Figure 1I-4. Die-set parallelism

GUIDE POSTS AND GUIDE BUSHINGS

Guide posts and guide bushings, in combination, establish and control the path followed by the moveable upper die assembly. In effect, alignment uniformity can be maintained only when the relative positions of the upper and lower die members remain unchanged during the working portions of the stroke.

Because deflecting forces are always present and can be minimized but not eliminated, the selection of guide posts and bushings must be based on their ability to resist both wear and deflection (Figure 1I-5).

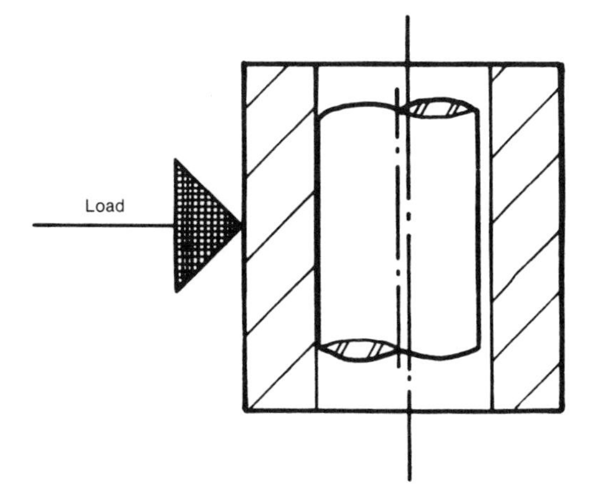

Figure 1I-5. Guide post/guide bushing deflection

Two basic guidance systems enjoy a marked degree of popularity: the ball-bearing bushing assembly and the more widely used conventional system. Each has advantages as well as disadvantages, and selection is usually a function of personal preference.

The ball-bearing bushing assembly, by virtue of its rolling balls, provides unusual ease of assembly. The interference fit or preload on the balls maintains the relative position of the bushing to the post, but the assembly is less resistant to deflecting loads. For a given bushing size, the post is more likely to bend because its diameter has been reduced in order to provide room for the balls. To compensate for this problem, Danly die sets maintain the basic guide-post diameter and increase the bushing size. The plate dimensions are correspondingly increased so that the available die space is unchanged. Rolling balls generate less friction than sliding members, but are more difficult to lubricate (Figure 1I-6).

The interference fit with the individual balls eliminates any possibility of a physical barrier being established between the components. Ordinary lubricants will break down due to extremely high pressures which rupture the film. It is necessary to utilize extreme-pressure additives that react with the various components to form chemical barri-

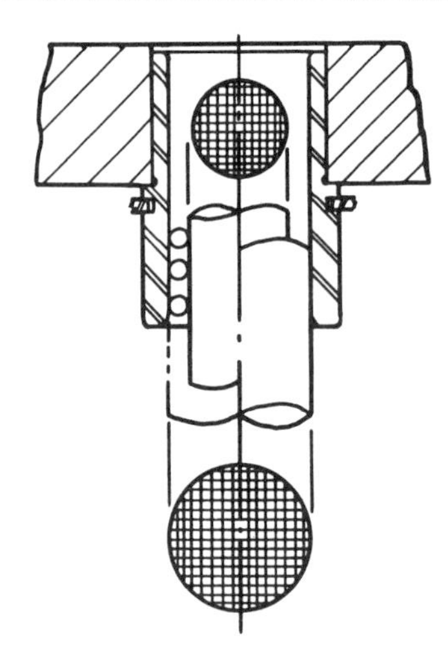

Figure 1I-6. Guide-post rigidity comparison between ball-bearing and conventional guidance systems

ers that prevent metal-to-metal point contacts (Figure 1I-7).

Ball-bearing bushing assemblies are recommended during the building phase, but the post and ball cage can be removed and replaced with the conventional demountable post for optimum performance.

In the conventional system, clearance is provided between the post and the bushing. This space is then filled with lubricant to provide a physical barrier between the moving members. The Die Set Standard establishes the permissible range of clearance without adversely affecting alignment ability and defines the configuration of the lubricant-retention grooves. Guide posts and bushings are prefitted to dimensions, within tolerance, and are furnished as matched units in each die set. By means of color-coded components and selective assembly, three classes of operating fits may be specified.

Guide posts and bushings are primarily subjected to frictional forces, which generate wear. Even deflecting forces are translated into wear factors because they tend to break down the lubricant and thereby increase the coefficient of friction between the sliding members. In selecting the best combination of materials for use in a specific application, the decision is dependent upon the anticipated operating conditions:

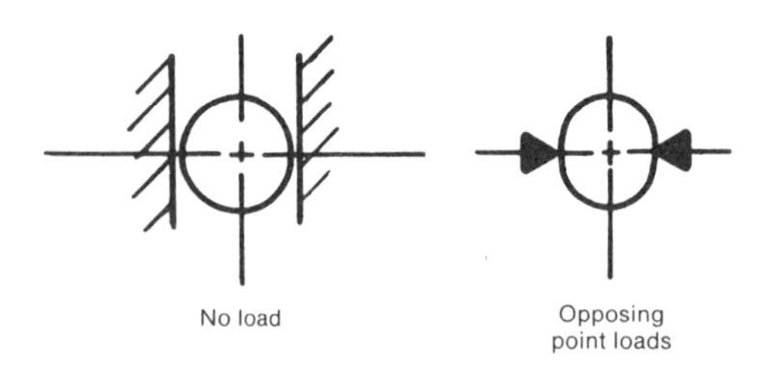

Figure 1I-7. Ball-bearing deformation

1. **Frequency, amount, and type of lubrication:** Lubrication must be performed on a regular schedule, and the amount must be adequate to replace the lubricant that has been displaced or used. When large deflecting forces are anticipated or encountered, extreme-pressure additives are required.
2. **Punch-to-die clearances to be maintained by the die set:** When clearances are extremely close, the guide post and bushing must be selected on the basis of abrasion resistance. Hardened steel posts and bushings are selected for the majority of applications.
3. **Condition of the press:** Specially processed bronze-plated bushings should be chosen when considerable ram play is anticipated. These bushings provide a bronze interface with a steel support and are used with hardened steel posts.
4. **Velocity of operation:** Press velocity is not merely expressed in terms of strokes per minute. In actuality, it is the product of twice the stroke multiplied by the number of strokes per minute. Bronze-plated bushings and chromium-plated guide posts are the best choice for high-speed operations.
5. **Abrasive impurities in the surrounding atmosphere:** When the surrounding atmosphere is charged with abrasive particles which cannot be economically removed, well-lubricated hardened steel guide posts and bushings will provide the best service. The hardness is such that the abrasive particles will have less effect on surfaces.

Surface wear due to friction, either sliding or rolling, is referred to as abrasive wear. Should the abrasive wear condition be permitted to continue without alleviation, the ultimate result will be seizure: the normally moving members will be literally welded together.

ABRASIVE WEAR

When relative motion exists between two parallel, lubricated surfaces, the protective film is subjected to stretching as well as compressive forces. As a result, there is almost never a complete protective film separating the surfaces. Minute voids in the lubricant film permit metal-to-metal point contacts to occur. These point contacts, along with hard foreign elements in the lubricant, induce an abrasive action on the two surfaces. The result is a removal and/or relocation of surface material caused by a small-scale welding action between contacting points. In its extreme form, this surface relocation action results in seizure.

A periodic visual examination of guide posts and bushings is recommended. In many cases, abrasive wear can be detected by the polished appearance of the surfaces, and corrective action can be initiated to prevent seizure.

Abrasive wear is the result of the frictional forces developed during the relative movement of the two members. It is extremely significant to note that friction cannot be present unless the two surfaces are forced together by the application of an external load. In effect, the conventional post-and-bushing combination is subjected only to deflecting forces; the ball-bearing bushing assembly is subjected to the initial load imparted by the interference fit plus the deflecting forces.

Guide-post and guide-bushing materials must be selected relative to the lowest frictional coefficients and the greatest resistance to deformation. The objective, of course, is to select material combinations that offer the greatest wear resistance with the least rate of wear. To determine which materials best satisfy the objective, a number of mechanical and physical properties must be considered:

1. **Relative hardness:** Where relative sliding action occurs, it is generally agreed that the greater the hardness difference between the two members, the lower the coefficient of friction. In the presence of hard foreign elements, however, a harder surface on both members will generally provide a lower wear rate. The ability of a material to bury within itself various amounts of foreign particles increases as its hardness decreases. This characteristic is particularly desirable because the abrasive particles are eliminated from the immediate region where the surfaces may be in contact. Die-set operational requirements, however, preclude the use of a very soft material for its embeddability characteristics alone. A layer thick enough to absorb a substantial quantity of particles would soon be plastically and permanently deformed by the lateral pressures imposed by deflecting forces.

2. **Elastic and plastic characteristics:** Hardened materials exhibit the greatest resistance to deformation. Their relatively high elastic limit permits them to withstand large lateral loads with a minimum of distortion. Soft materials may be permanently distorted when subjected to similar loading conditions. Their relatively low elastic limit may be exceeded, resulting in plastic deformation or displacement. Danly's experiments, conducted in a range of low operating pressures and velocities, conclusively demonstrated that hardened steel bushings are more wear resistant than those made of a cast alloy containing zinc, aluminum, and copper. In both instances, the mating surface was a hardened steel guide post.

3. **Presence and characteristics of surface films:** Lubricants tend to fail when subjected to large pressures and/or high velocities. When the lubricant film is ruptured, metal-to-metal point contacts occur. Extreme-pressure additives such as chlorine and sulfur combine with the base materials to form a chemical film or barrier which can substantially reduce the rate of wear. It is also possible to introduce a thin metallic film having a low coefficient of friction at the interface of the two surfaces. Danly's specially processed bronze-plated bushings exemplify this approach and exhibit an extremely low rate of wear.

4. **Ability to form a common welded joint:** As stated previously, seizure is an extreme form of abrasive wear. It is typified by considerable surface material removal and/or relocation and drastic increases in frictional forces, and is usually accompanied by an audible squealing sound. The point at which this action occurs and the extent of damage to the bearing surfaces are largely dependent on the physical properties of the materials involved. Because high local surface temperatures are generated, there is some molten flow of bearing material, generally resulting in a galled appearance. If the local heating produces a welding effect between minute quantities of the two materials, the strength properties of the resultant substance will influence the degree and direction of the galling. Materials, therefore, should be selected on the basis of least affinity for each other, and in the event of a welding action, the lowest strength properties in the resultant substance. An indication of the properties of the formed alloy will be given by its relative position in the periodic table.

GEOMETRICAL CONSIDERATIONS

Guide posts and bushings are secured to the appropriate plates by means of press fitting or the utilization of external toe clamps and fasteners. Components assembled by the latter method are known as demountable items.

The demountable components provide the best guidance system because all of the desirable geometrical requirements are maintained (Figure 1I-8). In addition, they can be easily removed and replaced to simplify die-maintenance procedures without disturbing the initial alignment.

The use of a wring or a light tap fit between the plate and the bushing or post, for a length of approximately 3/16 in. (4.8 mm), practically eliminates distortion in any of the components. The guide bushings are machine honed for accurate hole geometry and size. They are then mounted on arbors so that the shoulder seats can be ground flat and perpendicular to the guide posts. This combination of manufacturing accuracy and distortion-free assembly automatically

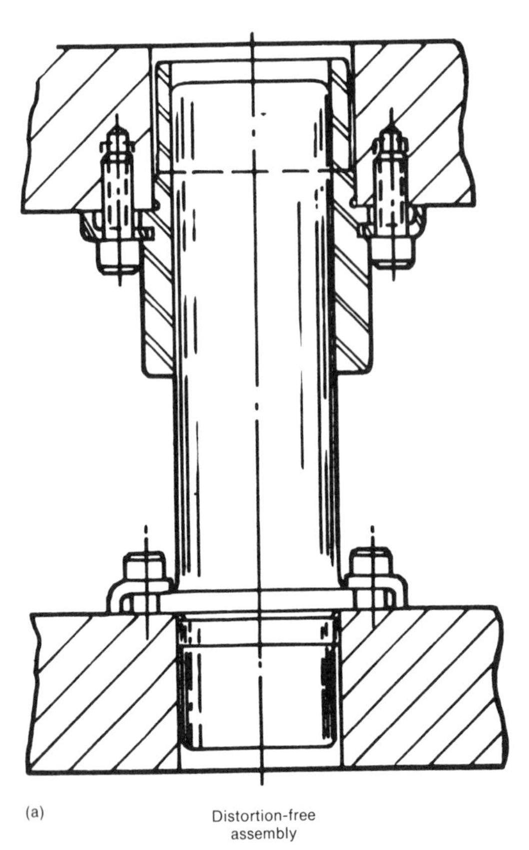

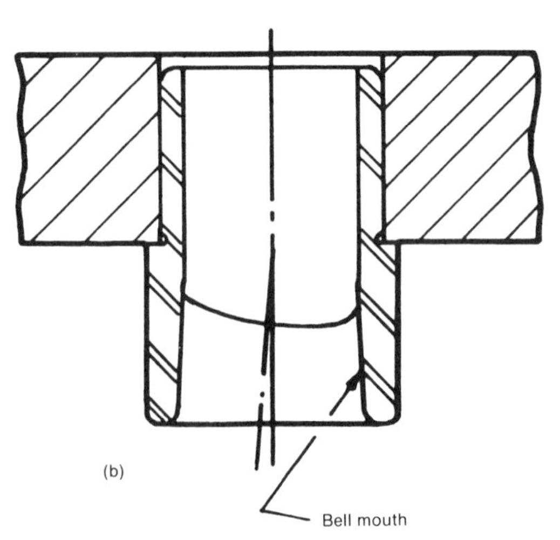

Figure 1I-8. Demountable (a) and press-fit (b) guide posts and bushings

squares up the posts and bushings to the die-set surfaces.

The holes bored in the die-set plates serve only to position the posts and bushings. Retention is accomplished by means of toe clamps and cap screws which develop greater resistance to stripping than press-fitted components.

In press-fitted assemblies, the plate holes must provide for the retention of the guide posts and bushings as well as their relative positions. The interference between the bushing and the hole in the plate distorts the bushing hole. The subsequent honing operation may bell-mouth or enlarge the entry end, resulting in loss of proper fit.

Press-hole distortion is an inherent risk in press-fitting operations. Consequently, the posts and bushings may not be square with the die-set surfaces. It may be necessary to tap the components to adjust the grip for squareness.

ABRASION AND SEIZURE TESTING METHODS

A single unit consisting of a guide post and bushing was installed in a fixture with a variable-speed motor drive that made available speeds of 50 to 430 strokes per minute. The bushing was split into halves to facilitate the application of variable surface loads. Accurately calibrated electronic strain gages were utilized to measure the magnitudes of the applied loads. Lubrication and cleanness conditions were maintained as uniformly as possible throughout the experiment.

The graph in Figure 1I-9 illustrates the effects of press speed and variable surface loads on the coefficient of friction between the sliding members.

ENDURANCE TESTING

A special die set was installed in an OBI press continuously operated at 320 strokes per minute. Variable surface loads were applied to the guide post and bushing assembly by means of a preloaded compression spring. Lubrication was introduced between the guide post and bushing at the beginning of each test run, and operation was continued until seizure occurred.

Following each seizure, the specimens were allowed to cool, examined for galling, relubricated if galling was not too severe, and reoperated to determine whether a healing characteristic existed.

This test demonstrated the superior wear resistance of hardened steel posts and bushings. This combination will maintain close tolerances for long periods of time when loads and speeds are below the point of seizure.

DIE-SET RECOMMENDATIONS

Catalog die sets that provide up to 25 by 14 in. (635 by 355 mm) of die space are offered in many guide post and bushing arrangements. They include many styles and types of guide post and bushing designs from press-fitted and demountable types to the ball-bearing bushing assembly. Both commercial and precision die sets are available for selection.

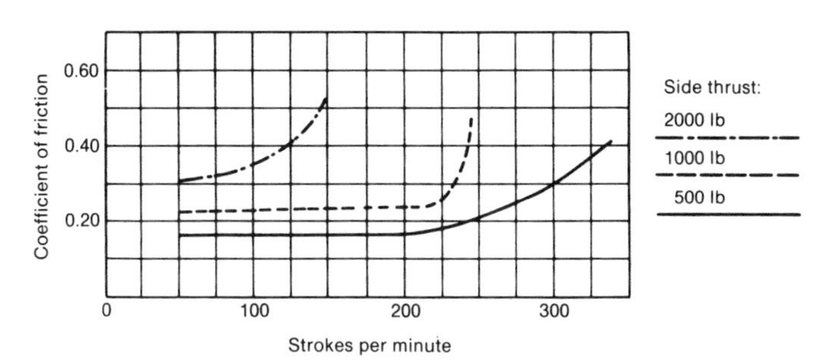

Figure 1I-9. Hardened steel bushing characteristics (stroke length, 1 in.)

The broad areas for selection do not impose any unusual die-design restrictions, and the advantages of die layout for catalog die-set usage are many:

1. The punch or die holder can be replaced if necessary. Danly manufactures each catalog set with interchangeability of all components.
2. Worn or damaged guide posts or bushings can be replaced without impairing the original alignment.
3. Components are stocked at strategic locations across the country, ready to be assembled without delay.
4. All steel punch and die holders are stress relieved to maintain dimensional stability during subsequent machining operations. It also helps to soften edges which may have become hard during the flame-cutting operation.

1J: Press Feeders for the Job Shop

A stamping job shop generally begins with the installation of a press. Either a reel or a stock cradle is added to handle coiled material. If labor costs are low, the material is pushed into the die by hand, but as labor costs increase, it becomes more and more important to automate the feeding operation.

There are four basic types of feeding arrangements: hand feeding, hitch feeds, roll feeds, and slide feeds. Hand feeding is most useful for short runs and those special cases where secondary sheets and precut lengths can be used to offset the higher cost of coil stock.

Of the automatic feeds, the hitch-type feed is the oldest and least expensive mechanism. In this type of feed, the material is advanced by a wedging action between a blade and a moving block. These units are best at feeding medium-weight material at relatively short progressions. Because of the low cost of some commercially available hitch feeds, they are often attached directly to a die where the setup can be of a permanent nature.

A popular type of permanently mounted equipment is the roll feed. In this unit, the material is advanced by the pressure of two opposed rollers, one or both of which are driven from the crankshaft of the press. These units are generally best for feeding rigid materials in moderate to heavy thicknesses. Thinner materials can often be handled by using a double-roll-feed combination in which a set of feeding rollers is attached to each side of a press and synchronized by a connecting link. Roll and mechanical slide feeds generally offer the longest life and highest speeds available today, but represent a permanent installation at prices ranging from moderate to high.

Air and hydraulic slide feeds are the fastest-growing types of feeding equipment. Because of the simplicity of air and its availability in press rooms, the air-operated punch-press feed is probably the most versatile and most widely used type of unit on the market today. Slide feeds fall somewhere between the hitch feed and the roll feed in cost, and have the capability of handling materials including the very thin and the very compressible. Air feeds are not limited to use in punch presses, but can be mounted on special machines in almost any position and can be actuated by mechanical, electrical, or air signals.

Because of the versatility of the air feed, we will examine various feeding concepts for the job shop utilizing air-operated punch-press feeds. The operating principles of air feeds are general and can be applied directly to hitch feeds and roll feeds. The problems that arise in the use of air-operated punch-press feeds are also encountered with other types of feeds. Note that if the proper precautions are followed, almost any type of feed, within its limitations, will give good accuracy, speed, and utility. Before considering a switch from one type of feed to another, always examine the application. The feed at hand is usually capable of doing the job if it is given the proper opportunity.

TYPES OF MATERIAL

Automatic feeds are normally designed to feed standard coil stock. If the stock is rigid enough to resist clamping or roller pressure and is self-supporting, it falls into the classification of a standard material. This is the type of material normally found in production shops. In addition to these materials, job shops are often called upon to feed materials not easily accommodated by the production installation.

The simplest special shape is wire. Almost all types of feeds will easily handle wire unless the wire diameter is very small — usually less than 0.015 in. (0.38 mm). The clamping and roller pressures relative to the area of the wire become so high at small diameters that deformation of the wire often occurs. The air feed, with its high clamping area, simplifies this type of feeding job. For small diameters, it is sometimes necessary to use telescoping tube guides to prevent buckling of the wire as it passes through the feed and the die. As the wire diameter becomes smaller, the job becomes more difficult and more delicate, but by no means impossible (Figure 1J-1).

Historically, thin materials have been difficult to feed. Again we are talking about thicknesses less than about 0.015 in. (0.38 mm), 15/1000 but here the deformation of the material is not a problem. Flexible inserts in the clamps can usually prevent scratching of highly polished materials, but again, depending on progression and the self-supporting features of the material, it may be advisable to install antibuckling guides or other types of support for the material as it passes through the feed. Small air feeds excel at feeding thin material.

The ease of grinding special shapes in the clamps makes air feeding of special formed material a simple operation. If the form is rigid, passageways can be machined to accommodate almost any shape. If the form becomes very high, it may be necessary to add special inserts or to machine special clamps — again, a relatively simple operation. Compressible and stretchable materials present a different type of problem. If a material is so flexible that it is distorted and stretched merely by being held in the hands,

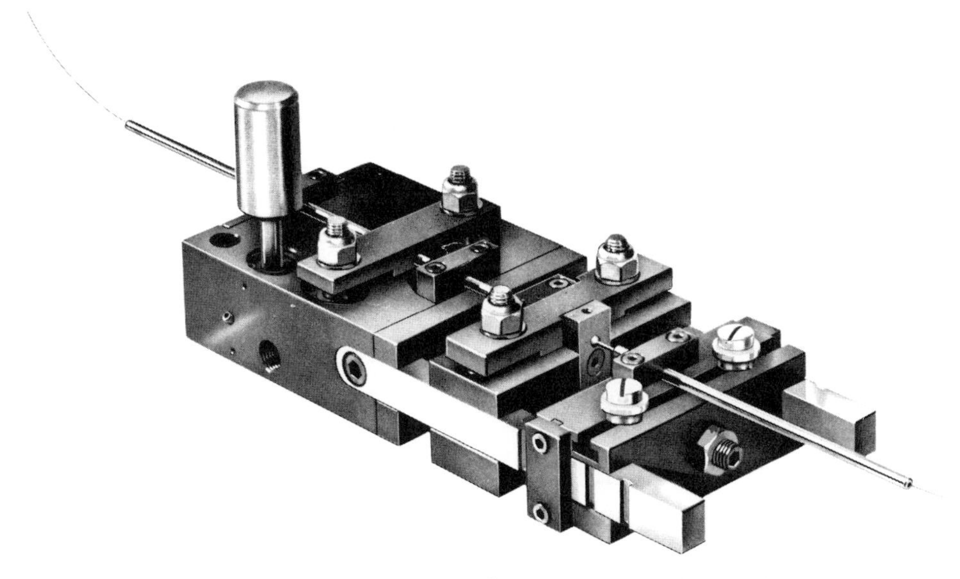

Figure 1J-1. Telescoping antibuckling guides for wire

it will cause considerable feeding problems. Nonetheless, this type of material must be fed, and probably the most practical solution is to bond the material, in a temporary manner, to some sort of carrier strip. This could be, perhaps, a paper carrier approximately the thickness of an IBM card (but, of course, in coil form).

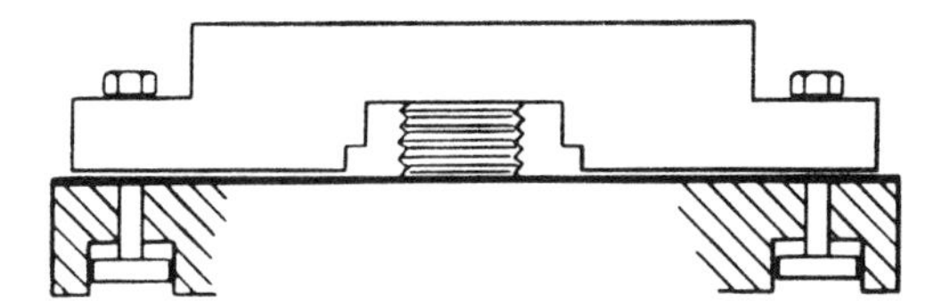

Figure 1J-2. Clamp design for compressible materials

Compressible materials are best handled by machining a cavity in a clamp slightly smaller in height than the thickness, and slightly larger than the width, of the material. The amount of difference depends on each individual application, but the basic idea is to let the clamp bottom on the feed yet slightly compress the material in order to utilize the friction of the material to resist the feeding forces. The ease of doing this type of job depends on the resistance of the material to deformation. Proper and imaginative techniques can simplify a difficult job (Figure 1J-2).

If more than one strip of material is to be fed, it is necessary to grip each piece individually. Again, the air feed makes this a simple job. There are many ways of gripping each individual strip. Among the favorites are use of a flexible insert in the clamp and use of a spring pin to grip each strip (Figure 1J-3).

Feeds that accommodate strips cut to precut lengths are not as common in the job shop. This type of feed is generally available at a very high price for companies that specialize in high production of special parts. Typical of such parts are the lithographed tops for bottles. Usually the press and the feed are available as an integrated package. If the job shop is called upon to handle strips, it must use its ingenuity, and, since hand feeding is often the best method, it should not be overlooked. Encountered in feeding of strips is the problem of what to do with the material between the feed and the die. A simple solution — but one that is relatively expensive and not always applicable — is to use two feeds, one mounted on each side of the die. The strip, of course, must be longer than the distance between the two feeds. Other methods involve either automatic or hand placing of one strip behind the other, perhaps using tape to fasten the strips together as they are running automatically.

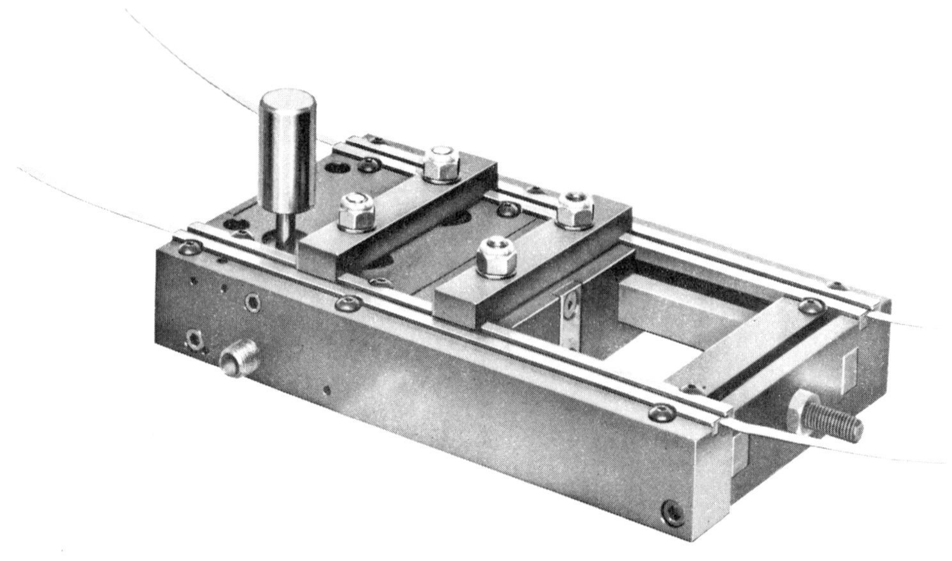

Figure 1J-3. Simultaneous feeding of two strips of material

ADAPTABILITY

Air feeds offer the utility of use with a wide variety of presses and machines. Their ease of installation and removal expands their versatility because they can be quickly changed from one die set to another. Often the actuating bracket can be left on the die and the feed can be moved from one job to another within a few minutes by removing two bolts and reinstalling the feed on another unit. The many available methods of actuation make air feeds quite simple to use on impact presses, electric presses, hydraulic presses, wire-forming machines, multislide machines, and special machines that bear little relationship to actual presses.

MOUNTING OF AIR FEEDS

Air feeds, such as Rapid-Air Precision Punch-Press Feeds, have almost no mounting limitations. Although they are generally mounted either directly on the die set or on the bolster plate of a press, they can also be attached to the feed vertically, sideways, or upside-down. One common application is to mount two feeds on one die or special machine and process and assemble two different parts. Consider also the possibility of mounting an air feed on a movable platform such as the punch holder of a die set. One ingenious fellow mounted two punch-press feeds, one behind the other, thus giving the progression capabilities of two separate increments, depending on which unit was signaled (Figure 1J-4).

AIR-FEED ACTUATION

There are two basic methods of actuation for air feeds: internal and external. Internal actuation, by means of the mechanical valve built into the feed, is normally used on average-stroke presses for normal conditions. It is the simplest and the preferred method of actuation.

If the stroke of the press becomes extremely short or long, or if special conditions exist, external actuation may be required. This consists of either electric or air actuation in which the mechanical valve is replaced with an electric or air valve mounted as close to the feed as possible. Of course this external valve must be triggered by a microswitch or air pilot valve generally mounted so as to be operated by a rotary cam on the crankshaft or a linear cam on the ram of the press. This method is most useful on special machines where a slide may not be adjacent to the feed, or in circumstances where long or short press strokes are being used. Although electric actuation is the most convenient method of external actuation, the cam-operated air valve offers a particular advantage on some special machines, especially where there is danger from electric arcs (Figure 1J-5).

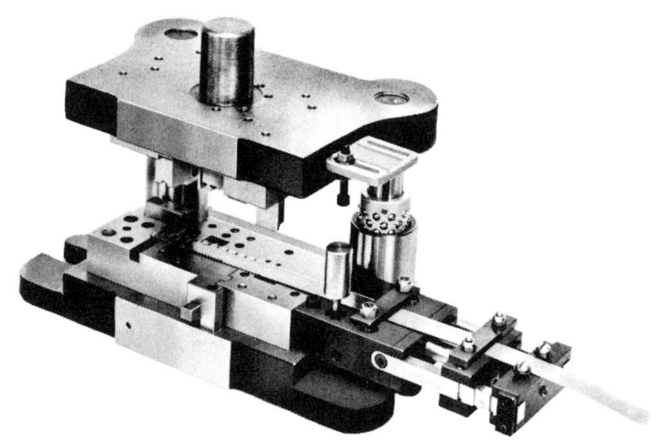

Figure 1J-4. Rapid-Air Feed mounted on die set

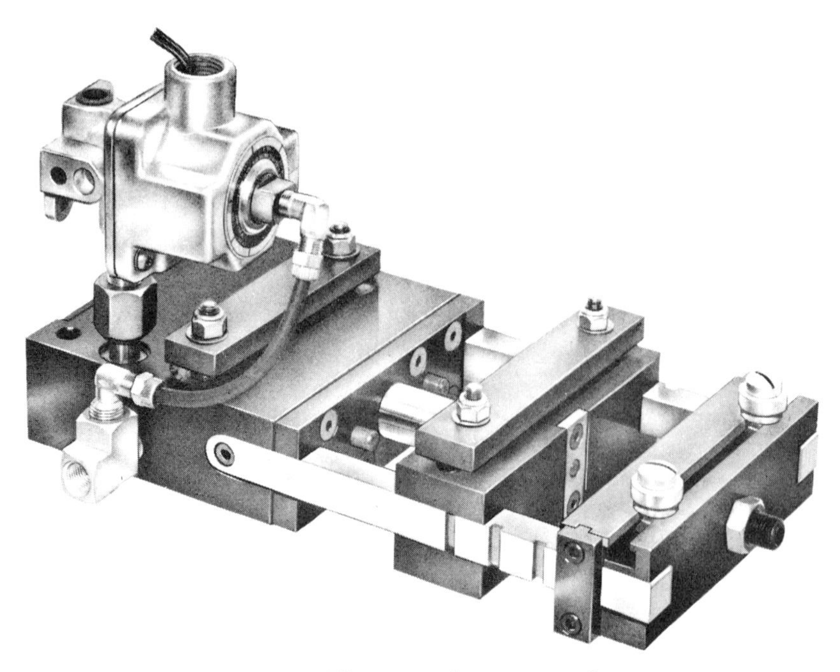

Figure 1J-5. Electric valve mounted on Rapid-Air Feed

Most air feeds allow some means of overtravel of the valve mechanism. Normally the valve switches in the first fraction of an inch of travel, and overtravel is provided to ensure that the feed has time for retraction at the bottom of the stroke. The bottom portion of the stroke is, in most instances, useless for feeding purposes, because the punches are perforating the material or forming the part. As the die opens, the feed can be triggered to push the material into the die. A general rule of thumb is to allow 50% of the cycle for feeding and 50% for retraction (Figure 1J-6).

The important item to remember regarding actuation of air feeds is to allow approximately 50% of the cycle for retraction. This is best accomplished by using the bottom half of the press movement for retraction of the feed. Obviously it is impossible to feed material in this position because the press is actually doing the perforating and cutting of the material. The mechanical valve attached to an air feed accomplishes this condition automatically, but the wide range of variable conditions available with electric actuation calls for greater care in the selection of the actual feeding and retraction signals.

This great versatility in mounting and actuation allows automatic feeding in machinery not normally adaptable to the use of standard feeds.

TYPES OF JOBS

The stamping job shop must have the flexibility to accommodate many different types of jobs. Job shops are called upon to run high-production and low-production jobs and must make a profit on both. In a typical application, two different parts are made from two pieces of material in a single die, and are assembled in the final station. Two or more feeds, mounted either at opposite ends of the die or at right angles, are used. The simplified mounting of air feeds again aids in this type of installation. Often a very

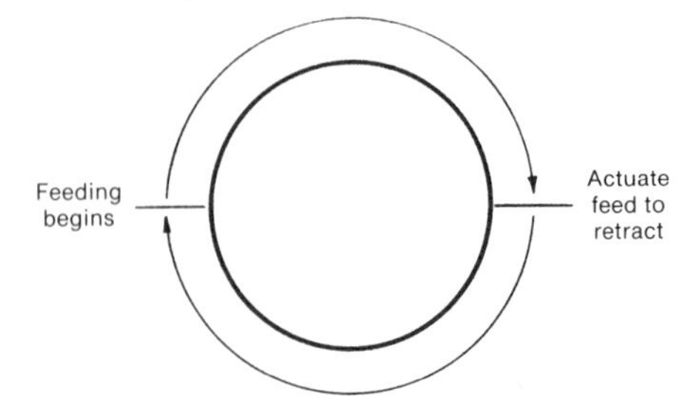

Figure 1J-6. Feeding cycle

high-production machine will utilize literally dozens of air feeds.

Often the job shop is faced with the realization that the completion of a part in a single die will be expensive for a short run. A solution is to run the material through one die, rewind it on a rewind reel, and then pass it through a second die for the secondary operation. Several good makes of rewind reels are available, but the major problem is generally the axial location of the material from one operation to another. The cumulative error will be enough to scrap the parts; therefore, a pilot is the only solution. The pilot can be placed either in the die or on one of the clamps. Again, the versatility of machining special shapes on the clamps of an air feed simplifies this otherwise rather difficult operation (Figure 1J-7).

Long progressions present a problem in that long-stroke feeds cost a great deal of money. A relatively simple solution for long progressions is to provide a multistroking device. At the end of the set number of progressions, the feed will stop and provide an electric signal for use in actuating the press. Although this approach is less expensive than buying an extremely long feed, it does have the disadvantage of reducing the available number of strokes per minute. A formula which we have found helpful for determining the speed of such a combination states that the number of strokes per minute is equal to $X/2N - 1$, where X is the speed of the feed at the maximum progression and N is the number of progressions. For example, if the regular speed for a 12-in. progression is 100 strokes per minute, and it is desired to feed three 12-in. progressions for a total progression of 36 in., the approximate available speed is $100/2(3) - 1$, or 20 strokes per minute. At this speed, the feed is generally operating more slowly than the press, but the job can usually be done at an economic cost.

Another popular use of a counting device is to feed a given number of times before actuating another secondary operation such as a separate cutoff. This is commonly done in the making of parts such as radiator fins. Instead of a counter, consider the use of a

Figure 1J-7. Hydraulic payout and rewind reel

microswitch. This is another example of a simple solution to what could otherwise be a complicated problem.

Another somewhat related problem is feeding of two separate and distinct progressions, either alternately or in some sort of fixed sequence. This is difficult with regular hitch feeds or roll feeds, but with a little ingenuity the air feed, with its reciprocating action between two fixed stops, can provide a solution. Either an air cylinder or a solenoid is used to insert a block of a fixed length between the regular stop on the feed and the sliding block. Generally, two progressions are required, but the solution could be expanded to accommodate a reasonable number of different progressions. The insertion of the fixed block must be timed to occur when the feed is in its forward position, and the actuating means must be fast enough to accomplish the job without interference. The control can be as simple as a push button operated by the press operator or can be electrically programmed. The choice again depends on the length of the job involved.

A situation rarely but sometimes encountered in the job shop is the problem of providing a continuously adjustable progression. This is relatively difficult. Although this procedure requires a great deal of work and diligence, it can result, if successful, in a great increase in production. Typical applications are the feeding of preprinted paper forms to be cut out into individual pieces, and the feeding of materials such as labels to be sewn into clothing. The technique begins with an electric eye that detects a light or dark spot on the part. The electric eye operates a solenoid or air cylinder that adjusts the fine-adjusting screw on the feed to a progression either slightly longer or slightly shorter than necessary. For example, if the electric eye is set to feed long on white material, the overfeeding will eventually allow a control spot to work its way into the range of the electric eye. When the control spot reduces the reflection from the built-in lamp to a minimum level, the electric eye provides a different type of signal which, through valving or solenoids, adjusts the progression of the feed to a shorter progression. The control spot will gradually recede from the range of the electric eye until the cycle again reverses. The degree of accuracy is relative to the type of electric eye and amplifying system used. This procedure is relatively complicated and, although not beyond the scope of the job shop, should be used only in high-production situations.

ACCURACY TECHNIQUES

The two main reasons for using automatic press feeds are to improve production and to provide uniform, accurate progressions. In some cases, accuracy is not important, but in others it can be the entire reason for using automatic feeds. All press feeds have accuracy limitations. Some are more easily adapted to overcoming these limitations, but accuracy generally can be improved in any feed by using the same techniques.

The simplest accuracy control is a positive or dead stop. When an automatic feed with a positive stop is used, the progression of the feed should be set for a distance slightly greater than the actual progression. This will ensure that the material always moves the required progression. In order to use this type of stop, the material must be cut off in the last station. The amount of material cut off controls the progression, and thus positive stops can be quite accurate on rigid material. However, this method is ineffective in controlling the progression of thin or compressible materials (Figure 1J-8).

The most common type of accuracy control is done with pilots. In this operation, a hole, generally round, is pierced in the material, and then, in a subsequent station, a punch with a rounded nose is inserted into the material, aligning it in the proper position. Pilots can be used for both axial and lat-

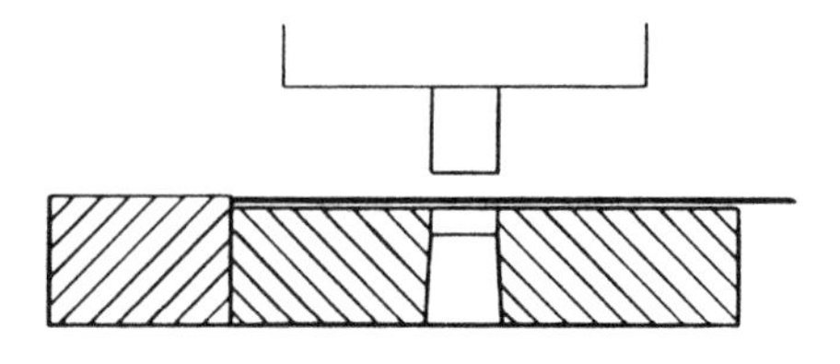

Figure 1J-8. Positive stop

eral control. In order to avoid uncorrectable axial errors, it is best for the pilot to engage the material in the station immediately following the piercing operation. When a pilot is used with an automatic feed, the usual suggestion is to feed the material slightly short and allow the pilot to pull it into the proper position. Thus, any feeding device should have the capability of pulling material through with just a slight amount of force. Again, pilots are effective on relatively rigid materials, but their performance drops considerably when they are used to feed thin or compressible materials (Figure 1J-9).

Disappearing stops are related to pilots in that they engage some edge or hole in the strip as it passes through a progressive die. They are generally less expensive than pilots, but do not offer the same accuracy control.

Perhaps the simplest means of controlling accuracy is to rely on the feed itself. Proper attention must be given to control of various factors that affect accuracy, but, with good conditions, repeatability much better than ±0.001 in. (±0.025 mm) can be achieved. We will now discuss various factors that must be checked in order to provide maximum accuracy.

The greatest accuracy is achieved when the feed is used only for feeding and is not made to serve some dual purpose. Most readers will be aware of the problems associated with nonpowered straighteners, but little attention is generally given to the effects of using nonpowered reels. The difficulty that arises when material is pulled directly off a reel comes from the inertia of the coil. At higher speeds, the jerk necessary to start the reel rotating is enough to cause either slipping or stretching of the material. For this reason, it is almost always best to pull the material from a free loop hanging between the feed and the reel or straightener. This loop should contain enough material for three or more progressions. Generally, automatic payoff devices are used to provide this loop, but in simple installations an operator can rotate the reel, or a weighted roller can be hung on the material between the feed and the reel (Figure 1J-10).

Another cause of apparent inaccuracy in feeding is the shear force generated at the cutoff station in a die. This shear force is generally not present when holes or forms are being pierced through the material, but is always present when parts are cut off before some restraining punch or pilot has engaged the material. The feed must then resist this force, and although in many cases it can, the best solution — not only for the sake of the feed, but also for die life — is to provide a

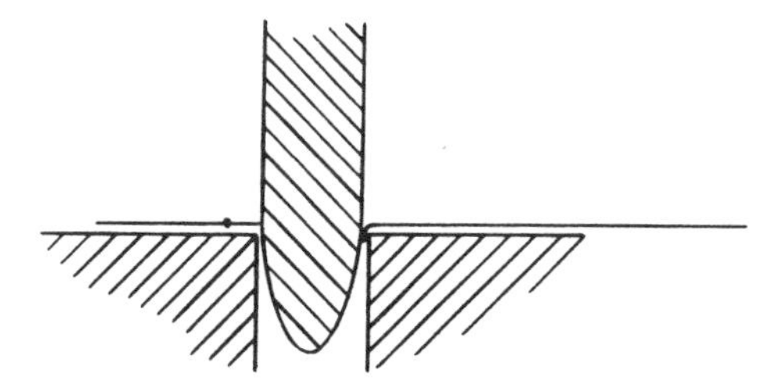

Figure 1J-9. Distortion of thin material by pilot

Figure 1J-10. Powered pedestal-type stock straightener

holddown clamp in the die that grips the material before the shearing action takes place. The heavier the material, the more pronounced the shear force (Figure 1J-11).

In many forming dies, the material is pulled up and down as a part is being formed. This means that the material must be pulled slightly through the feed and then pushed backward. This situation should be avoided in die design, but if it is present, the feed must be mounted as far away from the die as practical so as to maintain the integrity of the geometry to an acceptable level. Somewhat related to this problem is the pulling of material as a part is being drawn. Of course we are all familiar with the use of cutouts in a strip to reduce this problem (Figure 1J-12).

Mechanical interference often causes inaccuracy of an elusive nature. It is obvious that camber of the material will cause binding due to failure of the material to pass through the die itself. This type of problem can also occur as a result of improper alignment of the die and feed. Often this alignment is not square because slight differences in the shape or size of the material, or differences existing as the material passes through the feed, can sometimes cause the material to be pushed at a slight angle. The best solution is to observe the results in the die and adjust the feed accordingly. Although a powerful air feed can overcome some of these limitations, it is always advisable to provide the best conditions possible. The possibility of slugs in the die affecting accuracy should not be overlooked. Often a feed will push the slugs aside, but only after the restriction has affected the accuracy. Slugs can be particularly troublesome, and often very difficult to correct, depending on material characteristics. There are commercial devices available to aid in slug removal. Otherwise the standard diemaking techniques should be used to eliminate this problem (Figures 1J-13 and 1J-14).

Hitch feeds and air-operated punch-press feeds require that the impact at the end of the forward stroke be reduced to the minimum. The better air feeds have various devices such as speed-control valves and cushion pistons designed into the unit. Impact-reducing materials such as rubber and urethane can also be used as cushions. Generally, these techniques are quite simple, but if they are not followed, the material will continue to move forward due to inertia after the feed has stopped. Inertial slippage often will be accentuated by oily material. Although this problem can be reduced by overpowering the system with heavy clamps, this often leads to deformation of the material. A better solution is to provide a material with a high coefficient of friction. We have found leather to be one of the better materials for clamp inserts because it is slightly compressible and will not mark the stock. It

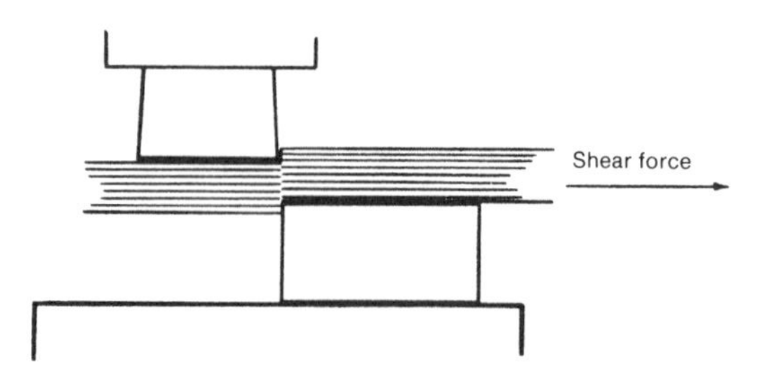

Figure 1J-11. Shear force on cutoff

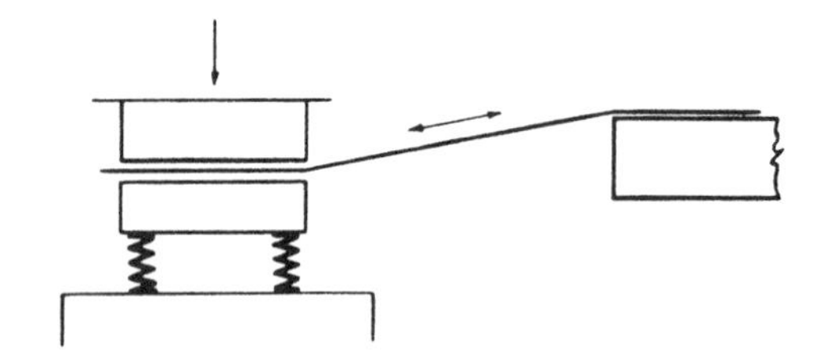

Figure 1J-12. Pulling of stock during forming operation

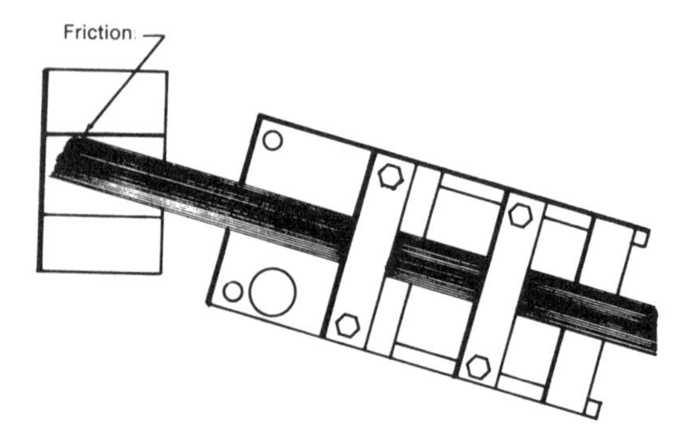

Figure 1J-13. Binding of stock between feed and die

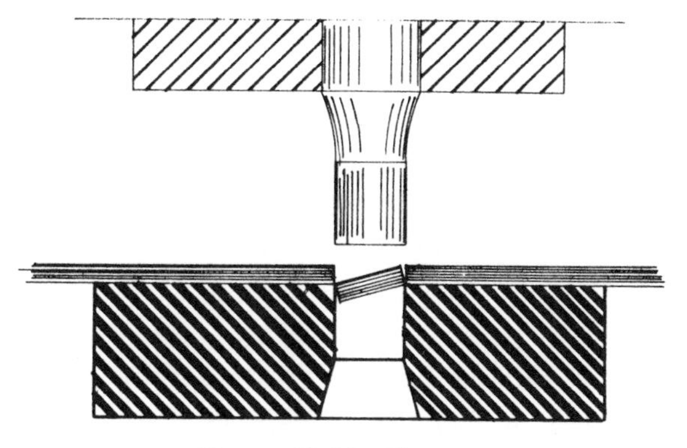

Figure 1J-14. Slug pickup

provides a high coefficient of friction even under oily conditions.

Some of the most difficult materials to feed by any method are stretchable materials. Often a double feed combination is the simplest way to accomplish feeding of such materials. One feed is mounted on each side of the press, with the exit feed set at a slightly longer progression. Actuation, of course, must be simultaneous, and a mechanical link often connects the sliding portions of each feed. Since stretchable materials do not lend themselves to feeding without buckling, special guides can be used. Consider also the possibility of using tension reels to keep the material taut as it passes through the working stations.

SPECIAL SPEEDS

High-speed feeding is becoming more and more popular because it is the most obvious and attractive way to reduce the costs of a part. Air feeds are capable of speeds up to approximately 500 strokes per minute at a ½-in. (13-mm) progression. Beyond this it is best to rely on the mechanical advantage of a roll feed or on one of the specialized high-speed feeds now being made available. These units are usually expensive and should be considered as part of a total package with a good, well-built, high-speed press. The problems encountered with such units are different, and for high production their advantages can be rewarding.

Strange as it seems, very low speeds are also sometimes required, typically for handling extremely delicate materials. Air feeds often are called upon to provide extremely slow, uniform feeding, such as feeding of solder into a furnace at a continuous, smooth rate. The air feed provides a very inexpensive solution to this problem. The low speed is usually controled by a separate speed-control valve. Again, the imagination of job-shop workers is constantly being called upon for inexpensive solutions to what could be complicated problems.

AIR CIRCUITS

Effective use of machinery requires that its operating principles and functions be thoroughly understood. Air feeds usually appear complicated, but only because the valves and moving parts are inside the unit, and thus not visible. There are a few simple principles the understanding of which will greatly increase one's ability to analyze problems and make effective use of air-feed equipment.

The air feed is divided into two separate air circuits. The first circuit is controlled by the main valve which operates both sets of clamps. The stationary clamp is used to prevent the friction of the slide block from dragging the material backwards as the feed retracts for a new progression. The movable clamp provides the greatest force for gripping and pushing the material into the die. The clamp portion of the circuit can be considered as follows. When the main control valve is depressed, the air is exhausted from the clamp circuit. As the pressure falls, the stationary clamp grips the material just before the movable clamp releases. As the main valve is allowed to come up, the action reverses and the movable clamp grips the material just before the stationary clamp opens. The clamps move in sequence by a balance between the areas as determined by the designer of the feed. If it is desired to add motion to this cycle, a pressure-sensing valve is added to the circuit. This auxiliary or pilot-operated valve allows a flow of air into or out of one side of the piston. The area of the pilot-operated valve is adjusted so that the feeding cycle takes place at the appropriate time; retraction occurs while the feed clamp is down. This general principle allows great flexibility in feeding, because the two separate circuits ensure that the clamping cycle takes place before the feeding cycle. It also aids in separating the two air circuits so that the pressures necessary for clamping are not greatly reduced by the air flow necessary for feeding.

MAINTENANCE AND LIFE

Any piece of machinery must be properly maintained for maximum performance and long life. There is no reason why an air feed cannot provide hundreds of millions of cycles. In fact, many Rapid-Air Feeds provide more than one hundred million cycles per year (probably more cycles than each piston of your automobile travels in one year). Needless to say, lubrication and air supply are the key factors to long feed life.

Many shops have a problem with water in the air lines. This is caused by compression of relatively moist air and subsequent condensation of liquid when the air is cooled by being passed through the pipes to its point of use in the shop. Most readers will be familiar with the stream of water coming from blow guns in the morning as the air system is started up. Although many extensive systems are installed to eliminate water, we generally find that for the job shop the simplest methods are the best. An after-cooler at the compressor will eliminate most of the water. There is a simple solution to this problem that should not be overlooked: turning the outlet tee in an upward direction and drawing the air from the top of the pipe. If you want water, it is usually best to get it from the bottom of the well (Figure 1J-15).

Most air feeds are built to operate for many millions of cycles before "O" rings or seals require replacement. In order to promote the longest life for moving seals, a filter should be used to keep the air as clean as possible. The lubricator should be set to provide a small amount of light machine oil for lubrication. Under good conditions, an O-ring life of at least ten million cycles can be expected. Under excellent conditions, O-ring life is considerably better.

There is no need to use "fancy" oils. One of the least expensive oils will work best. It should be remembered that motor oils are designed for automobiles and spindle oils are designed for precision spindles. Air feeds use inexpensive #10-weight hydraulic or machine oil. A periodic maintenance schedule will provide optimum service. Manufacturers' suggestions should be consulted for best results from maintenance schedules.

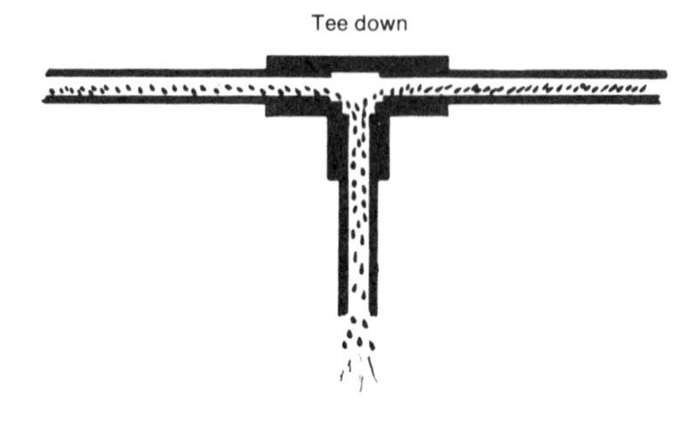

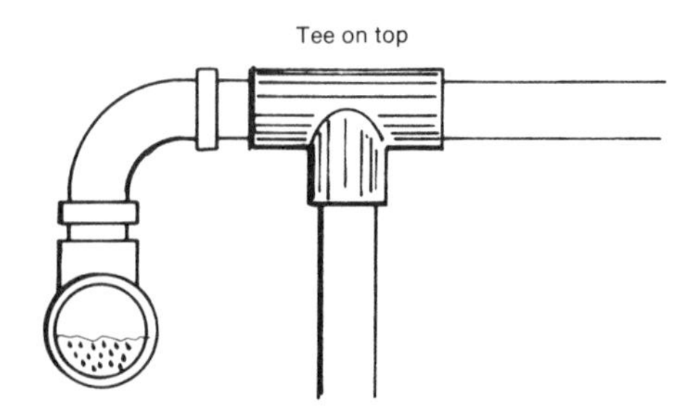

Figure 1J-15. Air lines with tee down and tee on top

GENERAL JOB-SHOP PROBLEMS

The job shop is called upon to make many types of dies. Some shops specialize in high production and others in very short runs. Therefore, the problems of the job shop vary so much that it is hard to specify general rules. There are, however, some universal suggestions that come up often enough to warrant comment — especially because these problems apply to automatic feeds of many different types.

In the straight cutoff-type die, the shear force can push the material backward. In fact, in hand feeding of heavier-gage materials, a considerable shock can be felt in the operator's arm and hand. Pressure pads should be used in the die to prevent this problem.

In dies where bending is performed, careful attention must be given to preventing projections in the strip from catching on cavities on the die. This will lead directly to buckling problems in the die and inaccuracy in feeding. Workers should be cautious of the friction generated by tight bends around forming mandrels. Feeding equipment is usually strong enough to overcome this type of friction, but, depending on the size of the unit and the nature of the stock, inaccuracies can result.

In forming and drawing dies, it is best that the level of material remain constant. Often this is difficult, and the best rule in such cases is to mount the feed as far away from the die as possible. It should be noted that this is exactly opposite to the normal recommendation.

A type of die not to be overlooked in today's production shop is the transfer die. This die is helpful in forming parts around arbors or in performing multiple bends. The part is generally carried in a strip to the next-to-last station, where a bend is made and the part cut off. The part is then pushed by an air cylinder to the final station, where final forming or assembly is performed. Transfer dies require ingenuity, but ingenuity is what job shops are all about.

In any job-shop operation, the length of the run and the setup are of primary importance. It would be convenient if dies and feeds could be mounted on presses permanently. Unfortunately, this would be rather expensive, because most jobs are simply not that long. Thus, setup time becomes extremely critical, and air feeds can serve a very useful function. Most air feeds can be completely set up in a matter of minutes and can be moved from press to press or die to die with very little difficulty.

1K: Metals Formed in Presses

Nearly all metals and their alloys can be formed in presses. Which metal is chosen for a particular application depends on the requirements of the part being made, the cost of manufacturing it, and the availability of the metal. Requirements of parts vary. The primary consideration for one part may be strength; for another it may be surface appearance. Most of the requirements of stampings used in making automobiles can be satisfied through the use of steel, copper and copper alloys, and aluminum.

Steel

Strength and durability are among the foremost requirements for stampings used in automobiles. Because of this, the most commonly used material for automotive stampings is the iron alloy, steel.

Steel is composed primarily of iron with carbon as the major alloying element. The greater the percentage of carbon, the stronger and stiffer the steel. High-carbon steels are generally unfeasible for use as workpiece materials in stamping operations because of their high resistance to plastic deformation. Most automotive stampings are produced from low-carbon steel, containing less than 0.15% carbon.

Two main varieties of low-carbon steel used are rimming steel and aluminum-killed steel. These steels derive their names from the processes used in casting the ingots. When steel is cast in ingot form, a boiling action takes place, resulting from addition of chemicals, that forces particles of pure iron to the outside of the mass, trapping impurities in the center. The violent action of the metal creates a collar or rim of still around the top of the ingot. This is rimming steel. Aluminum-killed steel lies quietly in the mold because of deoxidation that results from addition of aluminum. The impurities in aluminum-killed steel are spread uniformly throughout the ingot.

The impurities in rimming steel remain trapped in the center of the mass even after the ingot has been rolled. This provides steel strip and sheet with excellent surface finish. Most stampings used at the Ford Motor Company are made from rimming steel. It is relatively inexpensive and has strength properties equal to the demands of most automotive stampings. Its forming properties are suitable for most applications and it has excellent weldability.

The even spread of impurities in aluminum-killed steels results in steel strip and sheet that is highly ductile. This type of steel is frequently used where deep drawing qualities are desired. However, because of higher costs resulting from mechanical problems in producing killed steel, it is secondary to rimming steel in use.

Although low-carbon steels predominate in automotive stampings, a limited number of special parts are produced from stainless steels. These alloys, because of their fine surface appearance, are used to some extent for stampings such as window moldings and other trim details where "bright work" is desired. Stainless steels contain alloying additions of nickel and chromium. Because these elements are not abundantly available, and because of their high cost, they are not extensively used in automobile fabrication. They resist corrosion well and are suitable for forming operations if their carbon content is low.

Steel production methods influence the properties of the steel produced. The three common methods of steel production are the open-hearth process, the electric-furnace process, and the Bessemer-converter process. Most steel used in stamping operations comes from open-hearth furnaces. Properties of steel so produced vary with the degree to which the process is controlled. Generally speaking, open-hearth steel is conducive to press working, and its strength is sufficient for most stamping applications.

The highest degree of control of the quality of steel can be exercised in the electric-furnace process. Excellent purity and strength can be obtained. Excellent drawing properties are also obtainable. However, the nature of the electric-furnace process makes steel so produced relatively expensive. In addition, the quality of steel so produced is generally in excess of that required for stamping applications. As a result, this type of steel is used mostly for moving parts of automobiles and is not used extensively for stampings.

Steel produced by the Bessemer process has only limited application in stamping operations. Although such steel may be formed, it is less suitable for plastic working than open-hearth and electric-furnace steels. This is due to the presence of sulfur and phosphorus, which tend to remain in the steel throughout the process and impart exceptional stiffness.

Copper and Copper Alloys

Copper and copper alloys are generally used for stampings where the advantages of their special properties justify their higher cost.

In bending, forming, and drawing, copper takes the lead over all other metals in freedom from cracks, fractures, and wrinkles. This is due to its great ductility. However, its prime importance in automotive manufacturing is its properties of electrical and thermal conductivity. It is used in stamping radiator grids, to dissipate engine heat, and in electrical switches, relays, etc., because of its ability to conduct electricity.

Copper alloys usually have excellent corrosion resistance, but they are limited in automotive applications because of high cost and lack of strength.

Aluminum

Commercially pure aluminum is outstanding from the standpoint of workability. It can be easily formed into complex shapes, and it has properties that may make it suitable for use in many automotive stampings. The most interesting of these properties is its excellent strength-to-weight ratio. Aluminum parts may be designed to provide greater strength at less weight than steel. However, as in the case of copper, aluminum is expensive. Since low-carbon steels provide adequate strength for most stampings, aluminum has not been applied to general automobile fabrication.

BASIC METALLURGY

Any discussion of pressworking of metals must deal with a great variety of methods by which metal is transformed from flat strip or sheet into useful parts. This transformation is accomplished by four main types of operations: shearing, bending, drawing, and squeezing. To accomplish any of these actions, certain properties of the metal itself must be exploited or overcome. To understand the factors involved in working of metal, it is necessary to conduct at least a cursory examination of the structure and properties of metal. In this way it will be possible to determine the factors that permit metal to be plastically deformed and the factors that limit the possible extent of deformation.

Structure of Metals

To the naked eye, metal appears to be a solid, homogeneous mass. However, under microscopic examination, it is obvious that metal is composed of numerous intricate crystalline structures, each being held to the other by powerful electrostatic attraction.

These structures are formed during the transition of a metal from the liquid to the solid state. In the molten state, the atoms of a metal are in rapid and continuous motion. As cooling takes place, the motion becomes slower and slower. Crystals of solid metal begin to form at random throughout the mass. The crystals are composed of atoms that form the geometric crystalline shapes that are characteristic of the metal involved.

As cooling continues, atoms surrounding the various crystals join them, causing them to grow. The various growths increase in size until they abut against one another and solidification is complete. When a nucleus has completed its growth it forms a grain of metal. Neighboring grains differ in size and shape externally. Although the outlines of the grains are imperfect, the main body of the grain is truly crystalline. It is made up of small building blocks, each being a perfect unit crystal. The electrostatic forces that hold the atoms of the metal together exert greater force in some directions than in others. This leaves certain planes of weakness called slip planes. These slip planes are responsible for metal being able to be deformed plastically. They are oriented in many different directions in the grains of the same piece of metal.

When a load sufficient to cause plastic deformation is first applied, the slip planes situated most favorably will yield, causing deformation. The slippage along these planes will continue until arrested by grain boundaries or internal imperfections. Slip planes less suitably situated must then be used if deformation is to continue. Because they are less favorably situated, these new slip planes require greater pressure for slippage to take place. As deformation proceeds, the availability of slip planes decreases and the tendency to resist deformation increases.

When resistance to deformation increases, the metal is said to be strain hardened or work hardened. Thus, as metal is cold worked, its tensile strength and other properties are increased.

The process may be compared to a load suspended on a chain. The weakest link will break first. If the link is removed, the chain will support a greater load. This could continue for some time as each succeeding weakest link was broken and removed. Of course, deformed metal grains are not broken or removed; instead, they are merely so rigidly supported that further slip must occur in other grains. If sufficient pressure is applied after the slip planes are used up, the metal will break.

Mechanical Properties of Metals

Structures of various types of metal differ. However, all metals possess certain mechanical properties. The degree to which these properties are possessed by different metals varies considerably. As a result, different metals will behave differently under the press. Among the properties possessed in common by all metals are elasticity, malleability, ductility, strength, and hardness.

Elasticity, roughly defined, is the ability of a metal to resume its original shape after being deformed. The elasticity of any metal is limited. Cast iron, for instance, will undergo very little plastic deformation before it acquires permanent deformation or actually breaks. On the other hand, some steels may be deformed quite drastically and still snap back to their original shape.

To structural engineers, elasticity is important from the standpoint of how well metals will endure operational stresses. But, to the stamping process engineer, elasticity is a property that must be overcome in shaping a part to desired dimensions. When a part is bent or drawn in the press, it has a tendency to "spring back" when it is released from the die. Obviously, if the part is to have the desired shape, this springback must be compensated for. One method of compensating for springback is to overbend or overdraw the part by a percentage sufficient to cause the metal to retain the desired dimensions upon release from the die. Another method is to "strike bottom" in the forming operation. In striking bottom, the punch squeezes the metal against the die and rearranges its grain structure to cause a permanent set. Striking bottom is a sensitive operation. If stock is uneven in thickness, there is danger of overloading the press and breaking it. There is also danger of causing injury to the dies.

The point of pressure at which elasticity is overcome, and the metal acquires a permanent set, is called the elastic limit. All stamping operations work metal above the elastic limit. It is thus necessary for process engineers to be acquainted with the elastic limits of the various metals with which they deal.

Elastic limits are determined through tests in which metal samples are subjected to increasing loads until permanent deformation occurs. These tests may involve either tensile stresses or compressive stresses.

Malleability is another metal property of interest to the stamping process engineer. Malleability is a qualitative term describing the ability of a metal to withstand plastic deformation without fracturing. It refers particularly to deformation under compressive pressure, as in rolling, swaging, forging, and extrusion. Malleability varies widely among different metals. Copper is highly malleable. Steel has malleability in varying degrees depending on grain structure, percentage of alloying elements, and other factors.

As with elasticity, there is a definite limit to the malleability of a given metal. Above certain pressures, metal will break. Below certain pressures, no deformation occurs.

Ductility as a mechanical property of metal is similar to malleability. It is also a measure of the ability of a metal to withstand plastic deformation without fracturing. Ductility, however, permits plastic deformation in the direction of stretching. It is the reaction of metal to tensile stresses. Thus ductility is related to such metalworking operations as drawing and forming. Malleability and ductility, although similar, do not necessarily have the same values in the same metals.

The ductility of a given metal is measured by the amount of elongation and reduction in cross section that occur under measured loads. Tests to determine ductility are made in a special machine that exerts tensile stress on a test piece. The metal is usually stressed until it breaks. It is then measured for elongation and reduction in cross section, and a value for ductility is assigned. Another test utilizes a machine that presses a metal ball through the test piece.

Strength is another property of metal that has significant implications for pressworking. In general, strength is the capacity of a metal to support a load without being deformed. Strength is also related to the point of stress at which metal will deform, the point at which permanent deformation will occur, and the point at which metal will fracture or fail. The amount of pressure or stress required to cause a metal to fail is known as the ultimate strength. The strength properties of a metal must be known before predictions can be made as to the way the metal will behave under the press.

Hardness is another property of interest in the pressworking of metals. It is the ability of a metal to resist penetration or abrasion. The hardness of a metal has a definite relationship to the ability of the metal to be plastically deformed, and to the amount of pressure required for deformation to take place.

Several types of laboratory tests are used to gage the hardness of a metal. Brinell and Rockwell tests employ a steel ball or a diamond tip that is pressed into the metal under measured loads. Depth or diameter of penetration is related to the force employed to establish a hardness rating for the metal. Another test, the Shore rebound test, measures the distance of rebound of a diamond-tipped hammer striking the metal. The rebound is used as a basis for determining hardness.

Stress-Strain Analysis

The strength properties of a metal are determined in the laboratory through stress-strain analysis, which consists of applying a static load to a metal and recording the results. A stress-strain curve is developed to give a visual presentation of the metal's strength properties. This type of analysis can be developed for both compressive and tensile stresses. This analysis can be used to determine either the service potential of the material or the manner in which it can be expected to behave while being processed into a product.

When a static load is applied to a material, the load produces reacting forces throughout the material. This reaction is called stress. Although stresses are complicated in most stamping operations, they can generally be classified according to three major types: shear stress, compressive stress, and tensile stress. When a part is loaded and supports stress, it is deformed from its original dimensions. The deformation that occurs is called strain. The degree to which a part is strained depends on the magnitude and duration of stress. Strain is expressed as a percentage of deformation occurring, based on the original dimensions of the piece being tested.

As the unit stress is increased in testing, there is a relative increase in the unit strain. At first the strain is proportionate to the stress. On the stress-strain curve, this result appears as a straight line. As the load is increased, a point is reached where the direct strain increases at a greater rate than the stress. This is indicated on the curve by a cessation of the straight line and the beginning of a curving to the right. The point at which the line first breaks to the right is called the proportional limit. Closest to the proportional limit is the elastic limit. From the elastic limit, increases in stress will result in permanent deformation. When the elastic limit is exceeded, the strain is accelerated to the right as the load is increased.

At a certain point of stress, the metal begins to yield rapidly along slip planes. A relatively large amount of strain is experienced with little or no increase in stress. This point is called the yield point. Certain metals do not exhibit definite yield points on their stress-strain curves. In these cases, a definite permanent set is chosen and the amount of stress required to achieve the set is termed the yield strength. As the first well-situated slip planes are used up, the stress must be increased to obtain more strain. The strain curve again moves upward and to the right.

As greater stress occurs, more and more strain results. At length, the ultimate strength of the material is reached. This is the greatest load the metal will support before fracturing. Beyond this point, the strain increases rapidly under decreasing stress, and the metal fractures.

STEEL ROLLING OPERATIONS

Some of the properties of metals ready for the press are derived from the processes used to roll them into sheet and strip. Rolling operations are necessary for two general reasons. First, steel is originally produced in large ingots which must be reduced to proper thickness and size. Second, the properties of the metal are improved by refining grain structure, controlling directional flow lines, and securing desired surface finish. Metal is rolled in either a hot or cold condition.

Effects of Hot Rolling

Hot rolling is a steel-mill process done above the recrystallization temperature of the metal. At this temperature, deformation of the metal is accomplished with less pressure than in the cold state.

The crystalline structure of the metal is refined in hot rolling, with smaller crystals generally resulting. Because crystal growth occurs above the recrystallization temperature, hot rolling is generally continued until the metal's temperature is close to the upper limits of the cold working range. This is done to ensure that proper crystal size is maintained. Because reduction of metal is faster in the hot stage, metal is usually rolled hot until the final dimensional tolerances are approached. Another advantage gained in hot rolling is that any cavities in the ingot are welded shut and any slag inclusions are broken down. Surface impurities are burned off the ingot or slab at the beginning of the hot rolling process.

The finish resulting from hot rolling is relatively rough and porous. Thus hot rolled sheet is used only for stampings where surface condition is of minor importance — such as sheet for chassis frames. In the press plant, hot rolled steel is generally identified by a set of initials indicating the type of metal and the rolling process. For instance, a steel marked HRLCOH would be hot rolled low-carbon open-hearth steel. Steel marked CRLCOH would be cold rolled low-carbon open-hearth steel.

Effects of Cold Rolling

Cold rolling is another steel-mill process used to reduce sheet or strip to final dimensions, to impart certain desired mechanical properties, and to provide the required surface finish for exterior body panels and other visible surfaces of the final products. When metal is rolled cold — that is, below the recrystallization temperature — it becomes strain hardened. This increases the tensile strength of the metal and reduces the possibility of stretcher-strain markings appearing on the metal in subsequent processing. These markings, when they occur, give a lattice-like appearance on the surface of the part.

Effects of Flex Rolling

Flex rolling is a process used in stamping plants to remove age hardening and other effects of storage that tend to decrease the workability of the metal. This process subjects the metal to a flexing action between rolls, which helps to improve its workability. This type of treatment has also been found useful in preventing the appearance of stretcher strains.

SHEARING OPERATIONS

In pressworking of metals, four types of operations are involved: shearing, forming, drawing, and squeezing. Shearing operations include blanking, punching, piercing, perforating, notching, parting, and trimming.

Metal Shearing Theory

In all shearing operations, metal is worked above its ultimate strength. An increasing load stresses the metal in shear up to the elastic limit and beyond. As the elastic limit is exceeded, slippage occurs along the planes of crystals under stress. Deformation occurs according to the material. A reduction of the area under greatest stress proceeds until the ultimate strength of the material is exceeded. The surface crystals of the metal being cut are more severely stressed than those in the interior, because they are being forced to conform plastically to the profiles of the cutting tools. The cutting edges serve to localize the highest stress along one line in the surface material. When the strains in the material moving over the cutting edge reach sufficient intensity, fractures are initiated which follow properly oriented cleavage planes to meet similar fractures starting from the opposing cutting edge.

Thus, in shearing, the metal is stressed beyond its elastic limit, reduced in cross-sectional area, and fractured through the cross-sectional area when the ultimate strength is exceeded. Typical tooling for shearing operations consists of a punch and die, each in a suitable holder. The punch is ordinarily the moving member and usually enters the die.

Force Requirements for Shearing

The force required to effect shearing depends on several factors, the most important of which is the metal itself. Generally speaking, shearing of harder, less-ductile metals requires greater force than shearing of softer, more-ductile metals.

The sharpness of the cutting edges of the tools employed is another important factor. A sharp tool localizes the force more effectively than a dull tool. Thus sharp tools reduce force requirements. Dull tools, in addition to creating greater force requirements, cause secondary fractures in the metal, leaving rough, irregular edges. Thus, to ensure the greatest ease in shearing operations, the best quality of work, and protection of the press itself, sharp tools should always be used.

It is possible to reduce the maximum force necessary for shearing by setting one cutting edge at an angle with another, which reduces the amount of metal being sheared at one time. This practice is called giving "shear" to the cutting edges. This concept should not be confused with the shearing operation itself. An example of the use of "shear" is provided by a pair of household scissors. The blades are arranged to cut a small portion of material at a time, requiring less effort than if the entire cut were made simultaneously.

Another method of reducing force requirements in operations where a number of holes are to be punched in one operation is to "step" the punches. This means simply to arrange the punches so that each punch cuts at a different time. In this arrangement, the distance of punch travel is increased, but the total pressure requirement is reduced.

Another aspect of pressworking of metals that affects force requirements is uniformity of metal thickness. Some thickness variations occur in most materials. When thicker metal comes under the punch, greater force is required. Provision is made for thickness variations by selecting presses of sufficient tonnage to meet varying force requirements. Consideration is also given to variations in force requirements that are caused by age hardening. Allowances are made for these factors in order to provide protection of the press and the dies.

Approximate force requirements can be calculated according to mathematical formulas. The tonnage necessary to perform a shearing operation with a flat punch and die is determined by multiplying the length of the cut by the thickness of the metal multiplied by the shearing resistance of the metal. This formula is commonly stated as:

$$F = L \times T \times S_s$$

where F is force requirement, L is length of cut, T is thickness of stock, and S_s is shearing resistance. Shearing resistances of various metals can be found in materials strength charts; however, in most cases, a straight figure of 25 tons is used for shearing resistance.

Where "shear" is provided in punches or

dies, the force is reduced in direct proportion to the reduction in the length to be sheared at any one time.

It is to be remembered that shearing resistance may be increased 100% by thickness increases or severe age hardening. (Additional discussion of shearing science is in Article 2A.)

FORMING OPERATIONS

Forming includes a large variety of press operations. Among these are bending, forming, beading, curling, and flanging. These operations require working of metal above its elastic limit but below its ultimate strength. It is a characteristic of all these operations that deformation is confined to a small area adjacent to the bend line.

Bending

Bending operations are similar to the deflection of a beam under load. The metal is supported at either edge of the die and the load is concentrated at some point in between. As the load is applied, the metal is stressed in tension along the bottom grains at the point of the bend, and the top grains are stressed in compression. During the operation the metal is stressed above its elastic limit where it acquires a permanent set. Some elasticity usually remains, so that the metal tends to "spring back" as the pressure of the punch is released. This springback is compensated for by overbending or by bottoming the metal. Bottoming must be done with care, because thickness variations may cause overloading of the press or cause injury to the die.

One troublesome aspect of bending operations arises from the directionality present in rolled steel. Directional lines are set up in the steel in the direction of rolling, causing points of weakness running parallel to the directional lines. Because of these lines, cracking sometimes occurs when metal is bent parallel to the lines. This is especially true of hard-temper metals. To avoid this, bends may be designed across the grain of the metal.

Directionality is an exceptionally important factor in bending of heavy-gage work such as in automobile frame construction.

Most forming operations in automobile production have to do with the formation of flanges. Many straight bending operations are used in making flanges. Where straight flanges are made, the operations are fairly uncomplicated.

Contour Flanging

With the advent of streamlining and modern automobile styling, more and more situations have been created in which curved flanges are required. This type of flange is called a contour flange. Although contour flanges usually comprise the sides of a part, they are also used extensively in the formation of flanged holes.

The problems involved in making straight flanges are also common in contour flanging. In addition, contour flanging presents a number of unique problems. These problems arise from the stresses placed on metal in shaping of contours and curves.

Contour flanges are subdivided into three distinct types: stretch flanges, shrink flanges, and reverse flanges.

A stretch flange has a concave curvature. It is called a stretch flange because the material forming the flange must be stretched. This stretching is easily seen in forming of a flange for a hole. The metal used in forming the flange must be elongated.

A shrink flange has a convex curvature. In forming a shrink flange, the metal used to form the flange must be reduced in length. This relationship is easily seen in the flanging of a circular cake pan. The diameter of the finished flange is less than the diameter of the disk from which it was formed.

A reverse flange consists of at least one stretched and one shrunk portion.

Each of these types of flanges creates problems in metal control during forming. To obtain parts conforming to design, special blanks are usually developed. These blanks provide relief for the metal in areas of exceptional strain. In some cases the flanges are formed in rough blanks and then trimmed.

This is usually done when the stresses involved would result in cracking or severe distortion of developed blanks. Where part design permits, relief of the stresses involved in contour forming is frequently provided by notches or cutouts in the developed blank. These notches provide areas into which the stressed metal may flow and prevent buckling or wrinkling.

As in all stamping operations, the sequence of operations for producing a flanged part is very important. A hole pierced in a part prior to flanging may become distorted during flanging, a flange may interefere with tool access for subsequent operations. Each operation must be sequenced for the most effective production arrangement consistent with quality.

Beading

Rigidity and strength may be imparted to sheet-metal stampings through an operation called beading. Beading consists of pressing rounded troughs of uniform width into relative broad expanses of flat metal. Such troughs serve to strengthen the metal and to prevent vibrations or "oil-canning" sounds. Beads may be either straight or curved.

The formation of beads involves both tensile and compressive strains. Where the metal is bent over, the same forces are present as in bending. The metal in the trough is stretched.

Another term often applied to beading is "embossing." Embossing covers a broad area of stamping operations in which circles, squares, and other shapes are impressed into the metal. Some of these shapes may be for decorative purposes; others are designed to provide additional strength.

Pressure Requirements for Forming

Press tonnages required to accomplish the forming operations described above may also be calculated mathematically. To determine the required pressure (P), the length of the area being formed (L) is multiplied by the thickness of the metal (T) times the tensile strength (S_t) of the metal. The value thus derived is then adjusted by multiplying it by 0.25. The formula used in this calculation is:

$$P_s = L \times T \times S_t \times 0.25$$

DRAWING OPERATIONS

"Drawing" or "deep drawing" of metals refers to those operations which exert tensile and associated compressive stresses over a large area of the metal to obtain a stamping of desired shape. In forming, metal is plastically worked in a relatively small area. Drawing operations are more severe. It is difficult to arrive at a clear visualization of all that goes on in a drawing operation. Different types of stresses are being exerted in different directions at the same time. In fact, deep drawing comes closer to being an art and a craft than perhaps any other modern industrial operation. It is a craft dependent on the skill, ingenuity, and instincts of the die designer, the diemaker, and the die setter. The ability to draw metal objects is dependent on the skillful application of pressures to make use of the slip planes within the metal.

Movements in Drawing

The stresses at work in a drawing operation are extremely complicated. The punch descends on the metal and forces it into the die cavity, causing the metal to assume the shape of the punch. As this occurs, the metal is forced to move in several directions at the same time. The punch pressure is exerted in a manner that tends to tear the bottom out of the metal. This imposes tensile or stretching stresses in the walls of the part being drawn. At the same time, the edges of the blank are being pulled into the die opening, with the result that the metal at the edges is being compressed.

Pressure Requirements for Drawing

The pressures involved in performing drawing operations may be calculated to a fairly fine degree through the use of several mathematical formulas. However, selection of a press, based on tonnage requirements for a particular operation, does not normally require a precise knowledge of the total pressure involved. Estimation of pressure requirements can be done according to common sense and simple arithmetic.

In most drawing operations the metal is worked close to its ultimate strength. For this reason the ultimate strength of the material is commonly used as the basis for estimating pressure requirements. In fact, the pressure required for drawing a part is assumed to be the pressure required to "pull its bottom out." The pressure required to do this is equal to the area of the metal in the sidewall of the part multiplied by the ultimate tensile strength of the material.

The formula used to calculate a value corresponding roughly to the total tonnage requirement for a drawing operation (P_S) is:

$$P_s = 3.1416 \times D_1 \times T \times S_t$$

where D_1 is the diameter of the shell in inches, T is the wall thickness of the shell in inches, and S_t is the tensile strength of the metal in tons per square inch.

This formula applies generally to drawn parts of all shapes. Safety allowances may be added as indicated by past experience.

SQUEEZING OPERATIONS

The squeezing group of press operations includes a large variety of metalworking operations such as sizing, forging, swaging, coining, and extrusion. All of these operations require plastic flow of metal under compression with induced tensile stresses. The operations increase in severity with the amount of restraint to metal movement and the amount of metal displaced.

Metal to be squeezed can be considered an incompressible fluid. Thus any displacement must be accompanied by a corresponding flow of the displaced metal. Any restriction of the flow results in a tremendous increase in the pressure required to accomplish the displacement. This build-up of pressure is also enhanced by the tensile stresses induced in the metal, and often by the fact that the cross-sectional area of the metal is being increased.

Forging

"Forging" is the term used to define the plastic deformation of metal into a predetermined size or shape using compressive forces exerted through some type of die by a hammer, a press, or an upsetting machine.

Hot forging is done above the recrystallization temperature of the metal. At such temperatures, metal forms relatively easily. The raw material for forging is usually a bar or a billet. A piece of the desired weight is cut from the bar or billet, heated, and then formed by blows or pressure in the press. In some cases the end of a billet or bar is heated, shaped, and then cut off in the forging press.

Cold forging is done below the recrystallization temperature of the metal, although the metal may still be hot by normal standards. Among the benefits of forging are the desirable mechanical properties imparted to the metal. Greater strength and toughness are achieved than could be obtained by casting. Pressure requirements for forging vary greatly depending on the type of forging, the dies used, and other factors. This makes standard methods of calculating tonnage requirements ineffective. But, generally speaking, forging is done at very high pressures with practical experience being the best guide to press selection.

Cold Extrusion

Another form of metalworking done in presses is cold extrusion. This process employs a press with sufficient tonnage to force a billet of metal through a restricted orifice to form a greatly elongated part of uniform cross section. This action is usually accomplished quickly, although some extrusion is done by slow application of pressure.

Section 2
Stamping and Forming Operations

2A: Punching and Shearing Science

DEFINITIONS

Many pressworking operations are performed using the same "shearing" principles. Therefore, understanding the theory of shearing enables one to understand the following operations (see Figure 2A-1):

- **Shearing:**
  1. Cutting material with dies or blades.
  2. *Shear* is an inclination between two cutting edges used for the purpose of reducing the required shear force.
  3. A *shear* is a tool for cutting metal and other materials by a closing motion of two sharp, closely adjoining edges.
- **Punching:**
  1. A general term describing the process of die cutting a hole in material such as sheet metal, plate, or some structural shape.
  2. A *punch* is the male part of a die set and usually the upper member.
- **Perforating:** A more specific term used in the stamping industry for die cutting of holes in material.
- **Piercing:** Penetration of material using a sharp-pointed punch, leaving a jagged hole like a bullet hole.
- **Extruding:** Forming of a flange around a hole in sheet metal.
- **Blanking:** Cutting or shearing of material to a predetermined contour from sheet or strip stock.
- **Notching:** Cutting of various shapes from the corner or edge of a strip, sheet, or part.
- **Nibbling:** Progressive notching at a high rate of speed, making either a smooth finished edge or a scalloped edge.
- **Lancing:** Cutting into a workpiece without producing a detached slug. Usually combined with forming, such as in production of louvres.

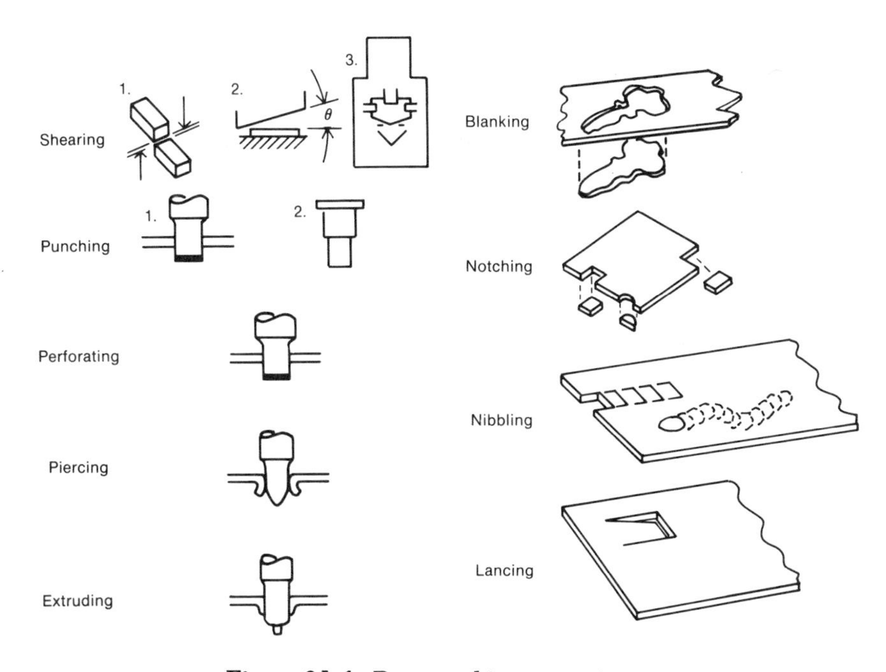

Figure 2A-1. Pressworking operations employing shearing principles. (See definitions in text.)

SHEAR ACTION IN METALCUTTING

The shear cutting or punching action results from a closing motion of two sharp, closely adjoined edges on material placed between them. The material is stressed in shear to the point of fracture while going through three phases (Figure 2A-2):

1. **Deformation:** As the cutting edges begin to close on the material, deformation occurs on both sides of the material next to the cut edge.
2. **Penetration:** The cutting edges cut or penetrate the material, causing fracture lines.

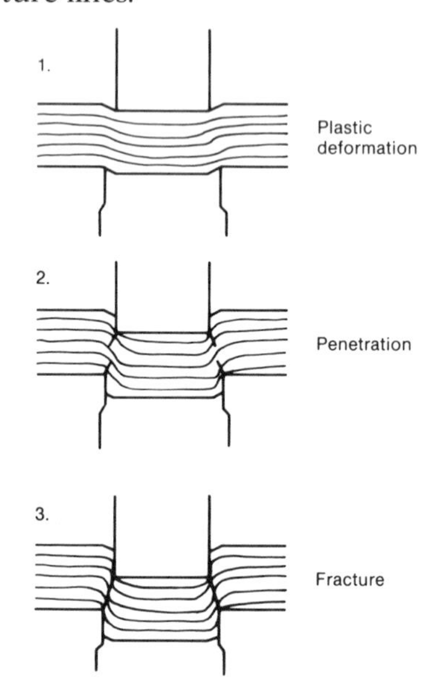

Figure 2A-2. Three phases of shear cutting

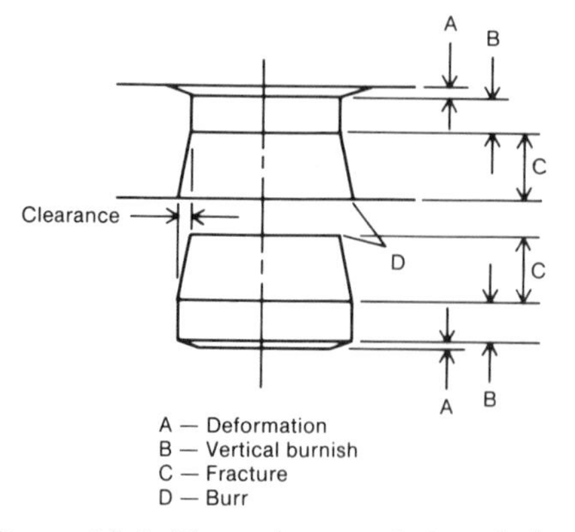

Figure 2A-3. Four characteristics of shear cutting

3. **Fracture:** The point where the upper and lower fracture lines meet. At this point the work is done, but in punching, the punch must continue to move through the material to clear the slug.

The shear cutting action produces four inherent characteristics found on both the parent material and the cut-off (or punched-out) part (Figure 2A-3). These characteristics are:

1. Plastic deformation
2. Vertical burnish-cut band
3. Angular fracture
4. Burr caused by the fracture starting above the cutting edge.

The amount of each of these four characteristics depends on:

- Material thickness
- Material type and hardness
- Amount of clearance between cutting edges
- Condition of cutting edges
- Firmness of support of material on both sides of the cut
- Diameter of hole or blank in relation to material thickness.

If all the preceding conditions are satisfied but the edge condition is still not acceptable, there are other pressworking methods that may be employed. One of the most common finishing methods is shaving, in which a small amount of material is removed to eliminate the fracture angle. Another is the fine-line blanking process, in which a special press or die set is used to compress the material in the shear plain during the cutting cycle and thus eliminate the fracture angle.

DEFORMATION

The type, hardness, and thickness of the material all have effects on the amount of deformation. Softer and thicker materials deform the most. Clearance between the cutting edges and the support of the material on both sides of the cut also have strong effects. Figure 2A-3 shows a hole punched with excessive clearance: note the deformation around the top surface. Supporting the material is generally no problem in hole punching, because the punch holds the slug or blank firmly against the punch face. However, in making a straight shear cut on a long bar, it is necessary to use a holddown to keep the bar from twisting around the lower blade. The combination of tensile and compressive stresses that occur in metal cutting causes this twisting. A support opposite the holddown produces the best results. Note that, without the support, the material bends before fracturing, and an angular impression from the blade is left on the end of the bar (Figure 2A-4).

PENETRATION

Penetration is the sum distance of the deformation and *vertical burnish* height. It is expressed as a percentage of the material thickness and is defined as the distance the punch must travel before the metal fractures. The percentage of penetration varies with the type and hardness of the material. As the material becomes harder, the percentage of penetration decreases (Figure 2A-5).

Penetration increases where holes are less than 1.5 times the material thickness due to the high compressive stress in the material cut zone.

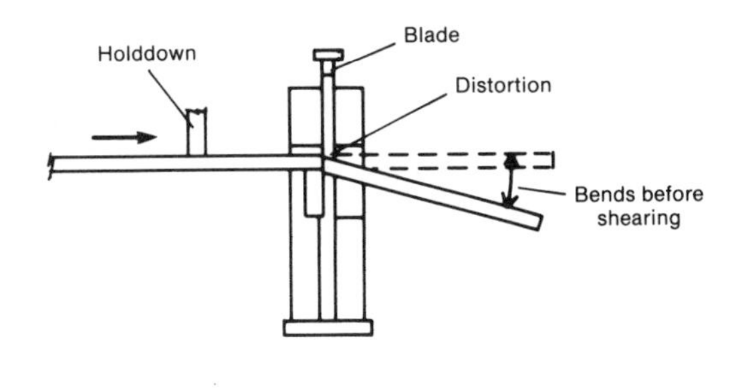

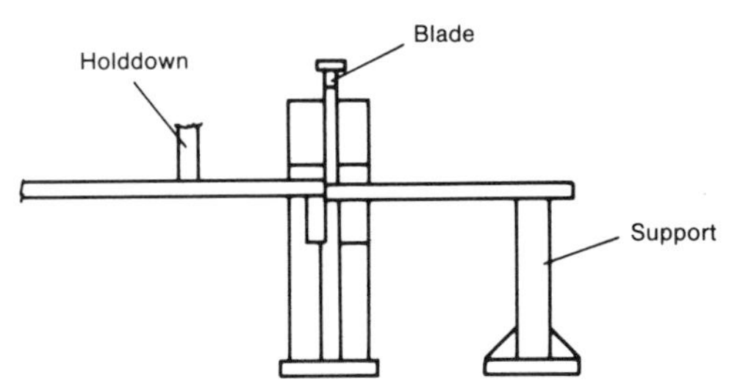

Figure 2A-4. Single-cut flat bar shearing with holddown and no rear support, and with rear support

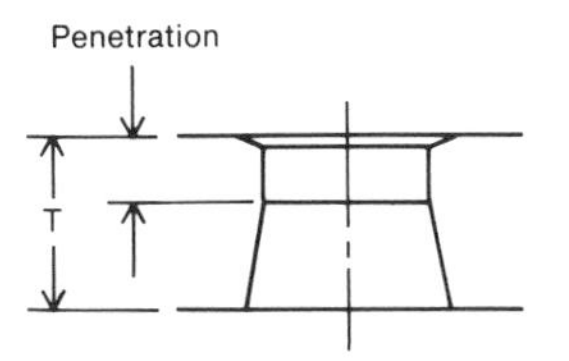

Material	Penetration, % T
Aluminum	60
Copper	55
Brass	50
Bronze	25
Steel, 0.10 C:	
Annealed	50
Cold rolled	38
Steel, 0.30 C:	
Annealed	33
Cold rolled	22
Silicon steel	30
Nickel	55

Figure 2A-5. Penetration as a percentage of stock thickness, showing variation with material hardness

CLEARANCE

The *angular fracture* and the quality of the cut, the punched hole, or the blank is greatly dependent on the amount of clearance between the two opposed cutting edges. Figure 2A-3 shows the clearance as being the distance between the mating cutting edges. Without proper clearance, the material will not fracture cleanly. Figure 2A-3 shows how upper and lower fractures should meet; if they do, a clean hole is produced with a minimum power requirement. With insufficient clearance, a defect known as "secondary shear" is produced, as shown in Figure 2A-6. The two fracture lines were not permitted to

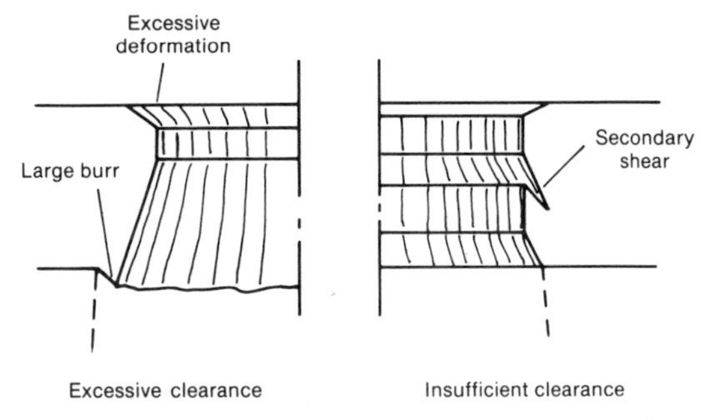

Figure 2A-6. Effects of excessive and insufficient clearances

Table 2A-1. Recommended Clearances for Punching

Material	Total diametral clearance, % of material thickness Minimum	Best	Maximum
Copper, ½ hard	8	12	16
Brass, ½ hard	6	11	16
Mild steel	10	15	20
Steel, 0.50% C	12	18	24
Aluminum, soft	5	10	15
Stainless steel	12	18	24

meet, leaving a ring of material that must be stressed to its point of fracture with a further expenditure of energy. The amount of secondary shear decreases as the clearance increases toward the proper clearance. Dull tools result in insufficient clearances, as well as burrs.

Excessive clearance between the mating edges causes extreme plastic deformation, a large burr, and a high angle of fracture, as shown in Figure 2A-6. Therefore, proper clearance may be defined as that clearance which causes no secondary shear and a minimum plastic deformation and burring.

Proper clearance varies with the thickness and type of material. As shown earlier in Figure 2A-5, the amount of penetration or fracture depends on the type and hardness of the material. Clearance is expressed as a percentage of the material thickness and must be qualified as to whether it means clearance per side or total diametral (overall) clearance. There are many opinions on what represents the best percentage of clearance. Proper clearances are best found by trial and error. No formulas or tables are available which will give exact clearances, but the values given in Table 2A-1 have worked well in short-run punching where it is important to select standard dies for punching materials in a range of thicknesses.

WHERE TO APPLY CLEARANCES

Where the clearance is applied must be considered in blanking or punching to close tolerances. Figure 2A-3 is a good illustration. The die is larger than the punch by the amount of clearance required to produce a clean fracture. If the hole is to be held accurately to size, the punch must be that size and the die must be oversize to allow proper clearance. Conversely, if the operation is blanking of a disk that has to be accurately held, the die must be that size and the punch must be undersize to allow proper clearance.

A punched hole will shrink very slightly, which must be allowed for if high precision is required. The amount varies with the type and hardness of the material but normally does not exceed 0.002 in. Punching or blanking consistently produces very accurate parts. (Additional discussion of die and punch clearance is in Article 2B.)

STRIPPING FORCE

The punching process requires two actions, punching and stripping. Forcing the punch through the material is simple in comparison with stripping or extracting the punch. A stripping force due to the resiliency or springback of the punched material grips the punch. Additional friction is created by cold welding and galling that occur on the punch surface.

Stripping force is generally expressed as a percentage of the force required to punch the hole. This percentage greatly changes with the type of material being punched and with the amount of clearance between the cutting edges (Figure 2A-7). Figure 2A-8 presents graphs showing the effects of these two conditions.

The surface finish of the punch changes

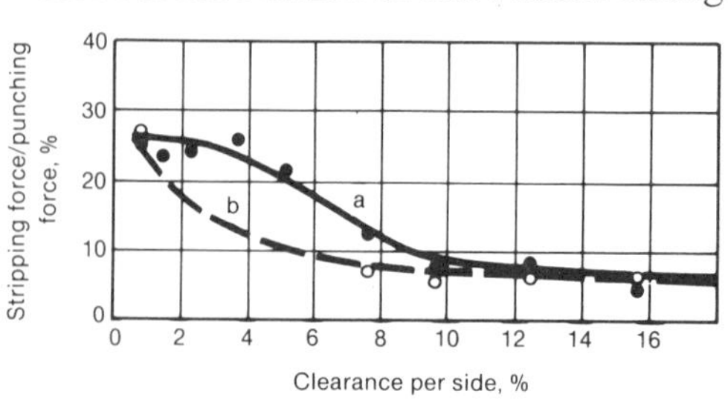

Punch diameter, 0.600 in.; no lubricant

a Thickness, 5/16 in.; tensile strength, 58,000 psi
b Thickness, 5/16 in.; tensile strength, 90,000 psi

Figure 2A-7. Variation of stripping force with clearance

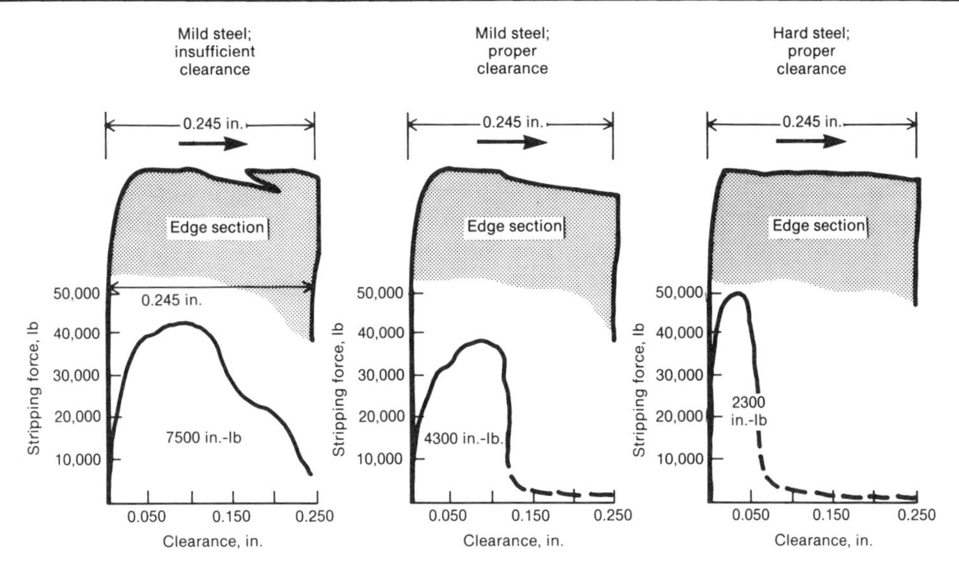

Figure 2A-8. Force curves showing effects of clearance and material hardness

with the number of holes being punched, and if a lubricant is not used, the effect of punch finish can be very severe, as shown in Figure 2A-9. The upper curve shows that dry punching produces a stripping force after seven holes of 25% of the punching force. After seven holes, the surface of the punch smoothens and the force drops (Figure 2A-10 shows the punch after each hole.)

A good grade of die cutting lubricant substantially reduces the stripping force, as shown in Figure 2A-9, (lower curve).

Stripping (Figure 2A-11) can be accomplished with a spring-loaded (urethane or die rubber) plate that holds the material firmly against the die or a positive stripper plate that is set a short distance above the material. With a positive stripper, the material is lifted with the ascending punch until it contacts the stripper and is freed of the punch.

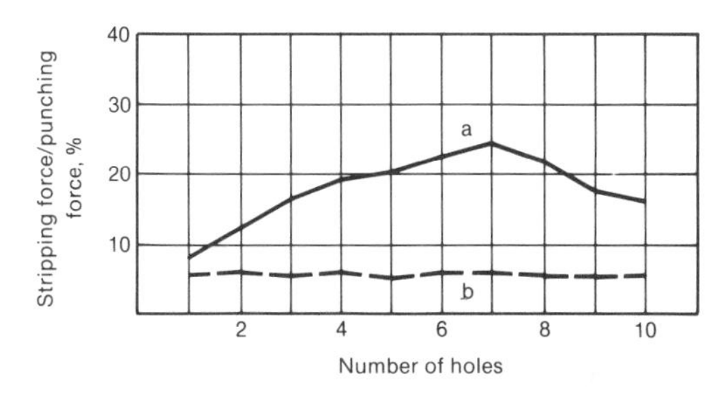

Figure 2A-9. Variation of stripping force with lubrication and number of holes. Mild steel, 1/2 in. thick; punch diameter, 0.600 in.; 10% over-all clearance

TOOL LIFE

Clearance has a great effect on tool life. Maximum life is attained only when clear-

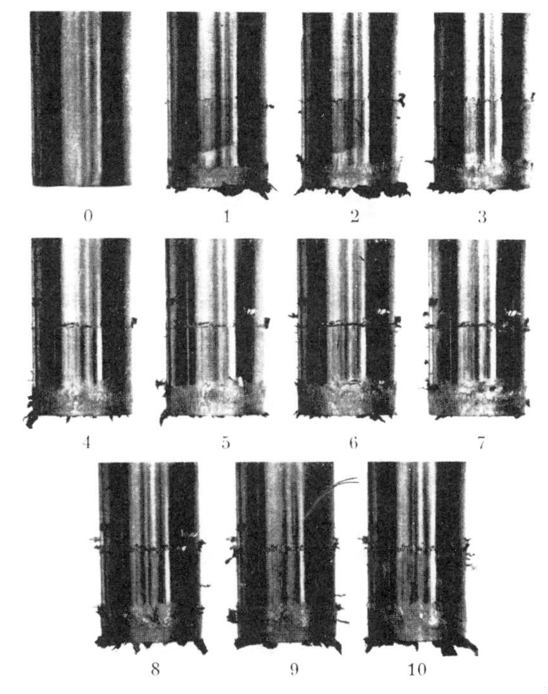

Figure 2A-10. Galling on surface of punch after punching each of ten holes

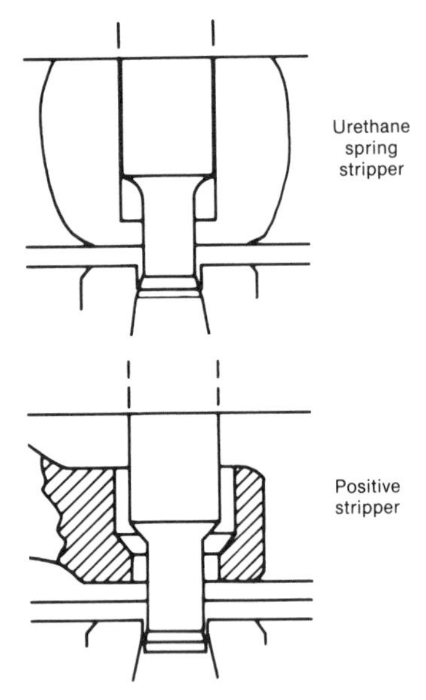

Figure 2A-11. Stripping with spring-loaded and positive stripper plates

ance is proper. Insufficient clearance greatly reduces tool life due to springback and cold welding of the material to the punch and die. In punching harder materials, insufficient clearance creates a much higher stress on the cutting edge and causes it to chip and break down.

Excessive clearance also has an adverse effect on tool life, caused by the material being stretched over the cutting edges, and causes the sharp edges to break down prematurely.

A good grade of die cutting lubricant greatly extends punch and die life.

Punches wear at twice the rate of dies because a punch has to pass through the material twice, in and out, whereas only the slug passes through the die.

SHEARING FORCE

The formula for the force F (area times shearing stress) required to shear, blank, or punch a given material, assuming there is no shear on the punch or die, is:

$$F = LTS \text{ (for any shape cut)}$$
$$F = DTS \text{ (for round holes)}$$

where L is sheared length, in inches; T is material thickness, in inches; S is shear strength of material, in pounds per square inch; and D is diameter, in inches.

The shear strength of the material, in pounds per square inch, is used, and values for some materials can be found in Table 2A-2. This represents the force required to shear or cut a 1-in.-square bar of metal (Figure 2A-12). The same rule is applied to bars of various shapes and for punching and blanking as shown. For punching and blanking, the perimeter of cut is multiplied by the thickness of the material to find the area (Figure 2A-13). Note in Table 2A-2 that the

Table 2A-2. Average Ultimate Strengths of Selected Materials

Material	Chart multiplier	Ultimate strength, psi Shear	Tensile
Aluminum:			
1100-O	0.19	9,500	13,000
1100-H14	0.22	11,000	18,000
3003-H14	0.28	14,000	22,000
2024-T4	0.82	41,000	68,000
5005-H18	0.32	16,000	29,000
6063-T5	0.36	18,000	30,000
6061-T4	0.48	24,000	35,000
6061-T6	0.58	29,000	41,000
7075-T6	0.98	49,000	82,000
Brass, rolled sheet:			
Soft	0.64	32,000	46,000
1/2 hard	0.88	44,000	65,000
Hard	1.00	50,000	78,000
Copper:			
1/4 hard	0.50	25,000	38,000
Hard	0.70	35,000	50,000
Steel:			
Mild A-7 structural	1.00	50,000	65,000
Boiler plate	1.10	55,000	70,000
Structural A-36	1.20	60,000	85,000
Structural COR-TEN (ASTM A242)	1.28	64,000	90,000
Cold rolled C-1018	1.20	60,000	85,000
Hot rolled C-1050	1.40	70,000	100,000
Hot rolled C-1095	2.20	110,000	150,000
Hot rolled C-1095, annealed	1.64	82,000	110,000
Stainless 302, annealed	1.40	70,000	90,000
Stainless 304, cold rolled	1.40	70,000	90,000
Stainless 316, cold rolled	1.40	70,000	90,000

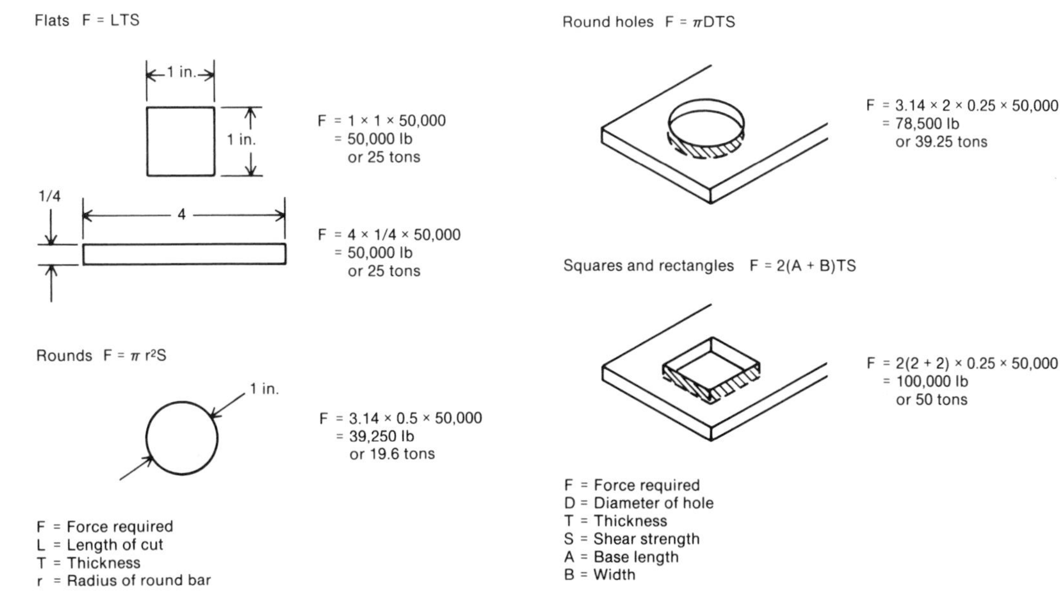

Figure 2A-12. Shearing-force calculations

Figure 2A-13. Punching-force calculations

shear and tensile strengths are not the same. Also, the yield strength of a material cannot be used for shear strength, since they are not the same. If the shear strength of a material is not given or known, it can be calculated by conducting a simple test on a hydraulic press by punching a hole.

The tonnages required for punching or blanking round holes in mild steel plate of various thicknesses are presented in Table 2A-3 for convenience. For example, punching a 1/2-in.-diam hole through 1/4-in.-thick mild steel requires 9.8 tons. Table 2A-3 can also be used for determining tonnages for other materials by multiplying the value in the table by the chart multiplier given in Table 2A-2 for the material in question. For example, the chart multiplier for aluminum alloy 6061-T6 is 0.58, which means that its shear strength is 58% of that for mild steel. Thus, punching a 1/2-in.-diam hole through 1/4-in.-thick 6061-T6 would require 9.8 × 0.58, or 5.68 tons. Tonnage requirements for punching holes greater than 1 in. in diameter can be calculated by adding the tonnages for two or more diameters the total of which equals the desired diameter. For example, the tonnage required for punching a 1½-in.-diam hole through 1/4-in.-thick mild steel would be the sum of the tonnages for punching a 1/2-in.-diam hole and a 1-in.-diam hole through 1/4-in.-thick plate, or 9.8 + 19.7 = 29.5 tons.

There are times when the shear strength of a material is unknown but the tensile strength is known. In such instances, the shear strength can be estimated by taking a percentage of the tensile strength, but this percentage varies with the type and thickness of the material. Figure 2A-14 gives percentages found by sampling three types of material.

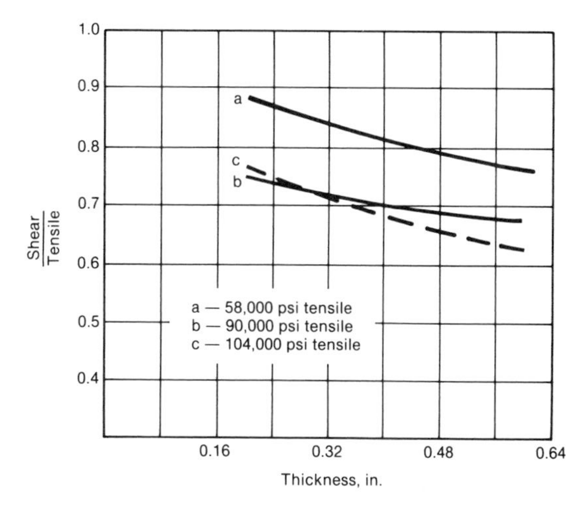

Figure 2A-14. Relationship of shear strength to tensile strength

Table 2A-3. Required Force for Punching Mild Steel Plate(a)

Metal gage	Thickness, in.	Force, tons, required for punching hole diameters, in., of:														
		1/8	3/16	1/4	5/16	3/8	7/16	1/2	9/16	5/8	11/16	3/4	13/16	7/8	15/16	1
20	0.036	0.35	0.53	0.71	0.88	1.1	1.2	1.4	1.6	1.8	1.9	2.1	2.3	2.5	2.7	2.8
18	0.048	0.47	0.71	0.94	1.2	1.4	1.7	1.9	2.1	2.4	2.6	2.8	3.1	3.3	3.5	3.8
1/16 or 16	0.060	0.59	0.89	1.2	1.5	1.8	2.1	2.4	2.7	2.9	3.2	3.5	3.8	4.1	4.4	4.7
14	0.075	0.74	1.1	1.5	1.9	2.2	2.6	2.9	3.3	3.7	4.1	4.4	4.8	5.2	5.5	5.9
12	0.105	1.0	1.6	2.1	2.6	3.1	3.6	4.1	4.7	5.2	5.7	6.2	6.7	7.2	7.7	8.3
1/8 or 11	0.120	1.2	1.8	2.4	3.0	3.5	4.1	4.7	5.3	5.9	6.5	7.1	7.7	8.3	8.8	9.4
10	0.135		2.0	2.7	3.3	4.0	4.6	5.3	6.0	6.6	7.3	8.0	8.6	9.3	10.0	10.6
3/16	0.187		2.8	3.7	4.6	5.5	6.5	7.4	8.3	9.2	10.2	11.1	12.0	12.9	13.8	14.8
1/4	0.250			4.9	6.2	7.4	8.6	9.8	11.0	12.3	13.5	14.8	16.0	17.2	18.5	19.7
5/16	0.312				7.8	9.2	10.8	12.3	13.8	15.4	16.9	18.4	20.0	21.5	23.0	24.6
3/8	0.375					11.1	13.0	14.8	16.6	18.5	20.3	22.1	24.0	25.8	27.7	29.5
1/2	0.500						17.2	19.7	22.1	24.6	27.1	29.5	32.0	34.4	36.9	39.4
5/8	0.625									30.8	33.8	36.9	40.0	43.0	46.1	49.2
3/4	0.750										40.6	44.3	48.0	51.9	55.4	59.0
7/8	0.875											51.6	56.0	60.2	64.6	69.0
1	1.00												64.0	68.8	73.8	78.8

(a) Chart multiplier: Values in chart above are for plate with a shear strength of 50,000 psi. For punching of materials with different shear strengths, it is necessary to use a multiplier (see Table 2A-2) for calculating the proper amount of force required to punch the hole. Example: To calculate the required force for punching a 15/16-in.-diam hole through ASTM A-36 steel (60,000-psi shear strength) 1 in. thick, multiply 73.8 tons by the multiplier for A-36 (1.20) to arrive at $73.8 \times 1.20 = 88.6$ tons. Recommended press size, 96-ton series.

EFFECT OF CLEARANCE ON FORCE AND POWER REQUIREMENTS

Clearance has little or no effect on the amount of force required for shearing or punching, provided there is no shear on the blades, punch, or die. Many studies have been conducted to support this theory, and the results of one are given in Figure 2A-15. The two curves represent mild steel of equal thickness, one with insufficient clearance and the other with proper clearance. The heights of these two curves are about the same, the one for insufficient clearance being slightly higher.

Take note of the areas under the curves, which represent the power requirements. Insufficient clearance greatly increases the work requirement (7500 in.-lb compared with 4300 in.-lb) as a result of the secondary shear. With insufficient clearance the shearing force is present throughout the material thickness, whereas with proper clearance the force drops off about halfway through the material. It is quite obvious at this point why tool life is shortened by insufficient clearance — the tool performs 75% more work.

REDUCING SHEAR FORCES

A progressive shearing action is commonly used to reduce the required force. This is accomplished by stepping or staggering of punches where more than one punch is used so that they do not cut at the same time, or by grinding an angle on the punch and die edges. In both methods, the work is done over a greater distance with less force, but the total work performed is the same as if there were no stepping or shear on the punches.

Figure 2A-16 illustrates three punches each stepped equal to half the material thickness. With proper clearance, in punching of mild steel, the maximum force required would be that required for just one punch. The work curve at the top of the figure shows the maximum force plus the fact that three times as much energy must be delivered from the flywheel of a mechanical press, which can be a problem for some presses. Most hydraulic presses deliver full tonnage throughout the stroke with sufficient energy.

If the die clearance for punching the three holes were insufficient, the maximum required force would be greatly affected, as shown by the dotted curve. The force requirement of one punch must diminish before the next punch contacts the material. (Figure 2A-15 illustrates the problem of insufficient clearance.)

Calculating the effect of applying a shear angle to the face of a punch is difficult due to many variables. Figure 2A-17 shows the effect of shear angle as related to the thickness of material being punched. For example, if the punch has an R of ¼ in. and the thickness

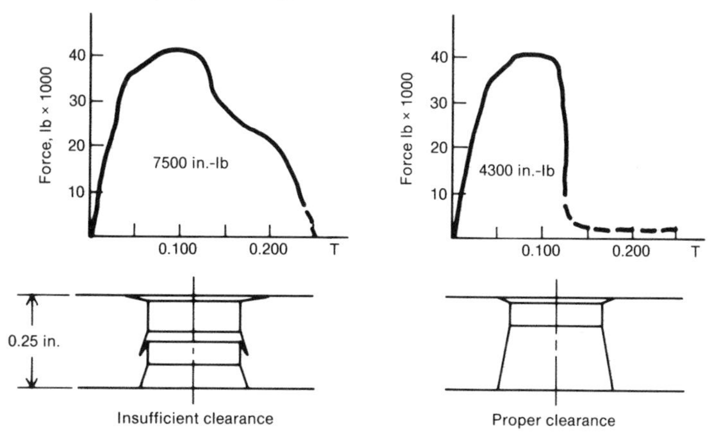

Figure 2A-15. Variation of work requirement (area under curve) with clearance

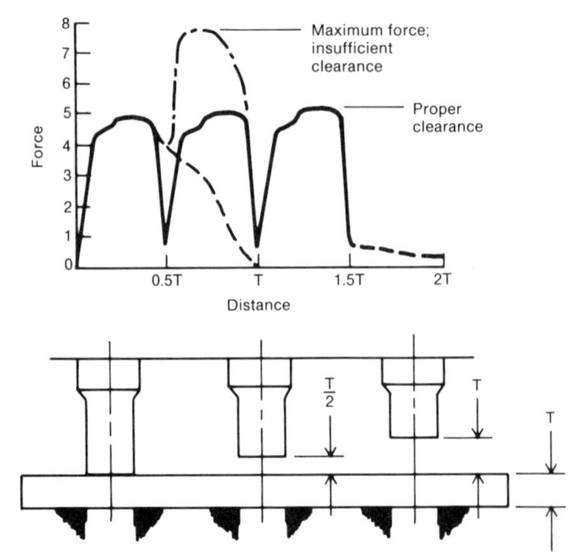

Figure 2A-16. Stepped punches

(T) of the material being punched is ¼ in., the calculated punching force using the formula in Figure 2A-13 would be reduced by 40%.

The effect of shear shown in Figure 2A-17 holds true only if a 15% over-all clearance is used. Less clearance increases the area being sheared, as shown in Figure 2A-18. Insufficient clearance can increase the punching force by as much as 50%.

An example of the necessity for proper clearance is where a 30-ton hydraulic press will punch a 2-in.-diam hole through ¼-in.-thick mild steel with a punch having a ⅛-in. shear (1/2T). However, with insufficient clearance, the required tonnage approaches the full tonnage for that of a punch and die without shear (40 ton), and the 30-ton press will not do the job. A die 1/32 in. larger than the punch should be used in this case to provide proper clearance.

There are many ways of applying shear to a blade, punch, and die. The standard practice on a squaring shear is to have the blade inclined from one side to the other, normally at a rate of 3/8 per foot. Figure 2A-19 illustrates various methods of applying shear to the punch and die. It is important to maintain balanced loading of the punch and die to prevent side thrusts on the punch that would cause the punch to crowd over and hit the die. This is the same problem encountered in punching part of a hole at the edge of a sheet.

Use of shear on a punch and die is an inexpensive way of stretching press capacity, provided the press has sufficient energy to accept the additional work. Many mechani-

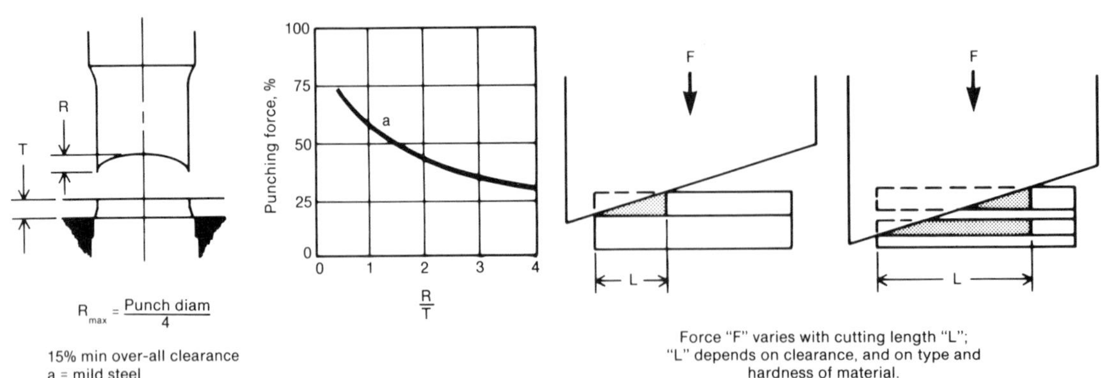

Figure 2A-17. Effect of punch-face shear on punching force

Figure 2A-18. Effect of clearance on force required for punching with sheared tooling

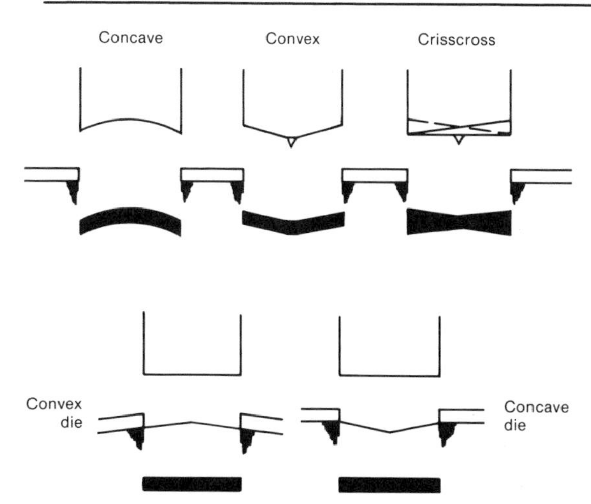

Figure 2A-19. Methods of applying shear to punches and dies

cal (flywheel-type) presses do not have sufficient energy to take advantage of shear. In fact, shear can decrease capacity if not properly applied.

DIAMETER-TO-THICKNESS RATIOS

Every good thing has its limitations, and punching is no exception. One limitation is that very small holes cannot be punched through very thick material — for example, a 1/4-in.-diam hole through 1-in.-thick mild steel. In this case the hole would have to be drilled, but where is the limiting point?

The old rule-of-thumb that the punch diameter must be at least equal to the thickness of the material has inexcusably cost industry thousands of dollars. One steel fabricator had the architect increase the hole-diameter specification from 13/16 to 15/16 in. so he could punch several thousand holes in 1-in.-thick beams. He consented to pay the extra cost for larger fasteners, which was far less than the cost of drilling. Unfortunately, he didn't know that with his portable press he could punch 13/16-in-diam holes through 1-in.-thick mild steel consistently.

When all the factors involved in the thickness-diameter ratio limitation are considered, it is possible to come up with a new, more realistic set of ratios for nonshock applications.

The diameter of the punch must be such that the punch's compressive strength is greater than the force required to punch the hole. This punching force can be found by multiplying the material's thickness by its shear strength (in psi), then multiplying by the length of cut.

Now let's see how this can be used to determine if a punch will endure when used in a hydraulic press. The following factors must be considered:

- A = Cross-sectional area of the punch, as determined by hole size and shape (Figure 2A-20)
- T = Thickness of material being punched
- S_s = Shear strength of material being punched
- S_c = Compressive stress in the punch
- L = Length of cut.

The compressive stress in the punch can be calculated from the formula:

$$S_c = \frac{T \times S_s \times L}{A}$$

The maximum allowable compressive stress (S_c) depends on the type and hardness of the tool steel from which the punch is made. A good grade of oil-hardened shock-resistant tool steel will withstand a compressive stress of 300,000 psi before breaking, and can be used safely at 250,000 psi with good tool life.

The curves in Figure 2A-21 are based on these punching strength values, and for a known shear strength (of the material being punched) the curves give the recommended thickness-to-diameter ratio.

The curve shown as a dotted line represents the ultimate strength (300,000 psi); the solid curve, the recommended working stress (250,000 psi). For example, for punching of mild steel with a shear strength of 50,000 psi, the recommended thickness-to-diameter ratio is 1¼ to 1; the ultimate ratio is 1½ to 1. Therefore, it is safe to punch a 1-in.-diam hole through 1¼-in.-thick mild steel.

Quite often a ratio between 1¼ and 1½ to 1 is used for mild steel, but punch life is shortened.

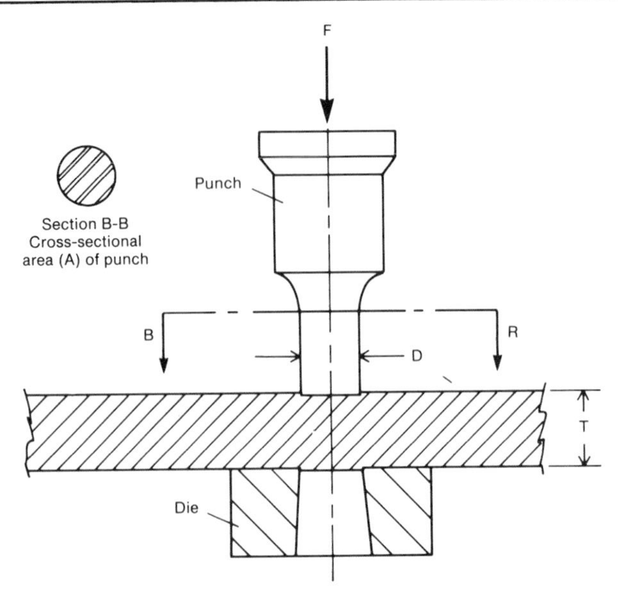

Figure 2A-20. Relationship of punch diameter to material thickness

Minimum Hole Size

The graph in Figure 2A-22 shows the minimum diameter of hole that can be punched through a given thickness of material. Three different materials are illustrated.

To use this graph, locate the thickness of material along the vertical scale and follow across horizontally to the lower edge of the shaded area for the material being punched. From this point of intersection, drop down to the horizontal scale and read the minimum recommended hole diameter.

The upper edge of each shaded area represents the breaking point for the punch. Working within the shaded areas will therefore result in shortened punch life. For example, in punching 3/4-in.-thick mild steel, 19/32 in. is the recommended minimum hole diameter; if a ½-in.-diam punch is used, it will fail.

A punch will fail in one of two ways when overloaded. If its elastic limit is slightly exceeded, the punch will expand as it is pushed through the material. A very high force is then required during stripping, causing the punch to break either at the punch end or under the head.

The second type of failure occurs when the compressive stress is greatly exceeded

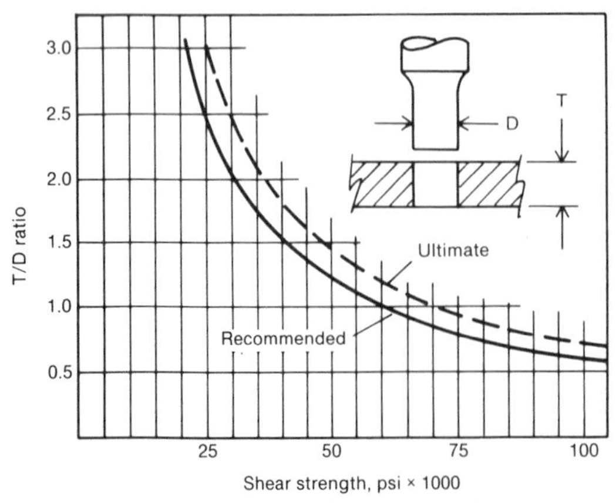

Figure 2A-21. Ratio of material thickness to punch diameter as a function of shear strength

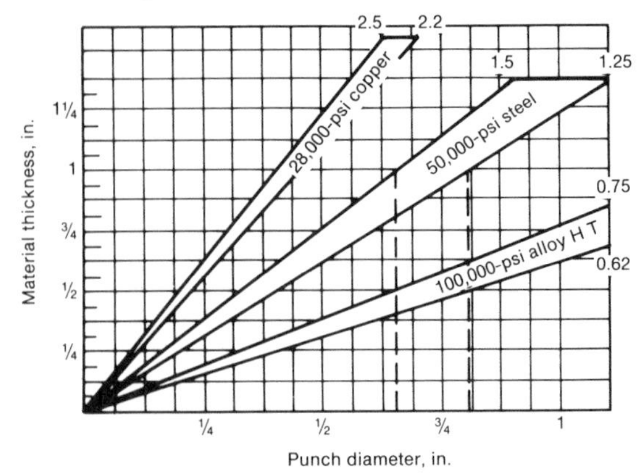

Figure 2A-22. Minimum punch diameter (hole size) as a function of material thickness

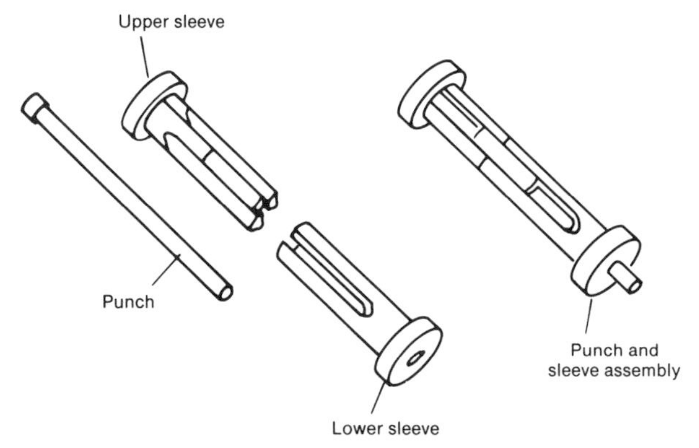

Figure 2A-23. Guided punch (courtesy of Durable Punch and Die Company)

and the punch simply buckles before penetrating the material.

Keep in mind that we have been talking about plain punches as used for most sheet metal, plate, and structural work. Special guided punches (Figure 2A-23) are available for piercing material up to ½ in. thick; these punches are supported to prevent buckling, and offer thickness-to-diameter ratios as high as 2 to 1 in mild steel.

LIMITATIONS OF PUNCHING

Punching can cause distortion. In some parts there is nothing that can be done to overcome it, but in many cases, steps can be taken to minimize it.

One common problem is the closeness of a hole to the edge of a part, which if too close causes bulging along the edge. Preferably, two times the thickness of the material should be allowed from the edge of the hole to the edge of the part.

Another frequent problem that occurs on strip and bar stock is camber due to off-center punching — for example, punching a row of 15/16-in.-diam holes on 3-in. centers in a 6-in.-wide, 3/4-in.-thick bar of mild steel where the common centerline is 2 in. in from one side. Figure 2A-24 shows the type of distortion that will result. The holes should be on-center where possible, and the bar should not be too narrow. Insufficient clearance will cause increased distortion, due to the increased outward forces produced.

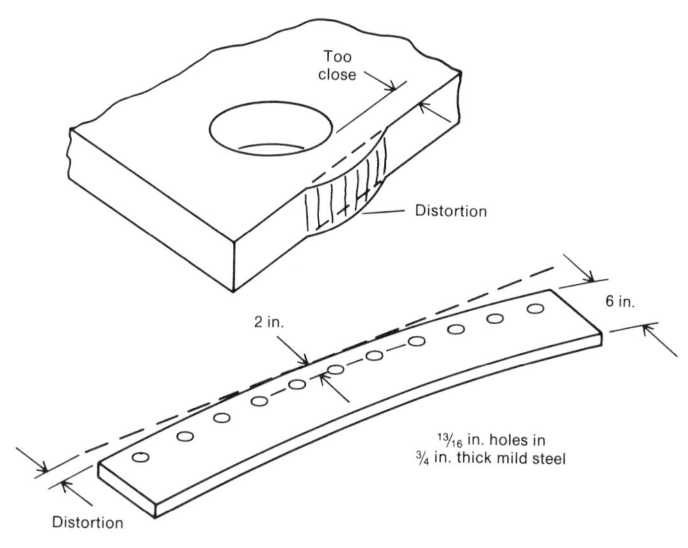

Figure 2A-24. Distortion resulting from off- center punching

2B: Die Clearances and Stripping Forces:

Effects on the Dynamic Properties of Metals for Blanking Operations

RELATIONSHIP OF DIE CLEARANCE TO STRESS-STRAIN CURVES

One of the most important factors in the design of punches and dies is the amount of clearance necessary between the punch and die. The amount of clearance, to a great extent, determines the life of the die, the interval of time between successive sharpenings of the die and punch, the quality of the work being stamped, the power requirements for the punch press producing the work, and the stresses built up in the press during a stamping cycle. Die clearance varies with the thickness, type, hardness, and complexity of the work. No exact information giving the correct amount of die clearance for a set of known conditions is available. The determination of die clearance has been largely a matter of guesswork and "rule-of-thumb" procedures.

The subject of die clearance and its many aspects covers an extremely large field, and prolonged and varied investigations are necessary to give complete explanation to the variety of phenomena involved.

Many static tests have been made for determining optimum die clearances, but very few tests have been run at speeds approximating actual mass-production rates because of the recording problems at such speeds. To record such instantaneous information would require an electronic device which would "pick up" and amplify two input signals at once and permanently record the information as it is being received.

A number of graduate studies at Purdue University have been directed toward investigating instrumentation for press operations. The die-clearance study described below

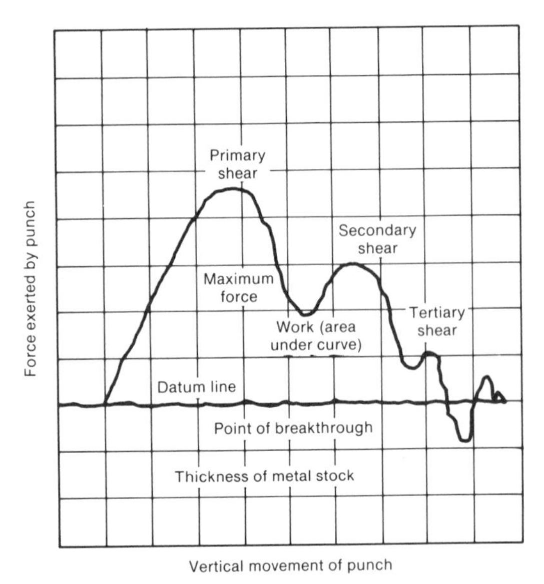

Figure 2B-1. Key for interpreting oscilloscope traces

proved to be a most interesting investigation.

A stress-strain indicator will record two electrical input signals plotting the force applied to the material and the distance through which the punch progresses. By examination of the load curve — i.e., a plot of force applied versus the position of the punch — it is possible to determine the dynamic properties of the metal being worked at that particular clearance (Figure 2B-1). By analyzing a series of load curves resulting from the use of different die clearances, an accurate determination of the optimum die clearance for a given set of conditions can be made (Figures 2B-2 and 2B-3).

The stress-strain indicator produces voltages which, when applied to an oscillograph, plot the load curve. Strain gages and a differential transformer are used as the input transducers. The load curves on the oscilloscope are recorded photographically.

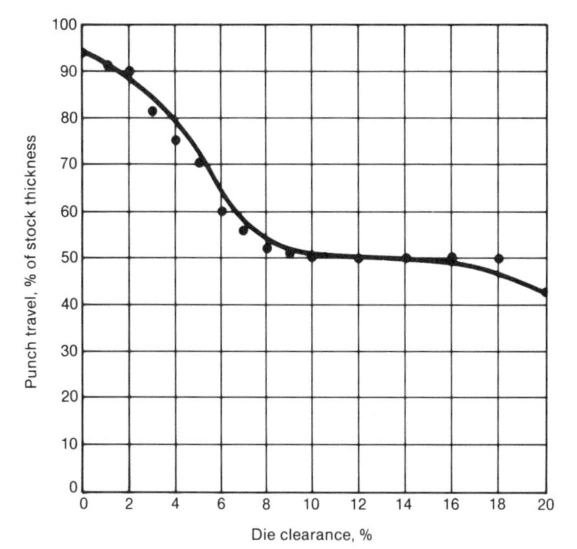

Figure 2B-2. Distance of punch travel through 1/8-in.-thick flat naval brass stock (hardness, Rockwell B62) for complete severance

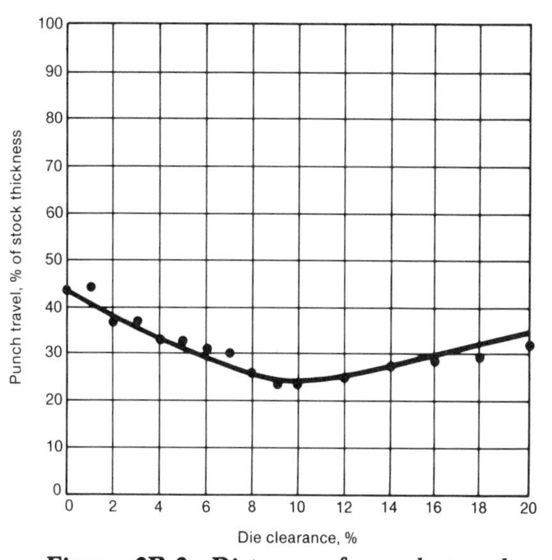

Figure 2B-3. Distance of punch travel through 1/8-in.-thick flat aluminum stock (hardness, Rockwell B73) for complete severance

The purpose of this study was to investigate the possibility, usefulness, and reliability of such an electronic approach to the accurate determination of optimum die clearances. The investigation was of a general nature in which various types of metals and clearances were used. The oscilloscope traces of the load curves were then analyzed for possible analytical relationships between die clearance and (1) maximum force required to stamp out the work, (2) work required to sever the part, (3) amount of "dish" of the resulting work, and (4) the distance of travel of the punch through the metal stock before complete severance occurred. From this study it was established that die clearance definitely affects the aforementioned factors and that such an electronic approach to the determination of an optimum die clearance for a given set of conditions is practical and useful.

INTERPRETATION OF STRESS-STRAIN CURVES

The functioning of a great many press-working dies may be likened to a tensile test for determining physical properties of a material — that is, the material is stressed at a concentrated point until rupture occurs. The pattern of this rupture — stress versus strain — closely approaches that of a metal slug being separated from its "mother" sheet by means of a blanking operation. The resulting stress-strain curve provides a wealth of information. The explanation of the procedure is in interpreting the stress-strain curves.

1. The integral of the curve represents the work required to blank out the slug. When the vertical-displacement conversion factor to stress and the horizontal conversion factor to strain are known, the small squares under the curve can be counted and the work determined directly.
2. The maximum height of the curve represents the maximum amount of force required to sever the blank from the sheet. The vertical-displacement calibration factor is known because the press ram was calibrated on a materials testing machine (refer to Figure 2B-1).
3. The amounts of secondary and tertiary shear can be determined from the areas under the second and third peaks, respectively.
4. The differential of the curve at any point indicates the rate of change in the load-

ing or unloading force at that time. If the curve is approaching the vertical, it means that the loading is applied almost instantaneously without any appreciable plastic deformation in the blanked slug, whereas a gradaully sloping curve indicates plastic deformation or "burring."

5. Likewise, if the curve drops off almost vertically, it means that the severed piece has been cut "clean." If the drop-off is gradual it means that a number of shear lines have overlapped each other and that, when the complete breakthrough does occur, the resulting cut is rough and ragged.

6. The point at which the trace dips down to or below the datum line is the point of breakthrough or sudden load reversal, as was the case with aluminum. In some instances, the propagation of a crack will move faster than the punch, causing the trace to fall below the datum line even though the piece has not been completely severed.

7. Any forces indicated beyond the horizontal distance representing the metal thickness is caused by the resistance of the slugs to movement between the die walls.

Results

The results of this test were highly satisfactory from the standpoint of a visual graphical record of the data secured. The data show a definite relationship between the die clearances and the dynamic characteristics of the materials being tested. Perhaps the major contribution of this investigation was to prove that the dynamic properties of metals are much different from the static properties in many instances. One author states that press-tonnage requirements are estimated by using dies with sharp edges and that the maximum pressure normally required equals the area adjacent to the perimeter of the cut. This is true for static conditions but is not applicable to dynamic conditions. Examination of the force-versus-die-clearance curve indicates that the dynamic force required for blanking is about 20 to 25% above the static force necessary to blank the same type and thickness of metal. This excessive amount of force required for blanking was apparent in all the metals used.

An increase in die clearance up to 8 to 12% gives work curves which show a decided decrease in the amount of work required for blanking. A clearance below 6%, for l/8-inch material, generally indicates excessive tool wear and excessive stresses in the press.

The nature of the blanked edge depends on the clearance. As clearance increases, the "finish" or quality of the blanked edge increases up to about 6 to 10%. Then plastic deformation occurs.

The "dish" deformation of the blanked slug tends to increase with increasing clearance. Dishing is much more prevalent in the softer materials; consequently, die clearance is more critical for softer metals. Die clearance has to be large enough to reduce work and dynamic forces, yet small enough to allow the product to fall within acceptable "dish" specifications. Die clearance can be excessive for the harder materials because of their ability to withstand "dishing."

Metal hardness is an important factor in determinng stress-strain characteristics. Because of the inability of the harder metals to absorb the inertial effects of the punch-press ram, the maximum force required for blanking is higher for harder metals than for softer metals of the same ultimate shearing stress. Figures 2B-4, 2B-5, and 2B-6 show forces versus die clearances for various materials.

DYNAMIC STRIPPING FORCES IN BLANKING

When a hole is punched, the material exerts a radial clamping force on the punch surface, and so a certain amount of force is required to withdraw the punch. This is called the stripping force.

The strength design of the stripper plate, which holds the material down during the upstroke of the punch, is based on the maximum value of the stripping force. Because experience formulas provide only a rough approximation method for the computation of this force, it is advisable to use a large safety factor. It is easy to design the stripper plate to be strong enough to withstand the largest stripping forces that occur.

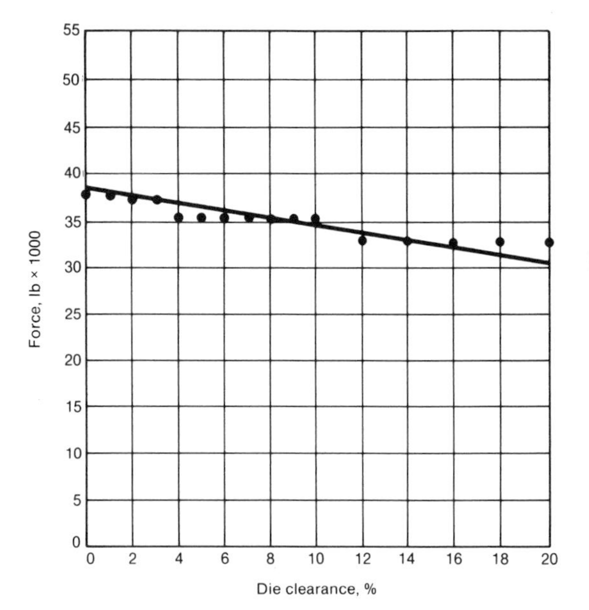

Figure 2B-4. Maximum force required for blanking a 1-in.-diam slug from 1/8-in.-thick flat naval brass stock (hardness, Rockwell B62)

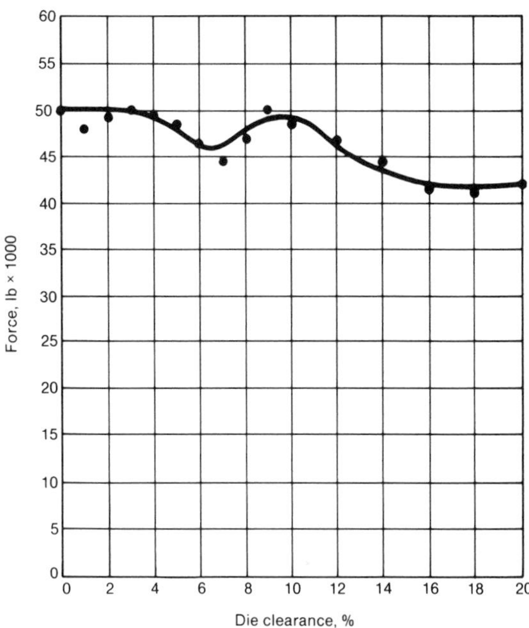

Figure 2B-5. Maximum force required for blanking a 1-in.-diam slug from 1/8-in.-thick flat stainless steel stock (hardness, Rockwell B85)

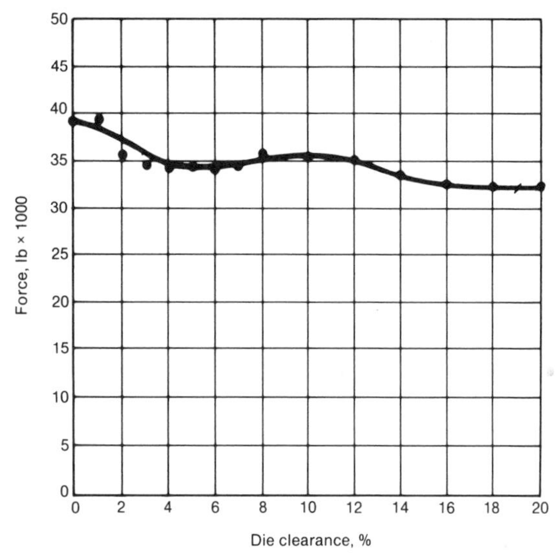

Figure 2B-6. Maximum force required for blanking a 1-in.-diam slug from 1/8-in.-thick flat aluminum stock (hardness, Rockwell B73)

This is as far as our present knowledge of the size and application of this force goes, and from this point of view, the stripping force is not very interesting.

There is, however, a more important aspect of this force: its characteristics provide important information about the punching condition, and about certain influential variables.

With an electronic recording method it is possible to measure the stripping force under actual operating conditions, thus eliminating the approximations of formulas based on experience. This investigation method results in a load curve, which is a plot of what is called the "dynamic stripping force" versus the position of the punch relative to the material being punched. Electric strain gages, attached to the stripper plate, govern the vertical travel of the electron beam of an oscilloscope; thus they record the size of the dynamic stripping force. The horizontal travel of this beam is related to the movement of the punch. This is achieved through a differential transformer with the core bar attached to the ram of the punch press. A 35-millimeter camera is used for the permanent recording of the load curves that appear on the oscillograph screen.

The primary purpose of this investigation was to design and build a test setup based on the recording method outlined above. Its

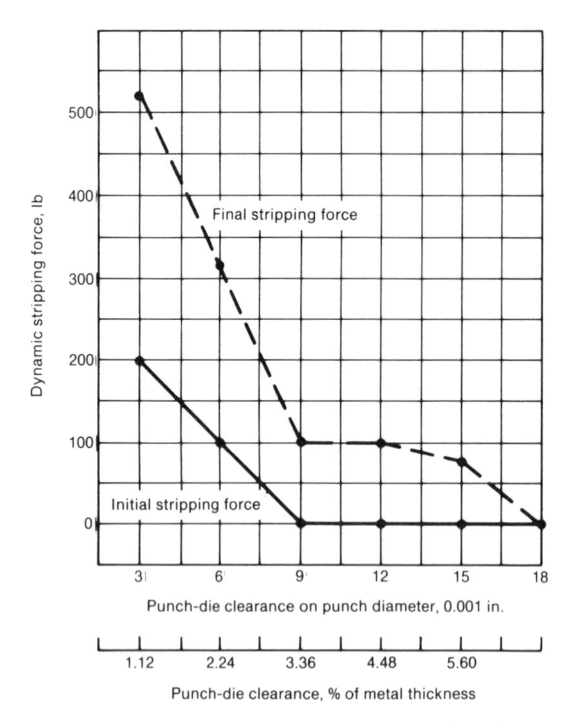

Metal: E – stainless steel. Hardness: Rockwell B86. Thickness of metal strip: 0.136 in.

Figure 2B-7. Dynamic stripping force as a function of punch-die clearance for stainless steel

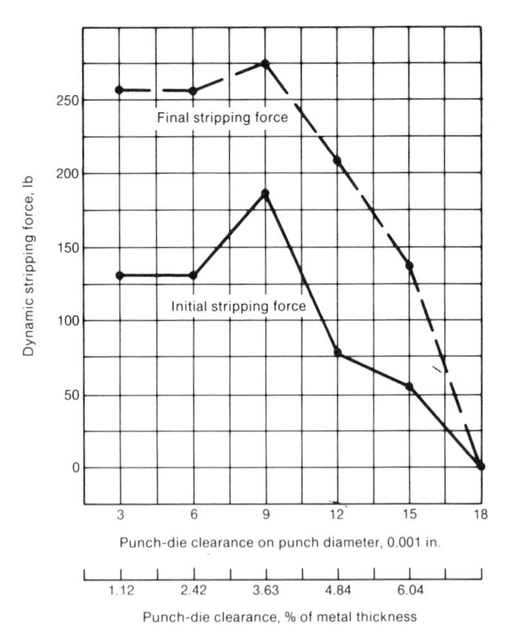

Metal: H – cold rolled steel (A). Hardness Rockwell B92. Thickness of metal strip: 0.124 in.

Figure 2B-9. Dynamic stripping force as a function of punch-die clearance for cold rolled steel

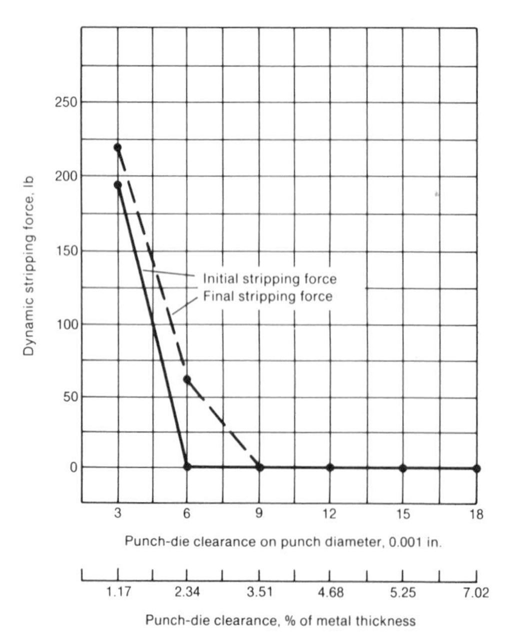

Metal: B – brass, 1/2 hard. Hardness: Rockwell B77. Thickness of metal strip: 0.128 in.

Figure 2B-8. Dynamic stripping force as a function of punch-die clearance for brass

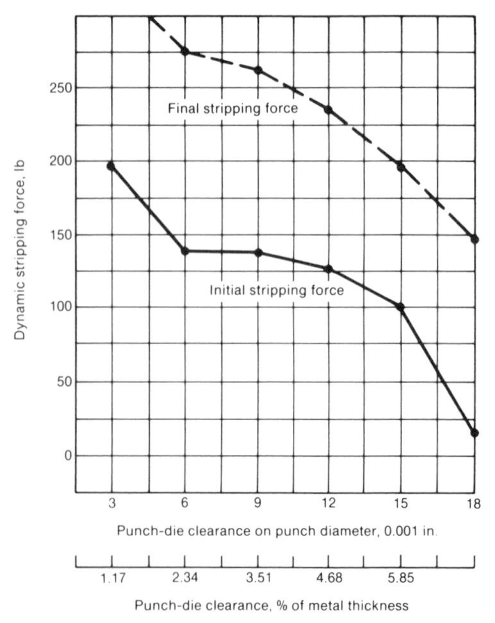

Metal: K – hot rolled steel. Hardness: Rockwell B65. Thickness of metal strip: 0.128 in.

Figure 2B-10. Dynamic stripping force as a function of punch-die clearance for hot rolled steel

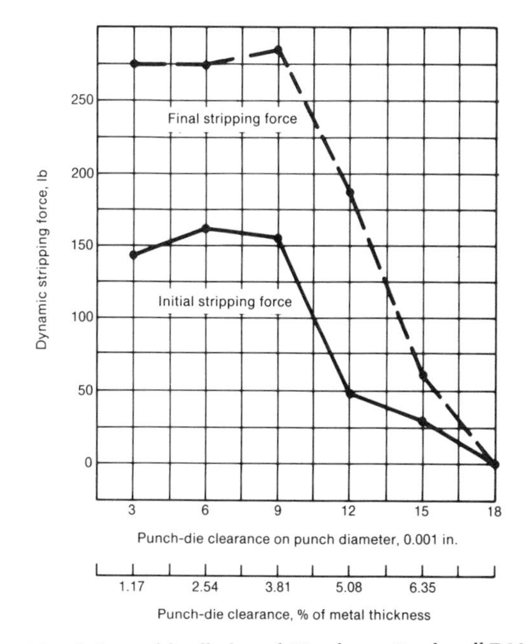

Metal: G – cold rolled steel. Hardness: Rockwell B93. Thickness of metal strip: 0.118 in.

Figure 2B-11. Dynamic stripping force as a function of punch-die clearance for cold rolled steel

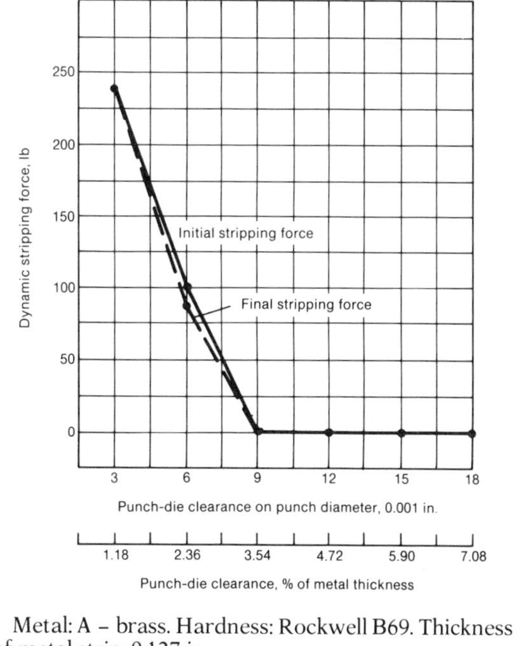

Metal: A – brass. Hardness: Rockwell B69. Thickness of metal strip: 0.127 in.

Figure 2B-13. Dynamic stripping force as a function of punch-die clearance for brass

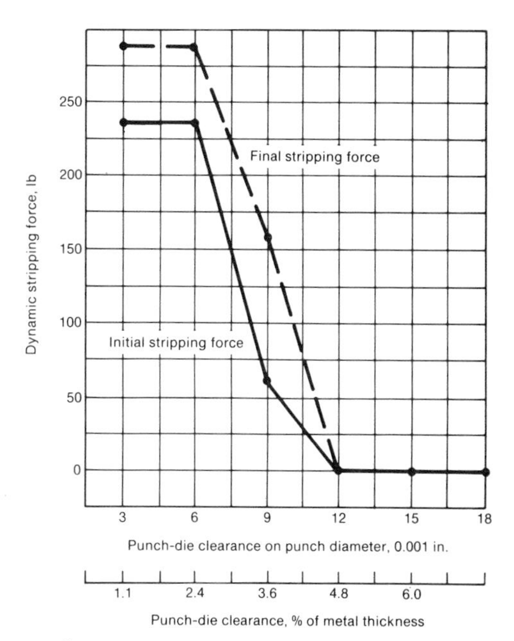

Metal: F – SAE 1020 cold rolled steel. Hardness: Rockwell B99. Thickness of metal strip: 0.125 in.

Figure 2B-12. Dynamic stripping force as a function of punch-die clearance for 1020 C.R. steel

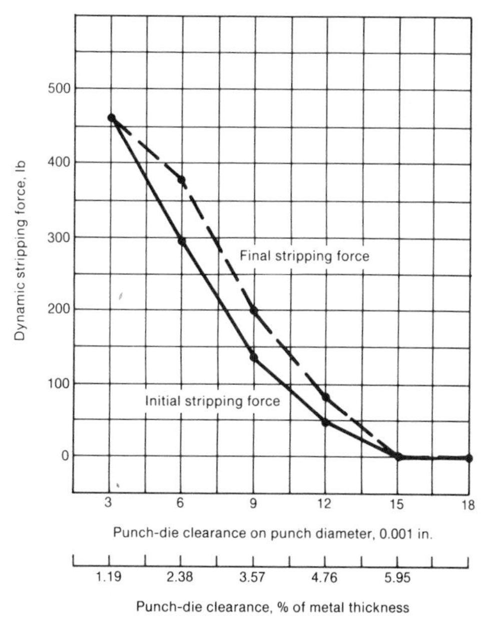

Metal: D – aluminum 24S-T3. Hardness: Rockwell B61. Thickness of metal strip: 0.126 in.

Figure 2B-14. Dynamic stripping force as a function of punch-die clearance for aluminum

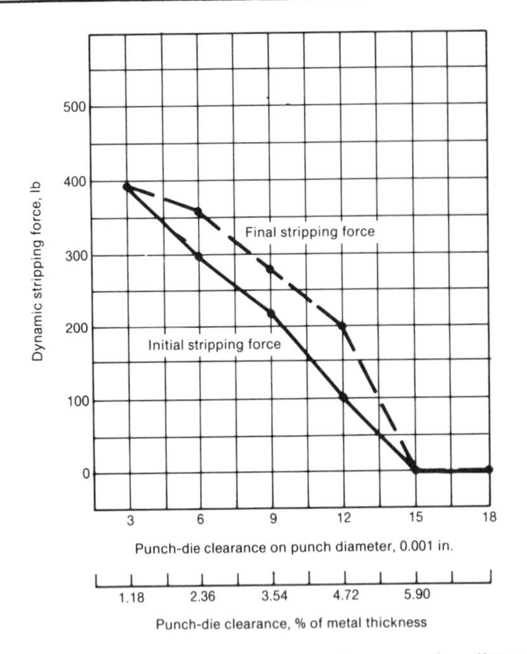

Metal: C – aluminum 24 ST. Hardness: Rockwell B64. Thickness of metal strip: 0.127 in.

Figure 2B-15. Dynamic stripping force as a function of punch-die clearance for aluminum

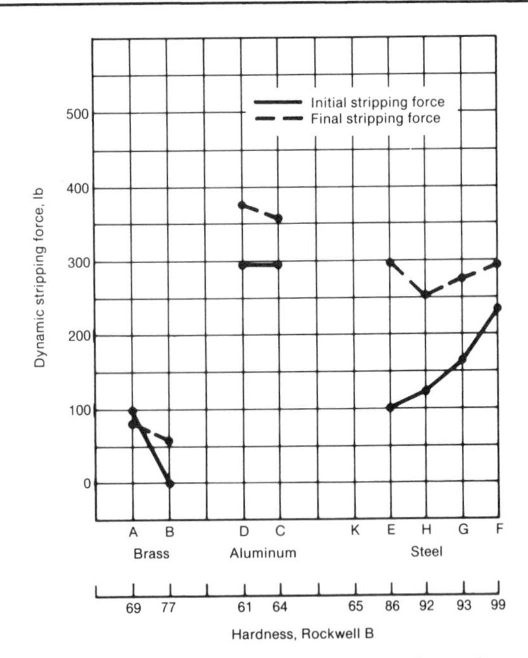

Figure 2B-17. Dynamic stripping force for brass, aluminum, and steel with punch-die clearances of 0.006 in. on punch diameter and 0.003 in. per cutting edge

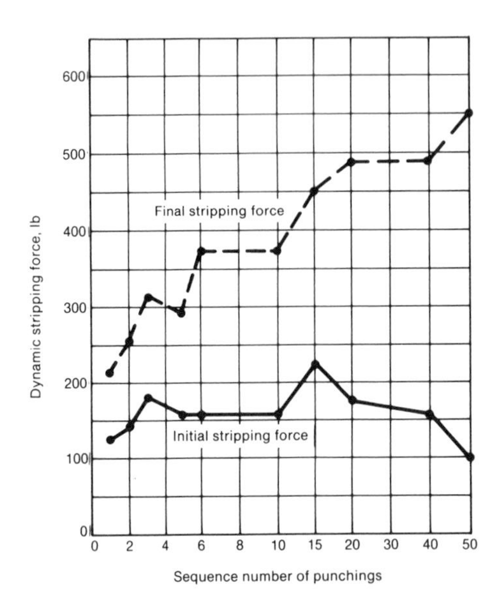

Metal: S – tool steel. Hardness: Rockwell B62.

Figure 2B-16. Dynamic stripping force as a function of punch wear for tool steel

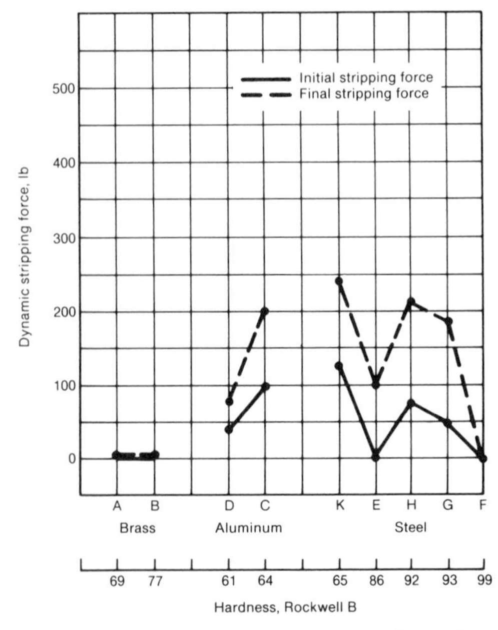

Figure 2B-18. Dynamic stripping force for brass, aluminum, and steel with punch-die clearances of 0.012 in. on punch diameter and 0.0096 in. per cutting edge

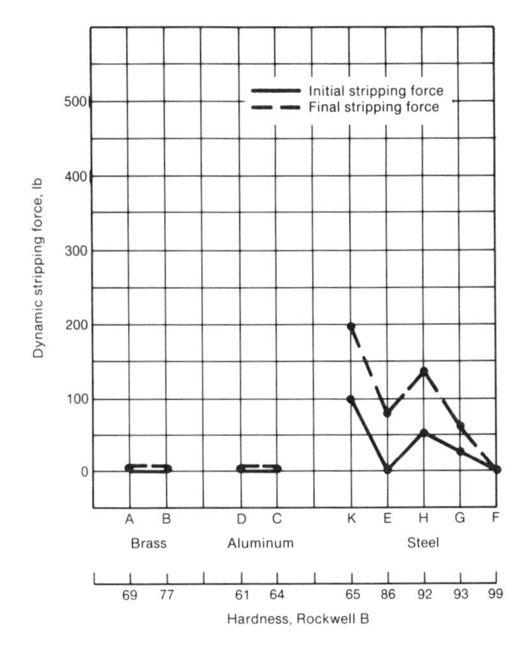

Figure 2B-19. Dynamic stripping force for brass, aluminum, and steel with punch-die clearances of 0.015 in. on punch diameter and 0.0075 in. per cutting edge

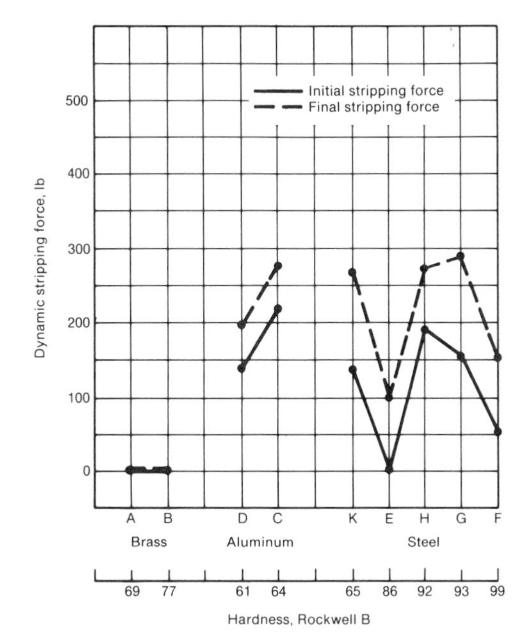

Figure 2B-20. Dynamic stripping force for brass, aluminum, and steel with punch-die clearances of 0.009 in. on punch diameter and 0.0045 in. per cutting edge

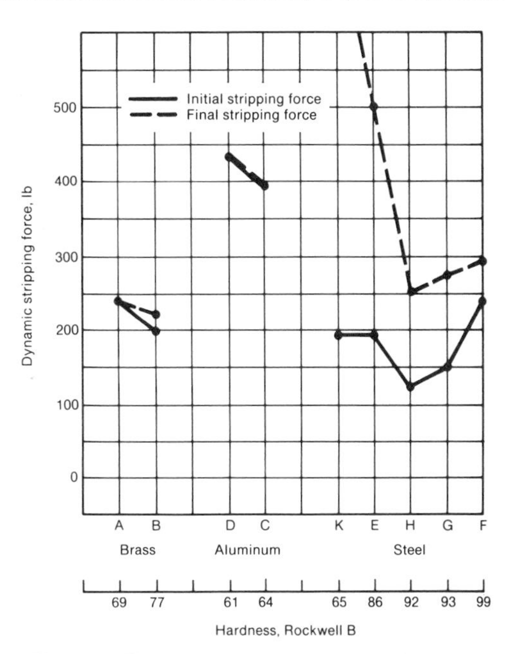

Figure 2B-21. Dynamic stripping force for brass, aluminum, and steel with punch-die clearances of 0.003 in. on punch diameter and 0.0015 in. per cutting edge

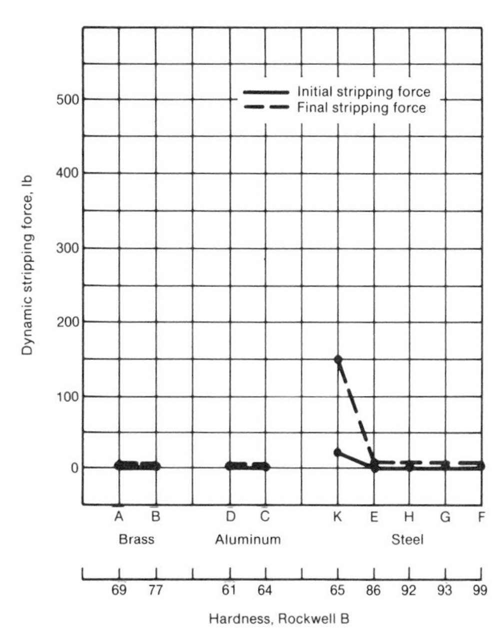

Figure 2B-22. Dynamic stripping force for brass, aluminum, and steel with punch-die clearances of 0.018 in. on punch diameter and 0.0096 in. per cutting edge

secondary purpose was to perform some limited investigations in order to determine its usefulness and reliability.

Two major investigations were performed:

1. For nine different materials the relation between the dynamic stripping force and the amount of punch-die clearance was established.
2. The effect of punch wear on the dynamic stripping force was recorded.

Results

The results exceeded the expectations, and a high degree of relationship between the characteristics of the recorded load curves and the amount of punch-die clearance was established (see Figures 2B-7 through 2B-22). The same was found to be true for punch wear, when it is caused by material buildup near the cutting edge of the punch.

Several pertinent points were found to be worthy of consideration and further research:

1. A rectangular shape of the load curve for the dynamic stripping force indicates a favorable punching condition. The more triangular the form of the curve becomes, the more material buildup occurs near the cutting edge of the punch, and the worse the general punching condition.

2. The characteristics of the insides of the holes are directly related to the size of the dynamic stripping force. It is possible to develop theoretical explanations for most of the characteristics of the dynamic stripping force.

3. A microscopical analysis of the area near the cutting edge of the punch reveals interesting aspects of punch wear. If this wear is caused by material buildup near the cutting edge, then the load curve clearly shows this condition with an increase of the terminal stripping force values.

4. A tentative theory can be developed which provides an explanation for the material buildup near the cutting edge of the punch.

5. This method of investigating punching conditions opens a wide field for further research along the same lines. All recordings of load curves for the dynamic stripping force should be supported by investigations of the insides of the punched holes and of the cutting edges of the punches.

2C: Blanking Operations

Blanking is the process of cutting or shearing, from sheet metal stock, a piece of metal of predetermined contour to prepare it for subsequent operations. The piece of metal resulting from this operation is commonly called a blank.

ANALYSIS OF PARTS

The starting point for setting up a given blanking operation is an analysis of the part itself. Certain factors concerning the part will have a definite bearing on the type of blank that should be selected and on the kinds of tools and equipment that may be used to produce that blank. To determine what these factors are and what their effects will be, it will be necessary to arrive at answers to the following questions:

1. What is the general classification of the part from a metalstamping point of view?
2. How big is the part?
3. How thick is the material?
4. Where, in the product, will the part be used?
5. What processes will be required to shape the part?
6. What is the production volume of the part?

A consideration of the types of stampings used in automobiles will readily indicate that the classification of the part will fall into one of two groups. It will be either a formed part or a drawn part. Determining which type of part is involved will narrow down the choice of which type of blank to use.

The size of the part will influence the selection of tools and equipment for blanking. Blanks for most formed parts generally can be produced by the use of dies. Some other means may be required for producing exceptionally large blanks due to difficulties in preventing distortion of large blanks in dies.

The thickness of the material may be a determining factor in the selection of the method of producing the part. Generally speaking, most heavy-gage material is formed rather than drawn. This will have a definite bearing on the type of blank to be used.

The point of use of the part in the product also has significance in blanking operations. A part that could possibly be shaped by forming will be drawn if it is to appear on an exterior surface of a car. This is done to ensure a wrinkle-free part. The reverse situation could possibly occur for an interior part. A decision to draw or to form will be necessary prior to selection of blank design and blanking methods.

The processes required to produce the part, in addition to determining the type of blank to use, will influence the choice of dies and other blanking tools. Also influenced will be the form of material from which the blank will be cut. The production volume of the part will have a significant bearing on how expensive a tooling arrangement for blanking can be afforded. This in turn will affect the type of blank and the form of material selected.

The prime objective in planning a blanking operation is to select a blank that will produce a part of prescribed quality, within reasonable cost, that can be produced in the quantity required to maintain production schedules. The blank must be designed to provide enough metal to shape the part and at the same time prevent excessive scrap loss due to trimming.

DETERMINING BLANK DESIGN

The factors involved in shaping a part must be considered in arriving at proper blank design. As noted earlier, either drawing or forming will be used to provide the shape of a part. Each of these shaping operations has its own set of implications for blank design.

Blanks for Drawn Parts

When a part is drawn, deformation occurs over a major portion of its surface. The metal flows and bends and sometimes stretches. To accomplish this severe deformation, the metal must be held around the edges to prevent wrinkling and tearing. In the design of the blank for a drawn part, metal must be provided at proper places to permit the part to be shaped to specifications. Metal also must be provided for holding. To make proper allowance for these requirements, the design of the blank must be carefully worked out. Working out the design usually involves a number of steps. First, a tentative blank design is made based on past experience; the amount of metal for a trimming allowance is calculated, as are the metal requirements for a binder surface and for the shape of the part. A blank is cut out by hand on the basis of this design and is tried out in the drawing die. The resultant part is analyzed for conformance to specifications and for excessive scrap loss. If the part does not conform, or if scrap is too great, a new blank design will be tried. This process will be repeated until a proper blank is obtained.

Blanks for Formed Parts

In forming, the deformation that occurs is usually confined to the area immediately adjacent to a bend line. The part is usually shaped without the metal flowing or stretching. The shape is provided by bending portions of the blank in various directions to achieve the desired over-all shape, in much the same manner that a boy would create a hat from a sheet of newspaper.

It would seem from this that, for most

formed parts, a blank could be laid out that could be bent to the shape of the part without the necessity of trimming. Actually, this is what is attempted in designing blanks for formed parts.

If the formed part has a complex shape, the actual design of the blank usually follows the steps taken in designing a blank for a drawn part. A trial blank is cut and placed in the forming die. Required alterations are made in the blank until the proper blank design is achieved.

For some formed parts, trimming allowances must be added. This happens when, because of problems in achieving proper part shape, extra metal is provided for holding, or when two parts are formed from one blank and must be separated by trimming (Figure 2C-1).

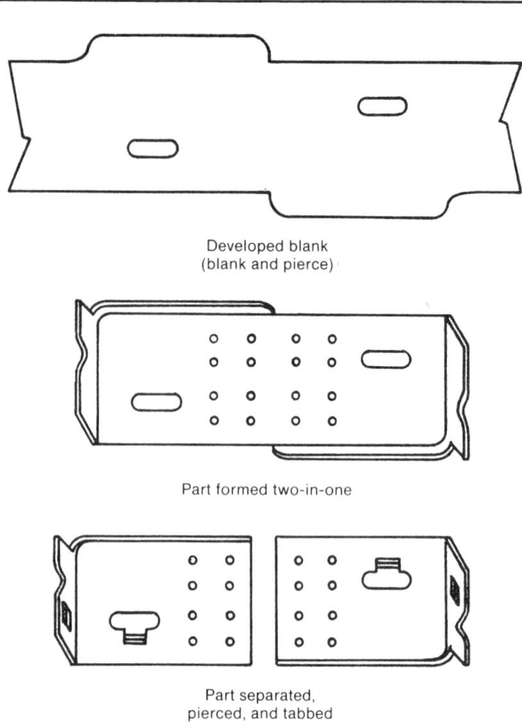

Figure 2C-1. Formed part: from blank to finished piece

The types of blanks used in both forming and drawing operations fall into four convenient classifications. These classifications are rectangular blanks, rough blanks, partially developed blanks, and fully developed blanks (Figure 2C-2 and Table 2C-1). This classification of blanks has permitted the application of some general rules to follow in setting up the blanking operation for a part. These rules deal with the type of a blank that should be used for a given type of part, the form of material from which the blank should be cut, and the method and equipment that should be used in making the blank.

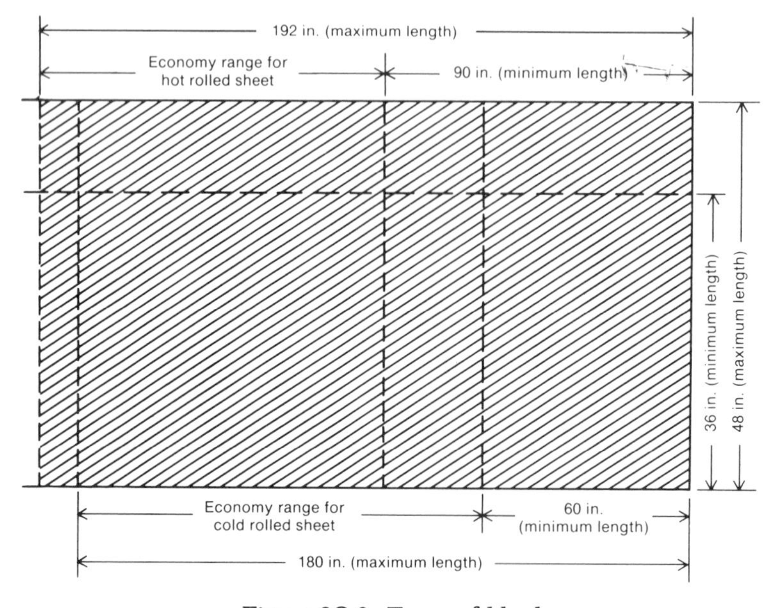

Figure 2C-2. Types of blanks

Table 2C-1. Summary of Application of Blanking Methods

Type of blank to be produced	Blanking method			
	Squaring shear	Push-through blanking die	Compound blanking and piercing die	Pierce, notch, and cutoff die
Rectangular blank	Good	Poor	Poor	Poor
Rough blank	Good. For straight-side symmetrical blanks; unsuitable for blanks with curved edges.	Fair. For small parts only.	Fair. For small parts only.	Good. For small and large parts.
Partially developed blank	Poor	Fair	Good. For small and large parts and parts with holes that can be pierced before drawing or forming.	Good. For small and large parts and parts with holes that can be pierced before forming.
Fully developed blank	Poor	Good. For small parts that do not require holes.	Good. For large parts and those requiring holes.	Fair. Precision sizing usually not obtainable with notch and cutoff die.

RECTANGULAR BLANKS

A rectangular blank is the least complex type of blank and the easiest to produce. As its name implies, it is rectangular in shape, having four straight sides. Its size is determined by the amount of metal required in the part and the amount of metal needed for trim allowances and holding. Rectangular blanks are usually restricted to production of drawn parts, although it would be possible to produce symmetrical, simple channeled formed parts from such blanks.

Only a portion of a rectangular blank appears in the finished part, because the outer edges of the blank provide surfaces for holding during drawing and are trimmed away. A fuel-tank half for an automobile is an example of a part made from a rectangular blank.

Why Rectangular Blanks Are Used

While the rectangular blank is somewhat limited in its application to various types of parts, certain definite advantages can be obtained through its use. These advantages result primarily from the type of equipment used in making rectangular blanks.

A squaring shear is the most prominent type of equipment used in making rectangular blanks. This machine consists of an appropriate frame and a moving cutting blade which acts on metal positioned over a lower cutting edge. A blank is produced at each stroke of the blade, which can be adjusted to operate at a very high speed. Thus a rectangular blank can be produced at a high rate of speed. In addition, a squaring shear can be set up with automatic feed and cutoff, resulting in a very low direct labor cost per piece.

The squaring shear can be adjusted to accommodate blanks of varying length with very little difficulty. A wide range of blank widths also can be accommodated. The ease of adjustment of the squaring shear makes it a very versatile machine. This makes rectangular blanks advantageous, because very little time is involved in setting up for their production.

One of the chief advantages of the rectangular blank is that no die is required. This results in a very considerable savings in the cost of establising a blanking operation.

One of the unfavorable aspects of the rectangular blank is the large amount of waste material that results from its use. Except in the case of a rectangular-shape part, such as a gas tank, a sizeable portion of a rectangular blank is lost to immediate production in the form of unusable scrap. On low-production jobs, the scrap loss is often offset by a reduction in die costs. In other jobs, however, especially high-production jobs, scrap loss is often one of the most important factors. For this reason, the choice of a rectangular blank should be weighed carefully in terms of the cost of scrap and its relation to the savings possible through use of the square shearing method.

Materials Used in Making Rectangular Blanks

The form of material used in producing rectangular blanks is an important consideration in setting up blanking operations. Two forms of material are generally used; sheet stock and coil stock (Figures 2C-3 and 2C-4).

Sheet stock may be used for producing rectangular blanks for low-, medium-, and high-volume requirements. Where this form of stock is used, manual operation of the squaring shear is usually involved. A package of sheets is delivered to the squaring shear by crane or lift truck. Each sheet is removed from the bundle and fed manually to

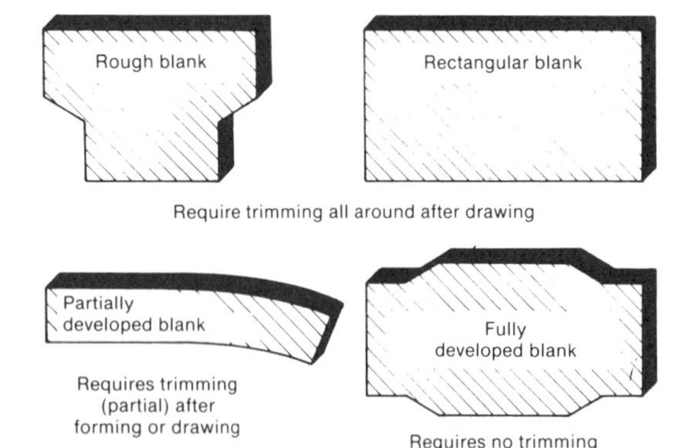

Figure 2C-3. Preferred sizes for ordering sheet stock

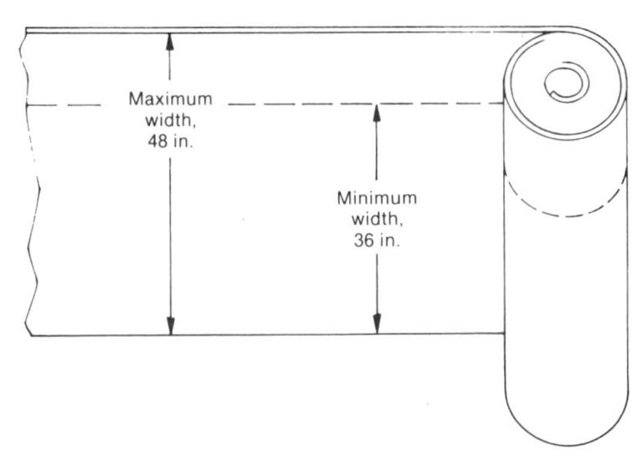

Figure 2C-4. Preferred sizes for ordering coil stock

the shear where it is cut to a predetermined layout.

Using the proper size of sheet for making blanks is important from the standpoint of material cost. First, a sheet should be of a size that will yield the most blanks with minimum scrap loss. A no scrap layout is most desirable.

Second, the size of the sheet should be planned with consideration of the variations in steel prices that are based on width and length of sheet. The most economical sizes of hot rolled low-carbon steel sheet range from 36 to 48 in. in width and from 90-1/16 to 192 in. in length. For cold rolled low-carbon sheet, the economical range of widths is the same as for hot rolled sheet. The length range, however, is from 60-1/16 to 180 in. The purchase price for sheet outside these size ranges is greater because of additional operations required at the steel plant. Narrower sheet must be supplied by adjusting the mills or by slitting wider sheet. Greater widths require special adjustment of the reducing mills. Lengths outside the economy size range cost more because of the increased effort involved in operating the flying shear which is used to cut the sheet.

Coil stock, when used, should also be ordered with consideration of the relation of price to size. Ordinarily, coil length does not affect price. Width, however, does, and the most economical width range is from 36 to 48 in.

Coil stock is used for producing rectangular blanks for high-volume requirements.

The coil is delivered by crane and clamped between the arbors of an uncoiling machine. The coil is passed to the shear automatically through feed rolls. The stock advances to a stop on the back of the shear and the cut is made by the blade, which is automatically activated by a limit switch. This type of operation is very economical from the standpoint of direct labor costs, because manual handling is greatly reduced and production is greater.

ROUGH BLANKS

A rough blank is a workpiece cut roughly to the size and general contour of the finished part, but having a margin of extra metal completely around the trim line. This extra metal is supplied for holding in subsequent operations and is cut off in the trim operation. A rough blank may be cut in a number of shapes. It may have straight sides and be in the shape of a triangle, diamond, trapezoid, hexagon, octagon, or even a "T". It also may have a combination of straight and contoured sides, giving it a nonsymmmetrical shape conforming to no particular shape classification.

Rough blanks are used for making drawn parts. The surplus metal outside the trim line permits the binder surface of the die and the blankholder to restrain the edges of the blank for desired metal flow and/or stretching action. In rare instances, a rough blank may be used for a formed part that presents difficult metal-holding problems, such as a part with a difficult contoured flange.

Why Rough Blanks Are Used

The chief purpose of using rough blanks is to obtain material economy. As compared with rectangular blanks, rough blanks achieve this purpose very well. Because the blank conforms roughly to the shape of the part, less metal is trimmed away as scrap. A rough blank, properly nested, will also require less metal per piece than a comparable rectangular blank. ("Nesting" refers to the layout of the blank in the sheet or coil stock.)

Another advantage of rough blanks is that they generally can be handled more easily through successive operations than can rectangular blanks. The binder surface of a rough blank is generally smaller, and it is usually free from extended protruding corners. This permits easier loading of the part in operations subsequent to the initial draw, and reduces the hazard of protruding corners being struck against press or handling mechanisms.

How Rough Blanks Are Made

Depending on the needs of the situation, one of two methods may be used in making a rough blank. For blanks having symmetrical shapes and straight edges, such as triangular, diamond-shape, and trapezoidal blanks, the squaring shear may be used. For nonsymmetrical parts, and for some symmetrical parts, presses are used with dies of the cutoff type. These dies are sometimes called progressive blanking dies.

A *plain cutoff die* (Figure 2C-5) cuts the blank in two or more steps in a manner similar to the squaring shear. The die set includes upper and lower shearing steels. In operation, the stock is moved forward to a predetermined stop, the press is cycled, and the part is cut off. Each cut creates the final edge of one blank and the initial edge of another blank. One of the principal differences between a progressive cutoff die with a die and press and a squaring shear is that the cutoff die often produces an irregular cut while the squaring shear always cuts on a straight line with simple shear blades.

The cutoff die is often used to produce two blanks simultaneously. This is made possible by incorporating two sets of cutting steels in the die. When this method is employed, one blank may be ejected off the end of the die and the other off the side of the die for effective disposal. Feeding must be arranged to advance the coil or strip as required.

The *notch and cutoff die* is also used for making rough blanks. This die is similar in operation to the plain cutoff die. However, two separate work stations are incorporated in the notch and cutoff die. In the initial work

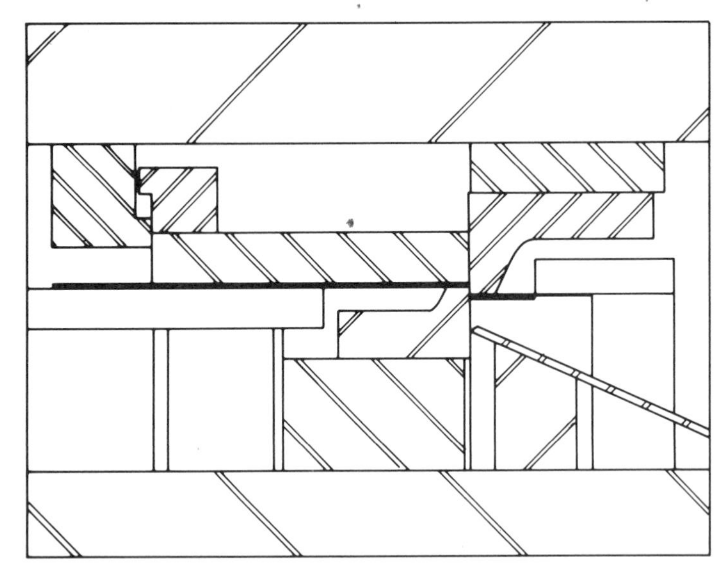

Figure 2C-5. Plain cutoff die

station, notching punches are mounted on the upper shoe with mating die sections on the lower shoe. These punches may be designed to notch triangles, semicircles, and many other blank-edge shapes that would be difficult or impossible to produce in a cutoff die alone. The notching tools provide only a part of the shape of a blank's edges; the remainder of the shape is supplied by the cutoff tools. The cutoff takes place in the last work station. The cutoff tool is usually lined up with the apex of the notch to ensure a clean cut.

Pierce, notch, and cutoff dies (Figure 2C-6) are used in some cases for producing rough blanks for formed parts. The major characteristics of this type of die will be discussed in relation to developed blanks.

The presses used for the foregoing progressive-type blanking operations are usually of the gap-frame inclinable variety for easy feeding of stock and quick disposal of blanks. For some large blanks, such as for fenders and quarter panels, where pressure requirements warrant, large presses of the straight-side type are used.

Materials Used in Making Rough Blanks

The principles cited for selecting an economical form of material for making rectangular blanks apply equally to rough

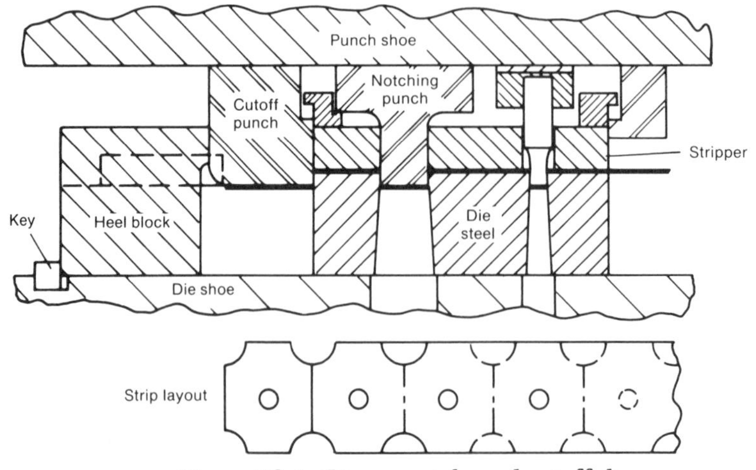

Figure 2C-6. Pierce, notch, and cutoff die

blanks. In fact, rough blanks require special consideration of layout for material economy. Since rough blanks are often nonsymmetrical, they must be planned and designed carefully to achieve material economy. The choice of strip or coil stock should be made with a view toward volume requirements and the reduction of offal (Figure 2C-7).

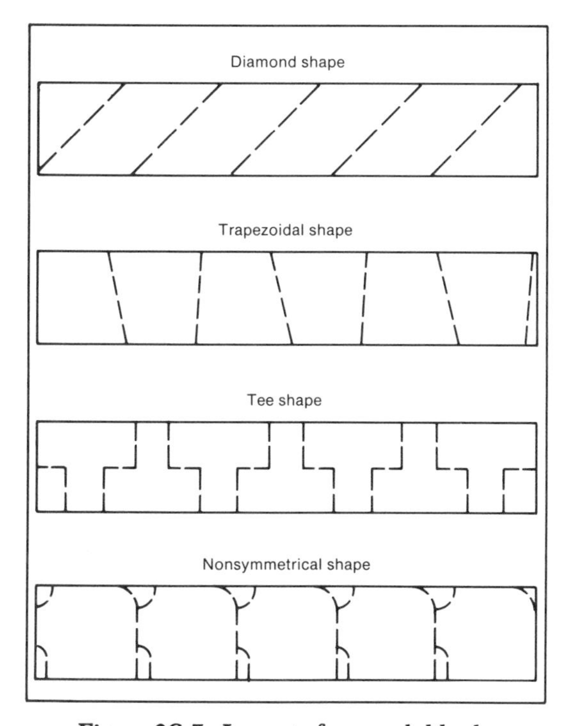

Figure 2C-7. Layouts for rough blanks

DEVELOPED BLANKS

A developed blank may be described as a blank of any shape, symmetrical or nonsymmetrical, that is precision sized so that no trimming is required after the part has been shaped. Developed blanks are used almost exclusively for parts that are formed. Practically all stampings used in chassis frame construction are in this category, as well as many internal parts of automobile bodies and front-end assemblies (Figure 2C-8).

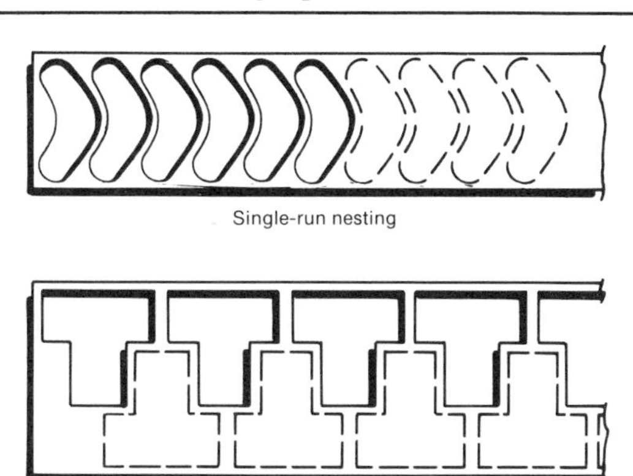

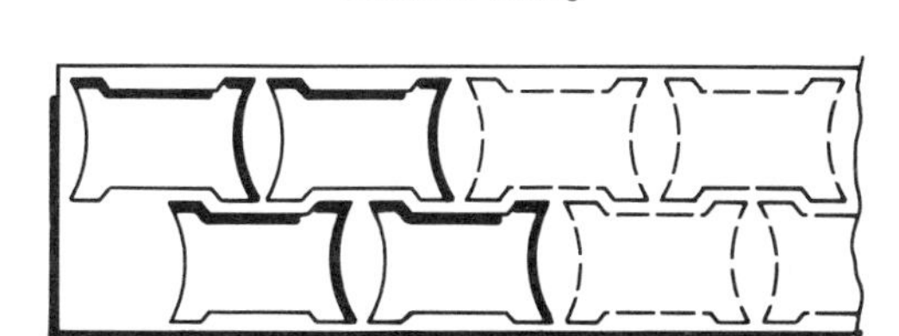

Single run 4 at a time

Figure 2C-8. Layouts for developed blanks

Why Developed Blanks Are Used

Developed blanks offer greater opportunities for savings in material and operating costs than any other type of blank used in stamping operations. There is no scrap resulting from trimming, and the need for expensive trim dies is avoided, where developed blanks can be used. Developed blanks cannot be used for drawn parts because of the requirement for extra metal for holding purposes in drawing.

Some scrap does result when developed blanks are used. This scrap is generated when the blank is cut from the parent stock. But, by proper nesting, scrap can be minimized. The scrap can also be used in production of blanks for small parts.

How Developed Blanks Are Made

A developed blank having mainly contoured edges is made in a push-through blanking die mounted in a suitable press. One advantage of this type of setup is the ease with which disposal of the blank is accomplished. The blank is merely pushed through an opening in the lower die shoe and the press bolster to a container below. In

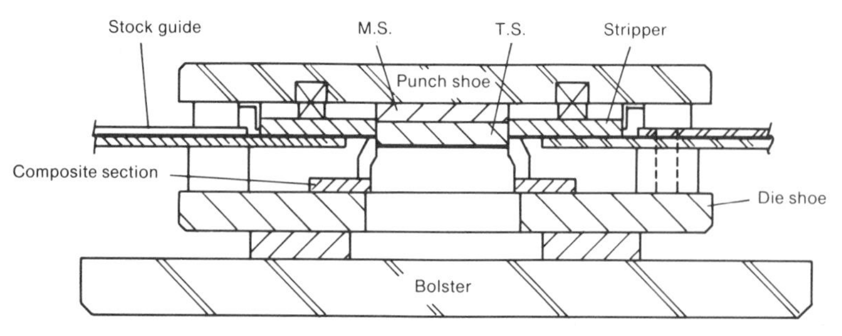

Figure 2C-9. Push-through blanking die

some instances, the blank is allowed to fall onto a conveyor which then carries it to a subsequent operation.

A *push-through blanking die* (Figure 2C-9) may be defined as one in which a sheet of metal to be cut is laid over a sharp-edged die opening, and a punch, with cutting edges to fit the die, is forced through the sheet metal into the die opening.

The punch assembly consists of a tool steel punch and a stripper mounted on a shoe with guide bushings. The stripper is provided to remove the scrap from the punch on the upstroke of the press.

The die assembly consists of a conventional die shoe which is equipped with guide pins to ensure accurate alignment. The guide pins are important in making a developed blank because of the precision sizing necessary. Careful alignment is also necessary to prevent damage to cutting steels. Additional components of the push-through die are the composite steels that comprise the cutting edges, backing plates, and stock guides.

Push-through dies are not considered practical for punching out blanks for large parts because a large die opening would allow the stock to sag and become distorted before the cut was made.

For low- or medium-volume work, a single push-through blanking die may be used with offal or sheet stock hand fed to the die. For high-volume work, double dies are often used to make two blanks, either separate or connected, with each stroke of the press. In some cases, dies for two different parts may be used in the same press at the same time. Coil stock and automatic feed mechanisms are usually used when production requirements are high.

A developed blank may also be made in dies other than the push-through type. These dies are the compound blanking and piercing die, and the notch, pierce, and cutoff die.

The *compound blanking and piercing die* (Figure 2C-10) performs more than one operation. Both blanking and piercing are completed in one stroke of the press and in one station in the die. The compound blanking die is situated in an upside-down position with the sized cutting section on the upper shoe. This upside-down position provides support for large parts to prevent buckling or sagging, permitting large developed blanks to be made. However, many small developed blanks are also produced in this type of die.

Combining piercing and blanking in one operation enhances the economy of a press operation. However, care should be taken that holes pierced will not become undesirably deformed in later forming operations.

The important parts of the punch assembly are the die section, the piercing-punch assembly, the stripper, and the knockout pad or pin.

The principal components of the die assembly are the tool steel punch and the stripper. The punch includes die buttons for receiving the piercing punches that are mounted on the upper shoe.

The actions that take place in a blanking and piercing die may occur simultaneously or in sequence. Piercing usually occurs first because the metal can be held more firmly before blanking takes place. In some cases, however, the blanking may be done first, with the face of the punch being used to hold the material while piercing takes place.

The *notch, pierce, and cut-off die* can also be used for producing a developed blank for a forming operation. This die is a modification of the progressive blanking die and includes the piercing function where holes produced will not be distorted in later forming operations.

Presses used in making developed blanks come in a range of sizes proportionate to the size of the blank being made. Most are of the straight-side type to ensure accurate alignment of cutting steels.

Materials Used in Making Developed Blanks

In making a developed blank, the choice of the form of material to be used is a highly important factor. The choice will be influenced by the size of the blank and the production volume of the part.

Offal stock is used for small, low-volume parts — that is, where requirements per 16-hour day are less than 1400. Offal is restricted to such work because of the difficulty in handling the unevenly shaped pieces. It is used only with blanking and piercing, blanking and forming, or push-through blanking dies. The progressive type of die cannot use offal for blanks because of difficulties in moving the part through such a die. The use of offal in all possible applications is desirable from the standpoint of material economy, but must be analyzed very closely.

Sheet stock is used for low-volume work, and for medium-volume work where the requirements are from 1400 to 2000 parts per 16-hour day, or for large parts. All sizes of blanks may be cut from this type of material. It may be used in push-through blanking dies or in compound blanking and piercing dies. It can also be used in progressive blanking dies. *Heavy-gage* sheet stock is used for high-volume work where requirements are over 2000 per day, because such metal is of too-heavy gage and coils would be too short. Stock of this type is used for making generator frames, chassis frame parts, and other heavy-duty applications.

Coil stock may be used for medium-volume work in a push-through, blanking and piercing, or progressive blanking die. Coil stock is used on all high-production stampings whenever it is practical to secure and use it in such form.

The layout of material for a developed blank is another important aspect of establishing an effective operation. The form of a developed blank is usually complex, and nesting it economically in a sheet or coil is sometimes more difficult than for a rectangular or rough blank.

One method of arriving at economical nesting is to prepare several paper templates of the blank and place them on the stock. By shifting them around on the stock, the nesting resulting in the least scrap may be determined.

If, in using strip stock, two nestings show equal promise, it may be necessary to lay out the entire strip to determine which nesting

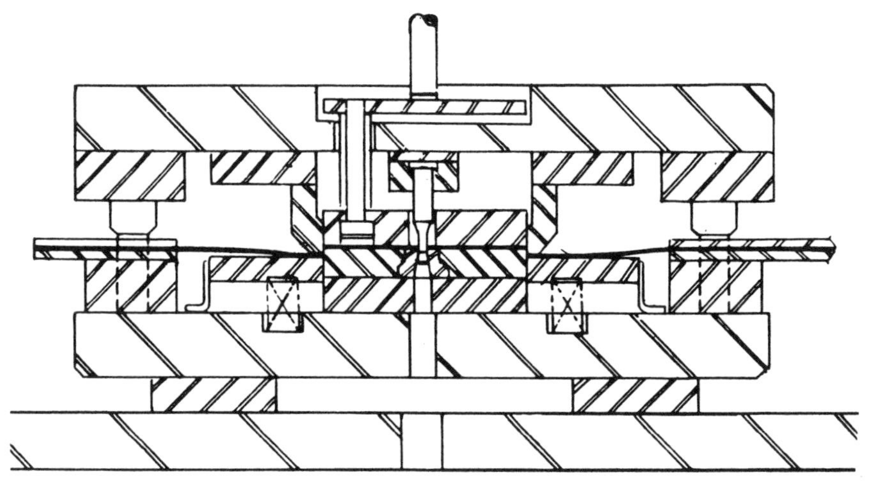

Figure 2C-10. Blanking and piercing die

will yield the least scrap at the beginning and end of the strip. End scrap is less important when coil stock is used because of the greater length of material. The method selected for cutting the blanks from strip or coil may also provide means for material economy. By feeding a strip through the die twice, it may be possible to obtain blanks from the scrap left after the first pass. Blanking in multiples (cutting more than one blank at a time) may offer additional material savings. It is generally practical to blank multiples from the widest stock possible. It is easier to feed wide strips of short multiples than it is to feed narrow strips of long multiples. Use of excessively wide strips must be avoided to escape payment of premium steel prices.

One method of determining the best nesting arrangement among several choices is to calculate the total area of stock required, in square inches, to produce each piece. The scrap involved is included in the calculation. To determine the area for each piece from a given layout, multiply the distance between the centers of two blanks by the width of the strip. A comparison of the figures obtained from analyzing several nesting possibilities by this method will readily reveal the best course to follow.

Certain basic rules in blank development for push-through die work should be followed in order to ensure production of blanks having clean-cut edges. These rules pertain to the amount of scrap that should be left between developed blanks to ensure a clean cut. For stock lighter than 11 gage, at least 1/8 in. of stock should be allowed between the blanks, and at the edge of the stock. For stock of 11 gage or heavier, at least 3/16 in. of stock should be allowed for scrap. This scrap allowance is necessary for a clean cut and to prevent the edges of the blank from being pinched to a sharp edge or to be drawn down into an excessive burr. Sometimes where edge conditions are less important and blanks are fairly symmetrical, no scrap allowance is necessary. If such is the case, blanks may be laid out adjoining one another.

In some cases, the grain direction of the metal may be a factor in laying out blanks. This factor is prevalent in electric-furnace and bessemer steels. There is no appreciable grain direction in open-hearth steel. Where definite grain direction does exist, however, blanks for forming operations should be cut in such a way that bends will be made across the grain by at least 30°. Bends made parallel with the grain direction will cause fractures at the radius of the bend.

Balancing the cuts in blanking operations is another important factor in ensuring effective blanking operations. Where cuts are unbalanced, shifting of the cutting steels is apt to take place. This will result in malformed blanks and excessive wear on cutting tools. This problem may be overcome, in some instances, by cutting blanks two at a time. The two unbalanced cutting edges will offset one another to create a balanced condition.

Wherever possible, blanking operations should be set up so as to safeguard the ease with which subsequent operations can be accomplished. This again may call for blanking of two parts at a time and leaving them fastened together to provide balanced conditions in forming or piercing operations that follow.

PARTIALLY DEVELOPED BLANKS

A partially developed blank (Figure 2C-11) is defined as a blank which is precision sized on two opposite sides of a part with the remaining two sides providing excess metal for holding purposes. This type of blank is used for drawing of parts where holding is required on only two sides. Roof rails are typical parts for which partially developed blanks may be used. There are only a few automotive stampings of this type, and hence partially developed blanks are rarely used. Where they are used, however, the complexity of trimming is slightly reduced.

A partially developed blank is usually made on a cutoff die, a notch and cutoff die, or a compound blanking die.

The general rules for selecting forms of material for partially developed blanks correspond to those cited for developed blanks.

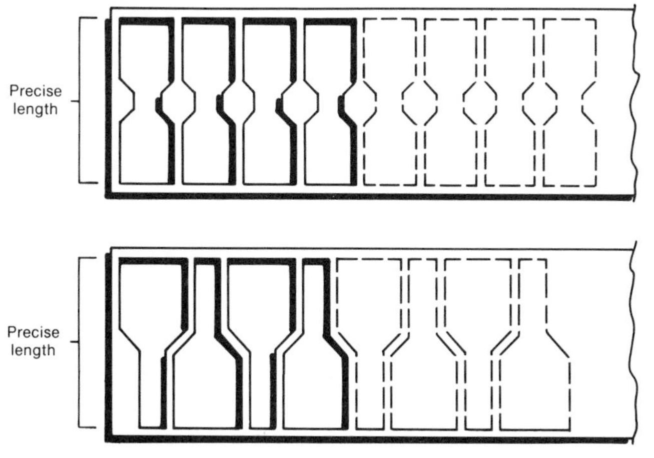

Figure 2C-11. Layouts for partially developed blanks

BLANKING CLEARANCES AND SHEAR

Certain mechanical problems that arise in cutting of metals also must be considered in establishing blanking operations. These mechanical problems include clearances between cutting steels and the pressure requirements for performing the shearing operation.

Clearances

In blanking, the die steels are built to the dimensional specifications of the blank, and any clearance required is machined on the punch. Clearances must be sufficient to avoid secondary shearing and close enough to prevent excessive burrs and deformation of blank edges. Pressure requirements for blanking are also affected by clearance. Usually, the greater the clearance provided the less the required pressure. For best quality in blank edges, cutting tools should be kept sharp.

Shear

Shear is ground on the die section in blanking operations. The sheet metal in a stamping operation always assumes the shape of the punch. Hence, if the shear were on the punch, the blank would be deformed. By providing shear on the die, the blank is produced flat and the scrap is deformed.

Shear is provided to reduce pressure requirements by permitting the cut to be made a little at a time. Shear provides the additional benefit of facilitating accurate blanking, because the reduction in pressure reduces the chances of distorting the part. Balanced or unblanced shear may be provided depending on the cut to be made and the precision required in the part.

SUMMARY

Blanking is the process of cutting or shearing, from sheet-metal stock, a piece of metal of predetermined contour to prepare it for subsequent operations. The nature of a blank, and the choice of material and processes used to produce it, will depend on certain factors concerning the part to be made. Hence, the starting point in setting up a blanking operation is an analysis of the part involved. Pertinent information to be obtained in this analysis includes whether the part is to be drawn or formed, the size of the part, the thickness of material, where the part will be used in the product, and the production volume of the part.

Determining the design of the blank is another step in developing a blanking operation. If the part is to be drawn, the design must allow material for the binder surface, the trim allowance, and the metal requirement for the part itself. For a formed part, the blank should be designed, whenever possible, to allow the part to be formed without the necessity of a trimming operation.

The blanks used for any operation will fall into one of four classifications: rectangular blanks, rough blanks, developed blanks, or partially developed blanks.

A rectangular blank, as its name implies, is rectangular in shape, having four straight sides. It is almost always used for making a drawn part. Only a portion of a rectangular blank appears in the finished part, because the outer edges of the blank provide surfaces for holding during drawing and are later trimmed away.

A rectangular blank is generally cut from the parent stock on a squaring shear. This process eliminates the need for a die. It is an economical method of making blanks because it is a rapid operation and involves a low direct labor cost per piece.

Rectangular blanks can be cut from either sheet or coil stock. When coil stock is used, automatic feeding is possible. Care should be taken to effect material economy in laying out rectangular blanks in sheet or coil stock. The size of sheet stock that can be used most economically varies from 36 to 48 in. in width and from 90$^1/_{16}$ to 180 in. in length for hot rolled steel. For cold rolled steel, the width is the same, but the economical length range is from 60$^1/_{16}$ to 180 in. Coil stock should be used, where possible, in the 36- to-48-in. range.

Rough blanks conform generally to the shape of the part involved, but have a margin of extra metal completely around the trim line. This extra metal is used for holding in drawing. Thus, rough blanks are used primarily in drawing operations.

Rough blanks having symmetrical shapes and straight sides may be made in a squaring shear. For rough blanks having contoured sides, dies of various types are used in conjunction with suitable presses. The dies used for making rough blanks include cutoff dies, notch and cutoff dies, and notch, pierce, and cutoff dies. Presses used in rough blanking include both gap-frame and straight-side presses of the single-action type.

Since rough blanks are often nonsymmetrical, they are designed and laid out carefully to achieve material economy.

A developed blank may be described as a blank of any shape that is precision sized to yield a finished part without using a trimming operation. A developed blank may be cut on a push-through blanking die, a compound blanking and piercing die, or a notch, pierce, and cutoff die.

Proper nesting of the blank is particularly important in making developed blanks. The best arrangement can be obtained by using paper templates of the part and arranging them in the most economical pattern on the material to be used.

A partially developed blank is defined as a blank which yields a part that requires trimming on only two sides. These blanks are made on cutoff dies, notch and cutoff dies, or compound blanking dies.

In making dies for blanking, clearances are machined on the punch and shear is ground on the die.

2D: Blanking and Shearing

The most common group of metalworking operations includes blanking, piercing, parting, shearing, notching, trimming, perforating, etc. In all of these operations, the metal is stressed in shear between approaching cutting edges until a fracture occurs. The equipment and maintenance engineer will wish to arrive at the maximum blanking or shearing pressure required for the various tools, the effect on this maximum pressure of shear on the punch or die (grinding of one or the other at an angle), and the power and flywheel energy needed for the work.

If the tools are ground flat and parallel and friction on the punch and die are neglected, the maximum pressure required to blank out a punching is normally the product of the shear strength of the metal and the cross-sectional area around the periphery of the blank. Thus the formula for maximum pressure (P) is:

$$P = LtS, \text{ or } P = \pi DtS$$

where S is the shearing strength of the material, t is the sheet thickness, and L is the length of the cut or the periphery of the blank. If the blank is round, L may be replaced by D, which is the diameter of the punching (or the sum of several diameters). The π is a constant and does not appear on the chart, because it is taken care of in the arrangement of the scales.

Figure 2D-1 is arranged from the above formula with its limits selected to cover the normal range of power-press (and squaring-shear) capacities. In reading this chart, any convenient ruler, sheet of paper, or other straightedge may be used. Lay the straightedge on the chart so that it connects the desired values of metal thickness and D or L. Mark or note the point where it crosses the centerline. Move the straightedge to connect this point on the centerline with the proper value for the shearing strength of the metal to be cut. The line through these two points crosses the shearing-pressure line at the tonnage required to punch the blank or hole. An example is shown in dashed lines.

The shear-strength values indicated at the left in the chart can be found in Table 2D-1. The data given may eventually be improved with closer control of materials, processes, and testing conditions. It is apparent, of course, that the harder any metal is, due to cold rolling or other causes, the higher is its shear strength or resistance to shearing.

As a flat punch progresses through a sheet of metal, it deforms the metal plastically to a point where the ultimate strength is exceeded and fractures start from the opposing cutting edges of the punch and die (see Figure 2D-2). If there is proper clearance between the tools, the fractures meet almost instantly and the fractured portion of the sheared edge appears quite clean. If the clearance is not sufficient, the first fractures do not meet, and secondary shearing takes place with a characteristically irregular and ragged appearance around the edge. When the cut is clean, a brightly burnished band around the edge of the blank indicates how far the punch had to penetrate before the fracture occurred. This is of interest in connection with Figures 2D-3 and 2D-4, because the work is entirely or nearly completed at this point.

Figure 2D-5 illustrates diagrammatically the effect of shear on the blanking load. The noun "shear" refers to grinding either the punch or the die at an angle, the other member remaining flat so as not to distort the product. The amount of the shear is the difference between the high and low points of the angular face, even though the punch may be ground down both ways from one or more high points.

The effect of shear is clearly to reduce the peak load of blanking by shearing a little work metal at a time instead of making the whole cut at once. The total energy required to do the job is not changed, being merely that a lower pressure is continued for a longer time. A small amount of shear (less

Table 2D-1. Properties of Metals(a)

	Theoretical yield points		Reduction in area (maximum annealed)	Shearing stress and per cent penetration to fracture				Tensile strength (nominal) annealed	Elongation in 2 inches (average) annealed	Elasticity Young's modulus	Modulus of strain hardening (tentative)	Bulk modulus volume	Annealing temperature (approx.) commercial	Forging tempera-ture (approx.)	Thermal coefficient of expansion	Specific gravity
	Minimum(b) (commer-cially annealed)	Maximum (severely cold-worked)		Annealed to soft temper		Cold-worked to temper noted										
	psi	psi	%	psi	%	psi	%	psi	%	psi	psi	psi	°F	°F	in./in./°F	lb/in.³
Aluminum, No. 2S commercially pure	(4-) 8,000	H 21,000	80	8,000	60	H 13,000	30	13,000	35-45	10,300,000	25,000	10,200,000	650		0.000,010	0.0963
Aluminum, No. 3S, Mn alloy	11,500	H 25,000	80	11,000		H 16,000		16,000	30-40	10,300,000	29,000		750			
Aluminum, No. 17S for heat treatm't	HT 30,000												630-650	400		
Aluminum, No. 52S, Cr alloy	21,000	H 37,000	80	14,000		H 36,000		29,000	25-30	10,300,000	38,000		650			
Brass, yellow, for cold working	(10-) 40,000	95,000	75	32,000	50	? 52,000	20	50,000	65	13,400,000	120,000	8,800,000	1100			0.303
Brass for forging, beta														12-1400		0.301
Bronze, Tobin alloy	25,000	120,000	53	36,000	25	1/4H 42,000	17	60,000	46		250,000					0.315
Copper	(10-) 25,000	62,000	65	22,000	55			33,000	68	14,500,000	81,000	17,400,000		16-1800	0.000,009	0.3195
Gold										11,600,000		23,200,000	6-800		0.000,008	0.6949
Iron, cast, gray	8,000 tension 30,000 compr.							23,000		12,000,000		13,900,000			0.000,005,5	0.260

Iron, cast, high-test	15,000 tension 40,000 compr.							45,000		20,000,000						
Lead	(1-) 3,000	4,000		3,500	50					2,470,000		1,100,000	below room temp.		0.000,017	0.4106
Monel metal	28,000	H112,000	63					70,000	30-50				11-1200			0.318
Nickel	21,000	H115,000	75	35,000	55			70,000	47	32,000,000		24,600,000	12-1300		0.000,007	0.3175
Silver										10,900,000		14,500,000	6-800		0.000,010	0.3791
Steel, 0.03C Armco	(20-) 35,000	70,000	76	34,000	60			42-48,000	48	30,000,000	100,000		13-1400			0.2834
0.15C	(32-) 55,000	110,000	60	48,000	38	61,000	25	57,000	35	30,000,000	150,000	23,200,000	13-1400	—	0.000,009,16	
0.50C	(55-) 70,000		45	71,000	24	90,000	14	85,000	25	30,000,000			13-1400	18-2100		
1.00C	90,000	145,000	20-40	115,000	10	150,000	2	120,000	10	30,000,000	200,000					
Steel, electrical high-silicon,				65,000	30											
Steel, stainless low-carbon for dwg. Enduro AA	50,000	120,000	60					80,000	25	30,000,000			13-1550	19-2100		
18-8	(35-) 50,000	165,000	71	57,000	39			95,000	57				13-1900	2100		0.285
high-carbon for cutlery, etc.			32					260,000	11				1450-1600	20-2100		
Tin	(2-) 4,000	4,500		5,000	40					7,260,000		7,260,000	below room temp.		0.000,015	0.2652
Zinc	(12-) 28,000		43	14,000	50	19,000	25	30-37,000	27	13,100,000		5,100,000	2-300		0.000,014,5	0.256

(a) Crane's *Plastic Working and Metals,* 3rd Edition, John Wiley & Sons, New York, 1941. (b) Figures in parentheses represent laboratory test minima which might prove misleading for estimating loads. All physical values are subject to some variations with testing methods and analysis of material.

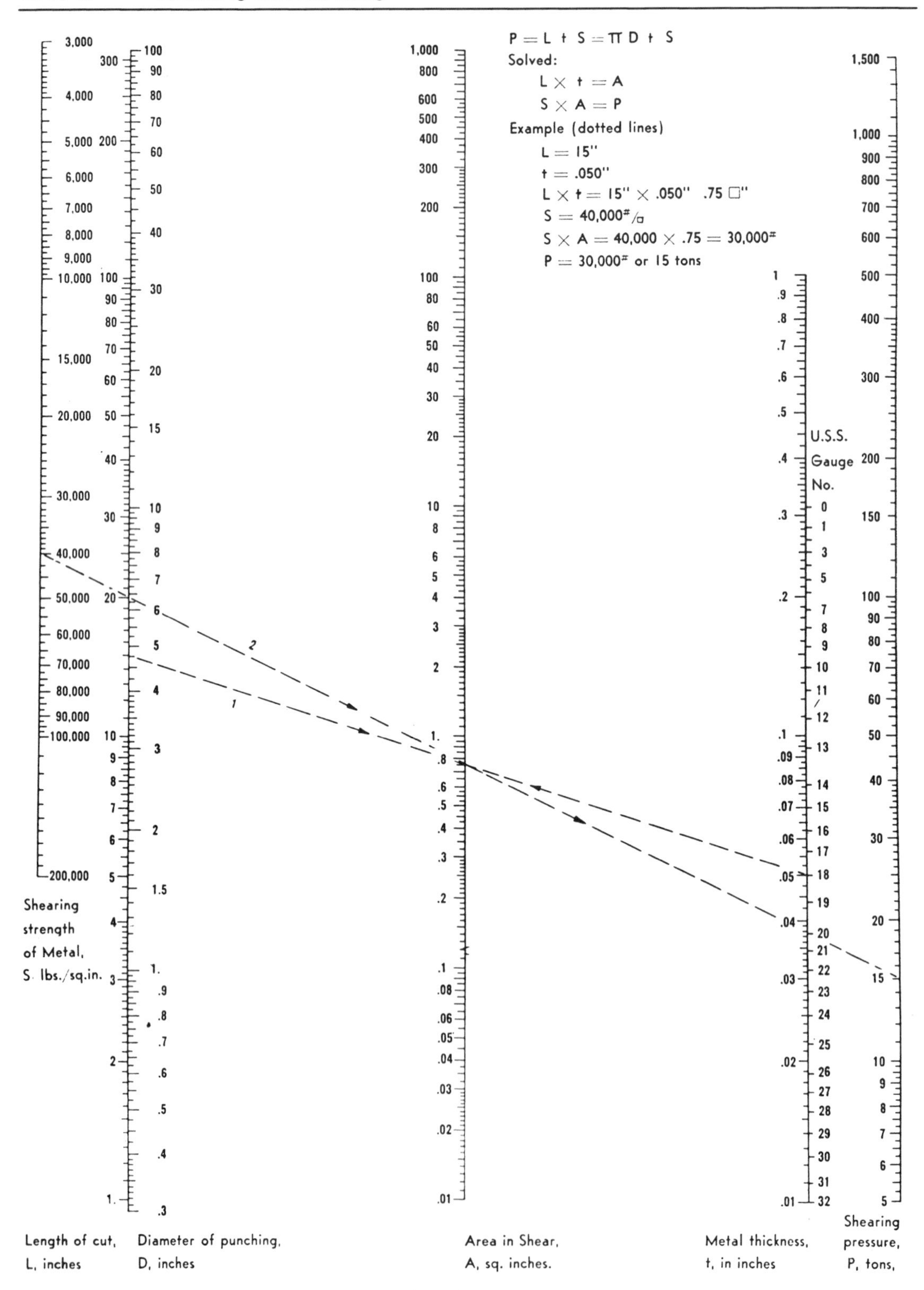

Figure 2D-1. Chart for determining blanking pressure

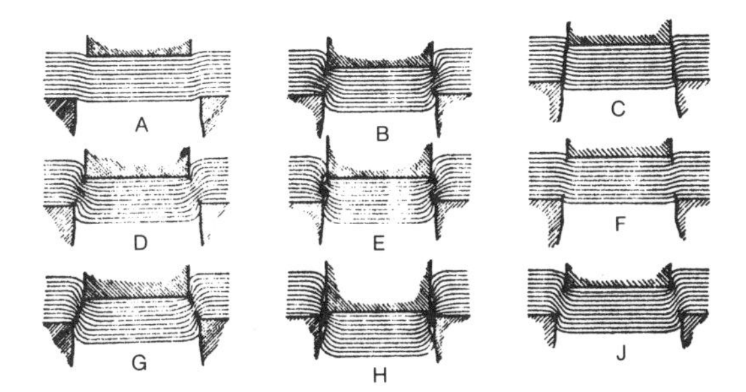

A, B, and C show ductile metal with ample clearance. D, E, and F show similar metal with insufficient clearance. G and H show hard metal with sufficient clearance. J shows effect of dull cutting edges.

Figure 2D-2. Progress of a punch through sheet metal, showing plastic deformation and fracture

than the distance a flat punch must travel to effect shearing) does not appreciably reduce the load, but does ease the snapback, especially of C-frame presses, which occurs when the whole load is released instantly by the fracture.

Figure 2D-3 is designed for use in approximating the extent to which the total or maximum load is reduced by shear on the tools. In the case of ample clearance and a clean fracture, the result obtained is high, because the calculation is based on the assumption that the working pressure without shear is equal to the maximum pressure recorded, throughout the working distance. Actually, however, the pressure rises gradually to the maximum and then usually falls somewhat to the point at which fracture occurs. The average pressure for the penetration distance, which would naturally be lower than the maximum pressure, would give an accurate result. An accurate pressure diagram is rarely available, and the chart in Figure 2D-3 is close enough for ordinary purposes. Note that it is always desirable to have considerable excess press capacity for blanking work for the sake of the tools.

In the case of insufficient clearance in the tools and a ragged fracture, secondary shearing occurs so that the pressure curve does not drop sharply. The amount of secondary shearing increases as the clearance becomes less and less. The square-root scale for per cent penetration is an arbitrary method of compensating for the effect of secondary shearing, but is ample for ordinary cases. In many cases, of course, the result will be high.

Figure 2D-3 is read from the top down, and, as indicated by the example in dashed lines, the third and fifth lines are used only as pivot points for the straightedge. The formula is:

$$P = \left(\frac{t \times \%\ \text{penetration}}{\text{amount of shear}} \right) \times P\ \text{max}$$

The thickness multiplied by the per cent penetration to effect shearing gives the actual working distance in inches. This value divided by the amount of shear on the tools in inches gives the proportion of the length of the cutting edge which is actually working at any instant (see Figure 2D-5). This value multiplied by the maximum pressure (without shear) from Figure 2D-1 gives the approximate maximum working pressure.

In checking power requirements or flywheel capacities, it is necessary to know the actual energy absorbed in doing the work. In the case of shearing, this is properly the working distance multiplied by the average pressure, or, graphically, the area under the pressure-distance curve. Because such curves are not easily obtained and therefore average pressure cannot be measured, we resort again to the approximate methods used above. Thus, in Figure 2D-4, the maximum pressure (P, from Figure 2D-1) is used in place of average pressure, and the working distance is taken as the product of the metal thickness, t, and the per cent penetration, using two scales for the latter. Then the energy or work, W, in inch-tons is:

$$W = T \times \%\ \text{penetration} \times P \times 1.16$$

The 1.16 adds a 16% allowance for machine friction. This can only be a general case, of course, because the arrangement and condition of machines vary widely.

It should also be noted that an allowance should be made for heavy stripper springs, when used, and for wall friction in pushing

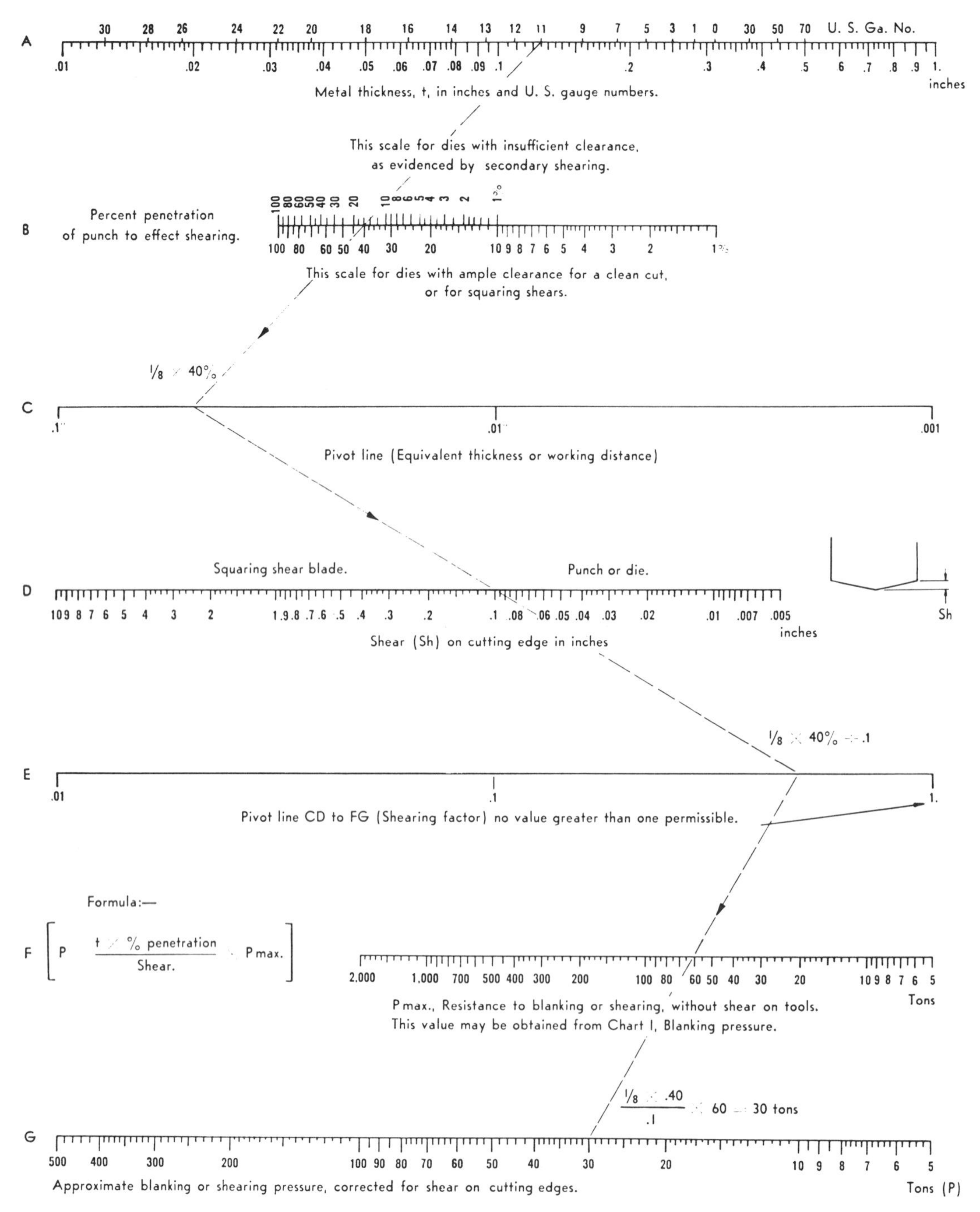

Figure 2D-3. Chart for determining effect of angular shear

slugs through the dies when there are long straight walls, as is occasionally the case.

In using Figure 2D-4, read thickness and per cent penetration first, to obtain the pivot point for the straightedge. Then set the straightedge to connect this point and the pressure value, to obtain the energy requirement.

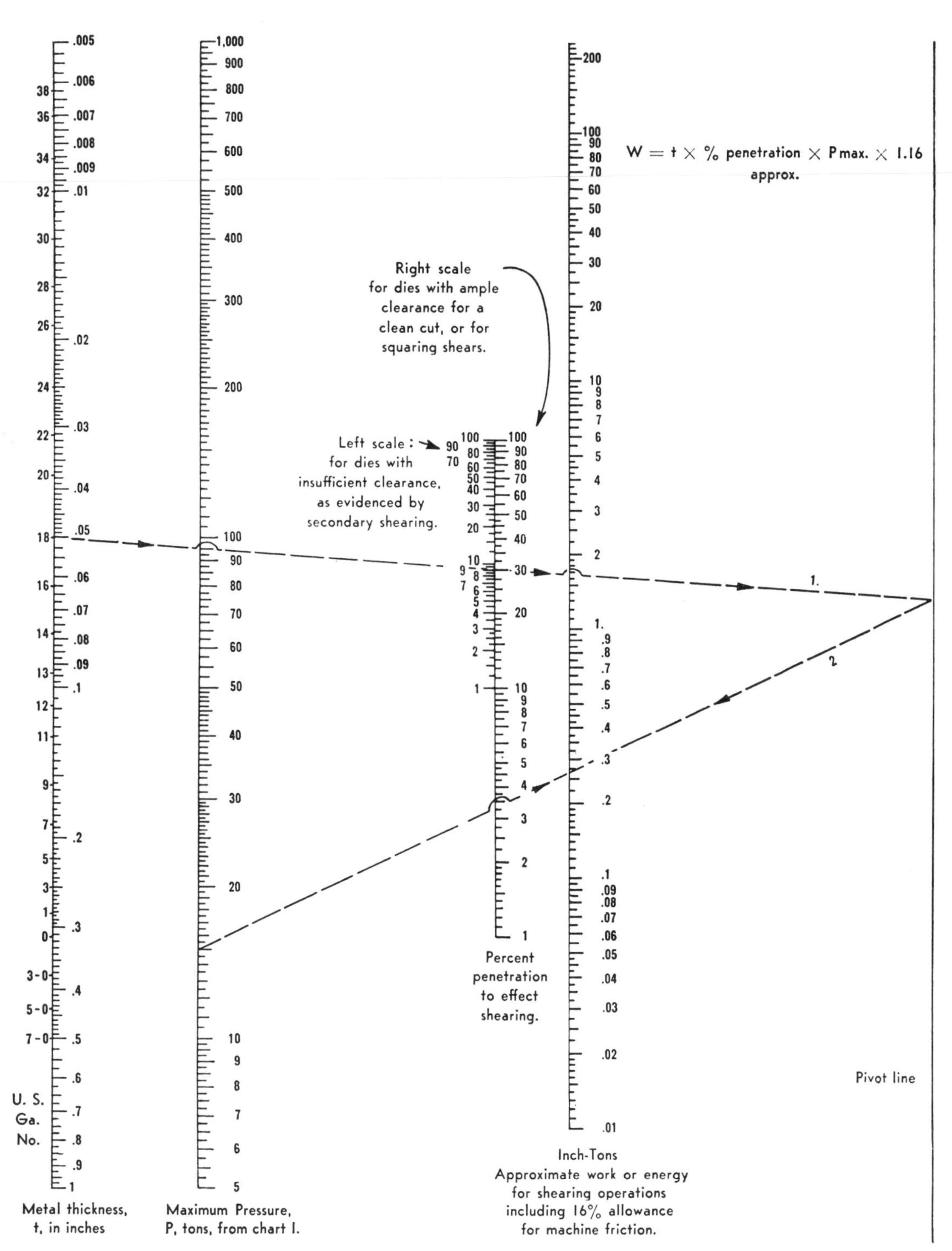

Figure 2D-4. Chart for determining energy required in shearing

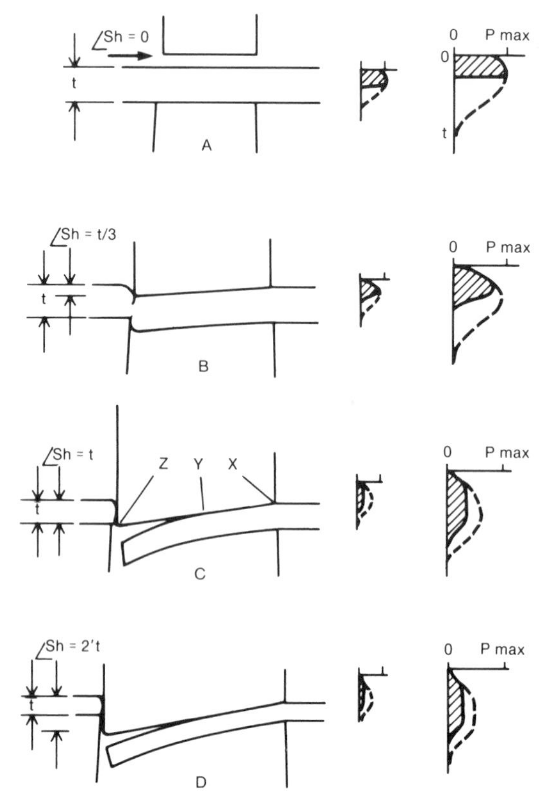

Different angles of shear on the punch (or die) reduce the amount of work it is doing at any instant.

Figure 2D-5. Effect of shear on blanking load

2E: Piercing Operations

Production of a sheet-metal stamping requires application of a variety of types of stamping operations. In the preparatory stages, blanking is used to cut out a workpiece of proper design and with sufficient metal to yield the desired part. This workpiece, or blank, is forced to assume the required shape by means of a forming or drawing operation. The specified edge dimensions of the part are obtained either in the blanking operation or in a trimming operation. But, in nearly all cases, a stamping requires more than sizing and shaping operations before it becomes a finished part. Holes, slots, tabs, or notches are incorporated in the part to bring it into conformity with specifications. The creation of holes or openings in sheet metal is done by the process of piercing.

Pierced holes serve many important purposes in the over-all design and use of a part. They provide means of fastening the part to other components of an assembly and sometimes provide means of making a component lighter in weight. They serve as locating points to facilitate subsequent stamping operations and provide access to oil holes, drain plugs, and other parts through firewalls and other panels of the car.

Although piercing is regarded as one of the least complex of stamping operations, the establishment of successful piercing operations at the Ford Motor Company requires careful planning by the process engineer. The great extent to which piercing is employed is, in itself, enough to support this conclusion. In addition, tools must be provided that will prevent excessive cost due to breakage. A good deal of standardization in tooling is called for. Operational sequences must be established properly in order to prevent deformation of pierced holes and the production of defective parts. In short, the success of a total stamping process is equally dependent on efficient piercing operations as it is on other types of operations.

TYPES OF PIERCING OPERATIONS

The various types of piercing operations employed in metal stamping (Figure 2E-1) derive their names from the kind of hole or cut that is to be made. Each of these holes or cuts is designed to accomplish a specific purpose in the design or use of the part.

Conventional Piercing

Piercing usually consists of driving a round flat-face punch through the part into a die section or button. The portion of metal under the punch is totally removed from the part. This piece of metal is called a slug. The resulting hole is normally used as a means of fastening parts together with rivets, bolts, or screws.

The majority of piercing operations are of this type, and the holes so produced cover a broad range of sizes. As a result, punches of various sizes are necessary to provide the tools required to meet piercing needs. This range of punch sizes has been modified by the application of punch-size standards. Care should always be taken to provide a standard-size punch wherever possible.

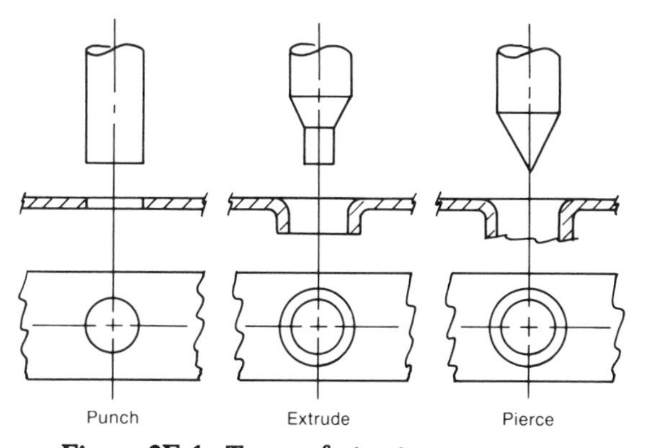

Figure 2E-1. Types of piercing operations

Piercing With a Pointed Punch

Another operation commonly referred to as piercing is the creation of a hole with a pointed or cone-shape punch. In this type of operation, no slug is removed. The hole results from a pushing apart and extrusion of the metal as the punch passes through it. This results in an extruded collar around the hole. The edges of this collar are ragged because of the nature of the break that takes place. The collar is sometimes tapped to provide a means of fastening another part with a bolt.

Pierce-and-Extrude Operations

A pierce-and-extrude operation consists of first punching a hole in the part and then extruding the edges of the hole to provide a circular flange around the enlarged hole. This operation calls for a punch of special design and special features in the die set. The end of the punch is flat, to break out the slug, and a beveled shoulder higher on the punch extrudes the hole edge and enlarges the hole as the punch passes through. Because the hole has been cut larger by the punch end, the edges of the extruded portion are cleaner and freer from cracks than in the case of the pointed punch. However, a certain amount of raggedness exists. Where a very smooth edge is needed, the hole may be reamed before extrusion. Extruded holes of this type may be used to provide spacing between panels; they may be used for strengthening holes in parts (such as hinges) that are mounted on rods or bolts; or they may be used as bosses or locators on which to mount other parts.

Slotting

Slotting is used to provide holes of special shape (that is, noncircular), such as elliptical holes or rectangular holes. These holes may be required on some parts to allow for adjustment of the fit of a part in an assembly. A slotting punch requires a special design. The cutting portion must be shaped in accordance with the shape of the hole to be produced.

Countersinking

Countersinking is also done with the use of a special punch. This punch is similar to that used in piercing and extruding. In countersinking, however, the distance the punch travels through the stock is less than in extruding. The punch is driven far enough to create a hole and a tapered flange around the hole. This depressed hole is used where the head of an assembly bolt or screw must be flush with or below the surface of the part.

Cutting and Lancing of Tabs

In cutting and lancing of tabs, the tab is usually the item of importance rather than the resutling hole. The tab is created by a punch which cuts all but one side of the opening produced, leaving the tab attached to the part. The operation consists of shearing a U-shape, V-shape, or rectangular cut in the metal and pushing the tab downward with an unsharpened edge of the punch. The tab may be used to locate the part for later processing or to facilitate fitting of the part in an assembly.

TOOLS FOR PIERCING

In a broad sense, the principal components of a set of piercing tools are the punch, the die, and the stripper. The punch creates a hole as it is forced through the material. The die consists of a steel block in which a hole matching the contours of the punch is provided. The edges of the hole provide the lower cutting edge and cooperate with the punch in shearing the material. The slug from the material usually drops through the hole in the die. The stripper consists of a plate with a suitable hole through which the punch point passes. It has a dual function in that it holds the material from slipping during piercing and keeps on holding it so as to strip the material from the punch as the punch retracts. The tool setup functions in the following manner. The stripper descends to hold the stock against the face of the die. The punch follows through the hole provided in the stripper, and is forced through the material into the die hole. On the upstroke, the stripper dwells and holds the material as the punch is withdrawn. The material must be held in this manner because it tends to cling to the punch surface. After the material is stripped from the punch, the stripper plate rises, allowing the part to be removed.

The foregoing description of the makeup and function of the piercing die is couched in the most general terms. Actually, many more factors are involved. The tools are generally much more complicated, the degree of complication depending on the operation involved.

The Punch Assembly

Actually, the term "punch" may refer to two different items in considering a piercing die setup. It may refer to a single tool for creating a hole, or it may refer collectively to an assemblage of components which may create several holes and even perform other operations.

The base for building a punch assembly is the upper shoe. This may be made from a mild steel plate, or it may be a casting. All other punch components are fastened to the shoe. Bushings should be provided in all shoes for guide pins to ensure accurate tool travel.

Hardened steel backup plates are attached to the shoe in areas where the individual punches will be placed. This is necessary to prevent the bases of the punches from being driven into the softer metal of the shoe during operation.

Another component of the punch assembly is the punch retainer. This is a device for holding the punch. Standards have been developed for types of retainers to be used with standard punches. A standard retainer is used whenever possible. In some cases, where there is insufficient space, or where several punches are necessarily mounted in a small area, special retainers may have to be used.

The business end of the punch assembly consists of the individual punches. These are held in the punch retainers by several means. The method of holding may depend on the size and shape of the punch and whether it is used singly or in combination with other punches. The service for which the punch is to be used also influences the holding method. For piercing heavy material, a more positive holding method is required than for piercing light material. This is true because of the heavy stripping load that exists in piercing heavy material. For an operation in which numerous parts are to be pierced, the holding system may be more elaborate and expensive than in an operation where total production quantity is low.

For high-production jobs, the ball-seat holding method may be employed. This system provides for the rapid changing of punches, an item of considerable importance in high-volume operations. In some cases, the stripping pressure may require the use of a shoulder-type holding system. In other cases, bridge-lock or solid-type holding may be best. (See Figure 2E-2.)

To accommodate the various types of piercing operations, a variety of piercing punches have been developed. Each type has certain special characteristics that make it suitable for the job to which it is applied.

The type of punch being used most widely is called the *regular-duty punch.* It is generally round in shape and is mass produced by a number of tool manufacturing companies. Its simple design and its abundance make it relatively inexpensive. It is of the ball-seat type and thus presents a minimum problem with relation to downtime for changing tools. It is used at the Ford Motor Company for making holes up to 1 in. in diameter when the stock thickness is 0.12 in. or less.

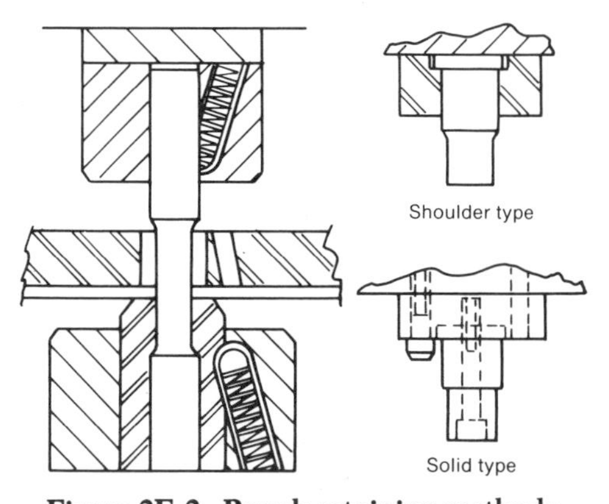

Figure 2E-2. Punch-retaining methods

For stock that is thicker than 0.12 in., and for holes 1 to 1½ in. in diameter, a *shoulder-type punch* is used. The heavier construction of this type of punch provides the added strength necessary to ensure reasonable tool life in this heavier type of work. The shoulder punch also permits a stronger holding method to resist the greater stripping pressure that exists where larger holes or thicker stock is involved.

A *solid-type punch* is used for piercing holes over 1½ in. in diameter. This type of punch is very heavily constructed and is bolted directly through the backing plate to the upper shoe.

Normally, piercing is confined to production of holes that are greater in diameter than the thickness of the stock involved. This is true because piercing of holes with diameters less than stock thickness results in excessive breakage of punches, unless special strengthening provisions are made in the punch. A punch thin enough to pierce a very small hole will not withstand the forces required to shear the stock on a sustained basis unless it is sufficiently reinforced. Thin punches tend to deflect or "whip" on contact with the stock, adding the probability of breakage. The reinforcement provided to enable such piercing to be done consists of several elements. The point of the punch is made as short as possible; the main body of the punch is made large and is brought as close to the point as possible; and a die cast soft metal sleeve is pressure cast to the punch to absorb the whipping and deflecting forces. This cast sleeve is called a whipsleeve, and punches so constructed are called *whip-sleeve punches* (Figure 2E-3).

Whipsleeve punches are used for piercing holes where the metal thickness exceeds the punch diameter. They are also used in producing slightly larger holes where whipping and deflection are a problem.

A *quill punch* (Figure 2E-3) may also be

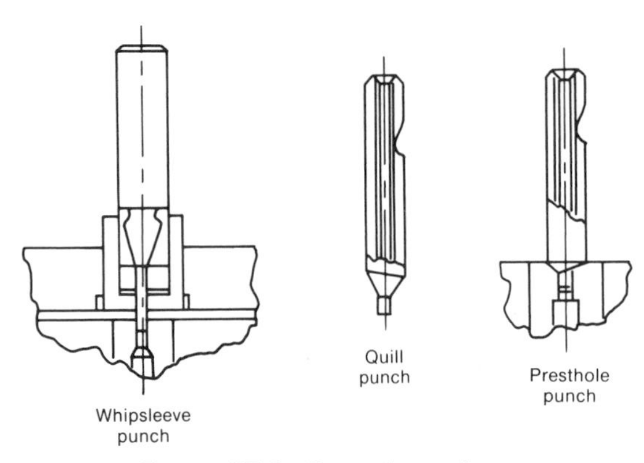

Figure 2E-3. Special punches

used to pierce holes smaller than stock thickness. This punch has a very short point and is encased in tool steel for strength. Hard stock may also be pierced with this type of punch.

A large number of special punches are used to create special types and shapes of holes.

A *blade-type punch* is used for piercing long, narrow slots. It consists of a steel blade mounted in a special retainer.

Another special type of hole is made by the *presthole punch* (Figure 2E-3). This punch is designed to pierce out a small amount of stock, countersink the edge of the resulting hole, and slit the countersunk edge. This provides a gripping device into which a screw or bolt can be assembled without using a backing nut.

Another special type of operation is accomplished by the use of a *developed punch*. This punch is used to create round holes in stock held at an angle, although the punch travel is vertical. To accomplish this, the punch point is ground into an oval shape.

The Die Assembly

The die section of the piercing-tool setup is also an assembly of several components. It generally consists of a lower shoe, die sections, die buttons, guide pins, and chutes or trays for the elimination of punched-out slugs.

The lower shoe, or die shoe, is a mild steel plate or a casting to which all die components are fastened. Guide pins are fastened to the shoe and are used to ensure accurate travel of the punch. Heel blocks are often added to absorb any tendency of the punch to deflect during the work stroke.

The die sections in a piercing setup usually perform the function of holders for the die buttons. In some cases, the die sections are drilled through and hardened to provide opposing cutting edges for the punch. This practice is usually not followed because of the simpler die maintenance achieved through use of die buttons.

A die button is a hardened steel cylinder with a hole designed to provide opposing cutting edges for a punch. The hole of a die button will conform to the shape of the punch with which it is used. Die buttons, like punches, are available in many standard sizes. Special die buttons are also available for use with punches of special design.

In establishing any piercing operation, consideration should be given to the elimination of slugs from the die. Slug chutes should be provided in the die to move the slugs to convenient accumulation points. In some cases, slug trays should be incorporated in the die to prevent the slugs from piling up on the floor under the press. Where slug trays are not used, adequate receptacles should be provided at the accumulation point.

The Stripper

After the punch has penetrated the metal, it must be pulled back. Because of elasticity, the metal compressed at the beginning of piercing tends to expand and grip the sides of the punch. This gripping causes the pierced part to rise with the punch. Thus, a mechanism must be provided to hold the part down while the punch is withdrawn. The holding of the part must be done properly to prevent distortion of the metal. The mechanism used for this task is called the stripper. The hole in the stripper through which the punch passes should be designed with the proper amount of clearance to permit free movement of the punch. In some cases, slender punches may be guided by bushings installed in the stripper plate.

Actually, a stripper may be a component of either the punch assembly or the die assembly. If it is a component of the punch it is called a spring stripper, because the stripping pressure is supplied by springs that hold the stripper to the stock as the punch begins to ascend. As the punch moves up, the stripper is picked up by the punch after the part has been freed.

If the stripper is mounted on the die it is called a solid stripper. The solid stripper is firmly mounted and does not move. It is generally used for large stampings or for stampings of heavy-gage material.

PIERCING SHEAR AND CLEARANCES

In piercing operations, one of the factors to consider is provision of safeguards against excessive tool breakage. In addition, the holes created must conform to certain quality requirements with respect to size and edge conditions. Edges should generally be free from excessive burrs. The provision of proper clearances and shear can aid considerably in reducing tool breakage and providing desired quality.

Shear

In hole-cutting operations, the metal cut out is the scrap. Hence any shear provided must be ground on the punch. The main purpose of shear is to reduce the load placed on the punches to accomplish the piercing.

Shear, to be of any advantage, should be ground on punches having diameters that are fairly large in relation to stock thickness. Shear on small punches may cause deflection of the punch and add to the likelihood of tool breakage. However, the benefits of shear may be provided where a number of small punches are being used by stepping the punches (Figure 2E-4). In this way, the pressure required to effect piercing is reduced by staggering the contact of the punches with the metal. The amount of punch stepping should be held to a minimum to avoid a jerky series of loads and releases as the punches progess through the metal. If the punches must enter the metal by 25% of stock thickness to effect shearing, each punch needs to be shorter than the next by no more than this amount.

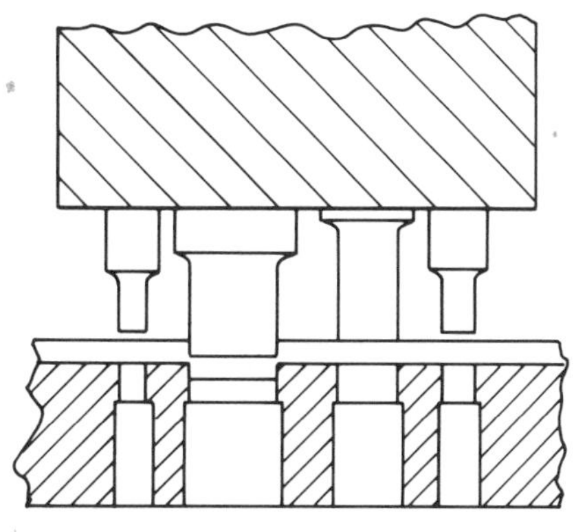

Figure 2E-4. Punches stepped to avoid breakage

Clearances

Tonnage requirements for piercing, and the quality of hole edges, are dependent in part on the provisions made for clearance between punch and die. The proper amount of clearance will result in a satisfactorily clean hole. Excessive clearance will result in rough, burred edges, distortion of material surrounding the cut, and pinched-off edge rather than a clean fracture. Insufficient clearance will result in unnecessarily high tonnage requirements because of the secondary and even tertiary shearing that takes place where insufficient clearance is provided. This greater force required under these conditions is apt to shorten tool life. In addition, particles of metal may be released during the secondary shearing. These particles may adhere to the punch surface and result in oversize or irregular holes. This adhering metal, called pickup, also increases tonnage requirements and may result in tool breakage due to deflection caused by a build-up of metal on one side of the punch. In addition to these undesirable effects, secondary or tertiary shearing leaves the internal surfaces of the hole ragged.

Proper clearances between punch and die are based on a percentage of stock thickness. This percentage is ground off the die hole, all around. The clearance varies with the type of material used. Generally, the harder the metal, the less the clearance. The range of clearances for most automotive stampings is considered to lie between 5 and 10% of metal thickness. The proper clearance is usually determined in die tryout.

SPECIAL FACTORS IN PIERCING

To establish successful piercing operations, thought should be given to certain special factors.

Proximity of Punches

Where a number of holes are punched close together, certain provisions should be made to prevent excessive tool breakage. This breakage may result from the crowding of metal that takes place in piercing. For instance, a small punch adjacent to a large punch may be deflected and broken as the result of the large punch crowding the metal in the direction of the small punch. This problem may be overcome by grinding the smaller punch shorter by an amount equal to the distance the larger punch must travel to shear the stock. In this way, the metal is already crowded before the smaller punch makes contact. Also, the smaller punch retracts first and thus escapes any side pressure resulting from springback when the large punch is withdrawn.

Sharpness of Tools

The sharpness of cutting edges in piercing operations has an important bearing on the quality of work done and the pressure required to do the work. A sharp edge localizes the severest stresses and causes the fracture to occur sooner and with less load. Thus, maintenance of tool edges provides protection against punch breakage. It also helps to ensure clean fractures.

Limitations of Piercing

Theoretically, the size and number of holes that can be pierced, and the thickness of stock that can be used, are limited only by the size and strength of the press. However, for producing specific parts in conformance to specifications, certain practical limitations should be observed.

The number of holes that can be pierced in one die is limited by the space required for assembly of punches and die buttons. The size of the holes to be pierced is another determining factor in the number of these items that can be arranged in a single die setup. While it is usually economical to combine piercing operations in a single die, care should be taken to avoid excessive die-maintenance problems caused by a weakening of the die by inclusion of too many operations.

There is usually no difficulty in piercing any thickness of stock used in automotive stampings. However, in piercing heavier-gage stock, care should be taken to provide punches sturdy enough to withstand the stresses involved. Proper guiding of punches is necessary to avoid excessive tool breakage.

Certain limitations should be observed in piercing close to the edges of stock or flanges. The general rule is to keep the outer edge of the hole at least two stock thicknesses from the edge of the material. With respect to flanges, at least three stock thicknesses should be allowed between the hole edge and the inside surface plane of the flange. The main purpose of this is to preserve adequate strength in the die button. The walls of the button must be thick enough to withstand the force exerted by the punch. If this thickness is reduced too much, in an effort to bring the die-button hole close to the edge of the flange, repeated failure of the button may result. If a hole is pierced too close to a flange breakline, prior to forming of the flange, undesirable distortion of the hole may take place during flanging. Where part design calls for holes in these limiting areas, a change in design may be necessary.

Most piercing is done with the punch traveling in a direction perpendicular to the horizontal plane of the stock. This provides the optimum conditions for punch strength and clean shearing. However, it is sometimes advantageous to pierce at some other angle (Figure 2E-5). Such a case might occur when piercing follows forming and the area to be pierced is at an angle to the horizontal. As the punch descends it first contacts the work on one side of the point. This, of course, tends to

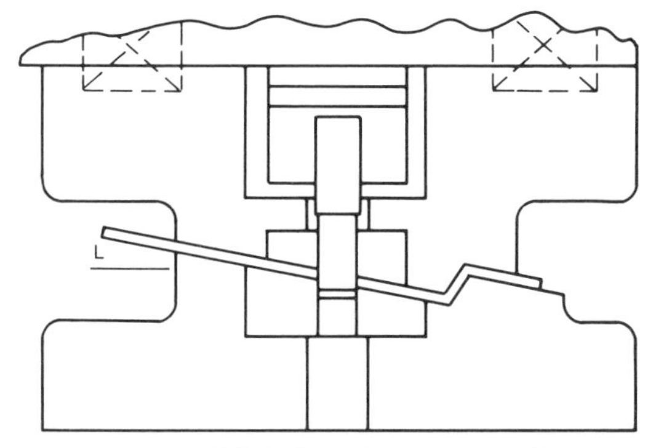

Figure 2E-5. Piercing at an angle

deflect the punch, requiring careful guiding. The greater the workpiece angle from horizontal, the more difficult it is to pierce with vertical punch travel. Actually, piercing in such a situation is limited to a part angle of 30° or less. If piercing is attempted at a greater angle, exorbitant costs will be incurred due to punch breakage.

The proximity of holes to each other is another factor which puts a limitation on piercing. This limitation applies only to equipping the punch and die assemblies with standard punches and die buttons. Holes actually can be pierced very close to each other; they may even overlap. However, when such piercing is done, special punches, punch holders, and die buttons are required, because the sides of the punches and the die buttons must be ground down to bring the hole-producing elements close together. The use of special piercing units may be justified on the basis of savings resulting from combining of operations. However, a comparison of tooling and operating costs for special units with costs of other methods should be undertaken to determine the best method.

CAM PIERCING

Piercing is a type of press operation that often can be successfully combined with other operations such as blanking, forming, and trimming. The economies that can be realized through combining of operations should always be considered when setting up a piercing operation. In addition, combining piercing with other operations is often a means of carrying out a greater amount of work with a limited number of presses.

In combining piercing operations, it is frequently unfeasible to pierce with a vertically held punch. This is true where the area to be pierced is in an angular position. In such a case, a cam-piercing mechanism may be used to bring the punch into right-angle contact with the part, the cam moving in from the side.

There are several types of cam mechanisms. The type selected for a particular application depends on such factors as space available for mounting the cam, pressure requirements for cam return, and the direction of piercing.

Spring-Return Cam

A spring-return cam (Figure 2E-6) consists of three main elements: the cam body, the driver, and the spring. The cam body carries the punch and is mounted on a base in such a way that the piercing punch may freely enter the die button which is mounted in proper position on the die shoe. The driver is a wedge-shape steel block mounted on the ram of the press. On descent of the ram, the wedge face acts on a similar surface of the cam body, causing the cam body to move in toward the die button. As this occurs, the cam spring is depressed. When the driver moves up, the spring expands to withdraw the cam body and its attached punch.

Air-Cylinder-Return Cam

The components and method of operation of an air-cylinder-return cam are very

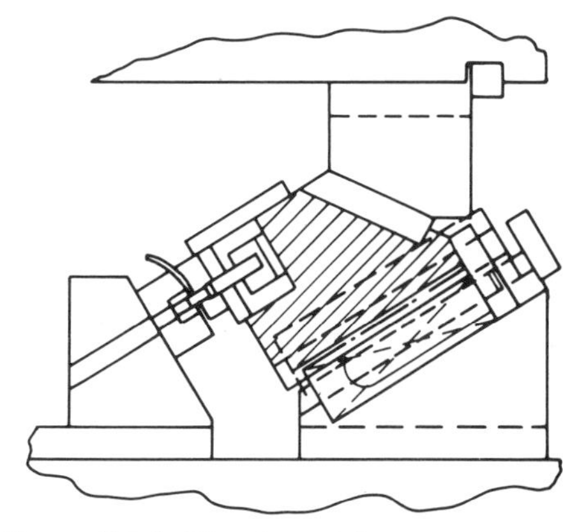

Figure 2E-6. Piercing with a spring-return cam

similar to those of a spring-return cam. The difference lies in the fact that an air cylinder is used instead of a spring to return the cam body to the open position on the upstroke of the press. Air cylinders usually work more satisfactorily than springs as return mechanisms. In addition, the air cylinder is considered safer, because springs sometimes break and may scatter pieces of metal around the work area, endangering employees.

Positive-Motion Cam

Some cam operations are accomplished with the use of a positive-type driver. This driver consists of a mechanism that engages the cam body throughout the work stroke, pressing the cam body inward on the downward stroke and retracting it on the upward stroke. No springs or air cylinders are used. This type of mechanism is called a positive-motion cam (Figure 2E-7). The cam body may be located on either the punch or the die. If the cam is on the punch, the driver is fastened to the die, and vice versa.

Internal Cams

In some cases, the location of the burr that is created on a part in a piercing operation is highly important. Sometimes, because of limited assembly clearances, a burr may not be tolerated on the interior surface of a part. The part may, however, be so positioned in the die where piercing is to be done that piercing in the normal way would induce burrs on the part interior. To avoid this, it may be necessary to pierce the part from the inside. This may be accomplished through the use of cams that work from inside the part when the part is in the die. These cams are linked to moving press mechanisms and carry the punch. The punch contacts the stock from the inside and forces a slug into a die button mounted outside the part on the punch, or perhaps on another cam mechanism.

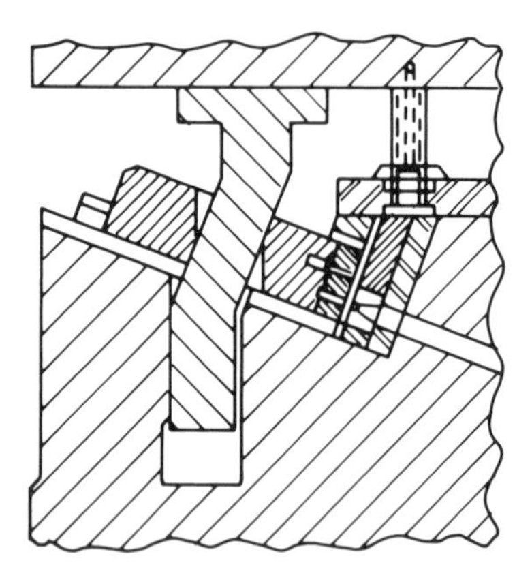

Figure 2E-7. Piercing with a positive-motion cam

PIERCING AND EXTRUDING

Quite often in press operations, a hole must be produced with an extruded collar all around it. This collar may be designed to strengthen the part, or to serve as a locating device or as a bearing for a shaft. To provide such a collar, a special type of die arrangement is necessary. The type of arrangement used will depend on the direction of the extrusion and the gage of the stock.

Pierce and Extrude Up

To accommodate other operations, the part may be positioned in a die in such a manner as to require an upward forming of the extrusion (Figure 2E-8). In such a case the upper shoe is equipped with a punch and a spring pad to hold the workpiece and keep it from shifting. A spring pad is also fitted to the lower shoe. This spring pad acts as a holding device during extrusion and strips the part when forming is completed. The lower shoe also contains a specially designed die button. In operation, the punch descends and the upper spring pad grips the metal while the punch pierces through the stock into the die button. As the punch continues down, the pressure of the lower spring pad is overcome and the pad moves downward. As the pad moves down, the free metal around the hole is extruded upward around the die button. The lower spring pad strips the part from the die button on the upstroke.

Pierce and Extrude Down

The position of the part in the die may require that the extrusion be formed downward (Figure 2E-9). In this type of operation,

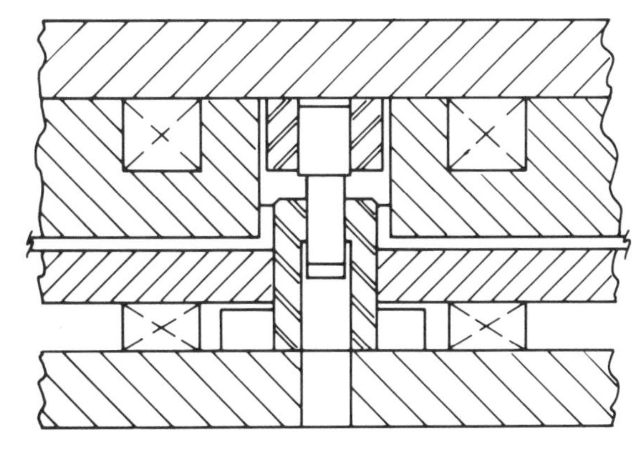

Figure 2E-8. Pierce and extrude up

the piercing punch is assembled in an extrusion punch and both are mounted on the upper shoe along with a spring pad. A lower spring pad is built around the die button, which is mounted on a plate that rides on an air cylinder or on springs. In operation, the upper pad grips the metal as the punch descends. The piercing punch pushes a slug out of the stock and continues down until the extrusion punch contacts the hole edges and wipes them down as the die button is pushed down. On the upstroke, the piercing and extrusion punches are withdrawn, and the die button rises to strip the extruded portion of the panel out of the die opening.

Extruding Heavy-Gage Metal

Heavy stampings are often designed with hole extrusions that can serve as sturdy housings for bushings or other shaft bearing units. When the material is of very heavy gage, the hole should be drilled rather than pierced. This is necessary because in heavy stock, cracks and fractures occur in extrusions formed from the edges of pierced holes.

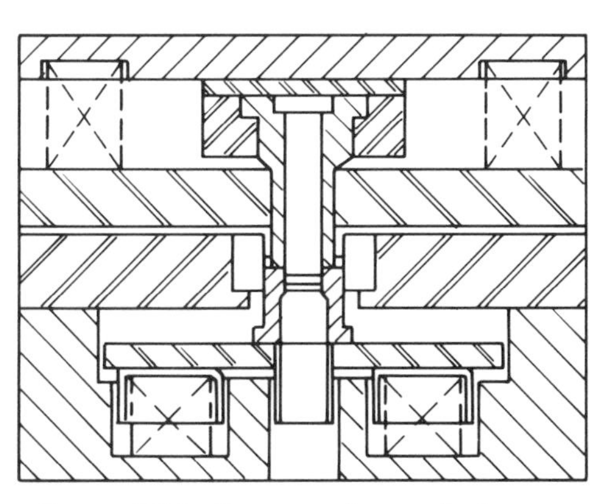

Figure 2E-9. Pierce and extrude down

COMBINED OPERATIONS

Piercing can be combined with practically any other type of stamping operation. In such combined operations, however, it is necessary to provide careful guidance between the punch and die.

Piercing is often combined with blanking when the holes will not be deformed in later operations. Dies for blanking are usually well guided and provide suitable protection for piercing tools. The same is true of trimming operations.

Piercing may also be performed in conjunction with drawing, forming, and restriking operations if hole distortion does not result and if the press tools are properly guided.

PRESSES EMPLOYED IN PIERCING

In nearly all cases, piercing is done on single-action presses. These presses may be of gap-frame or straight-side design. A triple-action press may also be used for piercing, the piercing being done with the third ram. In rare cases, a double-action press operation may incorporate a piercing punch which cuts through the metal just prior to bottoming to allow relief in an area of deep draw.

Presses selected for piercing should have adequate die area so that provision can be made to allow slugs to fall through the die base. Space for providing slug chutes under the die may be provided through the use of risers.

Presses for piercing should also be in good mechanical condition. This will help to avoid tool breakage due to variations or shiftings in the press stroke resulting from worn press parts.

Hydraulic piercing mechanisms are sometimes used for operations in which only piercing is performed. Such mechanisms consist of hydraulic cylinders which are mounted on a fixture base and which carry the punches. The number of cylinders is dependent on the number of holes to be produced. Die buttons are mounted in proper locations on a fixture which positions the part for the operation. In operation, the part moves into the fixture, usually automatically, and the cylinders are activated to effect piercing.

This type of mechanism is used in piercing the various holes in assembled automobile frames. By piercing all these holes in one fixture, at one time, the physical relationship of the holes may be held to closer tolerances. Performing the same type of operation in a press would be much more costly because of the greater expense of the required equipment.

Hydraulic piercing units are also used for piercing various hole patterns in fenders to accommodate various trim designs. They are also used in piercing rivet holes and air-valve holes in automobile wheels.

These mechanisms provide a means of piercing at low cost, are adaptable to quick changeovers, and are versatile in application. However, the use of hydraulic cylinders and hydraulic lines creates maintenance problems which are both costly and troublesome.

SUMMARY

The success of a total stamping process is as dependent on the provision of effective piercing operations as it is on other types of press operations. Nearly all stampings require piercing at some stage of their manufacture, and the number of types of piercing is quite large.

Most piercing consists of driving a flat-face cylindrical punch through the workpiece into a die section or button. A slug is removed from the metal as the punch passes through it. Piercing with a pointed punch removes no metal, but leaves a ragged-edge extruded collar around the hole produced. Othe types of piercing operations include piercing and extruding, slotting, countersinking, and lancing and tabbing.

The tools for piercing consist of a punch assembly, a die assembly, and a stripper, which may be a component of either the punch or the die.

The punch assembly consists of an upper shoe, backup plates, punch retainers, and punches. Punches come in a wide variety, including regular-duty punches of all sizes and special punches such as shoulder punches, solid punches, whipsleeve punches, quill punches, blade-type punches, presthole punches, and developed punches. Punches of the regular-duty type are usually of the ball-seat-release variety for quick changing.

The die assembly consists of a lower shoe, die sections or die buttons, and such features as slug chutes and trays. Die buttons often incorporate the ball-seat-release design.

In piercing, shear is applied to the punch if the punch is large enough in relation to metal thickness to make shearing of the punch worthwhile. On small punches, the benefits of shear may be provided by stepping the punches. Clearances in piercing are ground on the die because the punch controls the size of the hole opening, which is the critical dimension in piercing.

Certain special factors related to piercing should be considered in setting up piercing operations. Among these factors are the proximity of punches; the sharpness of tools; and piercing limitations, such as the number of holes that can be pierced on one die, the thickness of the stock, the distance from part edges and flanges at which piercing can be successfully accomplished, and the angle at which piercing may be done. Most piercing done at an angle to the horizontal is accomplished through the use of cam dies. Several types of cam piercing-die arrangements exist, such as spring-return cams, air-cylinder-return cams, positive-motion cams, and internal cams.

The process of piercing and extruding calls for a specialized type of piercing die. This type of die features spring pads on both the punch and die sections to control the extrusion of the hole edge after piercing has been done. Both upward extrusions and downward extrusions may be done. Each type requires special features in the die.

Practically all types of stamping operations can be accomplished in combination with piercing. Thought should always be given to effecting such combinations to achieve the resulting operational and tooling economies.

Single-action gap-frame or straight-side presses may be used for piercing. In rare instances, a double-action or triple-action press may be used for piercing that can be incorporated in a drawing operation. Another type of piercing mechanism that is now finding application is the hydraulic piercing machine. This machine is versatile, relatively inexpensive, and applicable to jobs requiring several holes to be pierced at one time. However, the hydraulic mechanism of this machine poses problems in maintenance.

2F: Trimming Operations

For many types of stamped parts, excess metal must be allowed for holding purposes during operations that shape the metal into the form of the part. This excess metal must, of course, be removed; such removal is done by trimming. Trimming is the removal of excess metal from a stamped part to allow the part to reach the finished stage or to prepare it for subsequent operations. For nearly all parts requiring drawing operations, trimming is necessary. Formed parts other than those produced from fully developed blanks also must undergo some trimming before they are brought to final size.

The broad scope of trimming operations at the Ford Motor Company is emphasized by the fact that 37% of the sheet metal delivered to the stamping plants is returned to steel mills in the form of scrap. Most of this scrap results from trimming operations. The remainder is generated in piercing and other types of shearing.

The generation of this amount of scrap indicates the importance of effective trimming operations in establishment of successful stamping production. Disposal of the scrap itself is a major problem. Also, proper processing of all types of stamping operations is necessary to obtain effective use of material.

ANALYSIS OF PARTS TO BE TRIMMED

Trimming of a part is usually performed immediately after a drawing operation. Therefore, it is necessary to consider what the part will look like after it is removed from the drawing die.

When the plaster for the drawing die is originally made from the mahogany model, the part is represented in its unfolded and untrimmed form. Flange breaklines are established, and extra metal for the flanges is allowed beyond these lines. The perimeter established by the unfolded flanges is designated as the trim line. This trim line then is the starting point from which the trimming operation evolves.

At this juncture the process engineer is in another area of decision making and must provide answers to many questions, such as:

1. Should the part be trimmed open side up or open side down?
2. Can the part be trimmed in one opera-

tion, or should it be trimmed in two operations?
3. Is it necessary to separate two or more pieces if parts are made in multiples?
4. Is any notching required?
5. Are both rough and finished trimming needed?
6. Should two trimming dies be built, or can a cam trimming die be used?
7. Are there any operations that can be combined with trimming?

A plan view of the plaster model will reveal that the shape of the trim outline which is going to require external cutting will be square, rectangular, oval, irregular but symmetrical, or irregular and nonsymmetrical (Figure 2F-1). A side elevation view may reveal that the binder metal is in a flat plane, has a single contour, or has multiple contours.

Internal cutting may also be required on the panel if it is necessary to separate parts (such as for the right-hand and left-hand sides of a car) or if large openings must be made somewhere in the panel for the purpose of allowing an extruded section to be made such as a window opening, a lightening hole, or an access hole.

The thickness of the metal to be cut may have some bearing on the trimming operation. In addition, consideration must be given to the possible creation of knifelike edges on the part that might injure employees who must handle the panels in later operations.

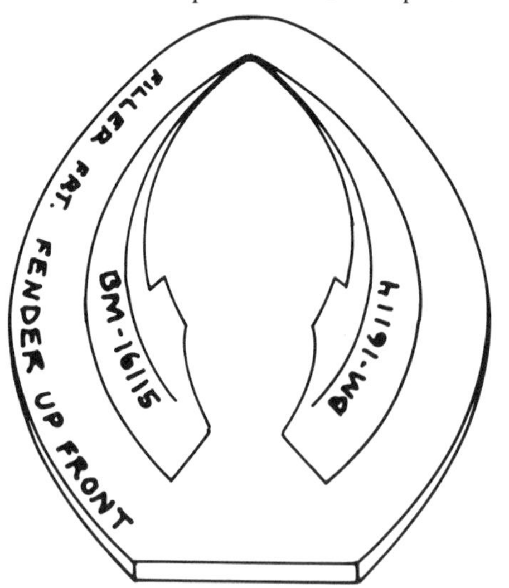

Figure 2F-1. Trim line on plaster die model

SELECTION OF TRIMMING DIES

Due to the wide variety of parts stamped for automotive fabrication, a number of different types of trimming dies have been developed. The edge requirements and size of the part will generally indicate what kind of die should be used.

Conventional Trimming Dies

The *flush trimming die* is the type of die most widely used in automotive stamping (Figure 2F-2). This type of die has two opposing cutting steels, operating much in the same fashion as a pair of scissors, that cut vertically on the downstroke of the press. The part is usually supported open side down on a solid pad mounted on the lower shoe; this pad is very similar to the type used on a solid forming die. There is usually a spring-loaded pad fastened to the punch. This pad holds the part firmly in position on the downstroke of the press. The upper cutting steels surrounding the spring pad contact the part after the pad has descended, and the scissors-like action takes place when the upper trimming steels pass by the lower trimming steels. After the press has completed its cycle, the part is ejected or lifted out of the die. Flush trimming results in a small burr around the edge of the cut because of the necessary clearance allowed between the punch and die.

The location of this burr is frequently of importance in flanging the part in later operations. The burr may be located on the top or the bottom of the flange depending on the position of the part in the die. If the part is positioned open side up, the burr will be on the underside of the flange. If the part is inverted, the burr will be located on the top of the flange.

After the press has completed its cycle, the metal cut away from the part, which is the offal, is usually in the form of a ring. This material must be disposed of in one way or another. If allowed to remain in ring form, it would require manual or mechanical removal from the die. To avoid this, scrap cutters are designed and built as integral units of the

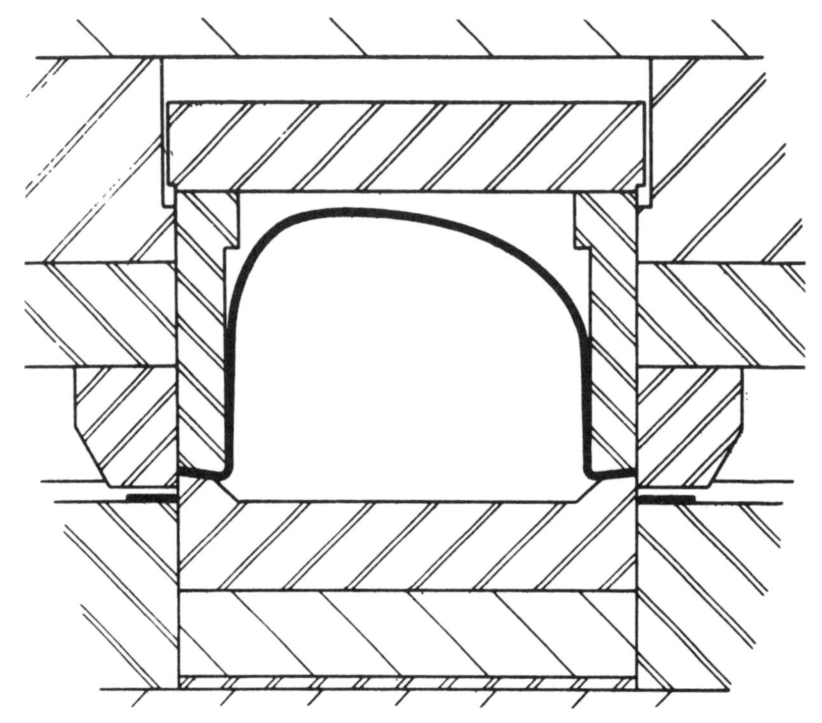

Figure 2F-2. Flush trimming die

lower trimming steels. These are knifelike blades mounted perpendicular to the trim line. Mating steels are also mounted on the punch. These scrap cutters are usually designed to part the material at distances from 12 to 18 in., as will be described later on. Provision is made for disposal of this stock in the design of the die and layout of the press department.

A *parting trimming die* is used for separating two or more parts formed or drawn from a single sheet of metal (Figure 2F-3). The cutting steels in these dies must be accurately aligned, and the parts must be firmly held during cutting to ensure that separation is accomplished at exactly the right place. Usually when parts are separated, a thin strip of metal is cut from between them.

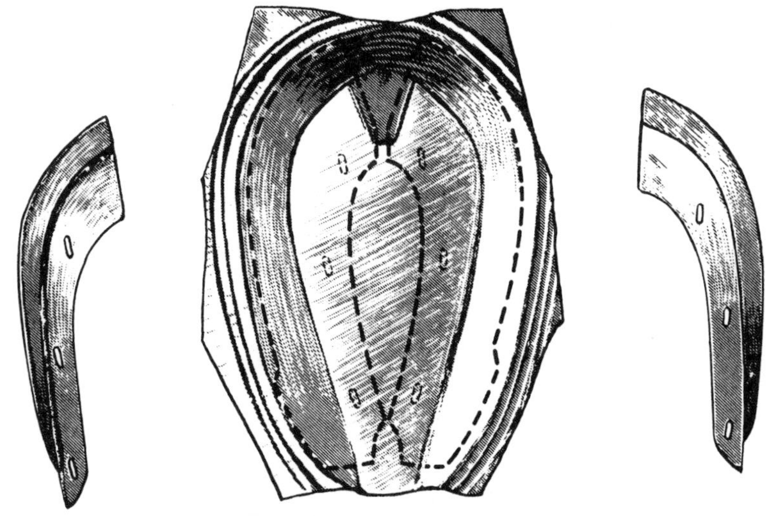

Figure 2F-3. Parting trimming dies used to separate parts drawn in multiples

Consideration must be given to the removal of this scrap metal. Parting is frequently combined with flush trimmming.

A *horn trimming die,* which is sometimes called a "hang-up" die, is a die generally mounted on a horn-type press which has an overhanging mechanism for separating the slide. The die is mounted on the front of the press. A portion of the part hangs over the die and then the press is cycled. Such dies are generally used when trimming of relatively short length is involved.

Cam Trimming Dies

A cam trimming die (Figure 2F-4) may be used where the operation cannot be performed entirely with a vertical stroke of the press. One or more cams are used to convert the vertical action of the ram into an action in a more suitable direction, such as 70°, 85°, 90°, or some other angle. The use of cams, while providing perhaps the only solution to a difficult trimming application, poses certain problems for pressworking. These problems should be thoroughly considered before a final decision to employ a cam trimming die is made. A cam die is justified if a saving in piece price can be achieved or if a limited number of presses are available for running the parts, which would necessitate the combination of cams on a flush trimming die. The design of the cam mechanism must be such that parts can be loaded and unloaded without interference. Provision must be made for easy removal of scrap generated by the cam trimming steels. Cam construction must be done with consideration of employee safety. Cams should be located where they are not in the way of the employee. Where cams move out from the press bed, guards should be installed to prevent injury to plant personnel.

The cam slide contains the cutting steel and is guided by steel gibs with hardened steel wear plates. Provision should be made for shimming of the wear plates to adjust the clearance of the cutting tools to compensate for wear.

A cam mounted on the die is usually driven by a wedge-shape driver attached to the punch. The action is obtained by a wedge-shape block on the punch, called the driver, which moves a wedge-shape block on the cam. On the upstroke of the press, the cam is returned by springs or by an air cylinder. In some cases, the advance and return of the cam are accomplished with a positive-type driver which is designed with several offsets to accomplish the desired action. Stop blocks are used to limit the back travel of the cam. The back travel must be great enough, however, to permit easy loading and unloading of parts.

A cam mounted on the punch is used for certain applications where the path of cam action is between perpendicular and 45% from the vertical action of the press. Such a cam will float on the punch and is actuated by a driver mounted on the die.

A *shimmy trimming die,* which is a form of cam trimming die, is often used in trimming cup-shape parts where exceptionally clean cut edges are required. In this type of die, when it is desired to trim the part open side up, the part is pushed into position by the ram of the press. A lower floating die, activated by positive-motion cams mounted on the punch, reciprocates from side to side and from front to back to trim the part. If it is desired to trim the part open side down, the floating section may be mounted on the upper shoe with the drivers on the lower shoe.

A *pinch trimming die* is frequently used for trimming cup-shape parts with only the vertical action of the press (Figure 2F-5). This operation is often combined with a drawing operation and may even be combined with blanking. In such a compound die, the punch first draws the part to the proper depth. After this, the punch continues its travel, and the flange is pinched off by the action of the cutting steels on the punch against the cutting steels on the die. Clearances between punch and die must be held to a minimum to ensure an even pinch-off. Although every effort is made to prevent the formation of ragged edges on the trim line, such an objective is seldom obtained with a pinch trimming die.

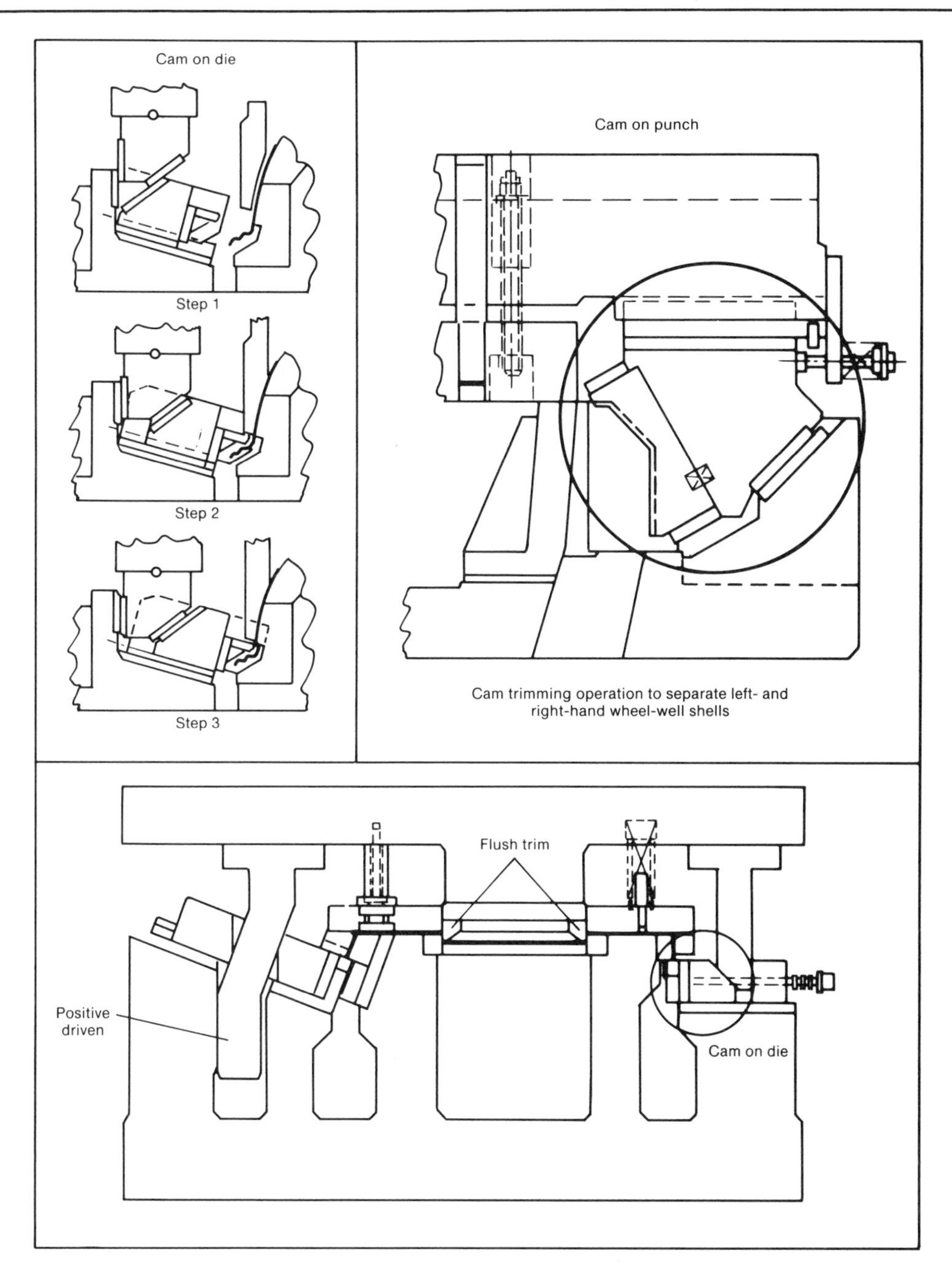

Figure 2F-4. Cam trimming dies

ROUGH AND FINISH TRIMMING

Rough Trimming

There are instances in which certain areas of a part may be trimmed away prior to completion of other operations. This is called rough trimming, because the edge produced will not be the final edge of the part.

Rough trimming is often done prior to a restriking operation. Restriking operations have as their purpose the final sizing of the curved areas of drawn parts. The rough trimming is necessary to prevent the scrap

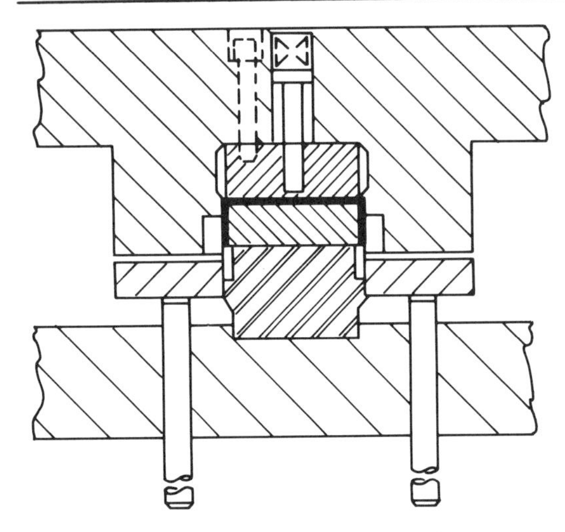

Figure 2F-5. Pinch trimming die for combined drawing and trimming

around the part from restricting metal movement or causing springback.

Rough trimming is used in other instances to permit convenient handling of a part after a first drawing operation. This is particularly true of large panels drawn from square-sheared stock, where large holding flanges represent a hazardous and awkward handling problem.

Rough trimming may also be necessary when a very large amount of scrap must be cut away. Cutting all the scrap in one finish trimming operation may cause problems in removing the scrap from the die. In such a case, a portion of the scrap may be trimmed off in one operation and the part finish trimmed in another operation. This may provide a more suitable arrangement and a solution to the scrap-disposal problem.

Finish Trimming

Finish trimming generally is done as the final operation in bringing a part to final size. The location of the rough trim line in relation to the finish trim line is of major importance in setting up a rough trimming operation. If insufficient metal is left for finish trimming, the finish cut will not be clean. An excessively narrow strip of metal will crush or draw down during trimming, leaving chips or hairlike burrs. In most cases, these chips or burrs will be deposited on the die surface, causing trouble in later operations. The chips will build up on the die, causing scoring of parts and injury to the die. Generally speaking, at least 1/8 in. of metal should be left for removal in a finish trimming operation to avoid burrs and chips.

CONSTRUCTION DETAILS OF TRIMMING DIES

The production of parts to specifications is the primary consideration in the construction of a trimming die. Before die design is begun, the shape and contour of the finished part (as shown by part print, prototype model, and die model) are carefully considered. The locations of holes and their proximity to the trim line are noted. If holes are too close to the trim line, distortion or breakage may occur unless provision is made for proper holding. The shape and contour of the part will also provide clues as to the amount of metal that will be provided for holding during operations preceding trimming and will indicate the amount of metal that will be left for trimming.

Die design must also be considered in relation to the specifications of the press to be used. A determination is made as to whether the press has the proper shut height and area to accommodate the die needed for trimming, and whether it will accommodate the stamping with the binder scrap attached. Consideration is also given to the possibility of designing a die that will be amenable to automation once it is installed in the press.

There are numerous details that must be worked out in the design and construction of a trimming die if proper performance is to be obtained. Although many of these details are of interest only to the die designer and diemaker, several should also be of concern to the process engineer.

One of the factors influencing construction of trimming dies is the extreme care that must be given to the guiding of cutting tools. While some side thrusts and unbalanced conditions may be tolerated in forming or drawing operations, positive guiding in trimming operations is imperative because of the necessity for proper mating between the upper and lower cutting steels. If cutting steels are not properly aligned, they will certainly suffer damage caused by the opposing edges striking together. To guard against such damage, guide pins are used to ensure accurate travel of the moving member of the die. Trimming dies for small parts may be adequately guided by two guide pins. For larger parts, more pins (up to four) may be required. These guide pins are supplied whether or not the cutting job to be done is evenly balanced.

In cases where unbalanced forces are present in trimming, heel blocks should be used to absorb the side thrust. These heel blocks consist of built-up steel bearing surfaces usually located at all four corners of the die. Ordinarily mounted on the lower die or die shoe, the blocks restrain the punch from moving sideways as it descends. This provision prevents tool breakage and helps to ensure a satisfactory edge on the trimmed part. It is important to locate heel blocks so that they will not interfere with loading or unloading of parts.

The selection of steels for trimming dies will depend on the complexity of the operation and the production volume involved.

The shoes for trimming dies are made of the same material as the other dies in the press lines. The cutting steels on both the punch and die are usually made of composite sections rather than from solid steel blocks, because it is easier to mount the sections at various angles on the mounting blocks and to remove them for resharpening.

As in all types of cutting, the pressure requirements in trimming operations can be reduced by grinding shear on the cutting edges of the tools. In trimming, the shear is ground on the punch. The shear may be either balanced or unbalanced. Where a long cut is to be made, it is generally advantageous to provide balanced shear to reduce the tendency of the blank to shift in the die.

Another consideration in setting up trimming operations is the clearances provided between the cutting steels. If insufficient clearance is allowed, the edges will be ragged. Secondary shearing will take place, adding to the load on the press and the wear on the cutting steels. Excessive clearance will cause the edge of the cut to draw down, leaving a burred edge. The same clearances outlined for piercing operations in Article 2E apply generally to trimming.

When parts are trimmed, there is a tendency for them to stick to the sides of the cutting tools. To prevent the edges of the part from lifting up and perhaps buckling as the punch ascends, strippers are provided. These strippers are usually spring-powered. Springs must be strong enough to strip the part from the cutting steels.

Trimming operations call for relatively close observance of tolerances. To ensure that tolerances are met, the part being trimmed must be accurately and firmly located in the dies. Special locators are placed over spacer plates where necessary for locating the part. If the shape of the part is not conducive to locating, pin locators or gages may be used in holes in the part. Gages should be adequate for holding the part firmly. Proper gaging (locating) is vital to accurate and fast trimming operations. Stripper plates frequently assist in holding the part by contacting the work prior to the cutting action.

SELECTION OF PRESSES FOR TRIMMING

Because of the requirements for exceptional control and relatively high precision, special care must be exercised in selection of a press for a trimming operation. The press as well as the die must be equipped to ensure accurate mating of the cutting steels. Older presses should be examined carefully to make certain they will provide good alignment.

Gap-Frame Presses

Gap-frame presses are frequently used for trimming small or medium-size parts in a wide range of volume requirements. Because of their accessibility, they are particularly adaptable to the use of automation for handling of parts. Such presses are usually of the inclinable type, and thus gravity can be utilized to facilitate unloading.

There are, however, certain features of

gap-frame presses that may serve to complicate trimming operations if press adjustments are not made. The feature of these presses that is most important is their gap construction. This gap tends to deflect, and spring open, under prolonged use and/or heavy loading. Where trimming is involved, this factor has important implications for the maintenance of the punch and die.

The deflection tends to cause the edges of the cutting steels to contact each other, resulting in breakage or undue wear. To counteract this disadvantage, tie rods may be installed across the gap to give the press greater rigidity. Another means of minimizing this problem is to employ presses of greater tonnage than is required by the cutting operation involved. This is called a press tonnage safety allowance and is a frequently used solution to the problem of deflection.

Guiding and stabilization of the press slide itself is another important factor. In addition to carefully aligned gibs on the press, guide pins and heel blocks should be provided in the die itself.

Straight-Side Presses

Deflection is more easily controlled in a straight-side press, where gibs at all four corners of the press are frequently used. The gibs tend to neutralize any side thrusts caused by uneven cutting pressures. Presses with two- and four-point suspension resist deflection from unequal loading. Nevertheless, heel blocks are often used in trimming dies run in four-column straight-side presses.

SCRAP HANDLING

The handling and disposal of scrap have an important bearing on the efficiency of trimming operations. In any trimming application, some scrap or offal is generated. Provision must be made for removal of this material from the area of press.

Implications for Over-all Press Operations

Removal of scrap in trimming operations has implications for the total effectiveness of all press operations. If adequate means of scrap disposal are not provided, an entire stamping operation could be easily bogged down. The removal of scrap often determines the speed of a trimming operation, and trimming speed often governs the entire press line. Thus, adequate facilities for scrap removal constitute one of the initial considerations in establishing effective stamping operations. In addition to the scrap involved in trimming, thought must also be given to the removal of scrap from blanking, piercing, and other types of cutting operations.

There are two aspects of the problem of scrap removal that stand out. One is the removal of scrap from the die itself, and the other is the removal of scrap from the area of operations. Several principles for solving these problems have been worked out. Consideration will now be given to some of the steps that should be taken.

Scrap Size

One of the first steps in effective scrap handling is to reduce the scrap to manageable size. Trimming involves cutting away portions of material that vary considerably with respect to size. The size of scrap will determine the ease with which it may be removed from the die and also from the press area. The maximum size of scrap that can be handled successfully is 24 in. in length. Smaller pieces are preferable. A length of 18 in. is considered to be optimum for most efficient removal from the press and from the work area.

Removal of Scrap From Dies

The methods to be used in removing trimmings and other scrap from dies should be planned concurrently with die design. This will permit the designer and the process engineer to make certain that adequate space is allowed for any die attachments necessary for removing scrap.

The die should be designed with sufficient height to allow the scrap to fall away from the die surface. This height will depend on the size of scrap and the provisions for receiving it outside the press.

In many cases, chutes should be incorporated in the die to carry the scrap to stock boxes or conveyors. The chutes should be designed to provide ample space for trimmings to fall through. If trimmings become wedged in a chute, the press will have to be shut down until they are cleared. In some cases, the design of the die will permit the scrap to fall directly into stock boxes without the use of chutes. It is necessary that the die design allow each piece of scrap to be completely removed from the die area, or at least to a space where it can be allowed to pile up for occasional removal. Operating speed is seriously reduced in cases where operators must clear away scrap after each press cycle.

Where scrap is large, or where it is trimmed from all around the part, scrap cutters should be provided to reduce it to manageable size. These cutters should be provided with shear to reduce load requirements. They should also be installed with consideration of possible changes in the binder surface of the die.

It is sometimes difficult to direct the fall of scrap to the proper area. This is particularly true of internal trimmings which fall through the die. To remove this type of scrap, small conveyors may be used. These may be of the belt variety, or, where trimmings are very small and tend to slip under a belt, vibration-type conveyors may be used.

Removal of Scrap From Work Area

The most effective means of removing scrap from the work area is to shed it through chutes to a conveyor that will move it directly to the scrap baler. This conveyor is usually situated under the press floor and away from the stamping operations. Where it is unfeasible to use this system, scrap bins or trailers can be used. These too, can often be located in a basement area where they will not interfere with work on the press floor. When these bins and containers are filled, they are taken away to the baler by industrial trucks and empty containers are substituted.

Use of Scrap and Offal

The metal cut away from parts in trimming and blanking operations is often referred to in press-room terminology as either scrap or offal. While there is no particular distinction between these two terms in standard dictionaries, each has a special meaning for the process engineer and each serves an important purpose in over-all company operations.

Offal is generally considered to be the type of trimmings that can be put to further use in the stamping plant. This material is often used for making small parts in low-, medium-, and in some cases high-volume production. Since offal usually is unevenly shaped, hand methods of feeding it to blanking presses are generally used. An operator, by using good judgment, can often obtain good blanks from a large percentage of a piece of offal. Offal is also used as blanks for some small parts that can be successfully blanked and formed from the material in one operation. A process engineer, by giving attention to the trimmings that must be cut away from some parts, can often find means of achieving significant material economy by using the trimmings for producing other stampings. It is estimated that in one stamping plant, 32 tons of steel are saved daily by using offal. This figure could be increased through effective processing.

Many trimmings, especially those from deeply drawn parts, are wrinkled and distorted, and it is impractical to use them in further operations. Other trimmings are too small or too oddly shaped to lend themselves to further use in stamping. Such material is baled and returned to the steel plant where it is used as the very necessary scrap charge in making new steel.

MATERIAL HANDLING IN TRIMMING

Automatic Loading

Trimming frequently sets the pace for a group of related press operations, and thus it is important that trimming dies be built to meet the production-volume requirements of the part. One method of attaining satisfactory speed in a press operation is to cut down the time required for loading the press. While loading takes place, the press is idle. Reduction in loading time can therefore yield a higher degree of utilization of press capacity. To reduce loading time, consideration should be given to the use of automatic loading devices. This possibility should be considered during construction of the die so that space can be allowed for any necessary attachments. For small parts, an automatic loading device may consist of a simple gravity slide that will bring the part to rest properly in the die. For larger parts, a shuttle conveyor may be used.

Automatic Unloading

Another means of reducing waiting time in press operations is to provide speedy automatic means of unloading parts. For small parts, air blasts should be used. For medium parts, an effective unloading system may use lifters to free the part from the die and a positive kicker to eject the part. Provision may also be made in some cases to use the incoming part as an effective tool for forcing the completed part out of the die. Where parts are very large, iron hands are frequently used to extract them. To expedite handling, belt conveyors or stock boxes should be used to receive the parts as they issue from the die.

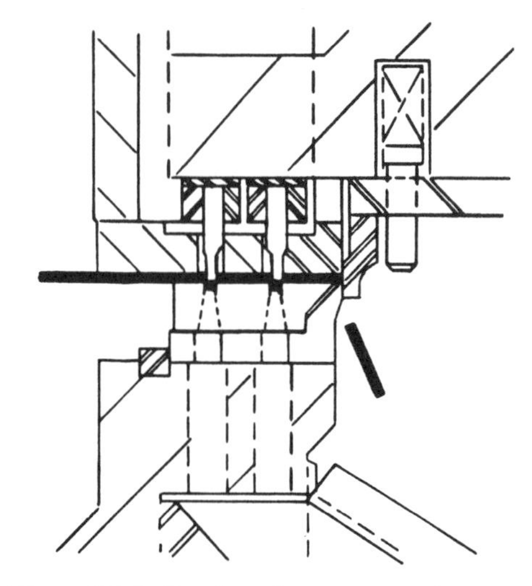

Figure 2F-6. Combined trimming and piercing

COMBINED OPERATIONS

Trimming operations are frequently combined with other types of cutting operations which must for various reasons follow the drawing of a part. Some forming operations also may be successfully combined with trimming.

Trimming and Piercing

Drawn parts frequently must be pierced after drawing in order for the holes not to be deformed. In these cases, it is economical to combine the piercing and trimming operations (Figure 2F-6). Holes may be pierced in the flange of a drawn part in the same die that trims the flange.

Trimming and Flanging

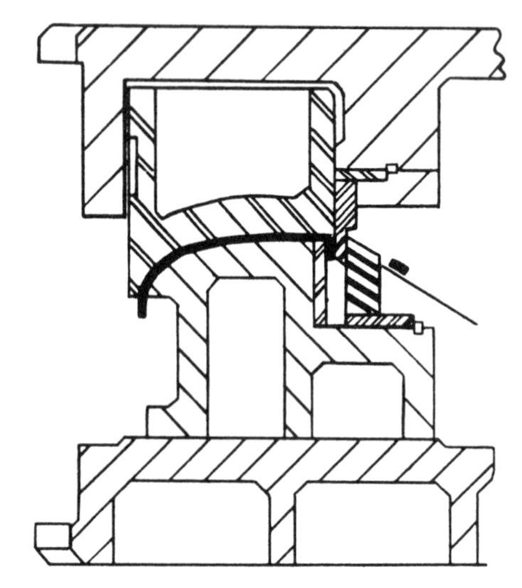

Figure 2F-7. Combined trimming and flanging

Trimming and flanging are also frequently combined (Figure 2F-7). The trimming is accomplished first as the punch depresses the part on a spring pad. When trimming is completed, the punch continues down and forms the flange. Trimming is also combined with many other types of operations.

SUMMARY

Trimming consists of removing excess metal from a part to allow the part to reach the finished stage or to prepare it for subsequent operations. Nearly all drawn parts require trimming to remove the flange material provided in the blank for holding purposes. Formed parts not produced from fully developed blanks also are trimmed.

In analysis of a part requiring trimming, the process engineer has numerous decisions to make. These decisions revolve around whether the part should be trimmed open side up or open side down, whether one or two operations are required, whether notching is required, whether rough trimming is necessary, whether cams can be used to limit the number of operations, and whether other operations can be successfully combined with trimming.

Selection of the proper type of die is another important factor in establishing successful trimming operations. Several types of trimming dies may be used, such as flush, parting, horn, cam, and pinch trimming dies.

Selection of presses for trimming operations is also important. Gap-frame presses may be used for some parts. Care must be taken in using this type of press to prevent deflection of shearing steels which results in excessive tool wear. This may be accomplished by using tie rods across the gap of the press to provide greater rigidity. In some cases, an oversize press may be used.

Straight-side presses present fewer problems in controlling deflection.

Scrap handling is a major problem in setting up trimming operations. The amount of offal and scrap generated by trimming can easily bog down an entire stamping operation unless adequate provision is made for scrap handling. An effort should be made to provide adequate means of removing scrap from the trimming die itself and also from the press area.

For removal of scrap from dies, consideration should be given to the inclusion of scrap chutes, conveyors, and/or stock boxes. Where scrap is large, scrap cutters should be used in the die itself to reduce the scrap to manageable size.

Conveyors, stock boxes, bins, and trailers should be considered in developing means to clear scrap from the area of press operations.

Consideration should also be given to the possibility of using suitable pieces of offal as blanks for small stampings.

2G: Bending of Sheet Metal

BENDING THEORY

The forces applied during bending are in opposite directions, just as in cutting of sheet metal. The bending forces, however, are spread farther apart, resulting in plastic distortion of metal without failure. The simplified sketches in Figure 2G-1 illustrate the forces applied during bending in V-dies, wiping dies, and U-dies. The latter two are more typical of high-production bending dies. The U-die is often referred to as a channel die. The spread of shear forces in a cutting die is equal to the clearance, which is usually about 10% of the sheet-metal thickness. The larger spread of bending forces is accomplished by using a clearance equal to the sheet-metal thickness plus the radii used on the punch and die steels.

Bending of sheet metal has the distinct characteristic of stressing the metal at localized areas only. This localized stress or pressure occurs only at the bend radius. The remaining or flat metal is not stressed during bending. The stresses of bending are illustrated in Figure 2G-2. The metal on the outside of the bend radius has been stretched or elongated, indicating that a tensile stress has been applied. The metal on the inside of the bend radius has been squeezed or placed under a compressive stress. Therefore, if failure or fracturing occurs during bending, it will occur at the outside bend surface. Any wrinkling will occur on the inside surface of the bend.

Neutral Axis. Because the sheet metal is stressed in tension on one surface and in compression on the other, a reversal of stresses must occur. At a certain line, there-

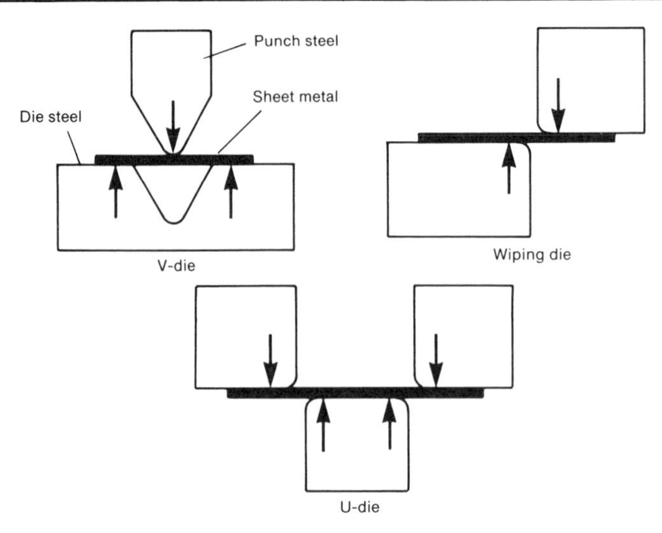

Figure 2G-1. Forces applied during bending

fore, the stresses are zero. This line of zero stress is called the *neutral axis* (see Figure 2G-2).

Before bending, the flat blank is of a certain length, and the length of the neutral axis is, of course, exactly equal to this original blank length. During bending, the outside surface of the sheet metal is increased in length, and the inside surface of the sheet metal is decreased in length, but the length of the neutral axis remains the same. Because the neutral axis is a true representation of the original blank length, it is used for blank-development calculations.

When the blank is first being bent, the neutral axis is near the center of the sheet-metal thickness. As bending progresses, the neutral axis shifts toward the inside, or compression side, of the bend. Normally, the neutral axis is measured as a certain distance

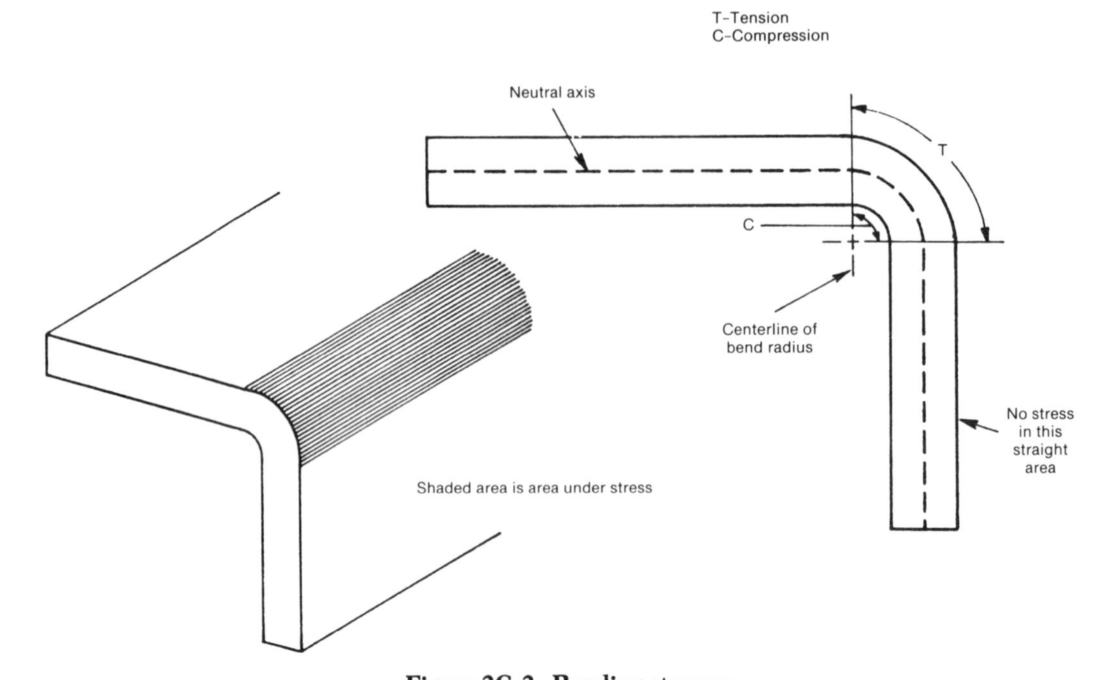

Figure 2G-2. Bending stresses

from the inside surface of the sheet metal at the bend area. Sheet metal thins slightly in the bend area, and the outside surface at the bend radius is not an accurate dimension. The inside surface bends tightly on the die-steel radius and is held to closer tolerances. Therefore, most parts are dimensioned with radii to the inside surface at all bends.

In bending of sheet metal, the distance from the inside surface to the neutral axis is usually about 40% of the thickness. Approximate positions of the axis for various thicknesses are shown in Figure 2G-3. Characteristics of the neutral axis are:

1. If the sheet-metal thickness is constant, the neutral axis shifts closer to the inside surface as the radius of bend is decreased.
2. If the radius of bend is constant, the neutral axis shifts closer to the inside surface as the sheet-metal thickness is increased.
3. If the radius of bend and sheet-metal thickness are constant, the neutral axis shifts closer to the inside surface as the degree of bend is increased.

Because the position of the neutral axis shifts for each variable listed, precise blank-size calculations are frequently difficult. Some alteration of blank dimensions may be necessary after bending the first parts in the die.

Metal Movement. During production bending, one area of the blank is usually held stationary by a pressure plate called a pad. The free blank area is then bent up or down to create the change in contour. The metal forced down or up by the punch steel moves through space to occupy a new position. This metal movement through space is often called swinging and is a characteristic common only in bending operations. Such metal movement does not occur in embossing, stretch forming, or drawing of sheet metal. Illustrations of bending motions are presented in Figure 2G-4. In design of a bending die, the swinging action must be predicted so

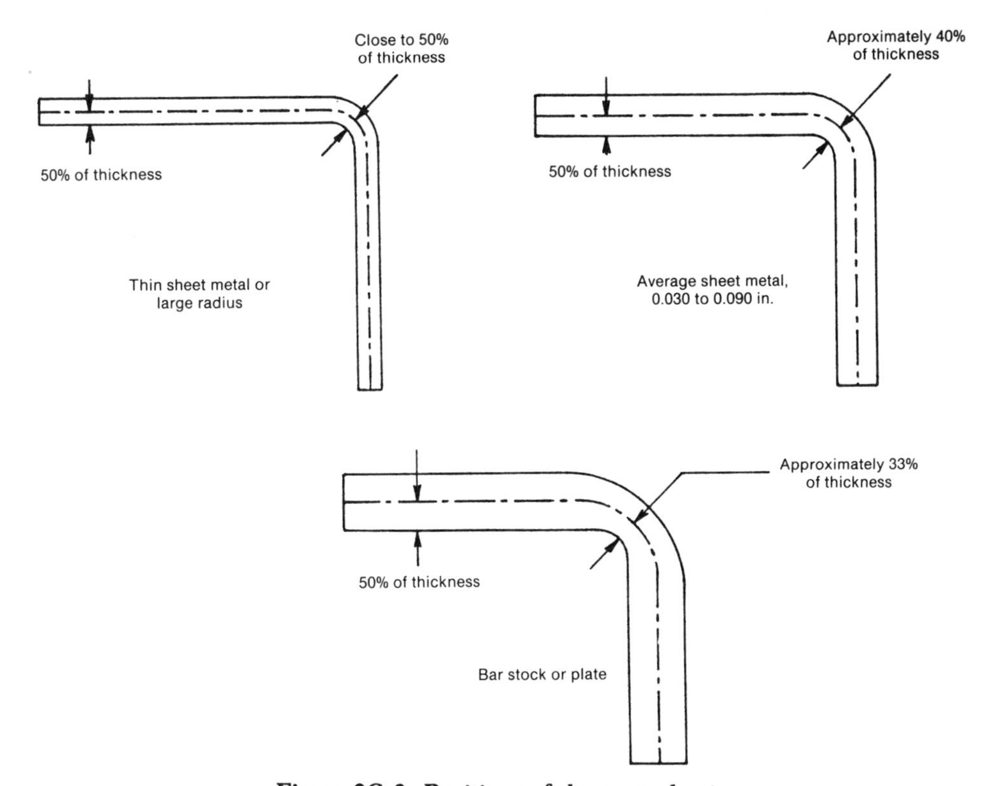

Figure 2G-3. Positions of the neutral axis

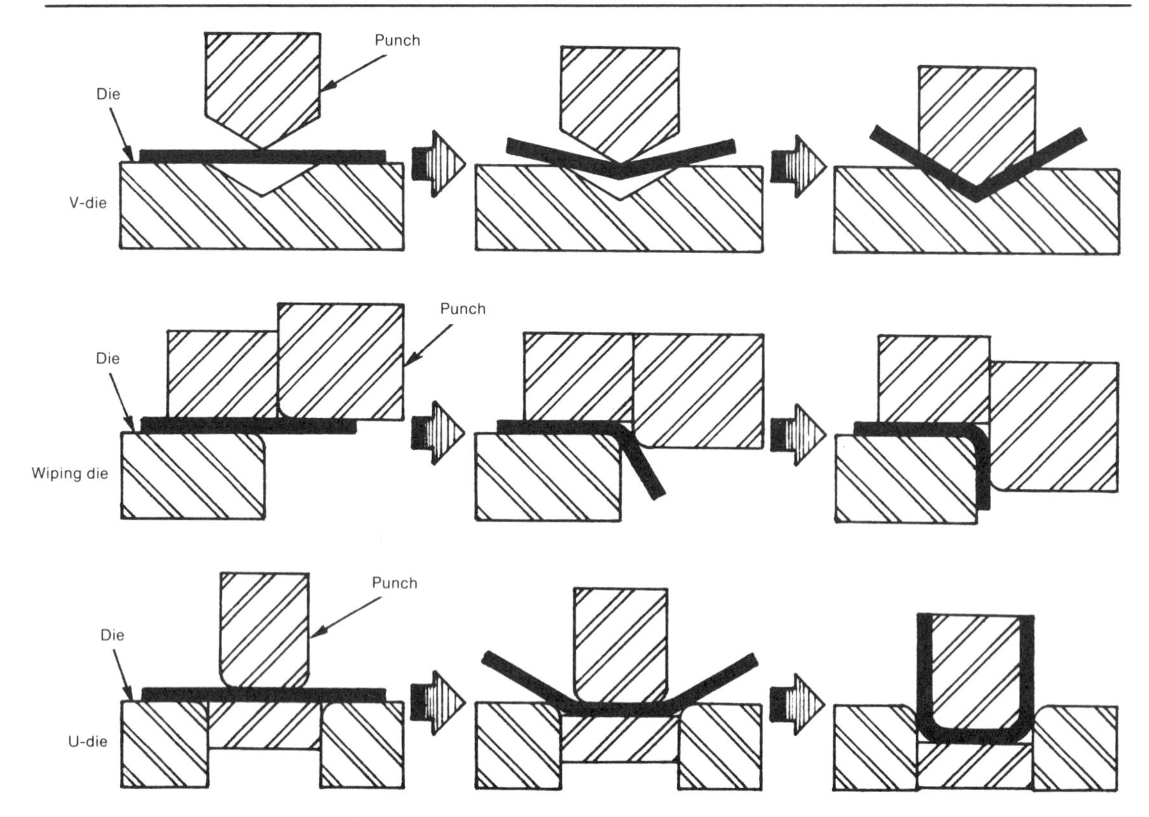

Figure 2G-4. Metal movement during bending

that no obstacles are placed in the way. Because of the metal movement, the larger area is usually held stationary and the smaller blank area is moved by the punch.

Springback. Variations in bending stresses cause springback after bending. The largest tensile stress occurs in the outside surface metal at the bend. The tensile stress decreases toward the center of the sheet thickness and becomes zero at the neutral axis. The pie-shape sketch in Figure 2G-5 depicts the changing tensile and compressive stresses in the bend zone. Since the tensile stresses go from zero at point 0 on the neutral axis to a maximum value at point X on the outside surface, the stress-strain curve developed by the standard tensile test may be used for an analysis of bending. For good bend design, the tensile stress at point X is less than the ultimate tensile strength, as shown. If tensile stess is greater than ultimate strength, the metal may fracture during bending.

The metal nearest the neutral axis has been stressed to values below the elastic limit. This metal creates a narrow elastic band on both sides of the neutral axis, as shown in Figure 2G-6. The metal farther away from the axis has been stressed beyond the yield strength, however, and has been plastically or permanently deformed. When the die opens, the elastic band tries to return to the original flat condition but cannot, due to restriction by the plastic deformation zones. Some slight return does occur as the elastic and plastic zones reach an equilibrium, and this return is known as springback. The springback forces about point 0 as created by the elastic metal are also depicted. Actually, plastically deformed metal has a very small elastic-return characteristic, which adds to the springback.

Variables and their effects on springback are as follows:

1. Harder sheet metals have greater degrees of springback due to a higher

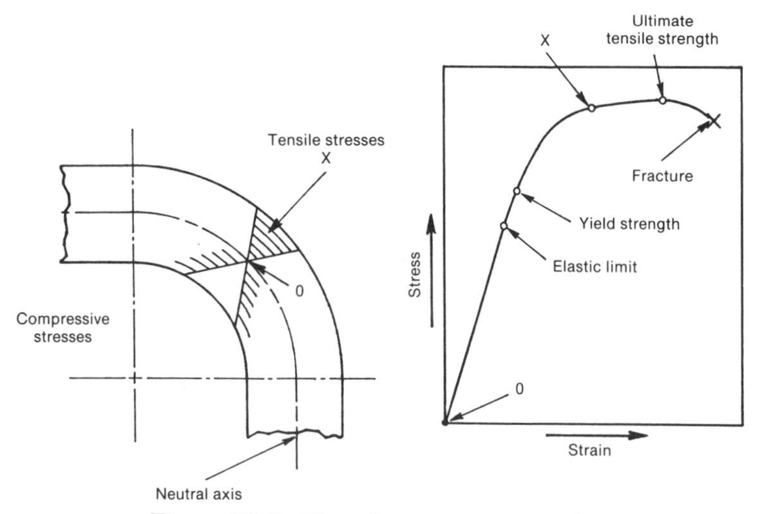

Figure 2G-5. Changing stress patterns in a bend

elastic limit and the resulting larger elastic band at the bend.

2. A sharper or smaller bend radius reduces springback by creating a larger plastic zone and could cause tearing, due to higher stresses in the outside surface.
3. As metal is bent through greater degrees of bend, the plastic zone is enlarged and springback is reduced for each degree of bend. Total springback is increased, however.
4. Thicker sheet metals have less springback, because more plastic deformation occurs, considering no change in the die radius.

Overcoming Springback. Several methods are used to overcome or counteract the effects of springback (Figure 2G-7). These are:

1. Overbending
2. Bottoming or setting
3. Stretch bending.

The sheet metal is often overbent an amount sufficient to produce the desired degree of bend or bend angle after springback. *Overbending* (Figure 2G-8) may be accomplished by using cams, by decreasing the die clearance, or by setting the punch and die steels at a smaller angle than required in the case of a V-die. When the clearance is re-

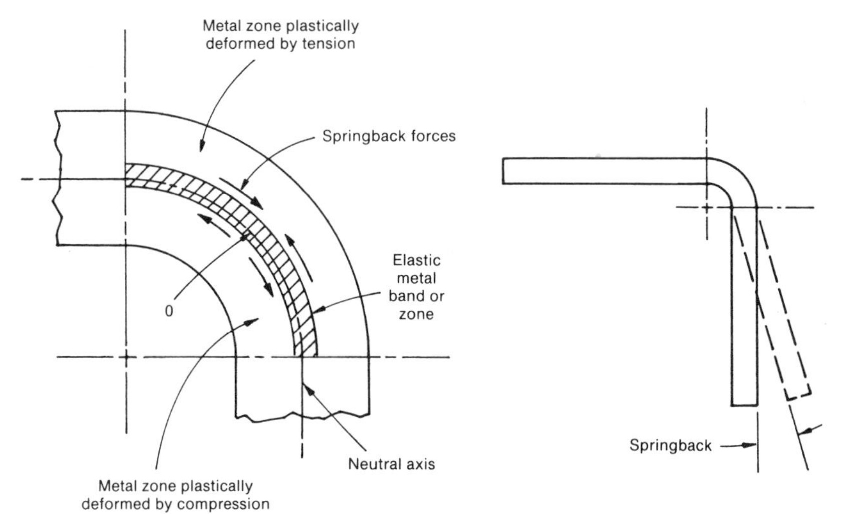

Figure 2G-6. Springback forces

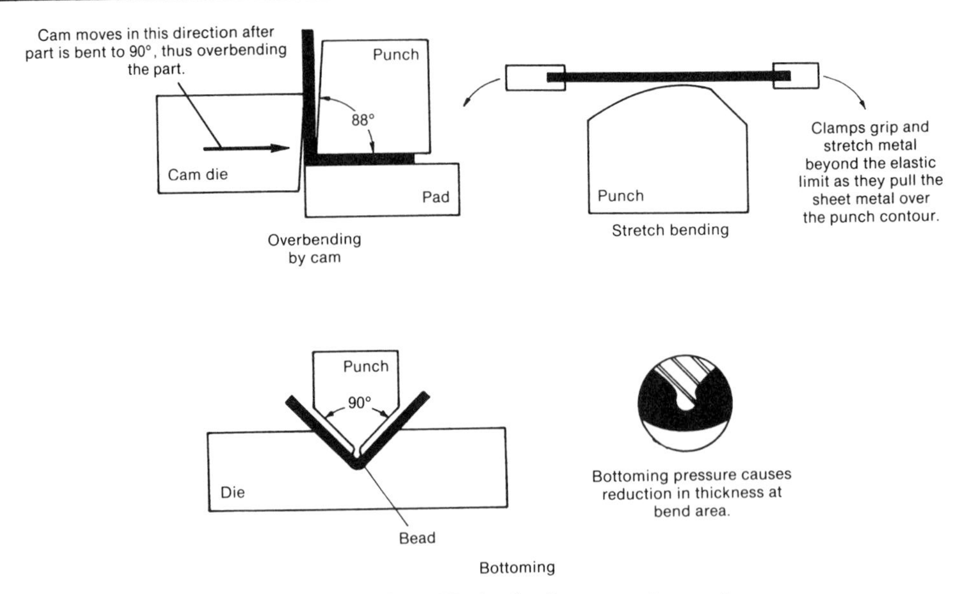

Figure 2G-7. Methods of overcoming spring-back

duced below the sheet-metal thickness, the burnishing action wipes the metal against an undersize punch or die steel.

Bottoming or *setting* consists of striking the metal severely at the radius area. This places the metal under high compressive stresses that set the metal past the yield strength. Bottoming is accomplished by placing a bead on the punch at the bend area. In a wiping die or U-die, the pad must *bottom* against the shoe or backing plate so that the punch may set the metal at the bend. It would be useless to bottom against the flat areas of sheet metal, because they are not stressed and do not cause springback. Also, bottoming against these larger areas would require extremely high press tonnages. Bottoming must be carefully controlled when adjusting the press ram, or the forces involved will rise at a rapid rate. Also, if two blanks are accidentally placed in a bending die that bottoms, press or die breakage may result.

Stretch bending consists of stretching the

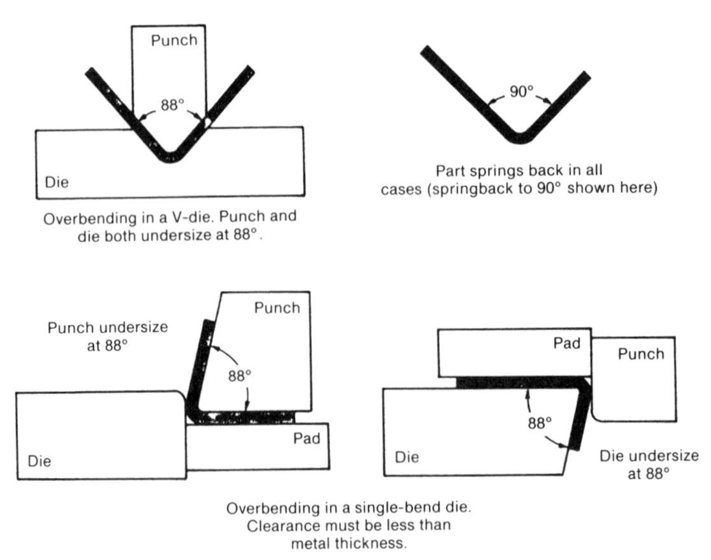

Figure 2G-8. Methods of overbending

blank so that all the metal is stressed past the yield strength. The blank is then forced over the punch to obtain the desired contour. This prestressing before bending results in very little springback. Only relatively large radii are bent by this method, beause sharp radii would take the prestressed metal beyond the ultimate tensile strength. The sheet metal must be uniform in strength. Any weak spots or defects will certainly cause failure. Figure 2G-9 illustrates methods of overcoming springback by bottoming and stretch bending. Stretch bending is most frequently done with a special hydraulic machine rather than with a die in a press. Hourly production rates are slower for such machines than for presses.

BENDING CALCULATIONS

Several formulas are available for computing the forces required for bending. These formulas rely on accurate measurement of the values substituted. Calculations are also necessary for predicting blank dimensions.

Blank Development.. When metal is bent, the length of the part when measured at the neutral axis is the same as the length of the flat blank. The neutral axis is located midway through the sheet-metal thickness from the inside of the bend, which is the compression side.

A common error in determining blank lengths is the failure to add or subtract the sheet-metal thickness when necessary. On part prints, the radius at a bend is shown at the inside surface of the bend. Therefore, when radii are on opposite sides, the sheet-metal thickness must often be considered in the calculations.

A general rule in blank development is to first divide the part into straight sections and bends or arcs. Then the length of each section is found. Often it is necessary to draw in right triangles to connect known to unknown dimensions. Trigonometry is then used to solve for an unknown side or angle. Another rule is that the legs of the triangle should always be drawn parallel to the dimension lines. The hypotenuse then is at the angle of the bend. With this arrangement, the sides of the triangle may be added to or subtracted directly from the dimensions shown on the part print.

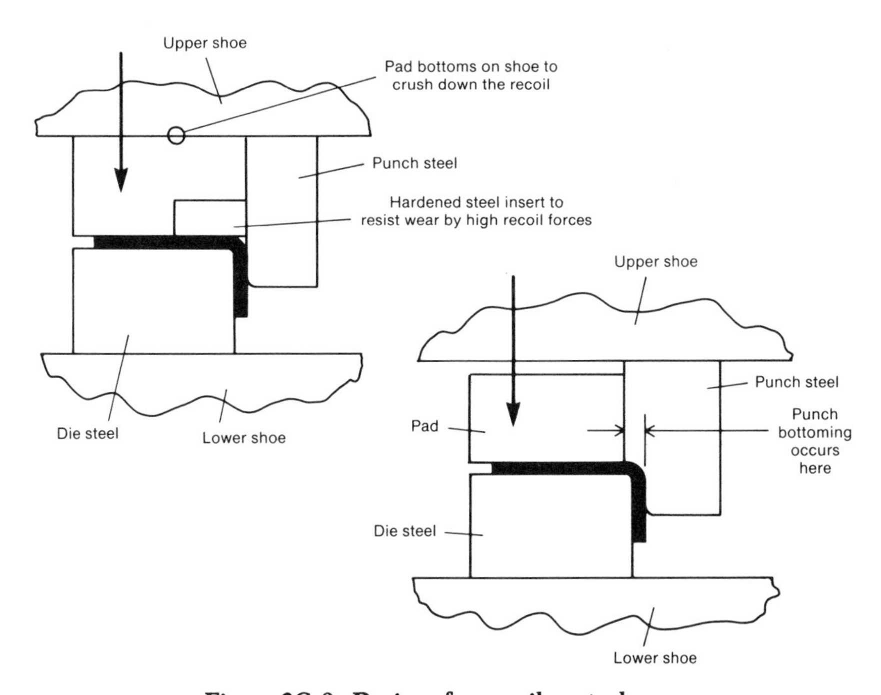

Figure 2G-9. Designs for recoil control

Bottoming Forces. To overcome springback, the bend area is often placed under high compressive stress to set the metal. This operation is called *bottoming* and is accomplished by placing a projection or bead on the punch steel. The force for bottoming is difficult to estimate. This force will increase due to work hardening of the metal if the yield strength is surpassed. Surpassing of the yield strength is all that is necessary to set the metal. The minimum force for bottoming could be estimated using the sheet metal's yield strength. The bottoming force would increase from this minimum value rapidly when press rams were adjusted downward too far.

Pad Forces. Most production dies of the wiping and channel styles require a pressure plate called a *pad* to counteract the bending force. In a wiping die with no pad, the sheet-metal blank would simply flip up vertically in the clearance gap and no bending would occur. The pad acts as a holddown plate in the die. Besides the holddown function, the pad must hold the metal flat against the tip die surface. A frequent difficulty in bending is that the sheet metal in the area next to the bend tends to rise or lift from the die-steel surface. To simplify the explanation, consider the pad as a knife-edge blade rather than a plate, as shown in Figure 2G-10. The lifting action is illustrated; the resulting corner condition at the bend is usually called *recoil.* Under some lighting conditions, a shadow occurs in the depression. A dark line can be seen to follow the bend and may be called a *ghost line.* On some parts requiring excellent appearance, such shadow lines are not permitted.

Although recoil cannot be prevented entirely, die designers use many methods to *control* recoil. The recoiling action causes the pad edge to wear rapidly, and hardened inserts are sometimes used in high-volume dies. The pad can be bottomed against the shoe, but very high tonnages can cause press damage. Bottoming against a lip on the punch steel is more effective and requires less force.

Several comments concerning pads will aid in the design of wiping dies:

1. To restrict recoil, the high pad forces are needed just as bending begins. Springs cannot be fully compressed this early and space is often limited. Most pads must be powered by air or hydraulic cylinders located in the press bed or lower die shoe.
2. Recoil problems are more frequent on long bends or flanges. Not only is the pad force high, but the pad must fit tightly for the entire length of bend. Recoil can occur due to the pad surface not being flat.

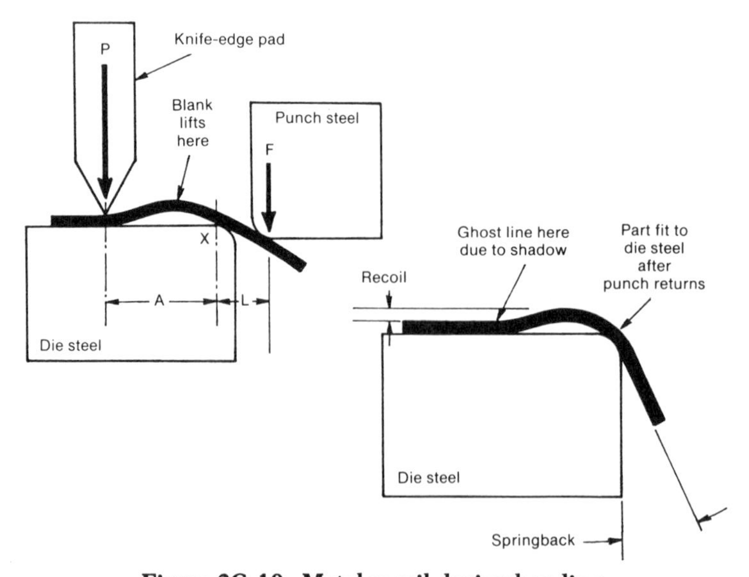

Figure 2G-10. Metal recoil during bending

3. It is vital that pads be well guided so that they do not cock or tip due to the unequal loading caused by recoil. An angular pad would permit some recoil despite high pad forces.

4. A third function of the pad is to prevent the blank from slipping toward the bending punch. The moving punch causes a dragging action on the blank. High pad forces create enough static friction to prevent blank slippage.

Very few experiments have been conducted to study pad-force requirements. Pad force is one of the largest variables between different companies doing presswork-ing of sheet metals.

BENDING OPERATIONS

Figure 2G-11 illustrates a number of bending terms. The characteristics of bending can be found in many operations for shaping of sheet metal, and each operation has a special name. Bending in a production plant may be called bending, flanging, hemming, seaming, curling, or corrugating.

Bending. Bends are made in sheet metal to obtain rigidity and to obtain a part of desired shape for performance of a certain function. Bending is commonly used to produce structural stampings, such as braces, brackets, supports, hinges, angles, and channels.

Flanging. Flanging is similar to bending of sheet metal except for one factor: during flanging, the metal bent down is short compared with the over-all part size.

Hemming and Seaming. The terms "hemming" and "seaming" are used in a manner similar to their use in the clothing

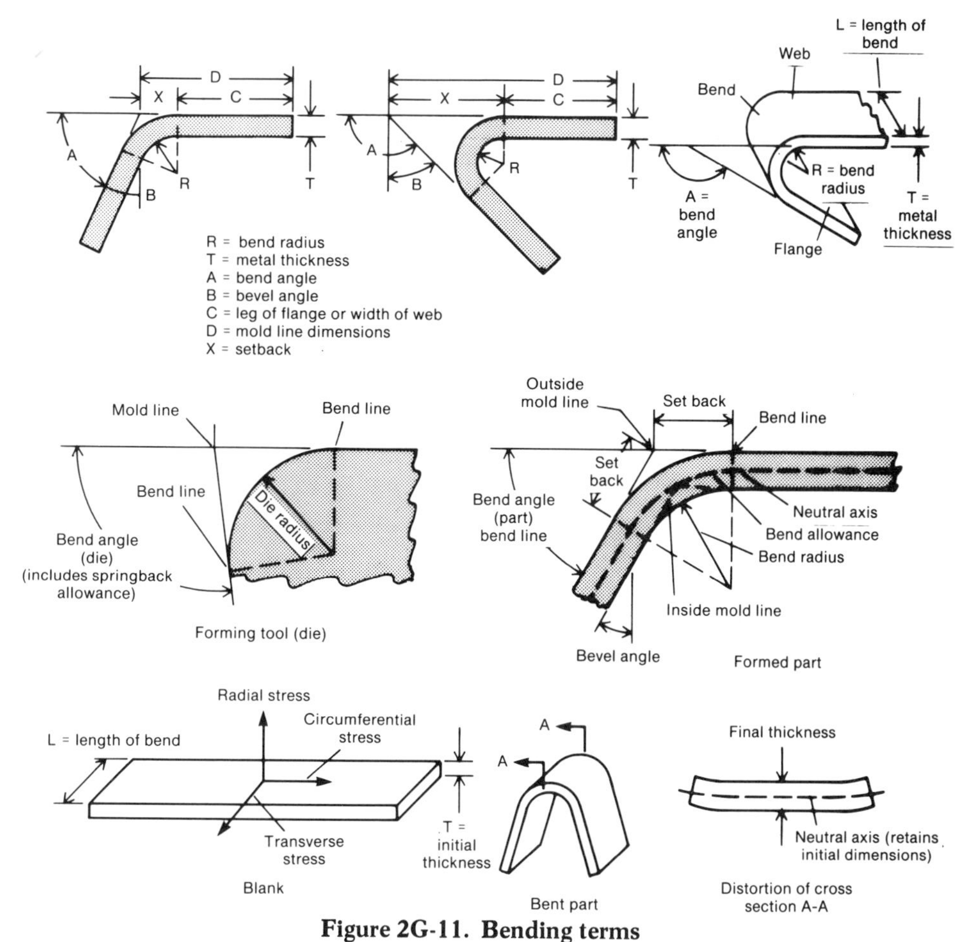

Figure 2G-11. Bending terms

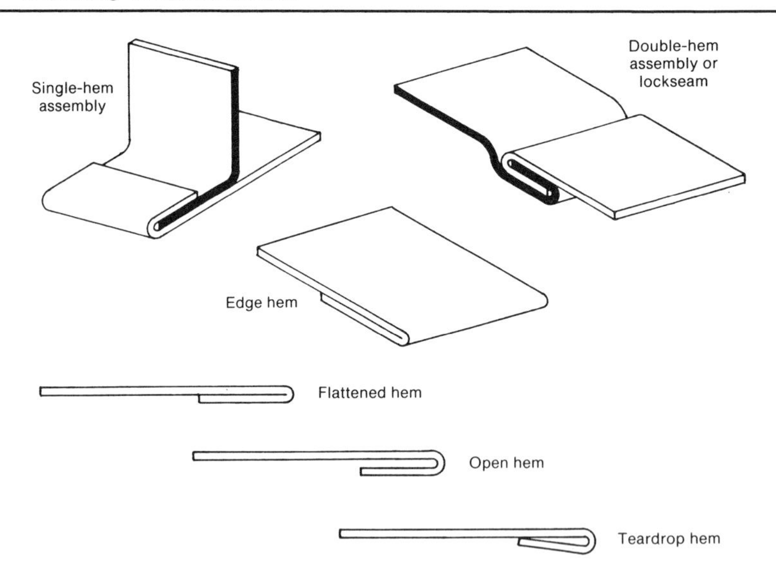

Figure 2G-12. Hems and seams

industry. A hem is a fold at the edge of the sheet metal to remove the burred edge and improve the appearance of the edge. The hem also adds slightly to the rigidity of the edge and improves wear resistance. Hems and seams are shown in Figure 2G-12.

Curling and Wiring. Curling and wiring are used to strengthen the edge of sheet metal. Actually, hems, curls, and wiring may be used on flat parts or on round parts such as cams or drums. During wiring, the metal is curled up and over a length of wire. The edge of the sheet metal is strengthened by both the wire and the curl formed in the metal. When the wire is not used, the operation is known as curling or false wiring (see Figure 2G-13).

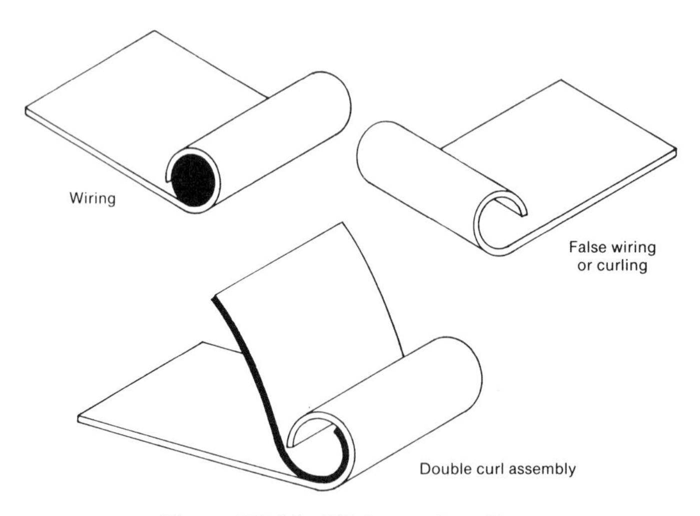

Figure 2G-13. Wiring and curling

2H: Drawing Operations

The drawing group of press operations is confined to the production of parts having compound curved surfaces. These parts range in kind from simple cup shapes to highly complex compound curved shapes. The form of the part is achieved by a plastic flow of metal around the contours of the punch. In each application of drawing, deformation of the metal occurs over a major portion of the part's surface. Hence, drawing is defined as a process in which a punch causes flat metal to flow into a die cavity to assume the shape of a seamless hollow vessel.

Drawing plays an extremely important role in the manufacture of automotive parts. The curves and styling lines of exterior body panels are produced, in most cases, by the drawing process. Such panels include hoods, fenders, roof tops, doors, quarter panels, and deck lids. Many interior panels, such as instrument panels and window moldings, also are shaped by drawing. The drawing process imparts certain properties to the metal of these panels without which extensive bracing and substructure would be required. These properties are stiffness and strength resulting from work hardening of the metal in the stretching process. Without this stiffening and strengthening, the broad expanses of sheet metal on the exterior of an automobile would buckle and vibrate. Some substructures (such as roof rails) assist in preventing these conditions, but the strength of many panels is relied on to prevent undesirable effects, such as "oil canning," caused by loose metal.

The mechanical principles involved in drawing are much the same regardless of the process used. A blank is placed on some type of binder surface and is forced to conform to the shape of the punch as the blank is pushed into the die cavity. The edges of the blank, in almost every case, are confined between the binder surface of the die and some form of blankholder. This confinement is for the purpose of controlling the movement of the metal into the die so that the blank will shape itself around the punch without wrinkles or tears in the area that comprises the part.

Several types of processes may be used to accomplish the drawing of a part. Each type of process has its special mechanical aspects. The selection of a process depends on which one can be used to produce the part most effectively and economically to meet quality and volume requirements. Further considerations in process selection include the requirements for metal holding, the shape and size of the part, and the depth of draw.

SINGLE-ACTION DRAWING

Many small and medium-size stampings are produced by single-action drawing, so called because it is done in single-action presses.

Range of Application

Among the chief advantages of single-action drawing is the speed of operation. The cycle time of single-action presses is relatively short, and, with proper tooling setup, excellent production volume can be achieved.

There are, however, certain limitations in the use of single-action presses in drawing metal stampings for automobiles. Depth of draw generally should not exceed 4 in., blankholding pressure requirements should not be great, and the width and length of a part generally should not exceed 24 and 36 in., respectively. Larger parts with deeper draws have been successfully drawn in these presses, but for larger parts the blankholding pressure requirements are generally too great. Narrow parts having lengths greater than 36 in. have frequently been drawn successfully in single-action presses.

Conventional Single-Action Drawing

In conventional single-action drawing, the part is drawn open side up. A punch is mounted on the press slide and the die is mounted on the lower die shoe.

The simplest type of single-action drawing is done with only a punch and a die. A blank is placed on the die and the punch descends to draw the part. No blankholder is used to control the metal. Such an arrangement is somewhat limited in application. It is used primarily on heavy-gage stock where the edge of the blank is strong enough to resist the tendency for wrinkles to form as a result of the compressive and tensile forces that develop in drawing. Most automobile stampings are made of lighter-gage metal which has a greater tendency to buckle and wrinkle at the edges during drawing. To eliminate the formation of wrinkles, it is necessary to supply a force for holding the blank edges and effecting control of the metal movement.

In conventional single-action drawing, this blankholding device is attached to the punch. The blankholder is often referred to as a pressure pad or draw ring. As the punch descends, this draw ring contacts the surface of the blank in the area of the binder surface. As the punch proceeds down, the draw ring, being spring loaded, presses on the blank edge while the punch shapes the part. Because the springs exert increasingly greater pressure as they are depressed, the force on the blank edge becomes greater and greater. As a result, only shallow draws are possible, because a point is soon reached where the blankholding pressure is too great to permit proper metal flow. This can be overcome by providing a spacer around the blank that permits only a certain amount of pressure to come to bear on the blank. The drawn part is either pushed through the die or stripped off the punch by the pressure pad. If the latter system is used, a positive knockout can be used to free the part from the die. Drawing tools in which the pressure pad is attached to the punch are limited in application because of the restriction of depth of draw caused by the necessity to use short springs and the difficulty of adjusting pressure.

Stretch Drawing

Among the forms of single-action drawing, stretch drawing, sometimes called stretch forming, is most frequently used.

Stretch drawing is distinguished by the position of the finished part in the press. The open side of the part faces downward because the punch is mounted on the lower die shoe. This arrangement permits deeper draws because space is available for longer springs or for air cushions. The die section is mounted on the upper shoe and is the moving member of the die set. For some parts made from heavy stock, the edges of the blank may be strong enough to resist buckling, which will permit stretch drawing without a blankholder.

In stretch drawing, a floating draw ring is also used for holding purposes. This holder is supported above or at the level of the punch by pressure pins extending through the die shoe and bolster to an air cushion or spring-pressure mechanism. The cushion or springs maintain a constant upward pressure on the draw ring.

In operation, a blank is placed on the draw ring, where it may be partly supported by the punch. The upper slide of the press descends, bringing the binder surface of the die in contact with the edge of the blank. The die continues downward with the edge of the blank being held between the draw ring and the binder surface. The force of the press slide overcomes the resistance of the binder ring, forcing it down around the punch. As the metal is contacted by the punch, it is forced to assume the shape of the punch as the die and draw ring continue downward. The metal is caused to flow plastically from between the die binder and the draw ring.

Certain advantages may be gained from using air cushions for supporting the blankholder. Pressure can be adjusted evenly and will be uniform throughout the press stroke. This is particularly important in drawing operations, where control of metal movement often decides the success of the operation. Where insufficient pressure is available, wrinkles will occur. Where excessive pressure is applied, cracks and ruptures will result. Air cushions provide part of the answer to this problem. However, air cushions are limited in the amount of pressure that can be applied. Their main use is in operations producing small parts or parts with relatively shallow draws. For large panels and deep drawing, some other type of blankholding device is required.

DOUBLE-ACTION DRAWING

The blankholding requirements for large panels, which comprise the majority of exterior body panels, necessitates the use of double-action drawing to achieve proper control of metal movement. This type of drawing is done on double-action presses which have a special press slide for the blankholding function.

Application

Double-action drawing is used in producing a wide range of parts. Most parts larger than 24 by 36 in., and those having depths of draw greater than 4 in. are drawn on double-action presses. In addition to this, many smaller, irregularly shaped parts that present difficult problems in metal control are drawn by this method.

In double-action drawing, the punch is mounted on the inner slide of the press and the blankholder is mounted on an outer slide. The part is usually drawn open side up. A blank is placed on the die and the blankholder descends to hold it flat against the face of the die. The punch follows the blankholder and forces the metal into the die opening. The same type of action takes place as occurs in the single-action press in that the metal is forced to flow from between the blankholder and the edges of the die to assume the shape of the punch. A die cushion, in this case, may be used to lift the drawn stamping out of the die.

Drawing operations must frequently be followed by restriking. Restriking operations are really forming operations and do not stretch or deform the metal to any great degree. Their main purpose is to sharpen radii and bring the part to the final dimensional specifications. This is frequently necessary due to the limitations of drawing operations. It may be impossible to obtain a required radius in an initial drawing operation. Hence, the part is struck again on another die to bring the radius to proper size. Restriking is also used to overcome the springback that occurs following initial drawing by overbending or bottoming. Restriking may also be used to obtain certain design features that cannot be incorporated in the initial draw, such as a bead in the edge of the part which would deter metal flow in drawing. Restriking is done on single-action presses and frequently is combined with other operations such as flanging or piercing.

Blankholding

Even with a separate part of the press dedicated to the task of blankholding, the problem of metal control in double-action drawing is a factor requiring careful study and a great deal of attention. Although the exterior panels of an automobile blend into a well-balanced and appealing appearance, the shape of an individual part is frequently unbalanced and irregular from a metalstamping point of view. The creation of the sweeps and curves of such panels calls for optimum control of metal flow in the drawing process. Thus, the pressures exerted on the blankholder must be properly distributed. The blankholder slide on a double-action press is usually adjustable so that the desired pressure can be brought to bear at all four corners or in particular areas. This procedure is satisfactory in some cases, but in others the localized pressure may result in fractures.

There are often areas of a part that require more inflow of metal than is necessary for other areas. In many cases this can be accomplished only by special adaptation of the blankholding shoes and the binder surface of the die.

Provisions for controlling unbalanced metal flow are usually provided by the inclusion of draw beads on the blankholder surface. These draw beads are also used to restrict metal flow to secure the required degree of stretch in a drawn part. The beads often reduce the pressure requirements for the blankholder slide, thus permitting the use of a lower-tonnage press. Another function of draw beads is to work the metal as it moves inward. It is theorized that this working warms the crystals of the metal and thus permits deeper draws.

Beads are usually used to deflect metal into or away from local areas or to retard the movement of metal into the die cavity. Two or more beads may be placed in a given area to achieve maximum control of metal movement. Positioning of beads is usually done in a die tryout, although modifications are sometimes necessary during a production run. When beads are used, a single bead is generally run around the die cavity, and additional beads are placed locally as required.

Placement of beads is often determined by the design of the die. Usually a bead is placed on the binder surface, with a corresponding groove in the upper blankholder surface. The size, spacing, position, and shape of the beads will vary according to the effect desired. Rounded beads are used where metal is to flow in. Square beads, with slightly rounded corners, are used to lock metal out. This latter type of bead is used where stretching is employed to eliminate loose metal.

TRIPLE-ACTION DRAWING

Triple-action presses are sometimes used for drawing parts that cannot be successfully shaped in one single- or double-action press.

Application

Triple-action presses are particularly useful for parts that require drawing in two directions. The third action of the press can draw the reverse shape while the blankholder and upper punch dwell. Parts requiring severe drawing around a planned opening in the interior of the part are also drawn on triple-action presses. In this case, the third action can be used to slit the metal in the scrap area of the part, thus providing relief for the metal to move into the drawn area. This will prevent tearing of the part.

A triple-action press is used to lance and form local areas in producing door panels at the Ford Motor Company. This die draws the outer panel of an automobile door, and the third action lances and draws the recess around the window opening. Because this latter draw is in a direction opposite to that of the first draw, the part would have to be drawn in two separate presses if the triple-action press were not used.

Stretching of Loose Areas

Triple-action presses are also used to provide the greater stretch requirements of local areas in parts. This is accomplished by forming beads or embossments in reverse of the initial draw. Such beads or embossments take up the loose metal in the local area. Obviously, this method of stretching metal cannot be used where it interferes with or adversely affects the engineering design of the part.

DRAWING AIDS

Sheet metal used in drawing operations must be properly conditioned if it is to be successfully drawn to desired shape. This is especially true of cold rolled steel, because cold rolling processes work harden the metal and increase its tensile strength. This hardening effect must be properly controlled to permit drawing.

Annealing

To prepare metal for drawing, annealing processes are employed, usually at the steel plant. Annealing consists of applying heat to metal under controlled conditions to remove the effects of work hardening. A work-hardened metal is in a state of unsatisfied equilibrium. Its crystal structure is distorted and it contains residual stresses tending to return the crystals to their normal symmetry. If the temperature is raised sufficiently, these stresses are relieved through three processes: recovery, recrystallization, and grain growth.

In the process of recovery, heating tends to realign atoms in their crystals, but does not cause a noticeable change in the crystal outline. Internal stresses are relieved without a detectable reduction in mechanical strength. The heating applied to accomplish recovery is relatively mild and of short duration.

In recrystallization, new grains form in

the metal to relieve the inner strains. Three factors control the amount of recrystallization that occurs: the temperature, the time at temperature, and the amount of previous cold work. By judicious manipulation of these factors, the desired crystal structure may be achieved. In this process, the metal softens and undergoes a reduction in mechanical strength and a recovery of its ability to withstand plastic deformation. This last effect has important implications for pressworking of metals.

Grain growth will result when higher temperatures are provided at longer intervals of time than are provided in the recrystallization process. It is believed that atomic activity of metals is increased in this process and that small grains are absorbed into larger grains as the inner strains are relieved. This process tends to return the metal to the state it was in before cold working was done. In conditioning of sheet metal for drawing operations, metal is seldom heated to the point where grain growth occurs to any great extent.

Closely controlled conditions are applied in annealing coils of sheet metal. The process is carried on in atmosphere-controlled furnaces that prevent oxidation of the metal. Metal coils are also covered to keep out harmful gases.

Annealing temperatures for the various types of metals can be obtained from any standard heat treating chart.

A common use of annealing in the stamping plant is torch heating of areas of metal blanks where drawing difficulties are expected.

Flex Rolling

Flex rolling is also used to condition metal for drawing. The light cold working effect of flex rolling has been found to decrease the effects of age hardening and to prevent the appearance of stretcher-strain markings on finished parts. This process is done in the stamping plant.

Lubricants

Another aid to drawing operations in the stamping plant is the use of lubricants. These lubricants, commonly called drawing compounds, facilitate drawing by reducing the friction generated between tools and metal. Their main purpose is to reduce wear of dies caused by friction.

Care must be exercised in the selection of lubricants so as to secure lubricants of proper consistency for the job involved. Some lubricants tend to tarnish or mar some types of metals, which increases the need for caution. Various devices have been developed for the application of lubricants. In many cases, they are sprayed on the metal as it enters the press. Roller devices are also used to spread lubricants on sheet-metal blanks as they are passed through the rolls.

DIE CONSTRUCTION

Although die construction falls primarily in the realm of die designers and diemakers, there are certain significant factors with which the process engineer should be familiar. These factors bear directly on the effectiveness of operations on a day-to-day basis.

Standards and Strength

To provide safeguards against failure of drawing dies in operation, certain standards for strength should be observed in die construction. Casting thicknesses should be as follows: walls should be 2.5 in. thick, ribs 1.75 in. thick, and punch and binder surfaces 3 to 4 in. thick. These are typical dimensions for drawing-die design. There may be some variation depending on the type of panel to be made and the type of stock to be drawn. Care should always be taken to provide dies with strength sufficient to withstand the mechanical forces involved in an operation.

Maintenance Provisions

In die construction, thought also should be given to the maintenance requirements of the die during production runs. To facilitate maintenance and repair, certain provisions can be incorporated in the die. Tapped holes may be provided for inserting crane hooks for lifting the die out of the press. Chain slots should be provided for proper handling of the die by cranes.

Part Handling

Provision should also be made in the design of drawing dies for lifting of the panel by some mechanical means when the die begins to open. This is done so that the part may be more easily removed from the die either by hand or mechanically. Wherever feasible, extractors or ejectors should be incorporated in the die to automatically remove the panel upon its completion.

SUMMARY

Drawing is a metalstamping process in which a punch causes flat metal to flow into a die cavity to assume the shape of a seamless hollow vessel. This process is used extensively in the manufacture of stampings for automobile construction. Typical parts produced by drawing are exterior body panels such as fenders, quarter panels, hoods, door panels, roofs, and deck lids. Also drawn are interior panels such as instrument panels and window moldings.

Three basic types of drawing operations may be employed: single-action, double-action, and triple-action drawing.

Single-action drawing is used in producing many small and medium-size parts. In single-action drawing, the blankholding function is generally performed by a draw ring mounted on springs or air cushions. This blankholding function is extremely important because it controls metal flow into the die and is employed to prevent wrinkles or tears from developing in the part.

In double-action drawing, the blankholding function is performed by a second press slide. This slide provides greater holding pressure and may be adjusted for desired metal control more effectively than a draw ring.

Triple-action drawing is sometimes used for accomplishing operations that cannot be achieved in one single- or double-action press. This is particularly true of parts that must be drawn in two directions. The third action can draw the reverse shape while the blankholder and upper slide dwell. The third action is sometimes used to eliminate loose metal in local areas by stretching it.

For proper drawing, metal must be properly conditioned. Annealing is used in the steel plant, and sometimes in the press plant, to provide metal with the ductile properties necessary for drawing.

Flex rolling is used in the stamping plant to condition coil stock for drawing. Lubricants are applied to sheet-metal blanks to reduce friction in drawing. This provides protection for punches and dies and also helps to protect the metal from tearing.

In constructing drawing dies and punches, care should be taken to provide sufficient strength. Provisions should also be made to facilitate maintenance of the die by incorporating devices for convenient handling of the die in and out of the press. Care should also be taken to provide for automatic loading and unloading of stampings, wherever feasible.

2I: Drawing and Reducing

Power-press operations involving cold drawing and redrawing of metal from flat blanks into cups, pans, and a multitude of other shapes are second only in breadth of application to the blanking and punching operations previously discussed.

The starting point in any drawing operation is to compute the size of blank required to make the desired shell. For most purposes the shell is relatively thin and the bottom corner radius is small, so that both may be neglected. The formula is then merely the equation of the surface area of the shell and the area of the blank:

$$\frac{\pi(D^2)}{4} = \frac{\pi(d^2)}{4} + \pi dh$$

where D is blank diameter, d is shell diameter, and h is shell height. For slide-rule use, this reduces to:

$$D = \sqrt{[d^2 + 4dh]}$$

Neither formula adapts itself to a chart which is easily read, and so it seems best to use the form:

$$D^2/4 = d(h + 1/4d)$$

In the chart shown in Figure 2I-1, then, it is necessary to add one-quarter of the shell diameter to the shell height, for use on the left-hand scale as a starting point. This is quickly done, however, and it is then possible to set the other end of the straightedge at the shell diameter and read the approximate blank diameter on the center scale. An example is shown dotted on the chart for illustration. Standard tables are also published, based on the same formula and covering quite a wide variety of shell sizes (see Table 2I-1).

If the bottom corner radius (r) is relatively large and should be considered, the second formula becomes:

$$D = \sqrt{[(d - 2r)^2 + 4d(h - r) + 2\pi r(d - 0.7r)]}$$

If the metal thickness (t) is relatively great, so that this also should be considered, the formula may be rewritten:

$$D = \sqrt{[(d - 2r - 2t)^2 + 4\,(d - t)(h - r) + 2\pi(r + 0.4t)\,(d - 0.7r - 0.3t)]}$$

This is really based on figuring the volume of metal required rather than the surface area. Volume or weight is also the only proper basis for approximating the dimensions of blanks for shells that are to be subjected to severe ironing (which reduces the wall thickness).

It is rarely possible to compute the size of any blank precisely or to maintain shells of perfectly uniform height in operation, because the thickening and thinning of the wall vary with the completeness of annealing; the heights of ironed shells vary with commercial variations in sheet thickness; and the top edge varies from square to irregular, usually with four more or fewer pronounced high spots resulting from the effect of the direction of rolling on the crystal structure of the metal. Thorough annealing, of course, should largely remove this directional effect. For all of these reasons, it is ordinarily necessary for the blank to be sufficiently oversize to permit a trimming operation. Common practice is to finish the drawing tools first and then determine the blank size by trial, before making the blanking die.

The trial method applies even more particularly to rectangular jobs than to round ones. The length and width of the blank can usually be approximated, because the sides are subjected to nearly straight bending. At the corners, the drawing action cannot be compared directly with that in round shells, because the unstrained sidewalls act as relief areas. The developed corner for a rectangular shell blank, as found by trial, is not bounded by a radius, but more nearly follows a 45° angle across the corner with a little radius at each end.

The chart in Figure 2I-2 is arranged spe-

Table 2I-1. Approximate Diameters of Blanks for Shells(a)

Diameter of shell	Approximate blank diameter for shell height of: ½	⅝	¾	⅞	1	1⅛	1¼	1⅜	1½	1⅝	1¾	1⅞	2	2¼	Diameter of shell
¼	.75	83	90	97	1.03	1.09	1.15	1.20	1.25	1.30	1.34	1.39	1.43	1.52	¼
⅜	.94	1.04	1.13	1.21	1.28	1.34	1.42	1.48	1.55	1.61	1.67	1.71	1.78	1.88	⅜
½	1.12	1.22	1.32	1.41	1.50	1.58	1.66	1.73	1.80	1.87	1.93	2.00	2.07	2.18	½
⅝	1.28	1.40	1.50	1.60	1.70	1.79	1.88	1.96	2.04	2.11	2.19	2.26	2.32	2.46	⅝
¾	1.44	1.56	1.67	1.78	1.89	1.98	2.08	2.16	2.25	2.33	2.41	2.48	2.56	2.70	¾
⅞	1.59	1.72	1.84	1.95	2.06	2.16	2.26	2.36	2.45	2.54	2.62	2.70	2.78	2.94	⅞
1	1.73	1.87	2.00	2.12	2.23	2.34	2.45	2.55	2.64	2.74	2.82	2.91	3.00	3.16	1
1⅛	1.87	2.02	2.15	2.28	2.40	2.51	2.62	2.73	2.83	2.93	3.02	3.11	3.21	3.37	1⅛
1¼	2.01	2.16	2.30	2.43	2.56	2.68	2.80	2.90	3.01	3.11	3.22	3.31	3.40	3.58	1¼
1⅜	2.16	2.31	2.45	2.59	2.72	2.84	2.96	3.08	3.18	3.29	3.39	3.49	3.59	3.77	1⅜
1½	2.29	2.45	2.60	2.74	2.87	3.00	3.12	3.24	3.36	3.46	3.58	3.67	3.77	3.97	1½
1⅝	2.42	2.59	2.74	2.89	3.02	3.15	3.28	3.40	3.52	3.63	3.74	3.85	3.95	4.15	1⅝
1¾	2.56	2.72	2.88	3.03	3.17	3.30	3.43	3.56	3.68	3.80	3.91	4.03	4.14	4.34	1¾
1⅞	2.70	2.86	3.02	3.18	3.32	3.46	3.59	3.72	3.84	3.97	4.08	4.20	4.30	4.52	1⅞
2	2.83	3.00	3.16	3.31	3.46	3.61	3.75	3.87	4.00	4.12	4.24	4.36	4.47	4.69	2
2⅛	2.96	3.13	3.30	3.46	3.61	3.75	3.89	4.02	4.16	4.28	4.40	4.52	4.64	4.86	2⅛
2¼	3.09	3.27	3.44	3.60	3.75	3.90	4.04	4.18	4.31	4.44	4.56	4.68	4.80	5.03	2¼
2⅜	3.22	3.40	3.57	3.74	3.89	4.04	4.18	4.32	4.46	4.59	4.72	4.84	4.96	5.20	2⅜
2½	3.35	3.54	3.71	3.87	4.03	4.18	4.33	4.47	4.61	4.74	4.87	5.00	5.12	5.36	2½
2⅝	3.48	3.67	3.84	4.01	4.17	4.32	4.47	4.62	4.76	4.89	5.03	5.15	5.28	5.52	2⅝
2¾	3.61	3.80	3.98	4.15	4.31	4.47	4.62	4.76	4.90	5.04	5.18	5.31	5.44	5.68	2¾
2⅞	3.75	3.93	4.11	4.28	4.45	4.60	4.76	4.91	5.05	5.19	5.33	5.46	5.59	5.84	2⅞
3	3.87	4.06	4.24	4.42	4.58	4.74	4.90	5.05	5.20	5.34	5.48	5.61	5.74	6.00	3
3⅛	4.00	4.19	4.38	4.55	4.72	4.88	5.04	5.19	5.34	5.48	5.63	5.76	5.90	6.16	3⅛
3¼	4.13	4.32	4.51	4.68	4.85	5.02	5.18	5.33	5.48	5.63	5.77	5.91	6.05	6.31	3¼
3⅜	4.26	4.45	4.64	4.82	4.99	5.15	5.31	5.47	5.62	5.77	5.92	6.05	6.19	6.47	3⅜
3½	4.39	4.58	4.77	4.95	5.12	5.29	5.45	5.61	5.77	5.91	6.06	6.20	6.35	6.61	3½
3⅝	4.51	4.71	4.90	5.08	5.26	5.43	5.59	5.75	5.90	6.06	6.21	6.35	6.49	6.76	3⅝
3¾	4.64	4.84	5.03	5.21	5.39	5.56	5.73	5.89	6.04	6.20	6.35	6.49	6.63	6.91	3¾
3⅞	4.77	4.97	5.16	5.34	5.52	5.69	5.86	6.02	6.18	6.34	6.49	6.64	6.78	7.06	3⅞
4	4.90	5.10	5.29	5.48	5.66	5.83	6.00	6.16	6.32	6.48	6.63	6.78	6.92	7.21	4
4⅛	5.02	5.22	5.42	5.61	5.79	5.96	6.13	6.30	6.46	6.62	6.77	6.92	7.07	7.36	4⅛
4¼	5.15	5.35	5.55	5.74	5.92	6.09	6.27	6.43	6.60	6.76	6.91	7.06	7.21	7.50	4¼
4⅜	5.28	5.48	5.68	5.87	6.05	6.23	6.40	6.57	6.73	6.89	7.05	7.21	7.35	7.64	4⅜
4½	5.40	5.61	5.81	6.00	6.18	6.36	6.53	6.70	6.87	7.03	7.19	7.35	7.50	7.79	4½

4⅝	5.53	5.74	5.93	6.13	6.31	6.49	6.67	6.84	7.01	7.17	7.33	7.48	7.64	7.93	4⅝
4¾	5.66	5.86	6.06	6.26	6.44	6.62	6.80	6.97	7.14	7.31	7.47	7.62	7.78	8.08	4¾
4⅞	5.78	5.99	6.19	6.39	6.57	6.76	6.93	7.11	7.28	7.44	7.60	7.76	7.92	8.22	4⅞
5	5.91	6.12	6.32	6.52	6.70	6.89	7.07	7.24	7.41	7.58	7.74	7.90	8.06	8.36	5
5⅛	6.04	6.25	6.45	6.64	6.83	7.02	7.20	7.37	7.55	7.72	7.88	8.04	8.21	8.51	5⅛
5¼	6.17	6.37	6.58	6.77	6.96	7.15	7.33	7.51	7.68	7.85	8.02	8.18	8.34	8.65	5¼
5⅜	6.29	6.50	6.71	6.90	7.09	7.28	7.46	7.64	7.82	7.99	8.15	8.31	8.48	8.79	5⅜
5½	6.42	6.63	6.83	7.03	7.22	7.41	7.60	7.77	7.95	8.12	8.29	8.45	8.61	8.93	5½
5⅝	6.54	6.76	6.96	7.16	7.35	7.54	7.73	7.91	8.08	8.25	8.42	8.59	8.75	9.07	5⅝
5¾	6.67	6.88	7.09	7.29	7.48	7.67	7.86	8.04	8.22	8.39	8.56	8.72	8.89	9.20	5¾
5⅞	6.80	7.01	7.22	7.42	7.61	7.80	7.99	8.17	8.35	8.52	8.69	8.86	9.02	9.34	5⅞
6	6.92	7.14	7.34	7.55	7.74	7.93	8.12	8.30	8.48	8.66	8.83	9.00	9.16	9.48	6
6¼	7.18	7.39	7.60	7.80	8.00	8.19	8.38	8.57	8.75	8.92	9.10	9.27	9.44	9.76	6¼
6½	7.43	7.64	7.85	8.06	8.26	8.45	8.64	8.83	9.01	9.19	9.36	9.54	9.71	10.03	6½
6¾	7.68	7.90	8.11	8.31	8.51	8.71	8.90	9.09	9.27	9.45	9.63	9.80	9.97	10.31	6¾
7	7.93	8.15	8.36	8.57	8.77	8.97	9.16	9.35	9.53	9.72	9.90	10.07	10.24	10.58	7
7¼	8.18	8.40	8.62	8.82	9.03	9.22	9.42	9.61	9.80	9.98	10.16	10.34	10.51	10.85	7¼
7½	8.44	8.66	8.87	9.08	9.28	9.48	9.68	9.87	10.06	10.24	10.42	10.60	10.78	11.12	7½
7¾	8.69	8.91	9.12	9.33	9.54	9.74	9.94	10.13	10.32	10.50	10.69	10.87	11.04	11.39	7¾
8	8.94	9.16	9.38	9.59	9.79	10.00	10.19	10.39	10.58	10.77	10.95	11.13	11.31	11.66	8
8¼	9.19	9.41	9.63	9.84	10.05	10.25	10.45	10.65	10.84	11.03	11.21	11.39	11.57	11.92	8¼
8½	9.44	9.66	9.88	10.10	10.30	10.51	10.71	10.90	11.10	11.29	11.47	11.66	11.84	12.19	8½
8¾	9.69	9.92	10.13	10.35	10.56	10.76	10.96	11.16	11.36	11.55	11.73	11.92	12.10	12.46	8¾
9	9.95	10.17	10.39	10.60	10.81	11.02	11.21	11.42	11.61	11.81	11.99	12.18	12.36	12.72	9
9¼	10.20	10.42	10.64	10.85	11.07	11.27	11.48	11.68	11.87	12.07	12.26	12.44	12.63	12.99	9¼
9½	10.45	10.67	10.89	11.11	11.32	11.53	11.73	11.93	12.13	12.32	12.52	12.70	12.89	13.25	9½
9¾	10.70	10.92	11.14	11.36	11.57	11.78	11.99	12.19	12.39	12.58	12.77	12.96	13.15	13.52	9¾
10	10.95	11.18	11.40	11.61	11.83	12.04	12.24	12.44	12.64	12.84	13.03	13.22	13.41	13.78	10
10¼	11.20	11.43	11.65	11.87	12.08	12.29	12.50	12.70	12.90	13.10	13.29	13.48	13.67	14.04	10¼
10½	11.45	11.68	11.90	12.12	12.33	12.54	12.75	12.96	13.16	13.36	13.55	13.74	13.93	14.30	10½
10¾	11.70	11.93	12.15	12.37	12.59	12.80	13.01	13.21	13.41	13.61	13.81	14.00	14.19	14.56	10¾
11	11.95	12.18	12.40	12.62	12.84	13.05	13.26	13.47	13.67	13.87	14.07	14.26	14.45	14.83	11
11¼	12.20	12.43	12.67	12.91	13.09	13.30	13.52	13.72	13.93	14.13	14.33	14.52	14.71	15.08	11¼
11½	12.45	12.65	12.90	13.11	13.35	13.56	13.77	13.98	14.18	14.38	14.58	14.78	14.97	15.35	11½
11¾	12.66	12.92	13.15	13.39	13.59	13.82	14.03	14.23	14.44	14.63	14.84	15.04	15.22	15.60	11¾
12	12.96	13.20	13.42	13.63	13.86	14.07	14.28	14.49	14.70	14.88	15.10	15.30	15.50	15.88	12

(a) Blank diameters calculated using the formula $D = \sqrt{d^2 + 4dh}$, where D is blank diameter, d is shell diameter, and h is shell height. Values given are approximate only; they do not include any allowance for stretching of metal. All dimensions are in inches.

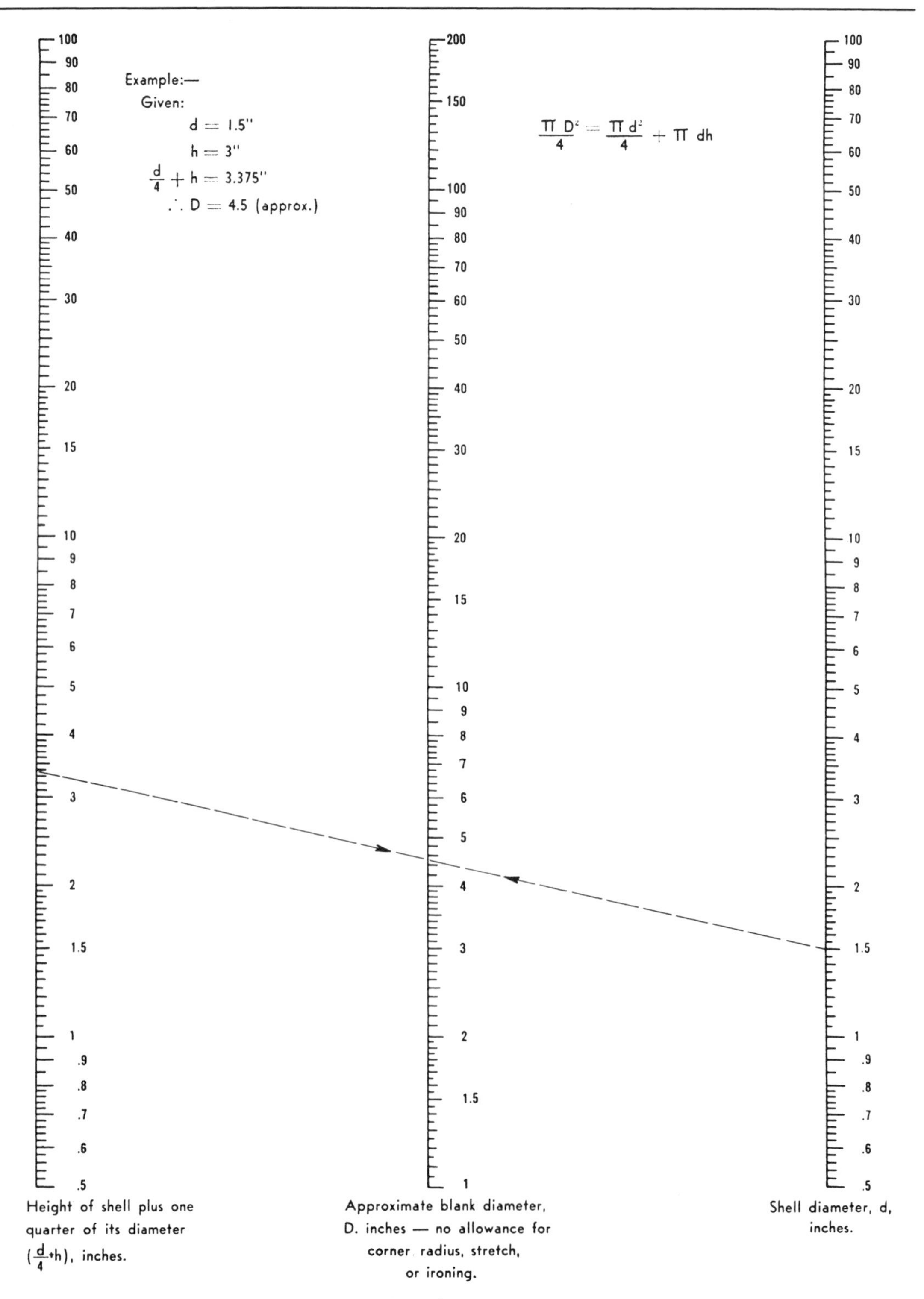

Figure 2I-1. Chart for determining blank diameters

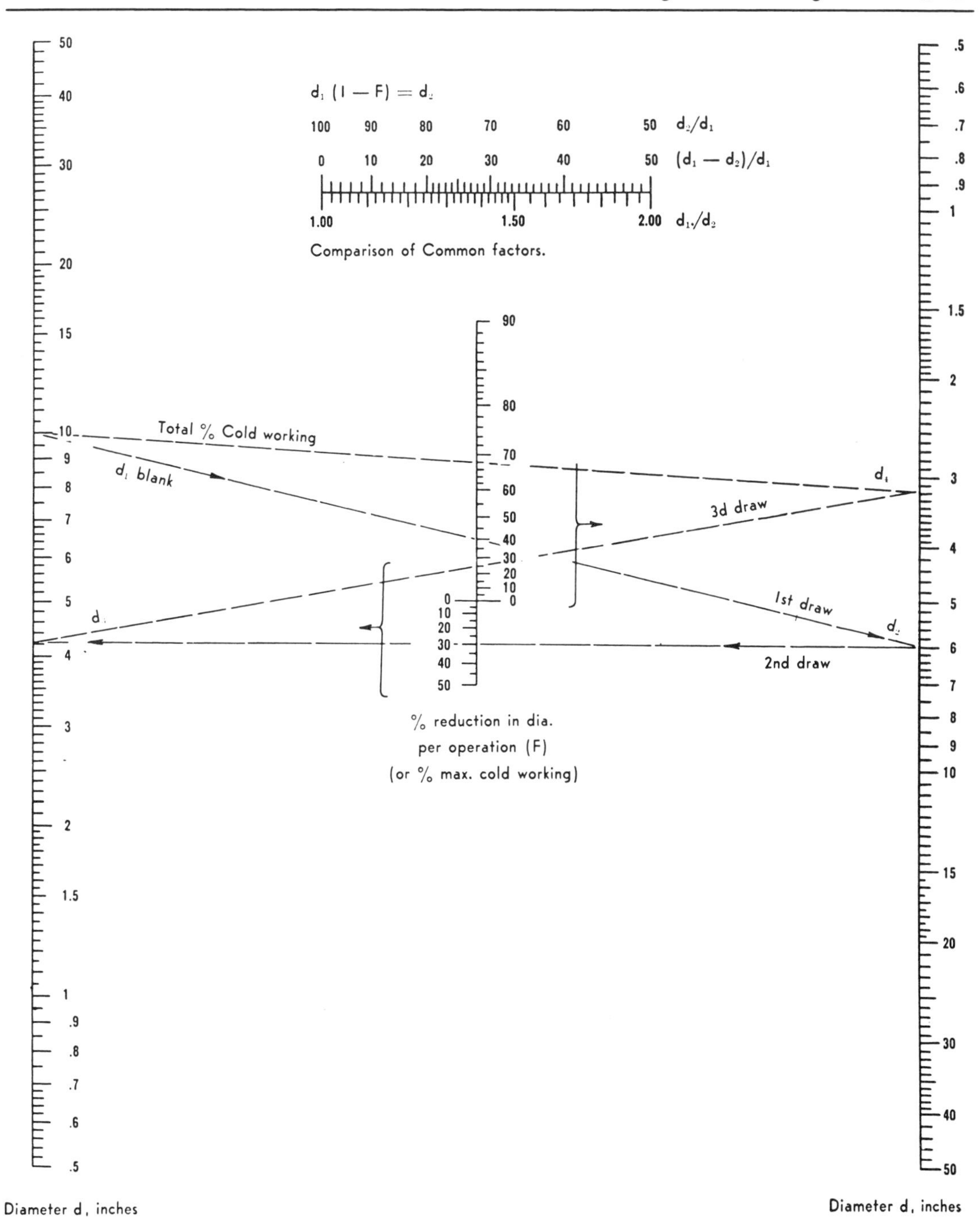

Figure 2I-2. Chart for determining reductions in drawing

cifically for convenience in figuring the number of operations in a series of reductions for a round shell, or for adapting stock drawing rings to a specific job. The scales on the chart are laid out such that they can be read progressively from left to right and then from right to left, etc., as indicated by the arrows. The example shown in dashed lines illustrates the selections of a series of operations for producing a $3^3/_{16}$-in.-diam shell from a 10-in.-diam blank. By following the dotted lines it can be seen that the blank diameter (d) is reduced 40% in the first draw to a 6-in.-diam shell, which is reduced 30% in the second draw to a 4.2-in.-diam shell, which in turn is reduced 25% in the third draw to the required diameter of 3.18 in. This series is conservatively chosen for most cases and should cause no trouble. Trial will also show that the job might be done in two steps assuming a limit 48% reduction to 5.18 in. in the first step and a double-step single-action reduction of 38% for the second, provided the metal is thick enough. Both of these operations would be troublesome to establish, however, and very close control of the material would be necessary. The upper dashed line in Figure 2I-2 indicates that the total reduction from 10 to 3.18 in. amounts to 68%. This is a measure of the amount of cold working to which the metal is subjected, and may be used in judging the need for annealing. Records are available that show total reductions of 67 to 68% between anneals in best-quality drawing steel. However, the initial condition of the metal must be taken into account as well as its rate of strain hardening.

As metal is pulled or squeezed into different shapes in press operations, it grows harder and less ductile. If the action is continued too long, the stress will exceed the ultimate strength of the material, resulting in cracking or tearing. Different metals will stand varying amounts of cold working. In an effort to measure and compare the "workabilities" of various metals, tests were made, values were computed and the charts shown in Figures 2I-3 and 2I-4 were drawn. These curves are offered tenatively, because there is still much room for experimentation and revision.

Only a portion of each curve is useful,

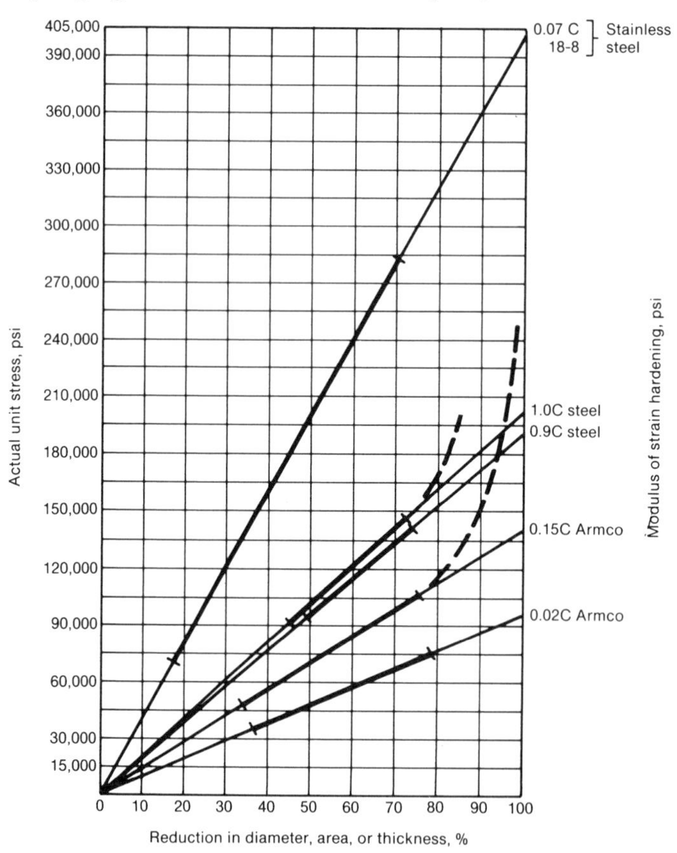

Figure 2I-3. Tentative curves showing commercial cold working range and rate of strain hardening for several steels (dashed portions indicate higher ranges)

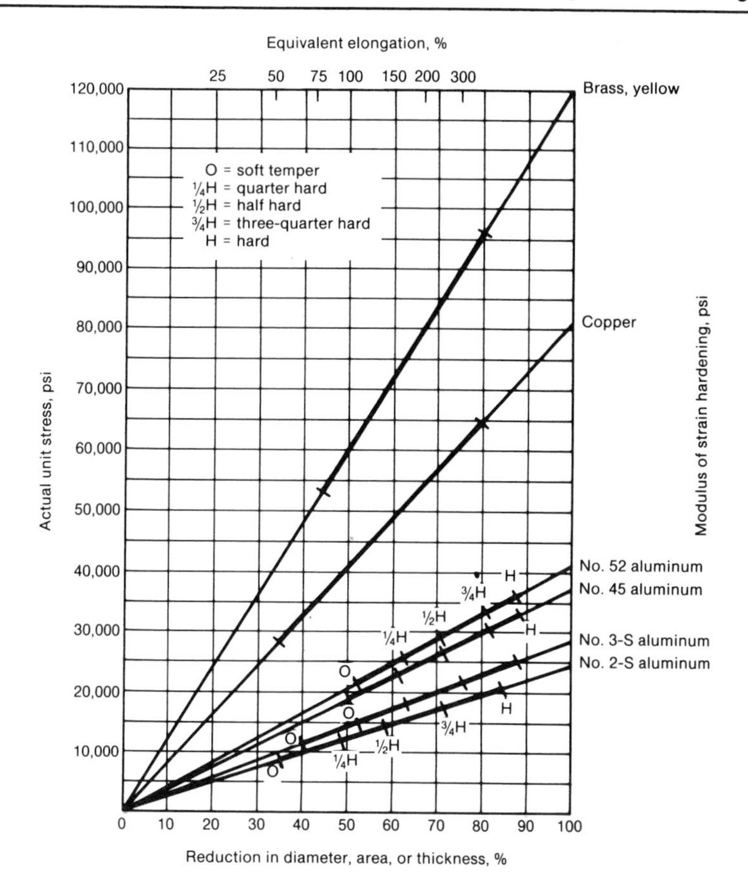

Figure 2I-4. Tentative curves showing commercial cold working range and rate of strain hardening for several nonferrous metals

and this we have shown as a heavy line. One end of the useful portion is the yield point of the metal in its softest condition — the condition after the best commercial anneal. The other end of the useful part of the curve is determined by the normal limit strength of the material. Between these two extremes lies the workable range which is of interest to sheet-metal engineers. The slope of the curve is a measure of the rate of strain hardening. The steeper the slope, the faster the metal hardens when worked.

To illustrate the use of these strain-hardening curves, let us cite an example. Suppose we have a disk of annealed copper 2 in. in diameter that is to be drawn into a cup 0.65 in. in diameter. The question is, can this be done without annealing? In Figure 2I-5 we have plotted the rate of strain hardening for copper, which has a modulus of 81,000 psi. The yield point of copper in the annealed condition is approximately 28,000 psi. The reduction is $(2 - 0.65)\ 2 \times 100 = 67.5\%$. We draw a line from the 67.5% reduction point on the base line (D) to the intersection of the copper strain-hardening curve and the 100% reduction line (E). We draw a horizontal line from point A to the point of intersection with our first line (C) and carry a vertical line from this intersection to the strain-hardening curve (at F). This is merely a graphical method of condensing the diagram; you will note that 100% reduction of the disk would be represented by the distance AB. A 67.5% reduction is equivalent to 67.5% of the length of this line, or AC. Line DE will cut in the same proportion any horizontal line drawn from the strain-hardening line. Thus the condition of the cup is shown by the point where vertical line CF meets the strain-

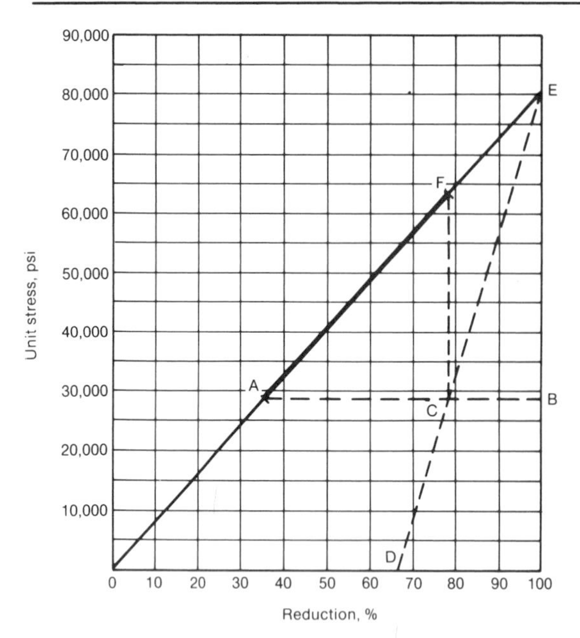

Figure 2I-5. Change in strain-hardened condition of a copper shell in a single operation

hardening curve at F. Here the yield point of the metal is 63,000 psi, which is very near the true ultimate strength of the material. Thus we probably would find any internal fractures starting around the edge of the cup, and if the edge is not to be trimmed, it would be advisable to draw in two or three steps and anneal at least once between draws.

Suppose the disk were only partially annealed. Then the initial yield point would be higher, and a 67.5% reduction would bring the final condition of the cup well beyond the ultimate strength of the copper, thus making one or more intermediate anneals necessary.

Figure 2I-6 is offered tentatively as a means of determining the maximum reductions permissible under various conditions. It is an initial effort at so concise a form and may be open to modification. The top limit of 50% reduction for a first operation seems to be substantiated both by practice and by the theory concerning the strains set up in drawing. The 30% limit for double-action reduction is dictated by practice and is modified by friction, corner radii, and whether the blankholder face is flat or at an angle to the shell wall. The 25% limit for single-action reduction is taken from practice and depends on the formation of wrinkles. Two-step dies in single-action reducing are not especially common, but seem to reduce the tendency to wrinkle and permit a higher limit.

The relative thickness of the metal is the factor which determines whether or not a blankholder is necessary or possible, so that in Figure 2I-6, the basis for comparison is t/d, or blank thickness divided by blank diameter. The lines which mark the boundaries of single- and double-action work are necessarily not sharply defined, because much depends on the radii or angles of various parts of the dies as well as on the material itself.

In order to compensate to a certain extent for the strain hardening of the metal, some die designers prefer to use a descending series of reductions where a number of operations are to be performed without an-

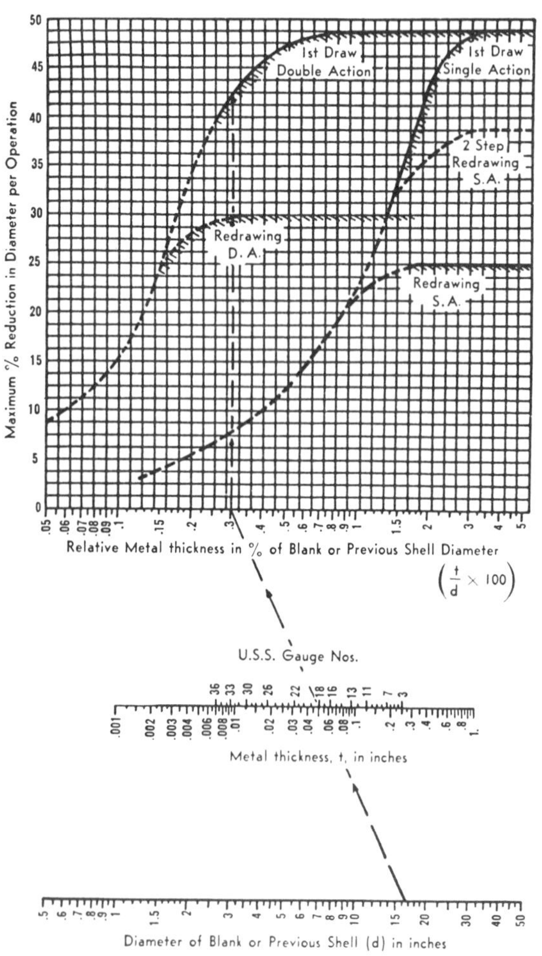

Figure 2I-6. Tentative chart for determining maximum reductions in diameter by various methods

nealing. Thus, for double-action reductions, the successive steps may be 30%, 25%, 20%, 16%, etc. For single-action redrawing, the series might start at 25% reduction. For very thin material, either series would start lower.

It is reasonable to suppose that the 67.5% reduction shown in Figure 2I-5 is excessive and would pull the bottom out. It would be more necessary to draw the shell in two or three operations. The final condition of the shell is the important fact — not the number of steps taken to get there. Suppose the disk is drawn into the cup in three steps, the first a 40% reduction, the second 30%, and the third 20%. These reductions are not additive because the reduction is a per cent change in the dimension at the start of each operation. Thus the second draw of 30% means a 30% reduction in the diameter of the shell after the first draw. It is a comparatively simple matter to show this graphically (see Figure 2I-7). The starting point (A) is the yield point of the disk. After the first draw, giving a 40% reduction, the yield point is at B; after the second draw it is at C; and after the final draw, at D. The sequence of drawing lines is indicated by the number 1, 2, 3, etc. If the cup were annealed after the first draw, A would be the starting point for the second draw.

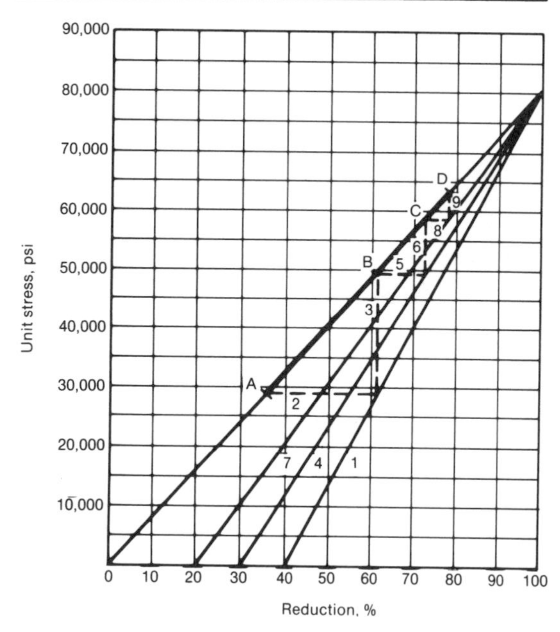

Figure 2I-7. Change in strain-hardened condition of same shell as in Figure 2I-5 if three operations are used to achieve the same total reduction

At the top of the chart in Figure 2I-2 is a scale for comparing various common methods of describing reductions, as a means of reconciling factors used in different plants. Thus the relation between a 10-in.-diam blank and a 6-in.diam shell may be described as a 40% reduction of the blank, a shell with 60% of the blank diameter, or a blank with a diameter 1.66 times the shell diameter. Figure 2I-8 shows a simple scale method of comparing reductions and extensions on a common basis.

In figuring reductions for oval shells, it is well to apply the rules for round shells based on a diameter equal to or slightly larger than twice the radius at the small ends of the oval. In drawing rectangular shells practice has established the possibility of drawing to a

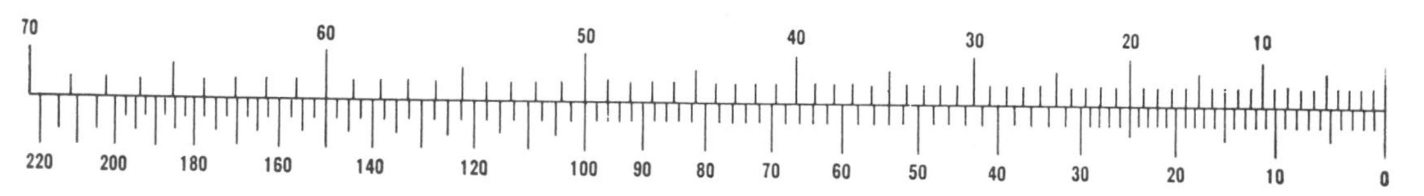

Figure 2I-8. Convenient means of comparing distortions in compression or tension as expressed in percentages

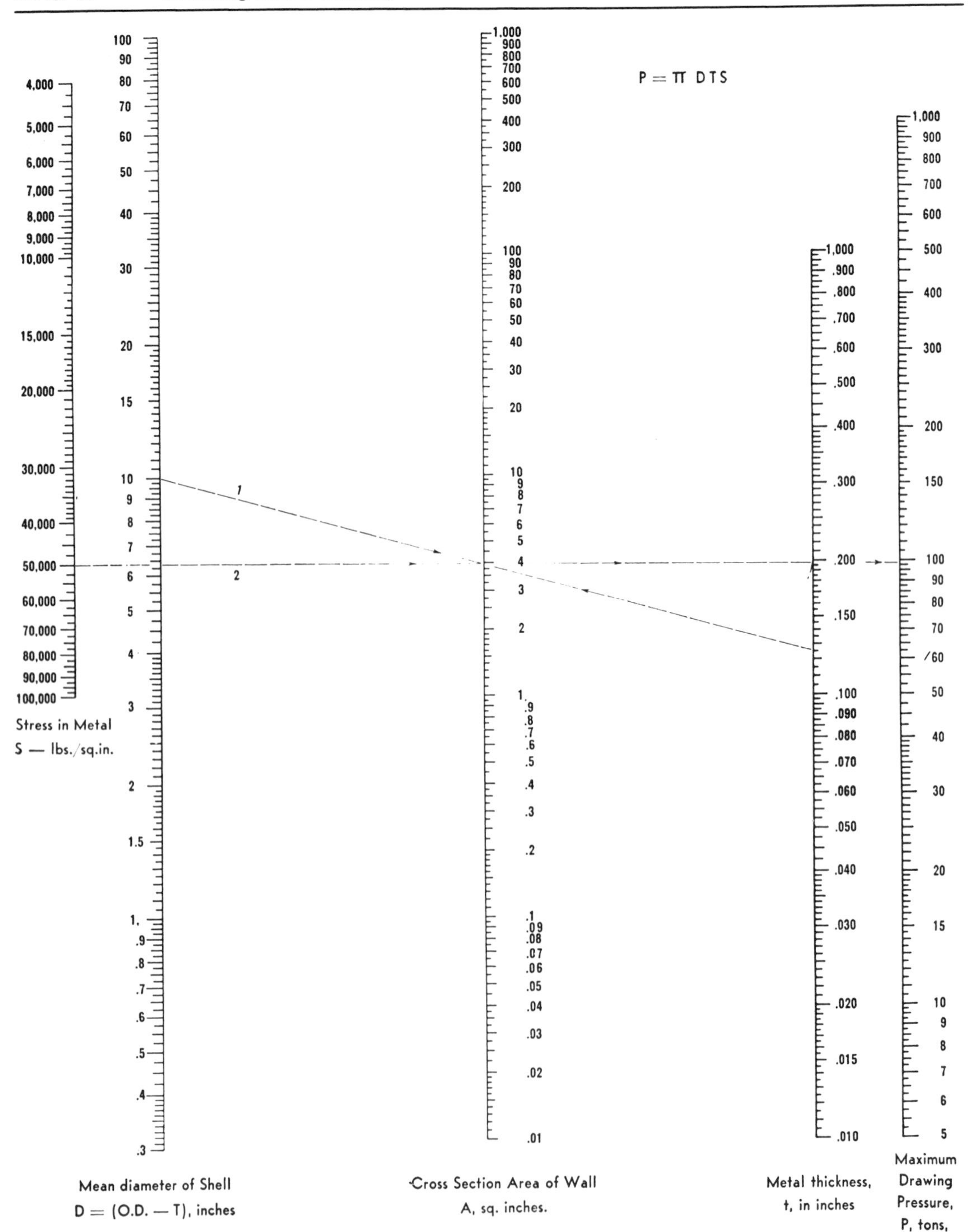

Figure 2I-9. Chart for determining drawing pressures

depth of 4 to 6 times the corner radius. As the corner radius becomes greater with reference to the length and width of the box, the jobs come nearer to the rules for round work. In fact, for very deep rectangular shells, the early operations are round in shape.

The chart in Figure 2I-9, designed for quick computation of the maximum working pressures in drawing operations, is based on a free draw with sufficient clearance so that there is no ironing or burnishing, and on a maximum reduction (nearly 50%). The formula actually gives the "load to pull the bottom out," or the tensile strength near the bottom of the shell. This must exceed the drawing load, or tearing will result. If there is ironing, however, the wall friction carries a considerable part of the added load. Other combinations of operations, such as blanking and drawing, drawing and piercing, drawing and stamping, etc., require that the added operation be figured separately, although it need not be added if it occurs before or after the actual drawing period.

In using the chart, the value given for the outside diameter of the shell will be close enough unless the wall is relatively thick, in which case it is better to use the mean diameter d – t (Figure 2I-10). Following the example shown by the dashed lines, note that a line between a diameter of 10 in. and a wall thickness of 1/8 in. (on the two inner scales) gives a wall cross section of nearly 4 in.² on the center scale. This value and the tensile strength of the material (a deep-drawing steel) on the extreme left scale, give the maximum drawing pressure on the right.

A formula for the drawing pressure P, in pounds, which takes into account the relation between the blank diameter D and punch diameter d (both in inches), is written:

$$P = \pi dtS\,(C - 1 + D/d)$$

where t is blank thickness in inches, S is maximum stress in the metal in psi, and C is a constant varying up to, say, 30 or 40% (0.30 or 0.40) to take care of bending and blankholding.

If the chart in Figure 2I-9 is being used, values of P may be corrected for low shells by multiplying by 1.40 × (D – d)/d, which is merely another form of the previous expression.

The work done in a drawing operation, or the energy required to do it, is roughly the product of the drawing pressure in tons and the depth of draw in inches, giving work in inch-tons. In actual practice, the work requirement is materially less than this due to the gradual rise in pressure. The true mea-

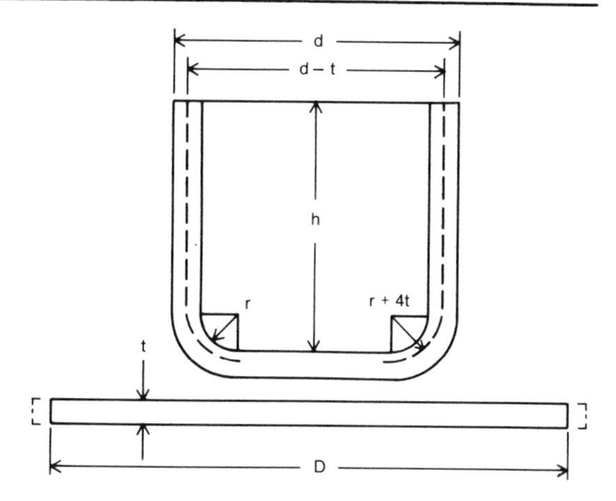

Figure 2I-10. Dimensions of thick-wall shell and blank referred to in blank-diameter formulas

sure of the work done is therefore the area under the drawing-pressure curve. For use in later checkups of flywheel capacity and motor drive, a formula for work, W, in inch-tons, may be expressed as:

$$W = P \times h \times C$$

where P is the corrected drawing pressure in tons, h is the shell height in inches, and the constant C may be taken at 60 to 80% (0.60 to 0.80) to allow for the gradual rise in pressure.

Drawing operations are done in double-action-toggle presses, cam presses, and single-action presses equipped with pneumatic, spring, or rubber drawing attachments. In toggle press construction, the blankholding pressure is taken on rocker-shaft bearings in the press frame so that the crankshaft sustains only the drawing load. The other types take both the drawing and the blankholding loads on the crankshaft, and allowances should therefore be made in computing press capacities. For deep drawing of round work, the blankholding pressure should be under 30 or 40% of the drawing pressure. For large rectangular work, the drawing load is lower than for round work, but the allowance for blankholding may run as high as 100%. In stretching shallow shapes such as casket tops, the metal must be gripped tightly around the edge, and blankholding pressure may be two or three times the relatively low drawing load.

2J: Progressive Dies

A progressive die incorporates several individual work stations where successive operations are performed on a part. Frequently, all the operations required to complete the part are incorporated in a single die.

Progressive dies are also called "cut and carry dies," "follow dies," or "gang dies" — labels descriptive of the operational aspects of such dies. Each work station performs one or more distinct operations on the part. The part is carried from station to station by the stock strip to which it is left attached throughout the operation. As the part moves forward, it is positively positioned in each work station until it is completed and cut out of the strip in the final operation.

FACTORS IN SELECTING PROGRESSIVE DIES

The operations performed in a progressive die could be done in a series of dies involving the use of several presses, but this would require a good deal of handling to move the parts from one press to another. Although this multiplicity of machines and material handling would seem to indicate that progressive dies should always be used, there are a number of factors that must be considered before the feasibility of using a progressive die can be established. Perhaps the most important of these factors is the part itself. Progressive dies are more or less restricted to the production of small stampings, including hubcaps and headlight shells. Handling problems often preclude the use of progressive dies for large parts.

Increase in Productivity

The primary purpose of using a progressive die is to increase productivity — that is, to achieve greater production at the same or less cost in manpower, materials, and machines. Thus, all the factors related to an operation must be considered to determine if a progressive die will meet these objectives.

Costs

The costs involved in the use of progressive dies are frequently a decisive factor in their selection. Progressive dies, due to their complexity, are generally very expensive to build. This expense must be weighed against anticipated savings in manpower, materials, and machines. The progressive die process and alternative processes such as line-die, multislide, or transfer processes must be compared to determine at what piece price each can produce a given part. If there is enough advantage in using a progressive die, the construction cost may be justified by a decreased piece price.

Production Volume

The production volume of the part to be produced is a most important factor in determining whether to use a progressive die. This matter ties in closely with the subject of over-all cost. Frequently, the die itself will be the most expensive item in a progressive operation. The cost of the die will ordinarily have to be amortized within the limits of the production run. For this reason, the use of progressive dies is generally restricted to high-volume parts. Except in rare instances, the cost of the die would prohibit progressive operations for low-volume parts. Some low-volume work may be done on progressive dies where only a few operations (two or three) are involved and the die construction is very simple.

Press Availability

Another factor involved in determining the selection of a progressive die is the availability of a press. Presses for progressive dies usually require wide beds to accommodate the multiple work stations of the die. The size of bed is, of course, dependent on the size of the part to be made.

Stock Requirements

Because of certain limitations presently inherent in progressive die operations, the material required to make the part is also a factor in determining whether to use a progressive die.

The progressive die utilizes a mechanical system to pilot the strip or coil stock into the various work stations. If the stock for a given part is too thin, it will wrinkle or buckle while being fed into the press. The wrinkles formed will result in scrapped production. As a result, progressive dies can be used only with stock that is of heavy enough gage to permit proper piloting.

Restrictions with respect to heavy-gage stock are also encountered in the use of progressive dies. These restrictions are related to the problem of supplying straight, well-leveled stock to the die. Roller-levelers or stock straighteners are usually used. But, if the stock is too heavy, proper results may be impossible to achieve. If wavy or uneven stock is introduced to the die, scrapped or damaged parts will generally be made. As a result there are restrictions on the thickness of stock that can be used in progressive dies.

Because progressive dies depend on a carrier strip to move the part through the work station, coil stock must generally be used. Some progressive operations may be done with strip stock, but coil stock is preferred for greatest efficiency. Single blanks cannot be accommodated in progressive dies although they are used for progressive work in transfer-type operations where separate die stations and transfer devices are employed. Therefore, where progressive dies are contemplated, the availability of coil stock should be ascertained well in advance of the beginning of die construction. The proper size of coil and proper gage of steel should be determined and ordered from the steel supplier.

STRIP DEVELOPMENT FOR PROGRESSIVE DIES

To determine the best coil size to use for a progressive die, a strip development should be made. A strip development is a visual representation of the operational steps a part will go through from start to finish in a progressive die. It provides a tentative picture of the position and relationship of each work station in the die and indicates the distance between each station. From this development, the process engineer determines how wide the strip should be (Figures 2J-1 to 2J-7).

Purposes of Strip Development

In addition to providing a means of determining coil size, numerous other impor-

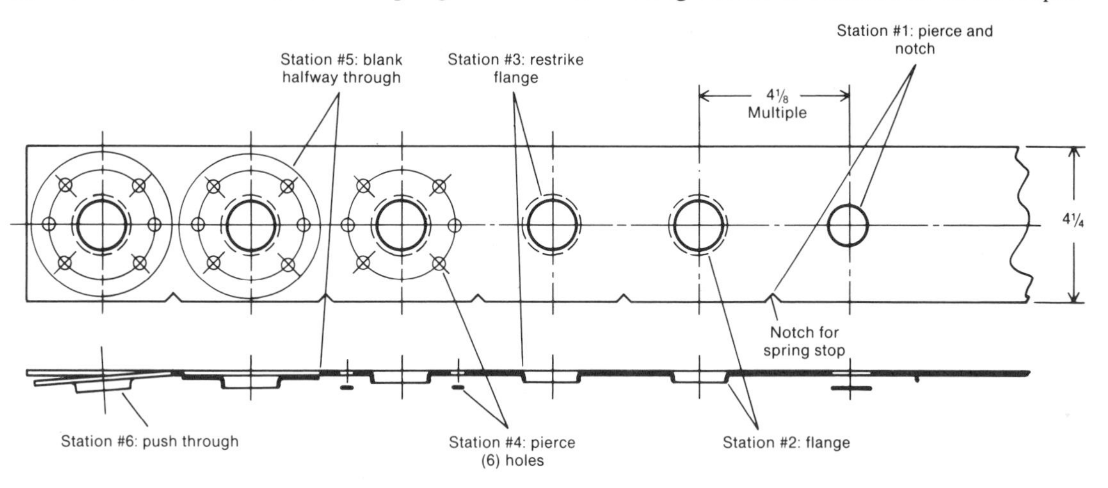

Note locating notch pierced in first station. Flanges formed prior to piercing of nearby holes.

Figure 2J-1. Strip development for a ring-shape part.

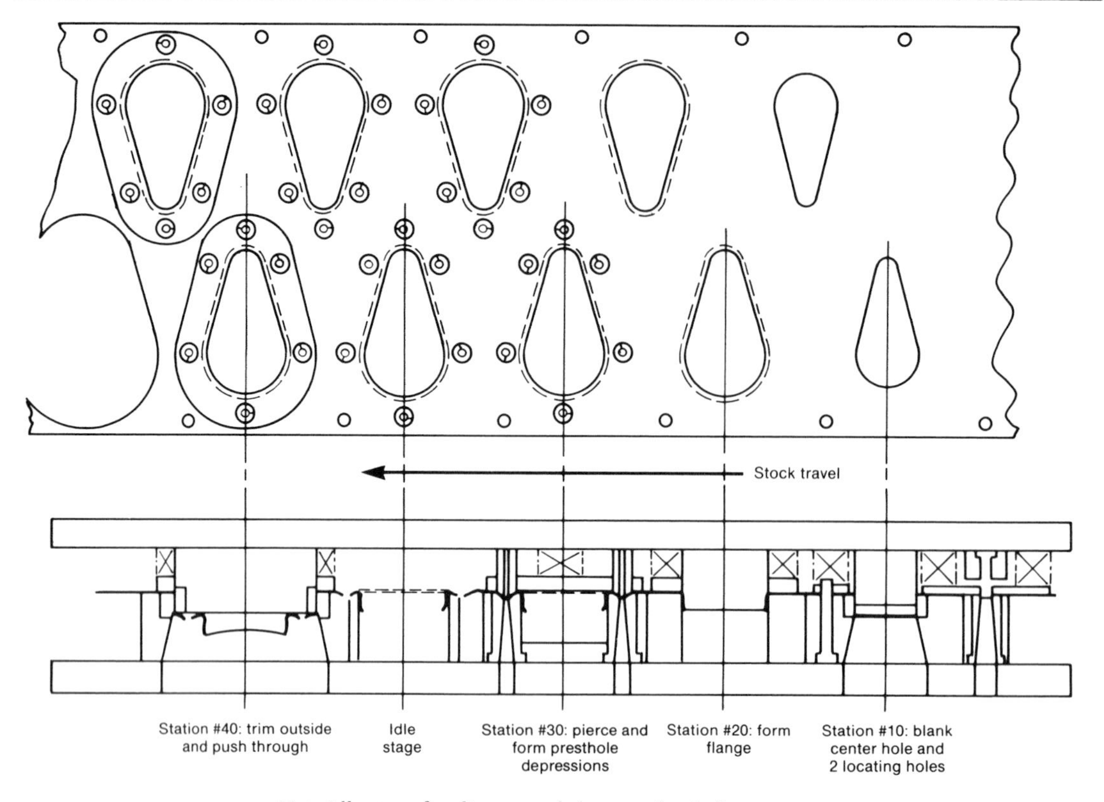

Note idle stage for die strength, layout of strip for material economy.

Figure 2J-2. Strip development for a ring-shape part (two at a time)

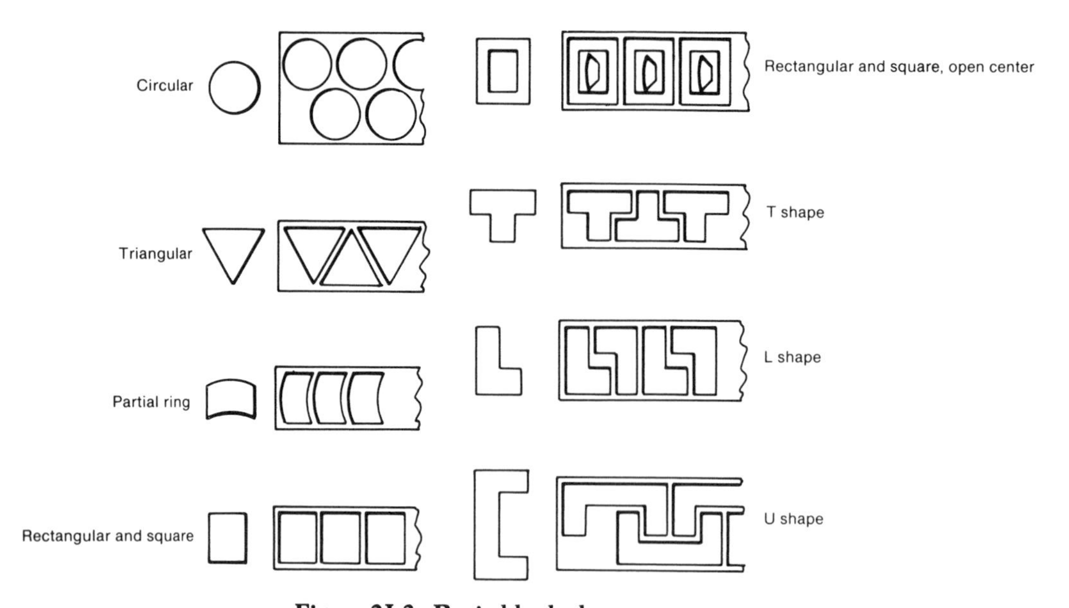

Figure 2J-3. Basic blank shapes

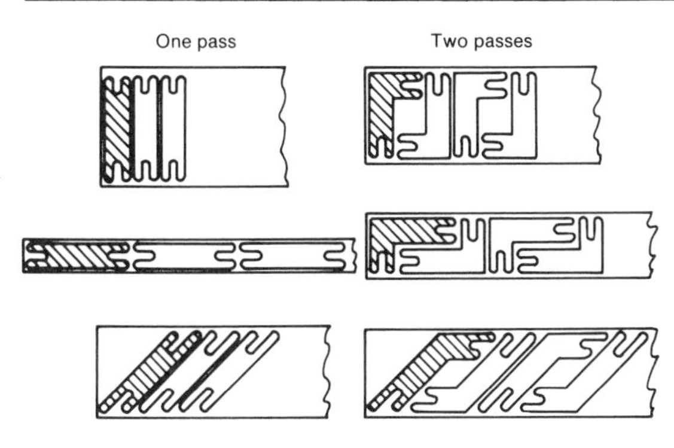

Figure 2J-4. Single-row blank layouts (shaded areas represent punches)

tant purposes are served by a strip development.

Practicality. The practicality of using a progressive die to make the part in question is one of the first considerations in using a strip layout. By obtaining a visual representation of the operations involved, it is possible to determine whether it is practical to use the progressive method. In some cases, an excessively, complex die design may be required. In others, the operations involved may be too difficult. In these cases, planning would revert to the use of line dies. Layouts that are too long for available presses, are too wide, or require too much tonnage are typical of the unfavorable conditions which may be uncovered in the study of a strip development.

Material Economy. A strip development also provides a means of arriving at the most economical use of material. Economy of material usage must be considered so as to obtain fully the possible reductions of manufacturing costs through use of progressive dies. The strip development should be studied with a view toward using the narrowest possible carrier strips consistent with requirements for producing parts of acceptable quality. It may be discovered that carrier strips outside the part are most feasible. On the other hand, some parts may be carried most effectively by a center strip. In many cases, experimental layouts indicate that the greatest economy can be achieved by running two parts at a time — a procedure which usually provides surer control of the part through the die. This procedure yields more balanced conditions with respect to the thrust of the ram. Often a right-hand and left-hand part can be run together. This gives a twofold advantage by eliminating separate manufacturing costs while contributing to a better die. The layout should be studied with a view toward reducing scrap production as much as possible. This may call for changes in the original strip layout. Such changes should be discovered early in the die-development process in order to procure the proper size of stock in time for the beginning of production. The objective for every part should be to obtain a "no-scrap" layout. This, of course, is a physical improbability for most parts, but adopting this objective will lead to the maximum scrap reduction.

Part-Design Changes. Study of a strip development may also indicate changes in part design. In progressive die work, some

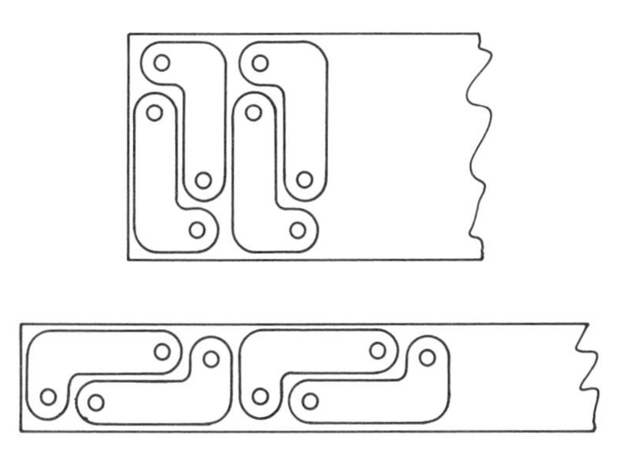

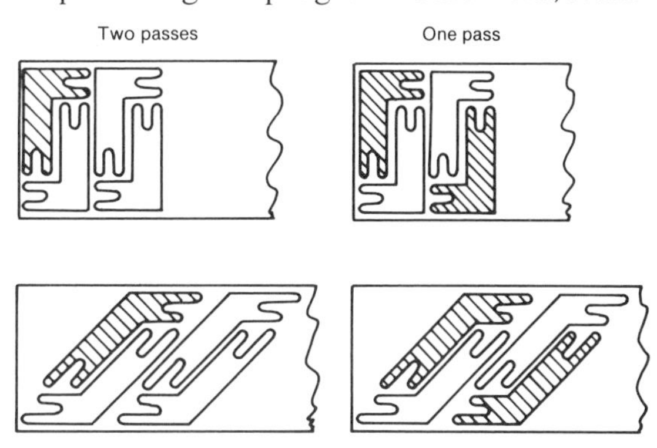

Figure 2J-5. Double-row blank layouts (shaded areas represent punches)

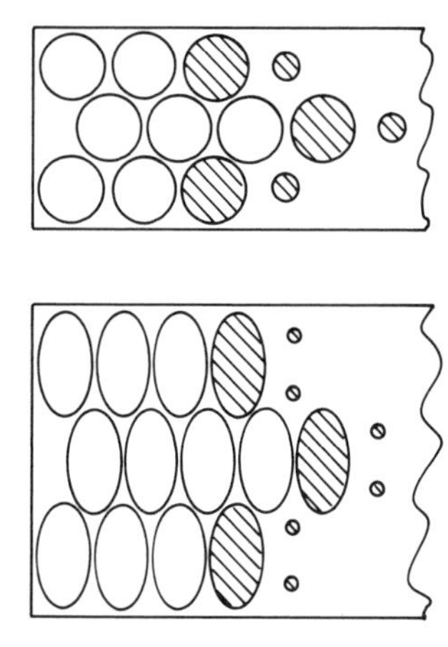

Figure 2J-6. Triple-row blank layouts (shaded areas represent punches)

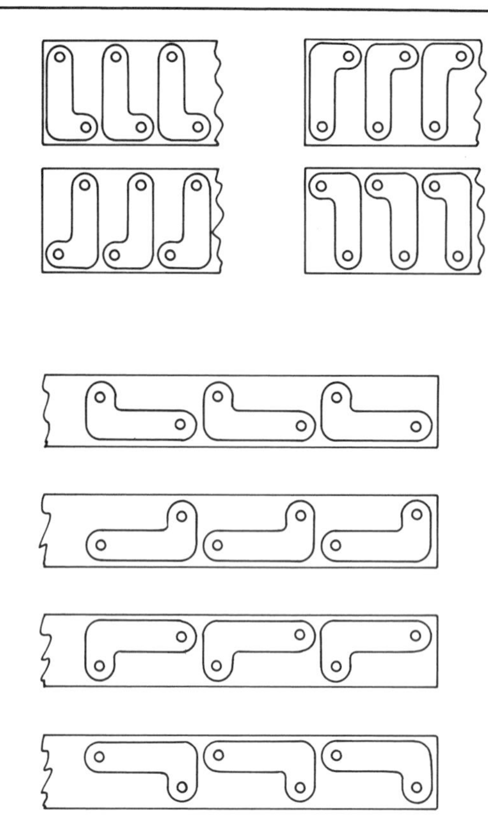

Figure 2J-7. Parts positioned for wide (top) and narrow (bottom) runs

notching and piercing is usually done in the first stations to allow for forming. When the part is cut out of the strip in the final station, the cutoff punch must line up with one or more of these previously cut notches. To avoid "slivers" and to obtain extreme accuracy in strip location, the cutoff line is usually designed to meet the radius of the notch. To obtain this desirable condition it may be necessary to change the design of the part to permit the cutoff line to line up with the notch.

Changes in part design may also provide an opportunity to achieve greater material economy through better nesting of parts in the strip. The full effect of any design change should, of course, be studied before such a change is suggested.

Operational Sequences. Sequences of operations that will produce the part most effectively are often determined through analysis of the strip layout. While rigid rules for sequencing operations cannot be formulated, there are principles that should be followed. The requirements for making the part within cost limitations will be the guiding factors in arranging the layout.

All holes that can be pierced without danger of deformation or mislocation should be pierced prior to forming. Some of these holes should be selected as piloting points for locating the part in subsequent operations. Where no suitable holes are available, it may be necessary to pierce pilot holes in the carrier or in the scrap. Holes should be as far from the center axis as the width of the strip will allow. This is particularly important in a strip with a center carrier. Pilot holes designed in this way will prevent mislocation of the part as it advances in the die. They will also prevent shifting of the part in forming operations.

The part should be blanked out following the creation of pilot holes. This should take place in the same station where piercing is done or in the station immediately following.

Forming follows the blanking operation. Where possible, all forming should be done in one station to avoid inaccuracies that are apt to occur from relocating a partially formed part in a subsequent station. All holes that would suffer deformation if made previously should be pierced after forming. Any restrike or additional forming operations should be done just prior to the final station in which the part is severed from the carrier strip.

Principles of Strip Development

In making the final development, certain principles should be observed as checkpoints to ensure proper die construction. Some of these principles apply generally to all types of dies. Special considerations must be made where the part design involves drawing or forming.

The following principles apply to all types of parts:

1. Pierce piloting holes and notches in the first station of the die. An important consideration in progressive die work is the proper gaging and location of the part throughout a sequence of operations. Means to facilitate location should be provided in the first operations. These piloting holes or locating points can then be used throughout the remainder of the operations to ensure that a part will be produced according to specifications.

2. Distribute pierced holes over several stations if they are too close to each other or to the edges of the die openings. While it is recommended that all holes be pierced in the same station, to better maintain tolerances, holes that are too close to each other are apt to cause damage to punches due to crowding of the metal. This can be compensated for in some instances by stepping the punches to permit larger punches to pierce first followed by the smaller punches. But if the number of holes to be pierced is great, better results may be achieved by spreading out the piercing operations. Holes that would be distorted by subsequent forming operations should be sequenced to follow forming.

3. Design the shape of blanked-out areas as simply as possible. The lines along which scrap pieces are cut out should be as simple as possible to reduce the problems involved in cutting them out. The simplest designs may permit the use of standardized commercial punches, thus obtaining maximum advantages with respect to tooling costs.

4. Consider the use of idle stations for strengthening the die and facilitating strip movement. Although at first glance it may seem that it would be best to perform work in each and every station of a die, there are many cases in which an idle station offers concrete advantages. At best, a progressive die involves cramming of numerous operations into a short space. The various operating tools are set in the main die body in die sections. The existence of numerous die sections tends to weaken the over-all die. By the addition of one or two idle stations, the die can be materially strengthened. Movement of the strip, and control of the metal, also can be facilitated by the inclusion of idle stations. There are often severe forces at work in adjoining work stations that tend to break the carrier strip. If an idle station is placed between two such stations, stresses are exerted over a greater length of strip, providing greater protection.

5. Provide for uniform loading of the press slide. In a progressive die, there are numerous forces at work. Pressures are mostly vertical, but in many cases sideways thrusts are encountered. These sideways thrusts should be balanced out as much as possible to prevent damage to punches and dies. Special emphasis should be placed on this factor because of the costly nature of the tools. In addition, damaged tools will result in inferior-quality parts. Balance may frequently be achieved through proper sequencing of operations.

6. Provide for efficient scrap disposal. The use of a progressive die is aimed primarily at increasing the productivity of a stamping operation. Ineffective scrap removal poses a serious threat to the attainment of this objective. In developing die design, space should be allowed for scrap chutes or other means of efficient scrap removal.

7. Provide efficient means of unloading finished parts. Parts to be drawn present special problems in the design of progressive dies. Proper provisions must be made to supply the metal required to shape the part. Special care must be taken to provide means of moving the part from station to station.

In the development of progressive dies for drawing, the following principles should be observed:

1. Allow adequate provision for lifting the strip from drawing punches. Where drawing is involved, the strip must be lifted

well above the drawing punch or out of the die cavity before it can advance. Because of this, space must be provided in the die to permit incorporation of merchanical lifters.

2. Provide adequate carrier strips and tabs to allow metal movement. As drawing takes place, the edges of the blanked-out portion of the strip are pulled in. This in turn exerts tensile stresses on the carrier strip. In fact, a center carrier strip may supply a portion of the metal required for the draw. This requires that carrier strip be of sufficient size and strength to avoid breakage and of proper design to allow metal to flow into the die cavity.

3. Drawing stages should be properly balanced. In progressive die design, attempts should not be made to draw too deeply in one station in an effort to reduce the number of draws required. Sound judgment must be used in determining the number of draw stations. It is better to spread the drawing operations out than to work too close to the allowable drawing limits. If excessive draws are attempted, the parts may rupture or tear. The tonnage load on the press may be too great, causing breakdowns. A progressive die with an extra station is less costly than a die of fewer stations that will not produce satisfactory parts.

GENERAL DESIGN FEATURES OF PROGRESSIVE DIES

Progressive dies are used for many small parts where tolerances and specifications are relatively severe. Production requirements of parts produced on these dies are high. Because of these factors, quality of design and structure in progressive dies must be of high caliber.

Strength

Progressive dies must be sturdily constructed to withstand sustained high-speed production. This calls for the selection of very strong materials for use in making the die. Areas of greatest stress must be properly built up and reinforced.

The design of progressive dies usually calls for die inserts and retainers and a sectional type of construction. This type of construction is necessary to incorporate the various types of operations that are united in a progressive die. Unless the various die components are properly distributed and strengthened, maintenance costs are apt to be very high.

To facilitate die repair, holes should be supplied in stripper plates and in other areas to facilitate removal of die sections. If such holes are not provided, it may be necessary to dismantle the entire die to remove a single section.

Punch retainers and die-button retainers should be backed up by steel plates to prevent the retainers from penetrating the die material. The die body itself should be made of steel if the demand for die strength is sufficiently high. Where cast iron is used in the die body, special steel wear plates and reinforcements should be added to prevent breakage or excessive wear.

Guiding

The precision requirements, plus the rapid cycling involved, require that the mating portions of the progressive die section be carefully guided. Precision guide pins with steel-backed bronze bushings should be used at all four corners of the die to help ensure exact alignment of mating surfaces. Stop blocks should be built into the die to prevent the punch from descending too far. These blocks should be placed where press operators cannot put their hands on them during operation. Guide pins should be long enough so that they are engaged at least 1/2 in. at the beginning of the press stroke.

All progressive die sets should be adequately heeled. Heeling of all die sets is a practical way of maintaining proper alignment of upper and lower shoes. The heels will offset any tendency of side thrusts to disturb die alignment.

In order to ensure that heels do the job for which they are intended, sufficient bearing surface must be engaged. The heels should be engaged a minimum of 1/2 in. at the bottom of the stroke.

Guide pins, stop blocks, and heel blocks

must be provided with suitable safety guards to prevent injury to press operators. They should be shielded wherever there is a possibility of someone getting caught in them. Sheet metal is generally used for shielding. Shielding also provides protection of the press and die by keeping out scrap and other materials that could cause difficulty.

Strippers

Progressive die sets usually employ stripper plates to free the die tools from the part being worked on. A stripper prevents distortion of the part and helps to ensure speedy operation. For best operating results, certain factors should be observed in the use of this tool.

The stripper's holes should be carefully designed to allow free movement of punches and other tools through them. Adequate clearances should be provided to prevent binding or excessive wear. Spring strippers should be well regulated with respect to travel. They should be supplied with springs that are sufficiently long to avoid excessive compression. Springs must be strong enough to depress all stock lifters that may be used.

Guiding of strippers is also important. Gibs should be used on two sides of the stripper to provide proper control.

Strippers should be made strong enough, and of a steel that is suitable, for the prevailing working conditions. If parts of a stripper break, they are apt to fall on die surfaces and result in injury to the die.

Lifters

The lifters used to facilitate movement of the strip through the die must also be constructed so as to withstand operating conditions. Breakage of lifters will often result in scored and damaged dies.

Ejectors

The full effectiveness of progressive die operations cannot be realized if swift, positive means of ejecting the part from the press are not provided. Presses using these dies are usually automatic in operation and run continuously until stopped by the operator. This rapid cycling provides only a short time for removal of the part.

Mechanical ejectors should be provided for removal of parts. Where there is a possibility of the part falling back into the die, air blasts should be incorporated to make ejection speedier and more positive. Air valves should be frequently checked for timing so that the part cannot fall out of the path of the blast.

AUXILIARY EQUIPMENT

The nature of progressive dies demands the choice of proper auxiliary equipment in order to ensure effective operations.

Coil Feeders

Most progressive die operations use coil stock. Coil feeders are used to pass the stock into the die. These feeders must possess the proper capacity for holding the required stock and should be so designed as to administer the proper amount of metal to the press at each cycle. Some coil feeders are merely holders for the stock, with automatic feed rolls being used to move the stock. Others are geared to the press cycle and actually feed in the stock.

Scrap Handling

An aspect of some progressive die operations not found in other types of operations is the requirement for handling a carrier strip. This strip issues from the press as finished parts are cut away. Some presses are equipped with devices that roll up this strip as it comes out of the press. In other instances, scrap cutters are mounted at the end of the die to chop the strip into easily managed lengths. Where the latter method is used, stock boxes or scrap chutes should be used to provide simple means for getting the scrap away from the press.

PRESSES USED WITH PROGRESSIVE DIES

In selecting presses for progressive die work, certain considerations are involved. Some of these considerations include the volume requirements of the part, the accuracy required, the size of the part, pressure requirements, and the relative difficulty of producing the part. Many types of presses are used for this kind of work. The choice will depend on how well the press meets the needs of a given situation plus the inescapable factor of press availability.

Open Back Inclinable Presses

Numerous progressive operations are performed in open back inclinable presses. The inclined feature of this press provides an often satisfactory solution of the problem of removing parts, because the parts tend to fall easily out of the back of the press. Where a multiple die is used to make two or more parts at a time, a choice must be made as to whether to let the parts running at the uppermost level of the die drop out of the rear or whether to use an air blast to blow them out of the front of the press.

Tie rods usually are used across the gaps of these presses where exceptional accuracy is required. This is done to prevent deflection of the press under load.

Because of pressure limitations, gap-frame presses are usually used on the smaller types of progressively shaped parts.

Four-Column Presses

For progressive work demanding high pressure and exceptional accuracy, straight-side four-column presses are frequently used. This type of press is generally superior in tonnage potential to gap-frame presses. In addition, the four columns absorb sideways thrusts to ensure better alignment of punch and die.

For both gap-frame and straight-side presses, the press bed must be sufficiently wide to provide room for the progressive die used.

Automatic Underdrive Presses

Automatic underdrive presses are designed specifically for use with progressive dies. These presses will be thoroughly discussed in a subsequent article. They have wide beds and provisions for easy application of automatic loading and unloading. At present, such presses are subject to limitations with respect to the sizes of the parts that may be produced on them. Nevertheless, their design lends itself well to the manufacture of stamped parts in progressive dies.

2K: Forming Operations

Among the techniques used to impart complex shapes to metal stampings is the process of forming. This process differs from drawing in the degree and type of deformation that takes place. In drawing, deformation occurs over a major portion of the surface of the part; a flowing of metal, often accompanied by stretching, is required to transform a flat piece of metal into the contoured shapes that are characteristic of drawn parts. In forming, the deformation of metal is confined to more specific areas of the part; changes in the surface planes are more angular than curved. Forming gives shape to a part by bending various portions of the metal to achieve the specified shape; there is seldom any extensive plastic movement of metal, and any stretching that occurs is confined to localized areas of the part. Forming, then, may be defined as a metal-stamping process in which complex shape is given to a part by means of deforming the metal by bending it. The stresses involved are above the elastic limit of the metal and below its ultimate strength.

The widespread use of forming is due to certain advantageous aspects of the process itself. Generally, forming is preferred to drawing in cases where either process can be used for a given operation. This preference is based largely on cost factors: forming dies are less costly than drawing dies. Material usage is generally more efficient in forming due to the fact that less material is required for holding purposes. Forming operations can usually be done on less powerful presses, with fast-acting single-action presses being used rather than double-action presses. The number of pieces per hour for forming generally exceeds that for drawing in comparable operations. In addition, the forming process is applicable to both light- and heavy-gage materials.

Forming, however, has certain characteristics that limit its application. Because of the relatively low amount of deformation possible in forming, parts with smooth curved surfaces and those with deep depressions cannot be formed but must be drawn. Forming operations also have a tendency to produce wrinkles in a part in the area of deformation. Thus, forming is seldom used in making parts for normally visible surfaces of automobiles. However, these wrinkles are not objectionable on many interior parts that are not exposed.

Another disadvantage of forming is the difficulty of obtaining a true shape due to springback. If a positively true profile is required, drawing is usually employed.

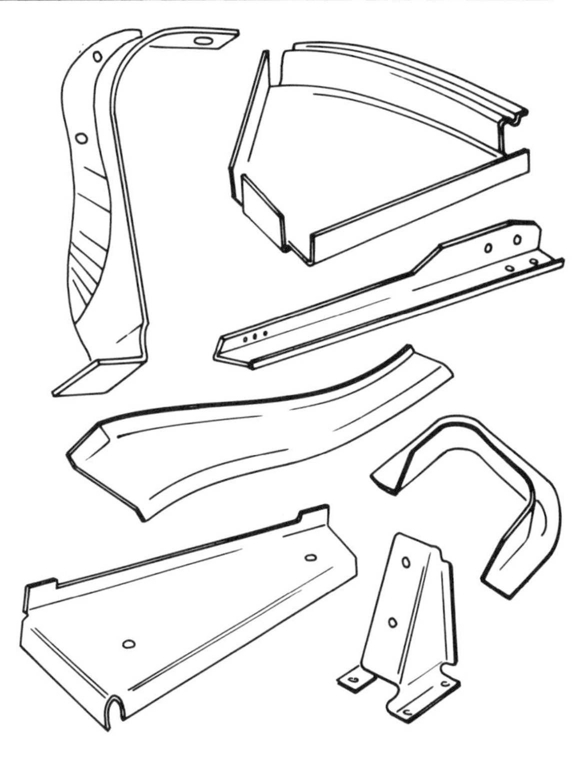

Figure 2K-1. Typical formed parts

ANALYSIS OF AUTOMOTIVE PARTS WHICH ARE FORMED

The range of complexity of shapes in formed automotive parts is very broad (Figure 2K-1). Many parts, such as body brackets, are quite simple in shape, the shape consisting of one or two simple bends. Other parts, such as deck-lid hinges, are highly complex, with many bends, contours, and flanges in various directions. The types of shapes involved, however, can be classified into six major groups (Figure 2K-2).

These classifications are plain channels, complex channels, simple plates, complex plates, simple contoured parts, and complex contoured parts, plus other unusual shapes. Parts are classified into these groups on the bases of the general configuration of the part and the complexity of the shape. Classification is used to aid the process engineer in determining the tooling required for a given part.

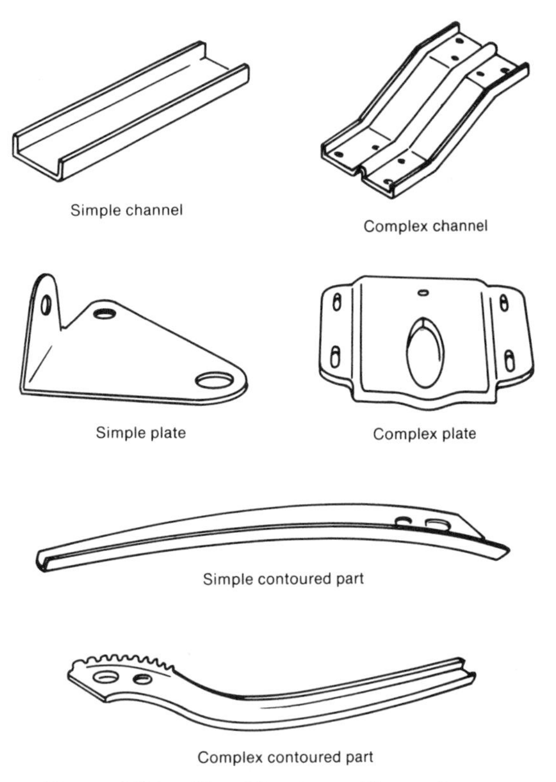

Figure 2K-2. Classifications of formed parts

Plain Channels

Plain channeled parts are characterized by a flat main body, called the web, with the sides of the part bent upward or downward. This gives the part the general appearance of a channel. Forming of parts of this description can usually be accomplished with the simpler types of dies and press equipment.

Complex Channels

Complex channeled parts have the same characteristics as plain channels with the addition of bends in the web of the part and/or the inclusion of curved edges or other more ornate shape characteristics. The tools required to produce these shapes are proportionately more complex than those used for plain channels. Floating die sections, pressure pads, or cam mechanisms are often included in the tooling for complex channels. There is also the possibility of two separate operations being needed to create bends that move in opposing directions.

Simple Plates

Simple plates are characterized by the absence of flanges or upturned edges. They are simple in shape, having only one or two straight-line bends. Such parts are usually of fairly heavy-gage stock. Simple plates are sometimes referred to as V-shape parts and are among the simplest types of parts to form. The tooling used is relatively simple. However, in forming simple plates, care must be taken to compensate properly for springback.

Complex Plates

Complex plates are also distinguished by an absence of flanges. However, they incorporate more bends than simple plates, and these bends may be made in more than one direction. The bends may be either horizontal or vertical in relation to the axis of the part. To achieve such bends, dies usually must be equipped with pressure pads or cam mechanisms for holding one portion of the part while another portion is being shaped.

Simple Contoured Parts

Simple contoured parts may have one or more flanges bent in the same direction. They also have curved surfaces. This curve is usually along one of the flanges rather than on the web of the part. The process of forming a curved flange is considerably more difficult than forming a straight flange because of the compressive and tensile stresses that are present. Greater metal control is required to prevent tearing or buckling of the flange. This control is generally provided by pressure pads in the die. Because flanging usually is done in only one direction in a simple contoured part, a single pressure pad is used.

Greater pressure is usually required to form contoured flanges than to form straight flanges due to the fact that the metal is being either stretched or compressed in the flange. If a shrink flange is being made, the metal is being compressed. If a stretch flange is being made, the metal is being stretched. The greater pressure requirements of this type of operation usually dictate the use of heavier presses rather than comparable straight flanging operations.

Complex Contoured Parts

Complex contoured parts are often characterized by the inclusion of reverse flanges. A reverse flange is a combination of a shrink flange and a stretch flange. Parts of the complex contoured type often have more than one of these flanges, and, in addition, one or more bends in the main body of the part may be included. This type of part requires the most complex type of tooling used in forming. Often two pressure pads may be used in one die, and in some cases several separate operations may be required.

PHYSICAL FACTORS INVOLVED IN FORMING OF SHEET METAL

Certain physical characteristics of sheet metal should be taken into account in forming operations. These characteristics are springback, directionality, and the balancing of press loads.

Springback

Springback is one of the principal physical factors that must be considered in forming operations. Springback is the tendency of metal to resume its original shape after forming. This factor is particularly present in forming becuase of the relatively low amount of metal deformation taking place. If springback is not compensated for in forming, the operation will not yield parts of the desired dimensions. Overbending of the formed area is one means of effecting compensation. It has been agreed by most press operators that a 2% addition to the angle of the bend will be sufficient allowance for springback in steel parts. Care must be exercised in applying this allowance, however, because thin steel tends to spring back more than thicker steel. Flanges with curved surfaces require greater stress in the metal and do not spring back as much as straight-line bends. It is difficult to find a rule that applies in each case. The knowledge and experience gained through trial and error seem to be the best aids in establishing proper overbends.

Another means of compensating for springback is to "bottom" or "strike home" with the punch in the area of the flange or bend. This compresses and sets the metal, eliminating springback. However, striking home sometimes results in heavy pressures being applied to the die and press because of thick areas in the stock. Striking home is unavoidable in some operations, but it is generally agreed that overbending is the safer approach to this problem (Figure 2K-3).

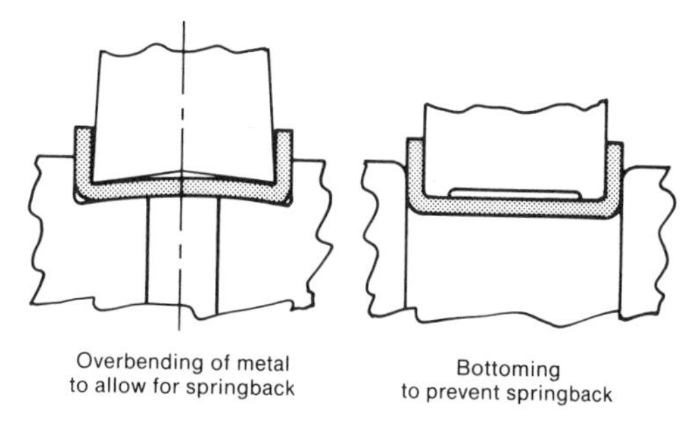

Figure 2K-3. Methods of compensating for springback

Directionality

Another physical aspect of metal that sometimes requires consideration is its directionality — that is, the direction of its grain structure.

In some types of steel, a very definite grain direction exists, in the direction in which steel rolling took place. Bends made parallel to this grain direction may result in cracking. As a result, the pattern of the part on the sheet should be arranged so that the bends or flanges will be formed across the grain. For many automotive stampings, however, this factor is not exceptionally troublesome. This is true of parts formed from open-hearth steel, for example, because such steel has no appreciable directionality in its grain structure.

Press-Load Balance

Another physical factor in forming has to do with the forces exerted by the press tools. As the punch descends on the metal, the first points of contact are in areas where bends will occur. If several areas of the part are involved, the contacts should be made evenly to prevent side thrusts and uneven loads on the metal. Uneven loads may force the metal to move and shift in the die, resulting in ruined stampings. Where uneven loads cannot be avoided, provisions should be made in the punch and die for holding the part firmly to prevent shifting in response to the work stroke. The types of devices used for this purpose will be discussed more thoroughly in the subsequent section on forming tools.

DESIGN OF BLANKS FOR FORMING

Layout of blanks for forming operations is an activity in which significant reductions in material cost can be achieved. Several types of blanks may be used, but the principal aim should be to eliminate the production of scrap and to eliminate the necessity of a trimming operation.

Developed Blanks

Because of the relatively small amount of metal displacement that occurs in forming, it is often possible to use developed blanks. Developed blanks result in minimum scrap loss, and no separate trimming operation is necessary.

Rough Blanks

In some forming operations, it is necessary to provide metal for holding purposes to enable the press tools to form the part. In these cases, a rough, or partially developed, blank is used. Extra metal is provided at the required locations for holding and must be trimmed off after forming.

Special double blanks may be used in forming where desirable advantages can be gained through their use. These blanks permit the forming of two parts at the same time. This provides proper balance of the thrust of the ram, preventing excessive tool wear. Double blanks sometimes can be either rough or developed. In addition to making the use of better dies possible, such blanks can also increase productivity.

TOOLS REQUIRED FOR FORMING

There are three major types of forming dies: the solid die; the single pressure-pad die; and the double, or two-way, pressure-pad die. The choice of which of these dies to use generally depends on the degree of complexity of the shape of the part. Other factors are the production volume and quality demanded of the part.

Solid Forming Dies

A solid forming die is so called because the punch steels and the die steels are solidly fastened to their respective shoes. There are no floating die members or pressure pads. Parts are solidly and positively forced into the shape of the punch and die (Figure 2K-4).

This type of forming die is used to produce simple U-shape parts, V-shape parts, and parts that require no more than one or two simple bends. Such a die may be used to produce plain channels, simple plates, and parts with shallow contours, where there are no flanges on the sides of the part.

Solid forming dies can be built quickly at low cost and are easy to repair. For these reasons, such dies are used to the greatest extent possible. Where a solid forming die can successfully shape a part, it should be used in preference to other types of dies. However, a solid forming die is limited with respect to the shapes it can produce; there is little provision for metal holding, resulting in wrinkles; and complex parts tend to shift, resulting in scrapped or poorly formed pieces.

Die Sections. The punch member is fastened to the top shoe and is machined to fit the inside shape of the part. It is usually made of one piece of steel; however, if the part is large, it may be made from a casting. The die section is attached to the lower shoe and is machined to fit the outside shape of the part. It can also be made of one piece. In

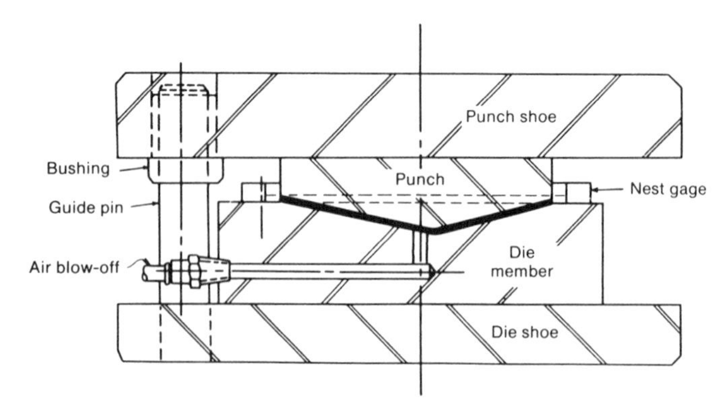

Figure 2K-4. Solid forming die

cases where the part is large, it can be made of two or more pieces or a casting.

Nesting the Blank. Nest gages are used to confine the blank in all directions so it cannot move or twist out of position. Gages usually fit against the outside edge of the blank and may be either pins pressed into the die or blocks of steel screwed and dowelled to the die. Inside gage pins may be used in pierced holes in the blank, although this results in slightly slower loading than when outside gages are used.

Guide Pins and Heels. Simple bending and forming dies do not require as much accuracy as do other dies, and in many cases guide pins need not be included in the die set. Guide pins are used, however, to make the die construction simpler, to make the die setting easier, and to keep the shoes in alignment while being transported and stored. Two guide pins are sufficient for small or medium-size dies, and four for large dies. Heels should be used if there is any unbalanced thrust which creates pressure on the guide pins.

Stripping the Part Off the Punch. The formed part may be of such a shape that it will stick to the punch and travel up with it unless stripped off. The usual types of strippers are hook strippers and internal positive knockouts. After the part has been stripped, it may be blown by a jet of air so that it falls clear of the die, or it may slide off the die if the press is inclined to the rear. If the part would otherwise stick in the die member, a spring plunger may be used to lift the formed part for unloading. This type of spring plunger may also be used to hold the blank against the punch member to prevent the blank from shifting sideways during the forming operation (Figure 2K-5).

An air blast may be used if the part cannot be unloaded from the forming block automatically. This jet of air is either directed at the part by an air nozzle, without any particular provision on the die, or else directed up through holes in the die member.

Single Pressure-Pad Forming Dies

A pressure-pad forming die is one in which a portion of the blank is gripped and held in place by a pressure pad while a stationary steel forms or flanges the remainder of the part.

Single pressure-pad forming dies are used for producing plain and complex plates and channels and, in some cases, simple contoured parts. Parts with shallow recesses also can be formed in such dies (Figure 2K-6).

This type of die has a floating die section which moves, generally downward, during the press cycle. The inclusion of such a feature adds to die costs. As a result, single pressure-pad dies are more expensive and take more time to build than solid forming dies. Maintenance cost is also higher. This expense is, of course, justified in view of the more complex shapes that can be produced.

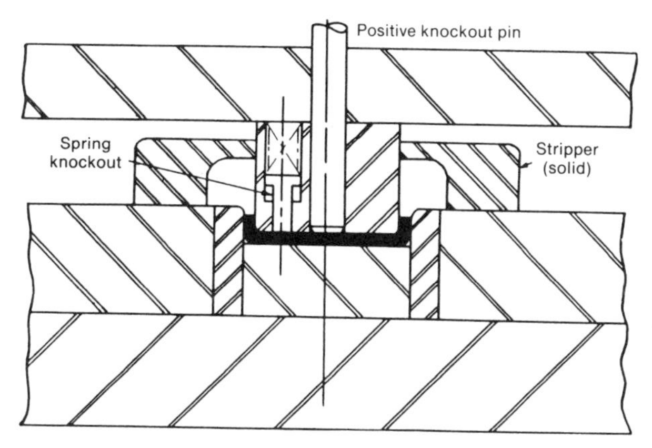

Figure 2K-5. Stripping methods

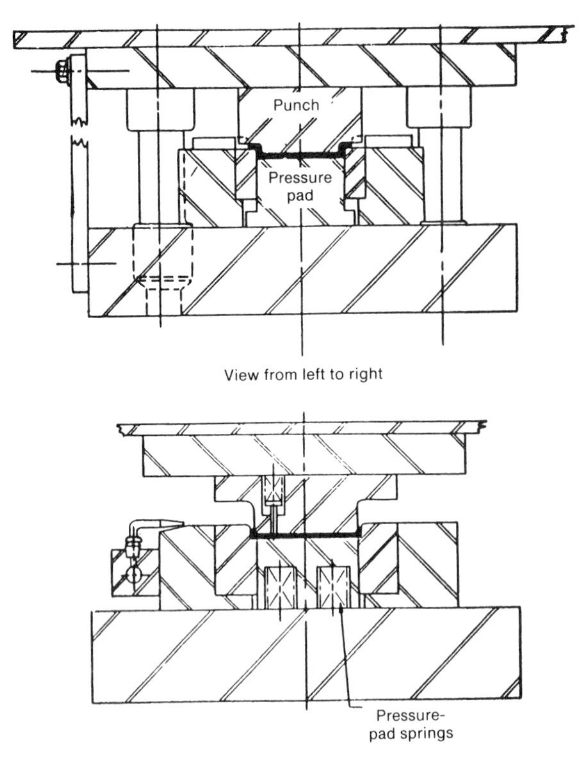

Figure 2K-6. Single pressure-pad forming die

Forming in single pressure-pad dies is done in one direction only. Flanges may be formed on two or more sides of a part, but only in the same direction. For forming that requires bends in several directions, a two-way pressure-pad die is used.

The mechanical action of a single pressure-pad die is as follows. The blank is placed between gages on top of the pressure pad, which is in the "up" position. The punch member on its downstroke contacts part of the blank and forces the pressure pad downward. As the pad travels down, the stationary die members on the bottom shoe form the remainder of the blank to its required shape. On the upstroke, the pressure pad keeps the formed part pressed against the punch until the pad reaches the top limit of its travel, which is a position flush with or slightly above the top level of any of the die members.

Because forming is done all in one direction, or one way from the pressure-pad surface, a pressure pad is required on only one shoe. This type of die can also perform the type of operation often called a flanging operation, because a flanging operation is a forming operation in which the edge of the blank is formed (Figure 2K-7). A pressure-pad forming die has an advantage over a solid forming die in that the blank does not shift as much, resulting in more accurate location of the edge of the part. Also, wrinkles from the excess metal are minimized in the region covered by the pad and are eliminated or at least minimized in the remainder of the part. A single pressure-pad die is, however, more expensive to build and is used only if a solid forming die will not produce quality parts.

The pressure pad usually floats on the bottom shoe and is operated by spring or air pressure. If springs are used, the up position of the pad is determined by retainers which limit the travel of the pad. When the pad is in its up position, the springs are still compressed, and this compression is called the initial compression of the springs. The initial compressed length of the pressure-pad springs is the length to which the springs

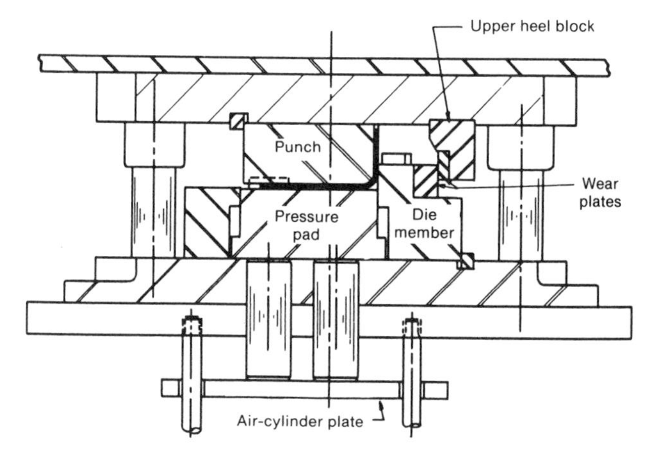

Figure 2K-7. Flange die

have been compressed from free length when the die is in an open position. The spring pressure corresponding to this initial compression is the pressure holding the blank when the forming operation starts. As the pad is forced downward, the springs are compressed to their final compressed length, which is the spring length at the closed die position and is equal to the free length minus the initial compression minus the pad travel. In general, the initial spring pressure is more important in a forming die than the final spring pressure because there is a greater tendency for the blank to shift at the beginning than at the end of the operation. Unfortunately, the spring pressure is least at the start when it is needed most, and then builds up to an amount considerably greater than required at the end of the stroke. The initial rather than the final compression pressure required for the operation usually determines the number of springs, and therefore there is an advantage in having as great an initial compression as possible. The final compressed length should be greater than the solid height of the spring.

The amount of pressure required for forming depends on the amount of sideways pull on the blank, tending to cause it to shift. This pull, in turn, depends on how large the radius of each flange break is, on whether the flanges have straight break lines, on whether the flanges are vertical or at an angle, and on how wide the flanges are. The longer the free length of the spring, the greater the initial compression with a given pad travel. It should be understood, however, that the pressures resulting from compression of a long and a short spring through the same absolute distance (rather than the same percentage of the total compression) will not be the same. The spring pressure builds up in direct proportion to the distance of compression. The advantage of a long spring is, therefore, not that it provides a greater final pressure, but that it allows the necessary pressure-pad travel and greater initial compression. If the press is equipped with an air cushion, the pressure pad of the die can be operated on by air pins extending down to the air-cushion plate. Because the air cushion has a stop of its own, air pins are used in lengths such that the air-pin travel ceases when the pressure pad is flush with or slightly above the die member.

Retainers are used with air pressure, but should have about 1/8 in. of additional clearance so that the pressure pad will not be forced up against them. The pressure pad is guided in its vertical travel by the stationary die members. The height of flanging steels for parts with straight break lines in both the plan and elevation views should extend 0.25 in. above the break radius for metal thicknesses of 0.06 in. or less. If a flange is being formed vertically, the punch wall is sometimes relieved on a 1 or 2° angle. This relief angle has the effect of reducing the amount of springback so that the flange is more nearly vertical. Flanging steels should extend around and spank the flange radius for short flanges with heights of twice the metal thickness or less. The pressure exerted on the sides of the forming steels as the pad descends necessitates rigid die construction.

Two-Way Pressure-Pad Forming Dies

A two-way pressure-pad forming die forms the part both upward and downward by pressure-pad action, and to do so requires a pressure pad on each of the top and bottom shoes. In an open die position, each pressure pad is flush with its die member, and the blank rests on the lower pressure pad and the lower die member (Figure 2K-8).

This type of die is used for the more complex types of formed parts. Plain and complex contoured parts, and complex plates, are generally formed in two-way pressure-pad dies. The costs of die construction, time required, and maintenance in operation are considerably higher than for the other types of forming dies. Because of this, the possibility of using a simpler die for forming a part should be fully exhausted before a decision is made to use a two-way pressure-pad die.

On the downstroke, the upper pressure pad contacts the blank on the lower pressure pad, and its springs start to compress because its spring power is less than the power of the lower pressure pad. As the lower pressure pad remains stationary, and the upper die member continues to travel down, the flange is formed downward over the corner of the lower pressure pad (which is acting as a punch). When the springs of the upper pressure pad have been compressed so that the pad has bottomed on the upper shoe, the lower pressure pad has to start descending, and as it does so the lower die member forms the upward flange. On the upstroke, the lower pad travels up first, because its springs are stronger than those of the upper pad; and then the upper-pad springs expand, stripping the part from the upper die member. The part can then be removed from the die.

Heels must be provided on the die, because the unbalanced forming thrusts tend to shift the top shoe sideways in relation to

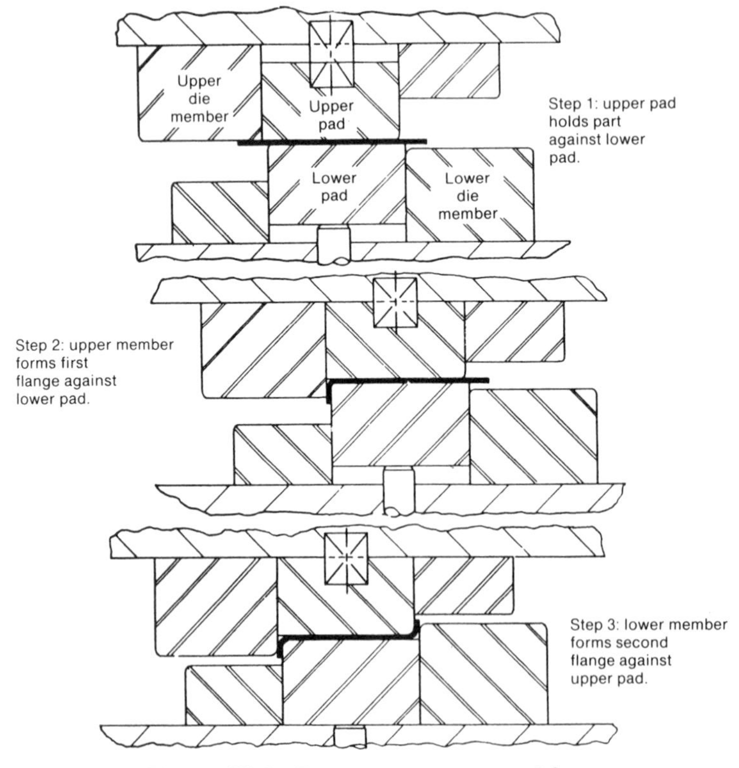

Figure 2K-8. Two-way pressure-pad forming die

the bottom shoe. The two pressure pads are guided by blocks on those sides not guided by the forming members themselves.

TYPES OF PRESSES USED IN FORMING

A rather broad image range of press-working equipment can be successfully used for forming operations. For small parts, rapid-acting gap-frame presses of the single-action type are often used. This type of press is sometimes used in conjunction with progressive forming dies for rapid production. Light-gage parts such as small brackets are typical of the parts formed on this type of press.

For medium to large formed parts, straight-sided, single-action presses can be used. Such presses can be used for radiator support brackets and interior body reinforcements. For heavy forming, such as frame rails, very large four-column presses of the single-action variety are used.

The selection of a press depends on pressure requirements, volume requirements, and the need for die room and shut height to accommodate tools and to facilitate loading and unloading of parts. For heavy-gage work, air cylinders may be used instead of springs for operating the pressure pads. In such cases, the press must be equipped with air cylinders in the base or on the superstructure.

SUMMARY

Forming is a metalstamping process in which a part is given a specified shape by means of deforming the metal by bending it. This process is widely used in providing flanges on such parts as hoods, fenders, quarter panels, and doors. It is also used as the primary shaping process in stamping such parts as support brackets, body reinforcements, frame rails, and many internal body and chassis components.

Parts given shape by forming may be classified into six major types: plain channelled parts, complex channelled parts, simple plates, complex plates, simple contoured parts, and complex contoured parts. By so classifying formed parts, a determination of the processes necessary to produce the parts is made easier.

In development of forming processes, certain physical factors must be considered. These factors are springback, directionality, balance of press load, and optimum size and shape of metal blanks for economical production.

The basic tools for forming consist mainly of three types of dies: solid forming dies, single pressure-pad forming dies, and two-way pressure-pad forming dies. The solid forming die is the cheapest and most simple to build. It is used in preference to other types of dies whenever it will successfully form the part involved. Pressure-pad forming dies are used where the metal-holding requirements for forming exceed those provided by a solid forming die. The single pressure-pad die is used for parts having bends in one direction. Two-way pressure-pad dies provide means for producing parts with bends in opposite directions.

2L: Contour Forming

Rolled or extruded preformed sheet-metal sections (L, T, I, U, etc.) sometimes must be edge-bent or edge-shaped. This is a rather troublesome process and is complicated because the problems and difficulties inherent in bending — and in forming — are increased by the additional trouble provoked by the deformation of webs or flanges, which are subjected to considerable additional stresses.

In fact, in convex forming (i.e., flanges directed toward the interior of the bend) there is excess metal at the inside of the bend with respect to the male-punch side, and in concave flanging (i.e., webs directed toward the exterior of the bend) there is a shortage of metal at the outside.

If the material is a rather thick-wall section and is annealed — i.e., is in a highly ductile condition — and if the webs (flanges, legs) are comparatively low, there should be no difficulty in forming the parts. However, if conditions are not so favorable, attempts at contour forming will present severe problems. The external portions, subjected to tension, will show cracks and tears (Figure 2L-1), and the internal portions, subjected to compression, will tend to buckle and exhibit formation of wrinkles or undulations (Figure 2L-2).

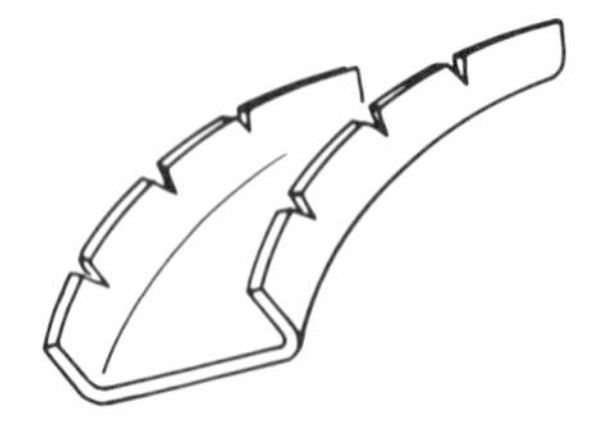

Figure 2L-1. Contour formed part, showing cracks in external webs

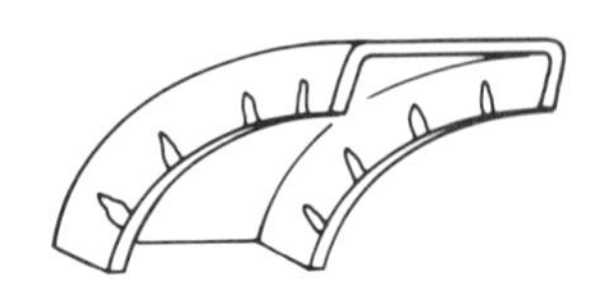

Figure 2L-2. Contour formed part, showing wrinkling in internal webs

CONTOUR FORMING METHODS

Depending on circumstances such as production quantities, workpiece sizes, dimensional tolerances, available equipment, and other factors, contour forming may be done either manually or with dies on punch presses (or press brakes) or on special machines.

One characteristic is common to practically all contour forming methods: the operation is almost always performed progressively, one short section at a time.

Taking into account all the characteristics of each case, the correct forming method is selected from among the following basic processes:

1. **Three-point air forming** (Figure 2L-3): This consists of pushing a preformed blank or an extrusion between two rollers or the rounded edges of a female die. In view of the high friction involved, rollers are usually preferred for air forming.
2. **Closed-die forming** (Figure 2L-4): Air forming does not ensure dimensional uniformity. Use of conventional (closed) forming dies (in punch presses) improves dimensional uniformity.
3. **Roll forming** (Figure 2L-5): Three-roller forming machines (horizontal or vertical) are sometimes used successfully for contour forming. The rollers are contoured according to the cross-sectional shape of the part.
4. **Wiper forming (compression forming)**: In this method, the blank is clamped firmly against a stationary formblock (punch), and then a shoe (roll or roll assembly) is wiped under pressure against the blank's outer surface, forcing it to assume the contour of the punch (Figure 2L-6). The same result may be obtained with a rotating formblock (not necessarily of circular shape) and stationary (hydraulically or otherwise actuated) wiper rolls (Figure 2L-7).
5. **Wrap forming** (Figure 2L-8): In wrap forming, the blank is gripped at both ends

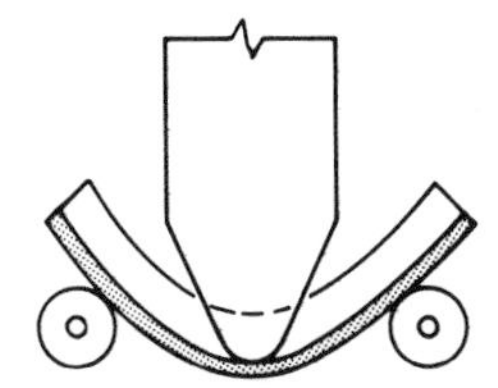

Figure 2L-3. Three-point air forming

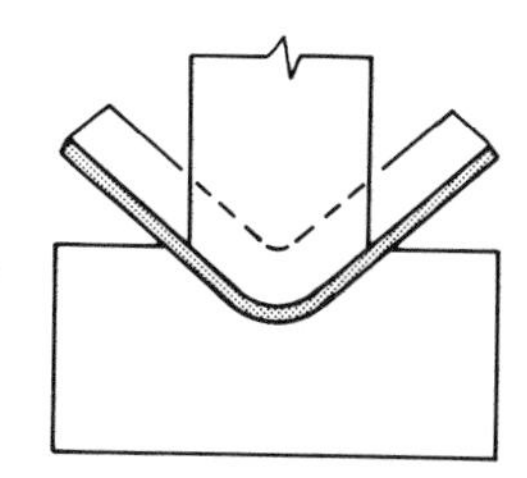

Figure 2L-4. Closed-die forming

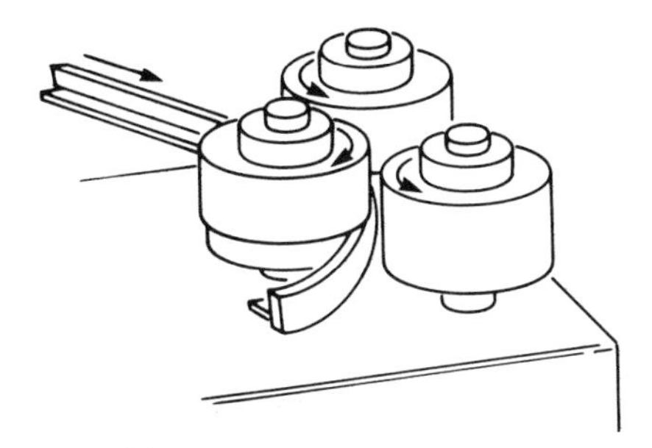

Figure 2L-5. Roll Forming

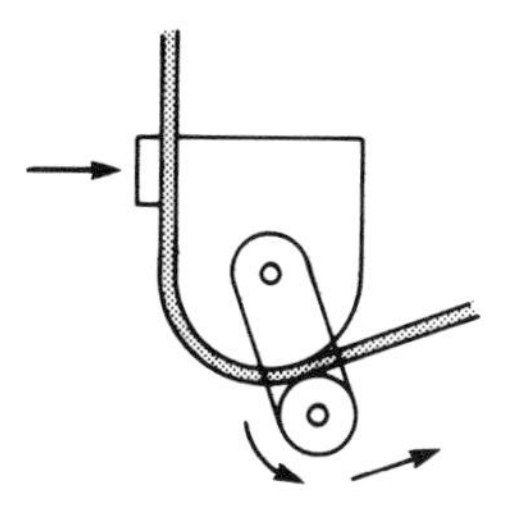

Figure 2L-6. Wiper forming with stationary formblock and movable wiper roll (webs not shown)

and then curved around a stationary formblock. Basically, wrap forming is also a compression forming process.

6. **Tangent bending** (Figure 2L-9): This process is essentially a wraparound operation. The basic difference lies in how the operation is performed. In tangent bending, forming is effected by means of pivoting tool members.

7. **Stretch-wrap forming**: This is the most

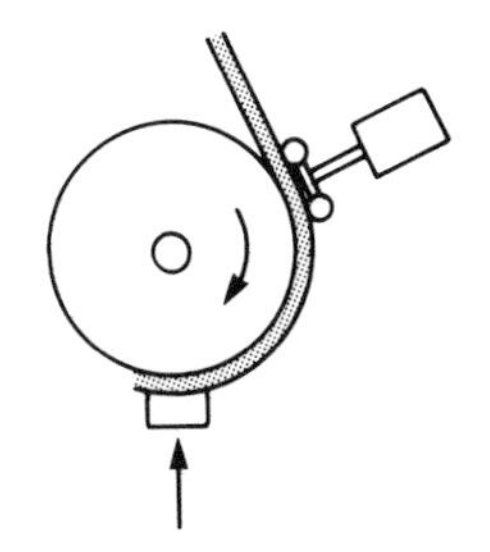

Figure 2L-7. Wiper forming with rotating formblock and stationary wiper roll (webs not shown)

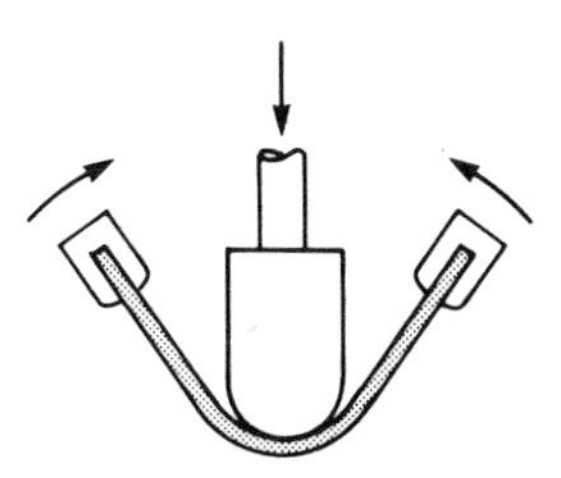

Figure 2L-8. Wrap forming (webs not shown)

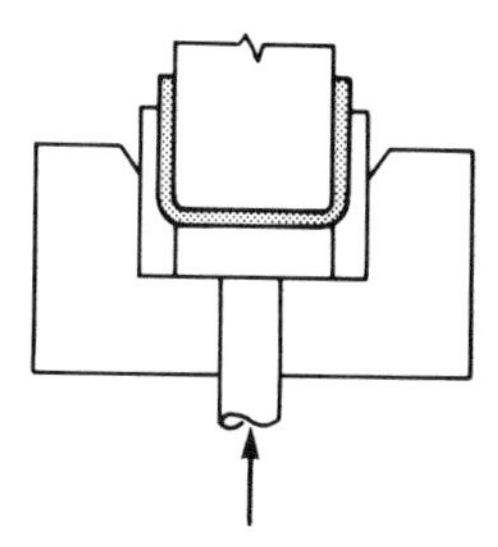

Figure 2L-9. Tangent bending (webs not shown)

sophisticated, most effective method developed for the most difficult cases. Here the straight blank is first stretched slightly beyond the yield point of the material, and then wrap forming (see item 5) is performed in such a way that a uniform tension is maintained on the blank at all times.

PART DESIGN AND PRODUCTION

In order to prevent problems (rather than having to remedy them when they happen), the designer and the manufacturing engineer must take preventive (rather than remedial) measures. The chief ones are as follows:

1. Select metals with stretching ability.

2. Select the correct temper. Soft tempers are preferred, but it should be kept in mind that softer materials have a higher tendency to buckle and thus to form wrinkles.

3. Thick-wall webs (within reason) cause less trouble than thin-wall webs.

4. Limit web height to the smallest possible value. If this is not possible, then notch the critical points where the maximum stresses are created (Figures 2L-10 and 2L-11).

5. Assign considerable rounding radii (Figure 2L-12). Minimum recommended values depend on material type and web height. Ratios of radius to height should be at least 3 to 4 for soft metals and at least 6 to 8 for hard metals.

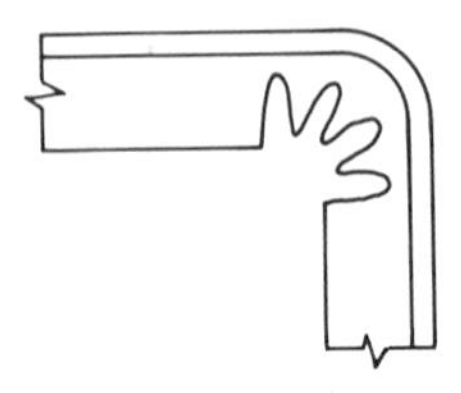

Figure 2L-10. Notched flange for avoiding wrinkle formation

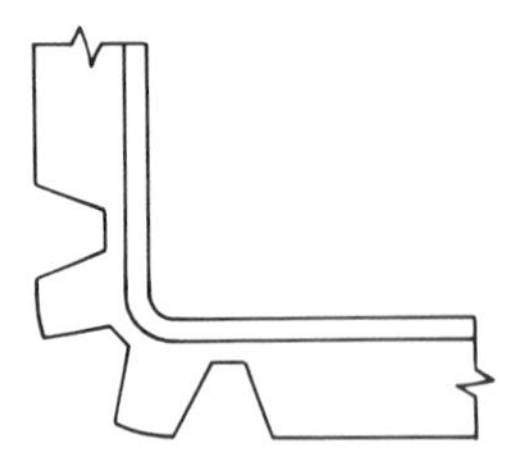

Figure 2L-11. Notched flange for avoiding cracks or tears

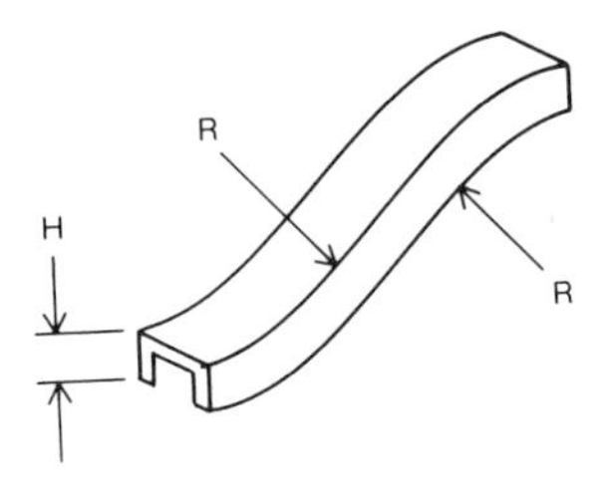

Figure 2L-12. Illustration of large-radius contours preferable for contour forming (minimum radii depend on stock quality and web height)

6. Smooth, burr-free edges improve formability.

7. Blanks should be annealed, to improve ductility and thus make the operation easier and smoother.

8. Contour forming sometimes is performed hot, especially in cases of very thick sections and low accuracy requirements.

9. Special tools (rollers, dies) with inserts or movable sections (as in thin-wall tube forming) are sometimes used in order to prevent collapse of thin-wall sections during forming operations.

SUMMARY

Contour forming involves many variables and unforeseeable complications. It requires prolonged developmental work before definitive part and tool designs can be established. Especially valuable in this field is the practical knowledge (and experience) of production personnel.

2M: Coining, Sizing, and Forging

Among the direct compression or squeezing operations, including stamping, sizing, swaging (or forging), coining, and extrusion, the working pressure required is based on the plan-view area or "projected" area of the job and the compressive resistance of the metal. The resistance varies with the relative amount of flow and freedom of flow of the characteristic operation as well as with the material and the amount of cold working (strain hardening) which it has undergone.

Figure 2M-1 involves a simple multiplication: pressure or press capacity, P, in tons, equals the product of the projected area of the piece, A, in square inches, and the compressive resistance of the material, S, in psi. In any calculation of required tonnage, proper allowance must be made for the character and arrangement of the job.

The required press capacity can also be determined by direct testing with the finished tools, performed in testing machines or hydraulic presses with suitable gages. In such cases, it is conservative practice to double the test pressure in selecting a mechanical press, due both to its much greater speed and to the inflexible character of a good stiff machine, a factor which contributes much to the accuracy of the finished product.

In selecting extrusion presses based on test results, the experimental pressures should be more than doubled because the increased speed of flow through the small orifice around the punch greatly increases the working load.

Somewhat less conservative ratings may be used with assurance for many operations of the squeezing group by using presses with hydropneumatic overload-relief cushions built into the press bed. Such combination machines take care of material-volume variations, double blanks, cold slugs (in forging), etc., without sacrificing any of the speed, operating, and maintenance advantages of mechanical equipment.

The working area is the projected area, in the plan view, of all surfaces of the work metal which are in contact with the die surfaces at the completion of the squeeze. Thus, note that the forging shown in Figure 2M-2(c) is relieved at two points to reduce the squeezing area; the die in Figure 2M-2(h) is relieved for the same purpose; and the area of the piece in Figure 2M-2(f) is increased by the area of the portion of the flash which is being squeezed.

Figure 2M-2 was prepared primarily to distinguish certain typical operations and to illustrate differences in the freedom of flow of the material. Considerable judgment must be exercised in this connection in estimating the surface pressure, which may be built up in completing the operation.

Stamping and embossing operations (Figure 2M-2a and b), in which the original thickness of the sheet is unaltered except by the relatively mild strains of bending and stretching, may be very easy operations, hardly belonging in the squeezing group. This is particularly true in case (a), where the die is relieved above and below so that it cannot strike solidly at any point. This operation therefore can practically be considered to be a drawing operation, and the required press capacity can be figured for steel at 15 to 20 tons per square inch of cross-sectional area along each line of the design. In case (b), the die "hits home" over the entire surface, but a sharp impression may be obtained with relatively low pressure because the operation is intended to be bending and not squeezing. A little carelessness in setup, however, or an oversize blank, will greatly increase this pressure. In fact, it may actually rise several hundred per cent, because the very large area relative to the metal thickness makes it almost impossible for the metal to flow out.

Sketches (e), (g), and (j) in Figure 2M-2 illustrate other "closed" dies in which dangerous pressures are possible if care is not exercised. In the sizing operation in sketch (c), the metal is entirely free to flow, but the die comes together on solid contact faces in order to make the accuracy obtained less

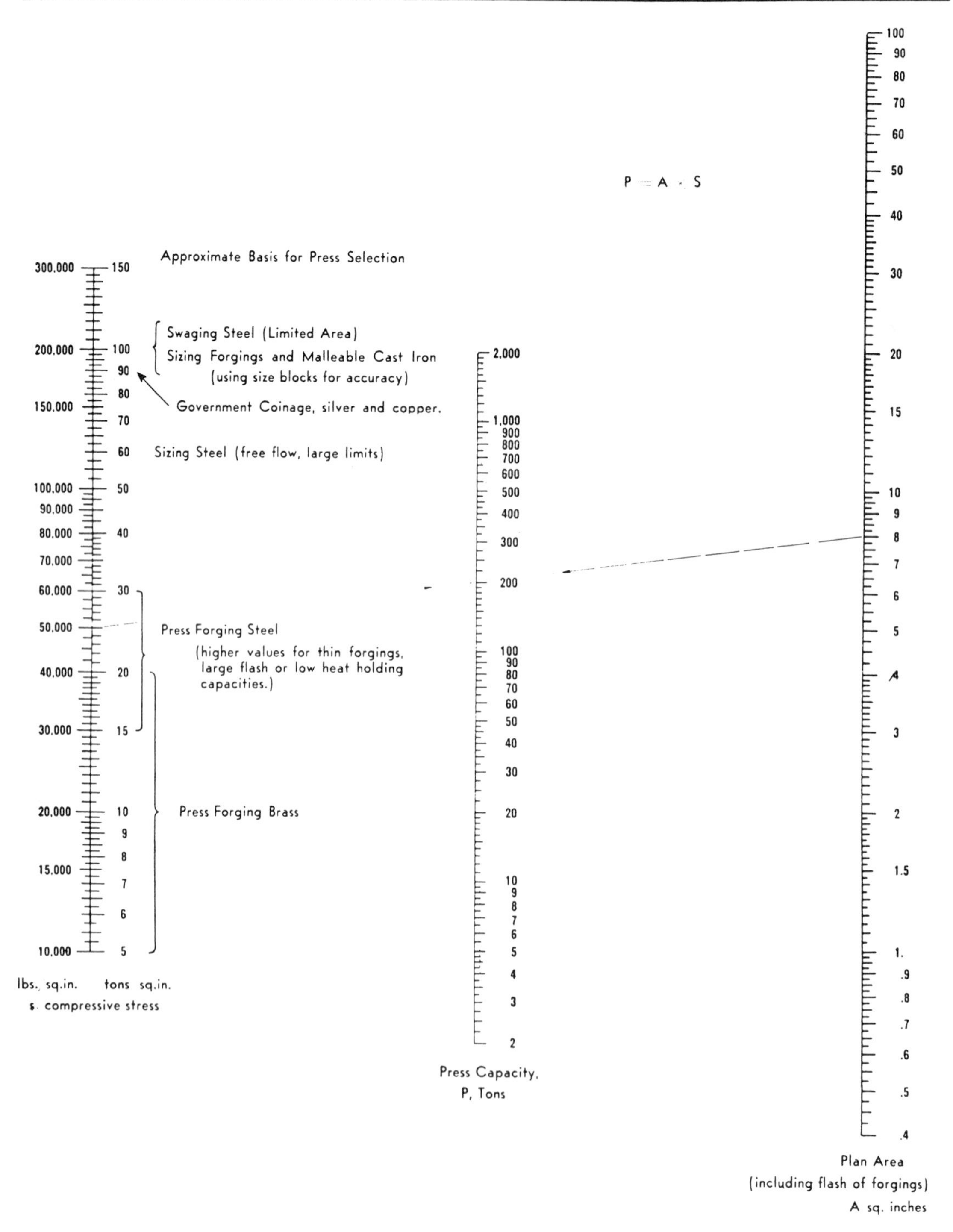

Figure 2M-1. Chart for determining required press capacities for coining, sizing, and forging

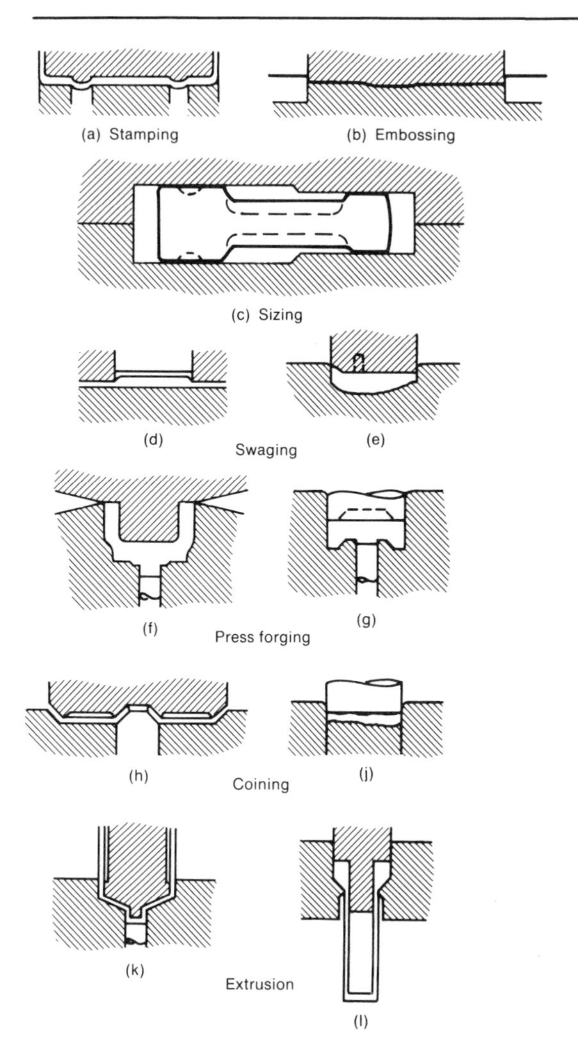

Figure 2M-2. Typical operations of the squeezing group, illustrating restriction of metal flow

dependent on the thickness and hardness of the original forging. Such surfaces usually should take at least as much pressure as the forging itself. Presses for this work are often selected on the basis of 100 tons per square inch (tsi), although the actual pressure on the surface of the forging may not exceed 25 or 30 tsi. In swaging, or cold forging, of steel (see Figure 2M-2d), the contact surfaces are usually unnecessary, but the pressures are higher due to more severe working of the material. Closed dies can be used with steel as well as with softer metals provided that proper consideration is given to press size and tool strength.

For coining (see Figure 2M-2j) in which metal is forced to fill a shallow impression (usually in a closed die), required tonnages may be figured at about 90 to 100 tsi. Some hard metals, as well as soft metals, can be coined.

Figure 2M-2(k) illustrates backward extrusion (as of collapsible tubes), and Figure 2M-2(l) shows forward extrusion. General tonnage figures of 50 to 100 tsi cover most extrusion of the softer metals and represent about the maximum that ordinary steels and tool constructions will withstand.

Hot forging pressures are figured conservatively at about one-third those of cold operations, although proper forging temperatures often bring the actual pressures relatively lower.

In figuring energy or work (average pressure times working distance) in squeezing operations, the distance the punch travels from its first contact with the slug or blank to the bottom of the stroke should include the distance the metal is moved and the amount of deflection in the press and tools (although part of the energy required to stretch the machine may be returned to the system, depending on speed and friction relations). The working pressure will begin at the yield point of the material times the area of contact and will rise gradually as the material strain hardens and the contact area increases. In the case of closed dies, there is likely to be a further sharp rise in pressure when the die fills, its duration and extent being determined by the amount necessary to stretch the press to get over bottom center. The material is practically incompressible.

The eccentric-shaft forging press is adaptable to many swaging and blanking operations as well as to the hot forging work for which it is named and in which its fast action is essential. Specialized types are now widely used for extrusion of collapsible tubes. Knuckle-joint coining presses are more compact and more powerful machines that are limited to short working strokes as in coining, swaging, and embossing operations.

2N: Ironing

In the blanking group of operations, metal is fractured in shear. In the drawing group, metal is worked primarily by the application of a tensile load on the sidewalls, although the resultant working stress in the flange is compressive. The group of operations that includes ironing, stamping, sizing, swaging, coining, extrusion, and press forging is distinguished by the fact that the primary working stress is compressive, and in all but the first case the blank or slug is squeezed directly between the punch and die.

Ironing, in which a drawn shell is reduced in wall thickness by being pulled through tight dies, is related to both shell drawing and wiredrawing. It is done to obtain a wall which is thin compared with the shell bottom; to obtain a uniform wall; to obtain a tapered wall, as in cartridge cases (Figure 2N-1); or merely to correct the natural wall thickening that occurs toward the top edge of a drawn shell. In this last case, the amount that the wall thickness will be increased by drawing must be discovered.

The theoretical maximum reduction in wall thickness per operation due to ironing is approximately 50%. In such a case, the cross-sectional area of the (unstrained) metal before ironing is about twice the cross-sectional area after ironing. Therefore, the area which is being worked in compression and which must yield is about equal to the strain-hardened area which is in tension and which must not yield. This would indicate that the practical limit should apparently be kept below 50%, although slightly higher reductions have been recorded.

The chart in Figure 2N-2 is arranged to show approximately the natural change in thickness accompanying a change in diameter. The results apply to the upper edges of drawn shells, because the wall thickness tapers from a maximum at that point to a minimum at the bottom corner, where it

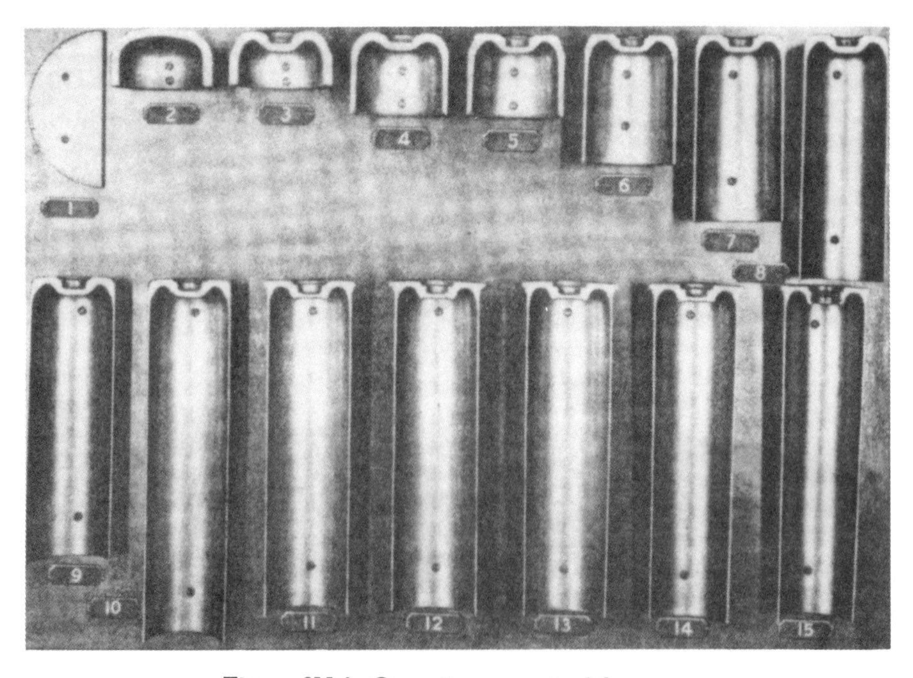

Figure 2N-1. Operations required for production of a cartridge case, including ironing (steps 4, 6, 7, 8, 9, and 10), coining of the end shape (steps 3, 5, and 11), and tapering (step 13)

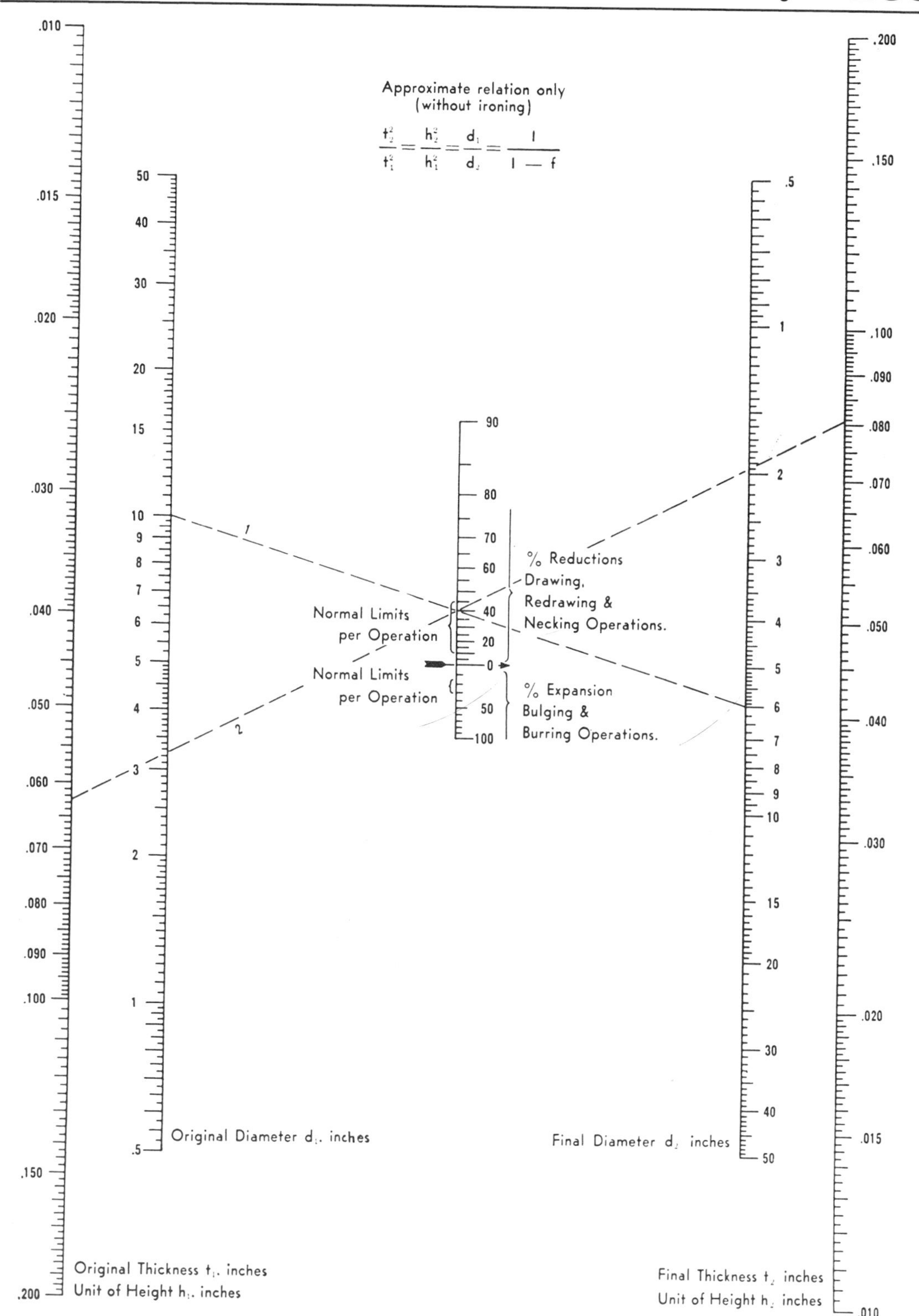

Figure 2N-2. Chart for determining approximate changes in thickness accompanying changes in diameter due to ironing

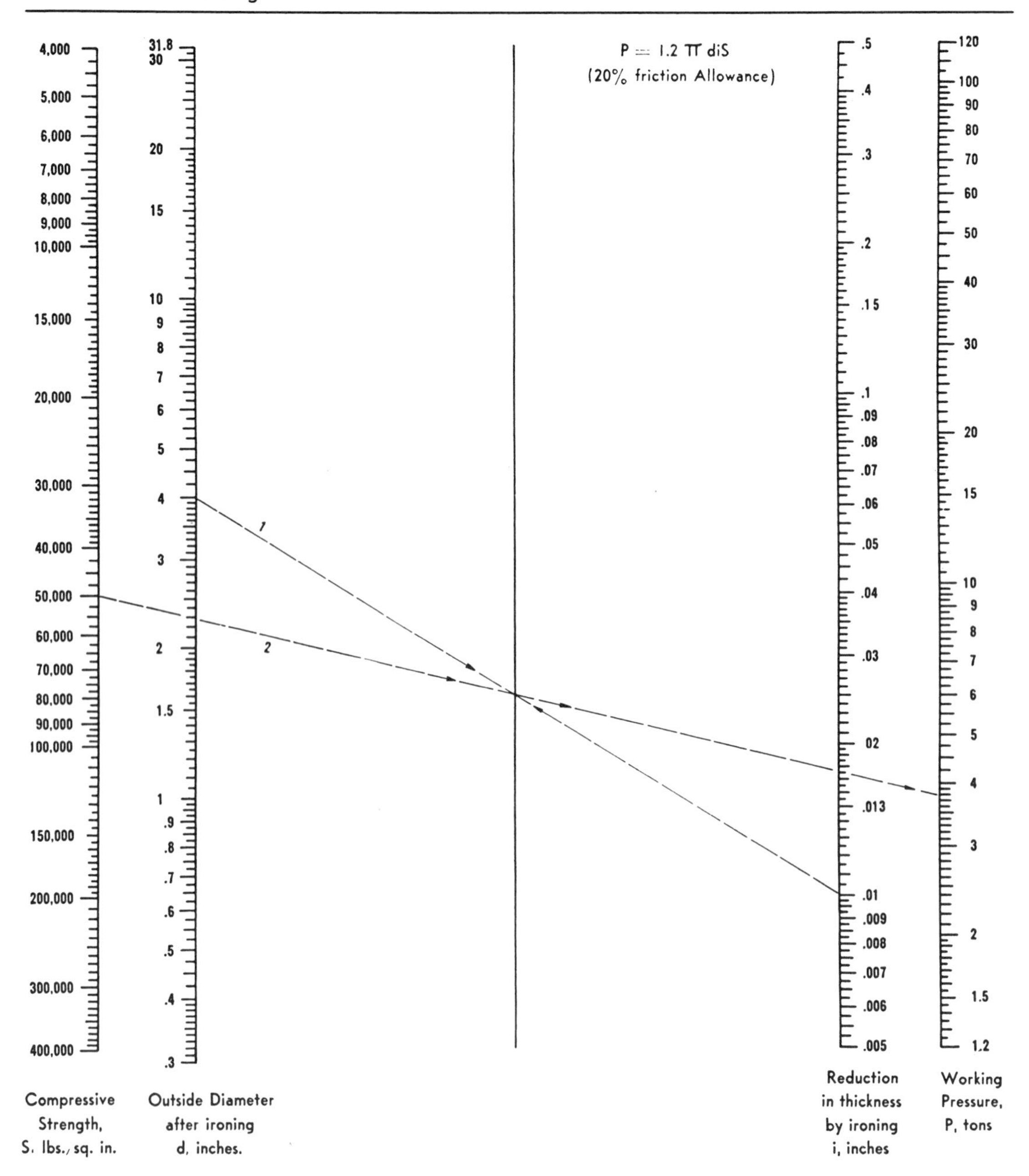

Figure 2N-3. Chart for determining pressures required for ironing

may be as much as 10 or 15% less than the original metal thickness. Even at the top edge, the metal thickness is likely to be a little less than the theoretical thickness given by the chart because of the thinning effect of bending over the drawing edge. A sharper radius or a deeper draw increases this thinning effect.

The chart in Figure 2N-2 is always read from left to right, drawing a line between the original diameter (of blank or shell) and the final (shell) diameter, on the two inner scales, to obtain the maximum per cent reduction or expansion on the center scale. A second line through this point, starting from the original metal thickness on the left-hand scale, will indicate the approximate maximum wall thickness at the upper edge of the final shell. An example is indicated in dashed lines. It is noted that this is based on a draw with sufficient clearance between the punch and die so that there will be no ironing.

If the clearance between punch and die is made equal to the metal thickness (0.0625 in. in the example), there will be an ironing load added to the drawing load (sufficient to remove 0.0805 in. – 0.0625 in. = 0.018 in.). If a parallel wall is desired, it will be necessary to iron down to the thickness of the thinnest part of the wall, which is appreciably less than the original metal thickness, depending on the sharpness of corner radii and the severity of reductions.

The chart in Figure 2N-3 offers a convenient means of approximating the pressure required in ironing. Referring to the example shown in dashed lines, note that the two inner scales are used first to establish the pivot point on the center scale. Thus a shell ironed to a finished diameter of 4 in., with a displacement of 0.010 in. of the total metal thickness, will require an ironing pressure of about 3.8 tons, assuming that the metal is spheroidized steel moderately strain hardened and therefore offering a compressive resistance of about 90,000 psi. This must be added, of course, to the drawing or redrawing load figured separately. The formula is $P = 1.2 \pi diS$, where P is the approximate required maximum pressure, in tons, d is the outside diameter after ironing, in inches, i is the reduction in wall thickness, in inches, and S is the compressive resistance of the metal under existing conditions of strain hardening, in psi. There is included a 20% allowance for surface friction in addition to the work of coining. This is an arbitrary figure which should cover well-polished dies and suitable lubrication. Lack of lubricant, tool-mark rings on the dies, or surface pickup will increase the friction load considerably.

If the wall of a shell is ironed thinner by the same amount for the entire length of the shell, then the work done is approximately the product of the length of the shell or of the ironed surface and the pressure required for ironing (from the chart). This is expressed as $W = Pl$, where W is work in inch-tons, P is pressure in tons, and l is length in inches. If a reducing operation accompanies ironing, then the drawing pressure should be added to the ironing pressure.

If the ironing operation is done merely to correct the natural changes in wall thickness due to drawing and reduces the thickest portion near the top of the shell to equal the thinnest portion near the bottom, the average ironing pressure will equal about half the maximum, and the formula will be $W = 0.5 Pl$.

2O: Automated In-Line Forming

Many varied types of metal-forming operations are used to form the bends, angles, flanges, wrappers, castings, shells, etc., that are required in the production of various industrial and consumer-oriented products. Some of these operations are tube bending, drawing, roll forming, press braking, wing forming, and tangent bending.

IN-LINE FORMING

Most people involved in producing products from metal blanks find production cost factors to be a constant source of worry, and are always concerned with the question of how to reduce costs while maintaining, or even increasing, production.

One answer could be automated in-line production, because in-line production operations, when automatic, are usually more productive than operations in which manual activity is involved in step-by-step production. The time spent in manually transporting parts between the various production steps is often a significant cost factor, and simply eliminating manual handling of the product often reduces costs and thereby increases profits.

A typical production line might consist of an uncoiler, followed by some type of straightening and shearing unit, a punching and/or notching unit, a roll former, and finally a metal-forming device. It is easy to see that if all of these individual operations are somehow connected automatically, then the productive output should be greatly increased.

What does the concept of in-line forming mean? It simply means that all positions are "go" and that there are no points of stoppage, or "bottlenecking," in the total flow of operations.

THE METAL FORMER: USES AND APPLICATIONS

Selecting only one segment of the in-line production line, the metal former is normally positioned last in the line, because its operations can most generally be performed only after all other operations have been completed.

A metal former can be put to excellent use in bending side or end flanges, by wing forming of U bends for cabinets and cases, or, in some instances, by tangent bending (a type of radius forming). This is often called the clamp-then-bend method, and it is ideal for bending of end flanges following formation of plain or configurative forms on the two long sides of the blank in a roll former.

It must first be determined that the total cycle times will prove to be realistic. Cycle times depend on the size of the piece-part, the number and shapes of desired flanges or bends, and the amount of time which will be involved in feeding and ejection of the piece-part. Roll forming is by far the faster of the two operations, and therefore the production rate of the roll former usually must be geared to the cycle time of the metal former. This is not to say, however, that actual production will slow down, because the continuity of automation will almost always substantially increase the actual number of parts produced per minute, hour, day, year, etc., even though the cycle time of the metal former may not seem fast to the naked eye.

Automating the entry into the machine, and the exit from the machine, of the piece-part will eliminate much of the loss in time normally involved in manual loading and unloading.

The metal former is designed primarily for straight-line forming of either right-angle or radius bends. The capability of forming compression or stretch flanges, without marking or marring the material, is an added bonus which is available in many instances for numerous applications.

By no means can all products be formed by this method, but the list of potential applications is longer than one might imagine. This method can be used in forming the cabinets, housings, wrappers, shells, liners, hems, and flanges for such items as: air conditioners (window and central), air purifiers, baseboard heaters, beverage and cold drink machines, boilers, commercial and home freezers and refrigerators, condenser coils and housings, camper shells, deep fat fryers, dehumidifier cabinets, deodorizers, desk pedestals, desk drawers, doors, dishwashers, dryers, electronic ovens, electrical boxes, furnace housings, file cabinets and drawers, grills, humidifiers, ice makers, incinerators, kitchen ranges, oven liners, baseboards, portable clothes dryers, range hoods, refrigerated food-display cases, room heaters, room dividers, safety deposit boxes, and coolers.

BASIC TYPES OF FORMING

Many varied capabilities are involved, and these can probably best be explained by explaining the five basic types of forming:

1. Fold action – fixed die
2. Fold action – hinged die
3. Wipe action – fixed die
4. Wipe action – hinged die
5. Fixed die – rack and pinion (gear)

Each of these five basic types of forming is necessary for particular applications and products; each, in its own particular way, fulfills a need.

Fold Action – Fixed Die Forming

The simplest, probably the most used, and also the most easily understood type is fold action – fixed die forming (Figure 20-1). This type of forming makes use of a bottom die block affixed to the bottom die plate, and a top die block affixed to the top vertically movable head. When the safety-protected palm button is depressed, the head will move in a downward motion, and, with the material to be formed having been correctly positioned between the top and bottom die blocks, the clamping action will firmly hold the piece-part in bending position.

Gaging will have already been accomplished, depending on whether the piece-part has been fed manually to a gage bar or pin stop, or on whether automatic feeding has been used.

When the safety-protected form button is depressed, the bending beam to which the bending die block is attached, bends at a preset angle to the stop block, thus forming the piece-part to the required angle.

Prior to the movement of the bending beam, the top head is securely latched to the bottom head, with the piece-part being firmly held between the top and bottom die blocks. This allows the required bend to be made correctly without slippage of the piece-part. This is strictly right-angle forming; however, acute or obtuse angles can be accommodated by means of correctly designed tooling. Overbending is somewhat limited due to the position of the top head and the length of the leg to be bent upward on the piece-part.

This type of straight-line forming can be done with one head forming one bend at a time, or, with a twin-head metal former, two bends can be made simultaneously.

Repositioning the piece-part under a single head and repeating the bending process, or moving the horizontally adjustable head of a twin-head unit to the desired location, can yield a substantially higher production rate if needed. A twin-head unit will allow varied dimensions between bends to be accommodated with little or no manual dexterity necessary. This, of course, automatically lowers the unit cost per part, thus increasing productivity and reducing costs.

The simplicity of both the single-head and twin-head metal formers sometimes tends to mask the built-in flexibility available.

Fold Action – Hinged Die Forming

The second type of forming is fold action – hinged die forming (Figure 20-2). This type of forming is commonly used when the product to be formed is designed with a flange, an offset, or a hem on that portion of the product that, when placed in the machine, is turned front or back.

In fold action – hinged die forming, the bottom plate and die block are in positions similar to their fixed-die positions, but the top die is hinged (often by means of a pin) so that, as the head is raised, the top die diagonalizes to a predetermined stop so that clearance for the flange, hem, or offset is available. This allows for easy removal of the formed part.

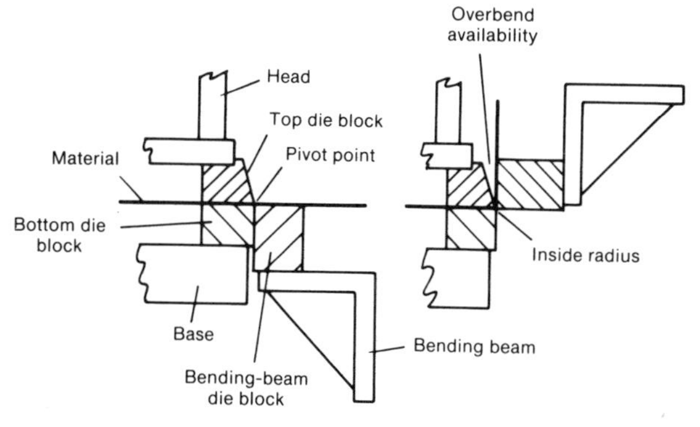

Figure 20-1. Fold action – fixed die forming

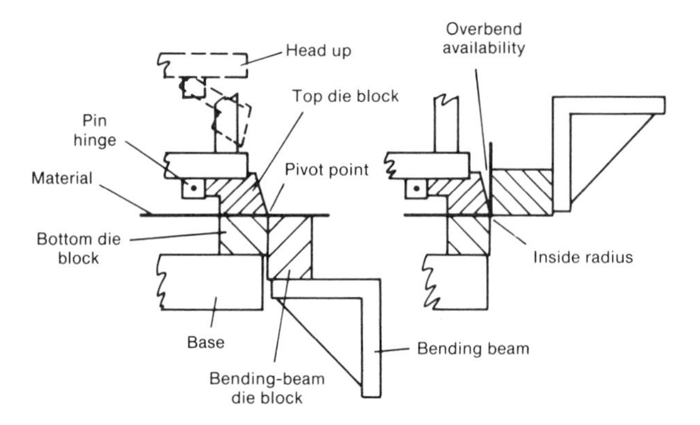

Figure 20-2. Fold action - hinged die forming

This method is used in U-forming. A twin-head machine can make both bends (as shown in Figure 20-3) with the capability of allowing dimension A to vary as needed, within the size limitations of the machine purchased.

Dimension A in Figure 20-4 also can be formed by using this method in forming the leg on a relatively shallow pan.

A complete box can also often be formed with a twin-head unit using multiple bending operations successively, as shown in Figure 20-5. This can also be termed a "wrap-around," because it will be "wrapped around" the head of the forming unit when finished. Varied products can be formed in this fashion, but again, the desired production rate, the need for changes in tooling or gaging, and the size of the product must all be considered in determining the correct method.

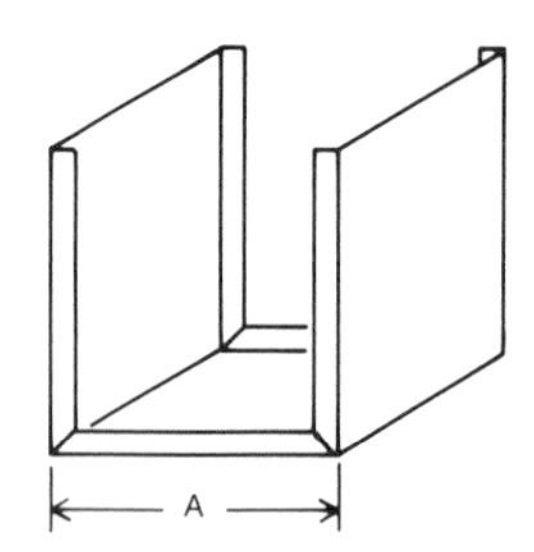

Figure 20-3. Formed section in which bends were made by fold action - hinged die forming in a twin-head machine. Dimension A is variable.

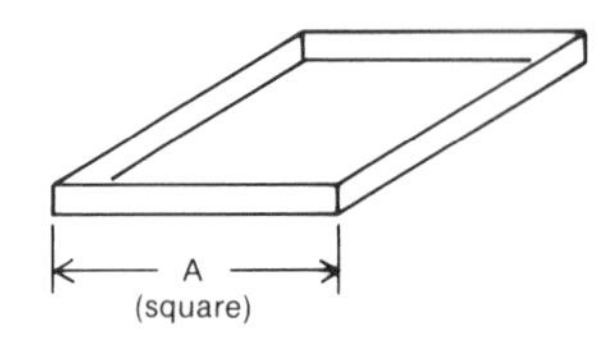

Figure 20-4. Shallow pan of which dimension A was formed by fold action - hinged die forming

When an inside radius bend of 1/8 in. or larger is desired, and the product does not require non-marring conditions, then wipe action - fixed die forming can be used (Figure 20-6).

The center point of the die radius is the pivot point on which the bending beam rotates. As the bending beam is activated, the bending beam die block literally wipes the material around a predetermined radius formed by the top die block, and some springback of material is normal to furnish the final required radius. Some marking or "burnishing" is normal with this method, and if it is objectionable it may be removed by grinding or some other method, provided the radius will be visible in its final location in the finished product.

Wipe Action - Hinged Die Forming

Wipe action - hinged die forming (Figure 20-7) also involves a type of wiping action around a fixed radius, and is used when the part has flanges, offsets, or hems on front and rear. It is similar in this way only to the pin-hinge action of a previously listed fold action method.

Again, the top die block is hinged so that, as the head is raised, the top die block diagonalizes to a predetermined location in order

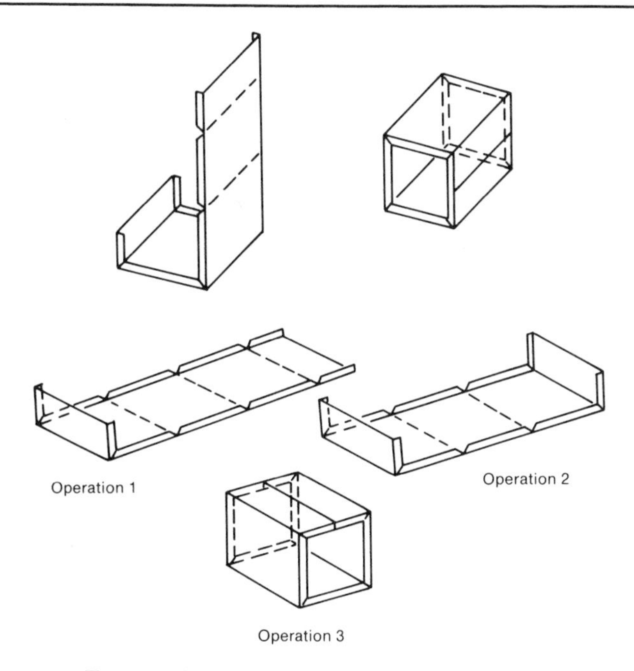

Figure 20-5. Complete box formed by fold action - hinged die forming in a twin-head machine

that the flanges have clearance so that the part can be removed from the metal former.

Fixed Die - Rack and Pinion (Gear) Forming

The final type of forming is probably the least understood, and yet in today's economy, and because of related happenings in the metal industry, it promises to have more potential usage than most of the other types. This type is called fixed die - rack and pinion (gear) forming (Figure 20-8).

Because much of the metal produced today is coated or painted, so that the various steps of production are accomplished on painted or coated metal, much different requirements must now be satisfied in every step of the production process. It seems that many products, especially in the appliance field, are intentionally manufactured from painted or coated coil, thus eliminating any need for final painting, and thus reducing not only the time required for painting but also the monetary requirements expended in the purchase and maintenance of painting lines. This, of course, makes it necessary for

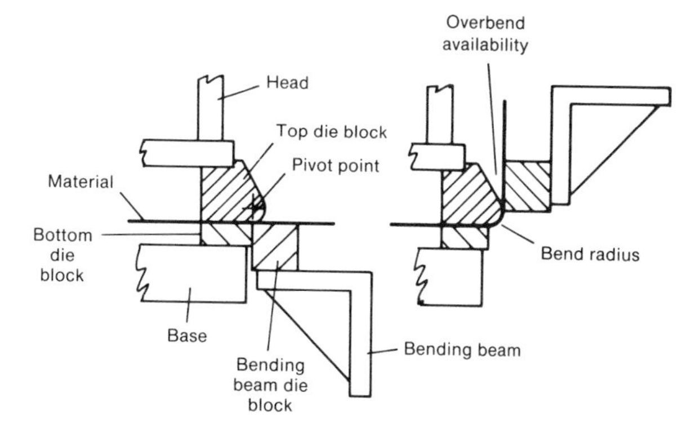

Figure 20-6. Wipe action - fixed die forming

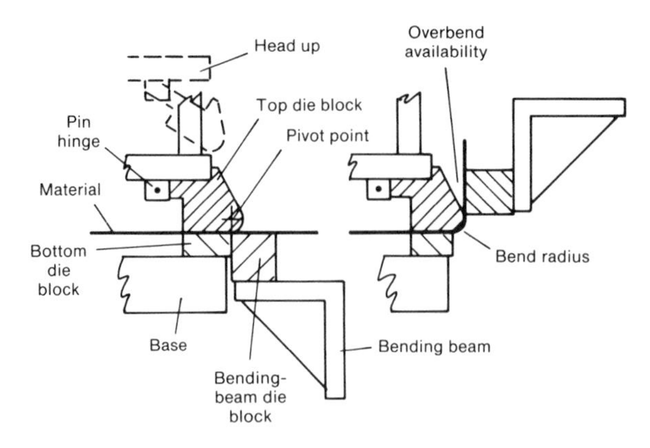

Figure 20-7. Wipe action - hinged die forming

all of the various production steps to be accomplished without marring or marking the paint coating, and this means that a wiping action will not prove satisfactory.

In addition, the present trend for many consumer-oriented products is toward more and more streamlined surfaces, shapes, corners, and other features, and thus fewer simple 90° corners are found, and more and more radius designs have become prevalent, such as on corners of refrigerators, dryers, washers, and freezers.

One way in which painted or coated metals can be formed into streamlined shapes is with fixed die - rack and pinion (gear) forming. This method is satisfactory for aluminum, painted metal, stainless products, vinyl-coated products, and many others of similar nature. This method is often called "tangent bending."

Again, the center point of the radius is the pivot point on which the bending beam rotates, but now there is a gear on the pivot point which has the same pitch diameter as that of the forming-die radius. This gear engages a rack, and the rack is mounted on a sliding plate. The bending-beam die block is also mounted on the sliding plate. As the bending beam is raised, the rack must mesh and rotate on the gear, which automatically keeps the bending-beam die block tangent to the radius die block. This allows the material to be formed around the radius without slippage of tooling on the material, and creates a mar-free product.

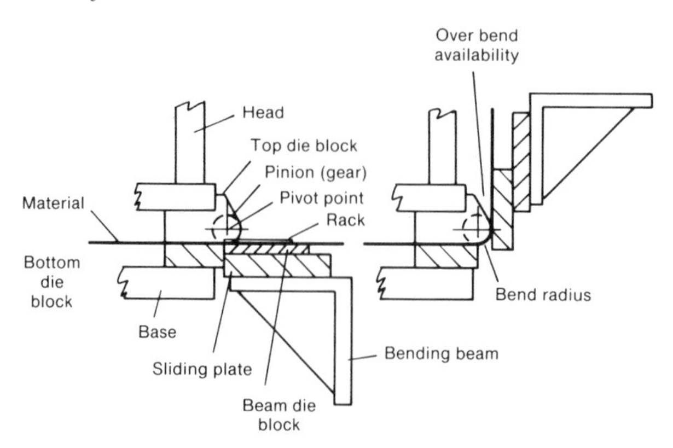

Figure 20-8. Fixed die - rack and pinion (gear) forming

2P: Extrusion

Closely allied to the coining, sizing, and forging group of press operations is the extrusion method of shaping metal, wherein the material is forced by direct compression to assume the desired shape.

The ever-increasing range of extrusion work now includes everything from collapsible tubes and continuous shapes in the softer metals (Figure 2P-1k and l) to heavy artillery shells cold formed from steel having finishing yield points of 120,000 psi.

Cold extrusion of steel has been made possible in recent years due to the availability of larger presses, dual surface protection, and better die construction and materials. In extrusion of steel, the cheaper grades may be utilized because the required physical properties of a given part can be obtained through working of the metal rather than by using more costly heat treated material. In general, the plain low-carbon steels adapt themselves most readily to cold extrusion. Higher-carbon steels should be in spheroidized form to improve freedom of internal movement.

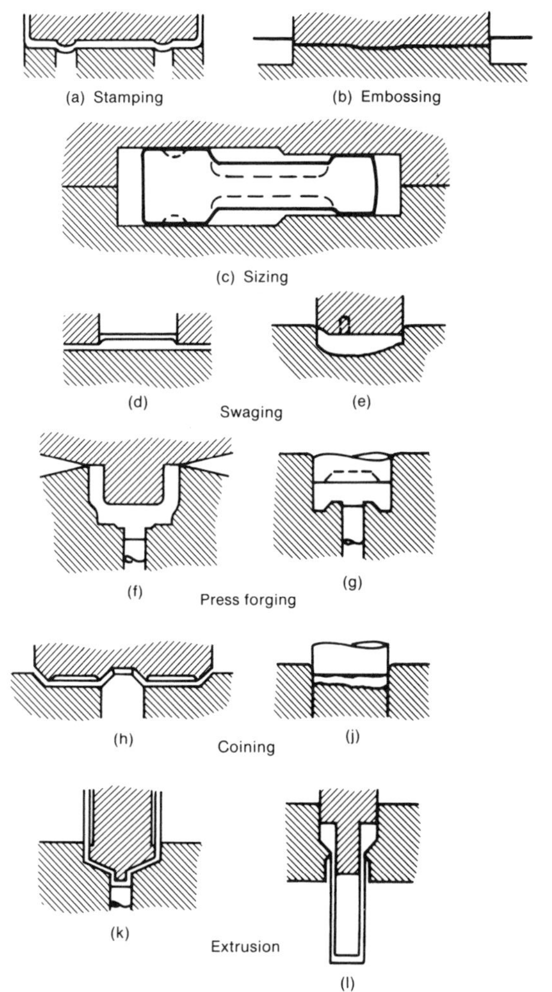

Figure 2P-1. Typical operations of the squeezing group, illustrating restriction of metal flow

Figure 2P-2 shows a typical example of a piercing or backward extrusion die in which the metal flows in a direction opposite that of the punch movement. Figure 2P-3 shows a sizing or coining die in which the billet is prepared for piercing to ensure concentricity, and in some cases to prepare the lower profile. This operation involves little extrusion or work hardening, and required tonnages may be computed per Table 2P-1 and the chart in Figure 2P-4. Figure 2P-5 shows a forward extrusion die in which the metal flows in the same direction as the punch, but at a higher rate due to the change in cross-sectional area of the part.

When steel is made to flow in steel dies

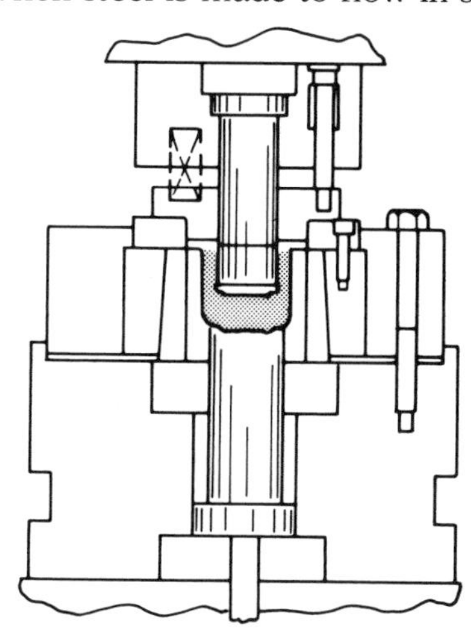

Note the centering ring for the punch and the carbide die cavity, preloaded in shrunk rings and supported on toughened load-distributing steels.

Figure 2P-2. Piercing or backward extrusion die

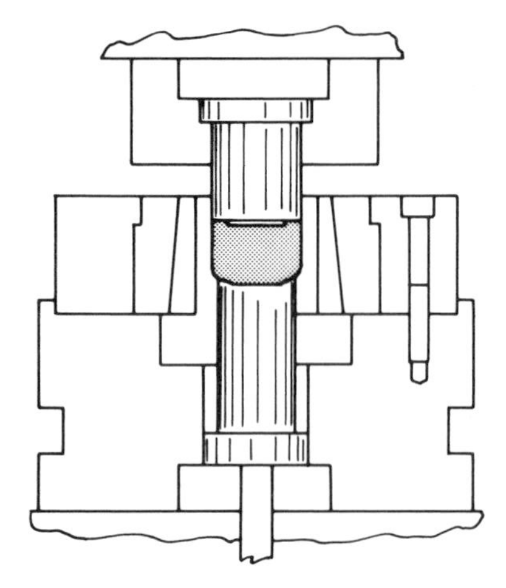

Figure 2P-3. Sizing die used to prepare steel billets for piercing

(or carbide dies), the stresses built up are so high that extreme care in die construction must be taken to support the highly stressed members against fatiguing deflections and to distribute the loads over a large portion of the slide and bed members of the press. The extremely high tool loading involved makes it desirable to draw or stress relieve the tool members at fairly frequent intervals to offset surface fatigue.

The use of carbide inserts at the points of excessive pressure is highly recommended. These inserts must be well supported on hardened blocks, and must be held in place by shrunken and tapered holders as illustrated in Figure 2P-2.

Extrusion dies can be classified into two general groups: "open" and "closed." In open dies the metal does not fill the die cavity, whereas in closed dies it does fill the die in order to "square up" in the corners and fill out all the details of the die. However, because closed dies are likely to cause excessive pressure, they are often relieved at one or two places to compensate for variations in slug volume.

The pressure required in extrusion work is dependent on several factors: the yield point of the material before extrusion, the yield point after extrusion, and the friction or resistance to flow that the nature of the die imposes. Also, the speed of extrusion creates additional loading on the tools. It is necessary, therefore, to make a study of each die to determine the tonnage required.

When there is little restriction to flow and low speed, the press load can be based on the yield point after work hardening. With high speed and restricted flow, the press load may increase to three times the finishing yield point, and pressures up to 330,000 psi have been recorded. Figure 2P-4 and Table 2P-2 are useful in estimating the work-hardened yield point.

The pure metals can be worked at pressures lower than those required for alloys, because in alloys certain groups of molecules tend to keep the slip planes together, thus making movement difficult. Figure 2P-6 illustrates this phenomenon, and it can be seen that alloying of gold and silver in equal amounts creates the strongest alloy, whereas, each metal by itself is relatively soft.

The following examples will serve as guides in calculating press loads. Figure 2P-2 illustrates a fast piercing operation (also known as backward extrusion) on a steel billet 4 in. in diameter, and with a reduction in bottom thickness of about 40%. The punch is 2¾ in. in diameter (area of 5.9 in.^{2}). Because this is an open die, and the billet is subjected to severe work hardening and flow, a unit pressure on the punch of 130 tons per square inch (from Table 2P-1) is reasonable. The total press load then becomes 770 tons.

An actual application similar to the above example, but with a 65% reduction in cross-sectional area and a 77% reduction in bottom

Table 2P-1. Extrusion Pressures for Common Metals

Material	Pressure(a)
Pure aluminum, "extrusion grade"	40 to 70
Brass, soft	30 to 50
Copper, soft	25 to 70
C-1010 steel, "extrusion grade"	50 to 165
C-1020 steel, spheroidized	60 to 200

(a) These values, in tons per square inch, are subject to considerable variation due to composition and structure of material, restrictions to flow (area reductions and tool profiles), and severity of work hardening (reduction in thickness).

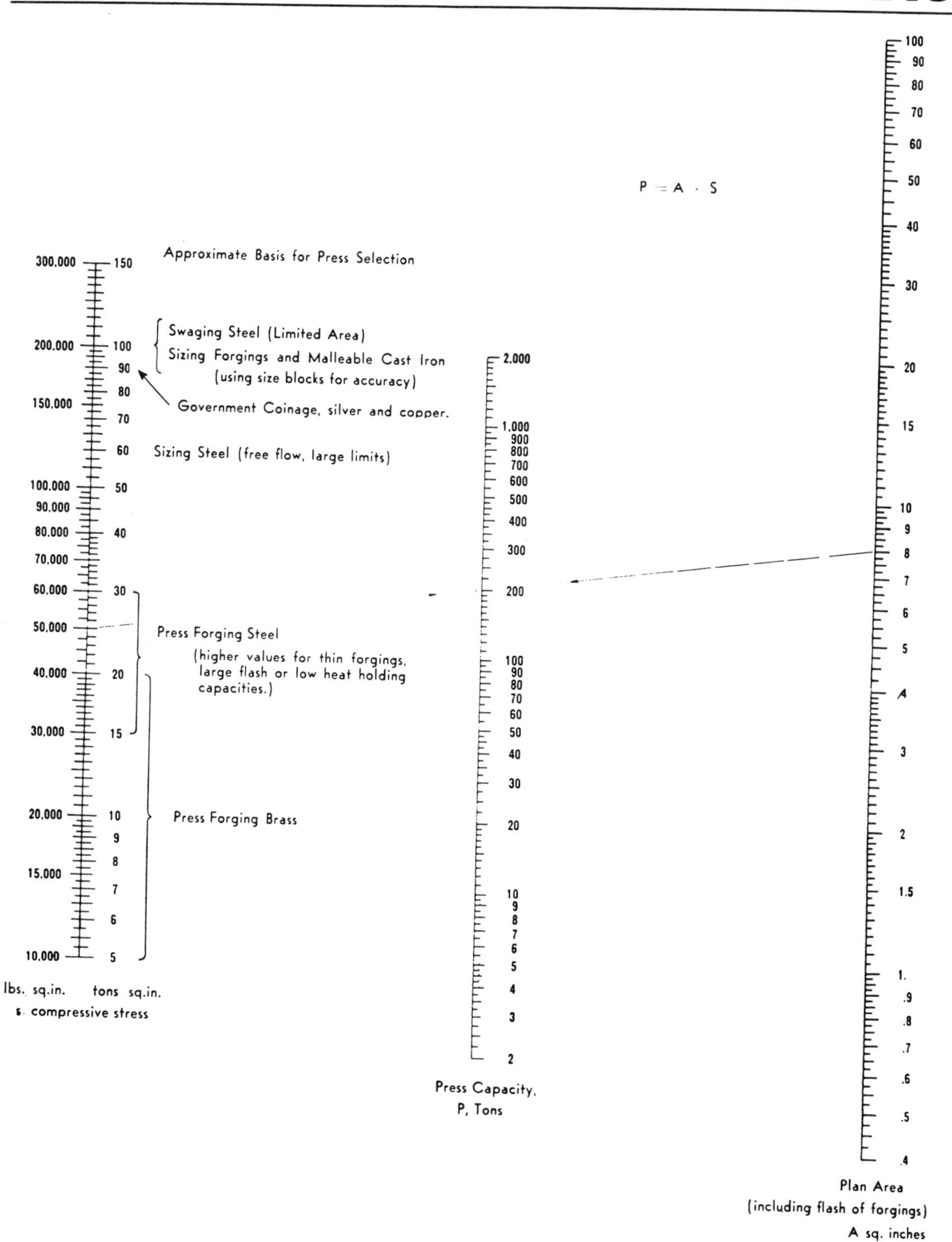

Figure 2P-4. Chart for determining tonnages required for coining, sizing, and forging

Table 2P-2. Properties of Metals(a)

	Theoretical yield points		Reduction in area (maximum annealed)	Shearing stress and per cent penetration to fracture				Tensile strength (nominal) annealed	Elongation in 2 inches (average annealed	Elasticity Young's modulus	Modulus of strain hardening (tentative)	Bulk modulus volume	Annealing temperature (approx.) commercial	Forging tempera-ture (approx.)	Thermal coefficient of expansion	Specific gravity
	Minimum(b) (commer-cially annealed)	Maximum (severely cold-worked)		Annealed to soft temper		Cold-worked to temper noted										
	psi	psi	%	psi	%	psi	%	psi	%	psi	psi	psi	°F	°F	in./in./°F	lb/in.³
Aluminum, No. 2S commercially pure	(4-) 8,000	H 21,000	80	8,000	60	H 13,000	30	13,000	35-45	10,300,000	25,000	10,200,000	650		0.000,010	0.0963
Aluminum, No. 3S, Mn alloy	11,500	H 25,000	80	11,000		H 16,000		16,000	30-40	10,300,000	29,000		750			
Aluminum, No. 17S for heat treatm't	HT 30,000												630-650	400		
Aluminum, No. 52S, Cr alloy	21,000	H 37,000	80	14,000		H 36,000		29,000	25-30	10,300,000	38,000		650			
Brass, yellow, for cold working	(10-) 40,000	95,000	75	32,000	50	? 52,000	20	50,000	65	13,400,000	120,000	8,800,000	1100			0.303
Brass for forging, beta														12-1400		0.301
Bronze, Tobin alloy	25,000	120,000	53	36,000	25	1/4H 42,000	17	60,000	46		250,000					0.315
Copper	(10-) 25,000	62,000	65	22,000	55			33,000	68	14,500,000	81,000	17,400,000		16-1800	0.000,009	0.3195
Gold										11,600,000		23,200,000	6-800		0.000,008	0.6949
Iron, cast, grav	8,000 tension							23,000		12,000,000		13,900,000			0.000,005,5	0.260

	tension 40,000 compr.							45,000		20,000,000						
Lead	(1-) 3,000	4,000		3,500	50					2,470,000		1,100,000	below room temp.		0.000,017	0.4106
Monel metal Nickel Silver	28,000 21,000	H112,000 H115,000	63 75	35,000	55			70,000 70,000	30-50 47	32,000,000 10,900,000		24,600,000 14,500,000	11-1200 12-1300 6-800		0.000,007 0.000,010	0.318 0.3175 0.3791
Steel, 0.03C Armco	(20-) 35,000	70,000	76	34,000	60			42-48,000	48	30,000,000	100,000		13-1400			0.2834
0.15C	(32-) 55,000	110,000	60	48,000	38	61,000	25	57,000	35	30,000,000	150,000	23,200,000	13-1400	—	0.000,009,16	
0.50C	(55-) 70,000		45	71,000	24	90,000	14	85,000	25	30,000,000			13-1400	18-2100		
1.00C	90,000	145,000	20-40	115,000	10	150,000	2	120,000	10	30,000,000	200,000					
Steel, electrical high-silicon,				65,000	30											
Steel, stainless low-carbon for dwg. Enduro AA	50,000	120,000	60					80,000	25	30,000,000			13-1550	19-2100		
18-8	(35-) 50,000	165,000	71	57,000	39			95,000	57				13-1900	2100		0.285
high-carbon for cutlery, etc.			32					260,000	11				1450-1600	20-2100	0.000,006,7	
Tin	(2-) 4,000	4,500		5,000	40					7,260,000		7,260,000	below room temp.		0.000,015	0.2652
Zinc	(12-) 28,000		43	14,000	50	19,000	25	30-37,000	27	13,100,000		5,100,000	2-300		0.000,014,5	0.256

(a) Crane's *Plastic Working and Metals*, 3rd Edition, John Wiley & Sons, New York, 1941. (b) Figures in parentheses represent laboratory test minima which might prove misleading for estimating loads. All physical values are subject to some variations with testing methods and analysis of material.

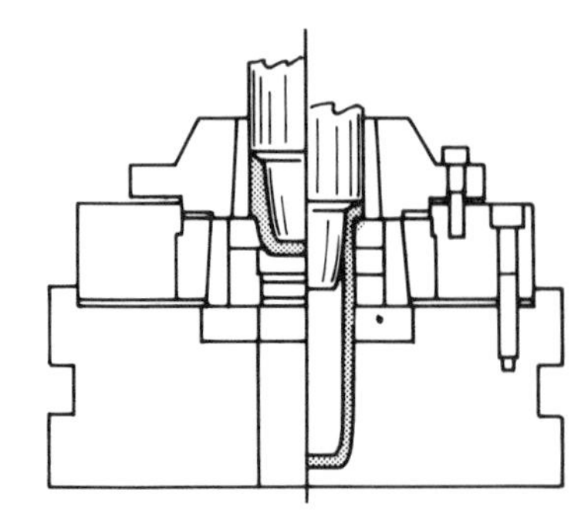

Note the distance the bottom of the shell has moved relative to punch movement. The secondary guide ring is added to help maintain straightness.

Figure 2P-5. Typical forward extrusion die

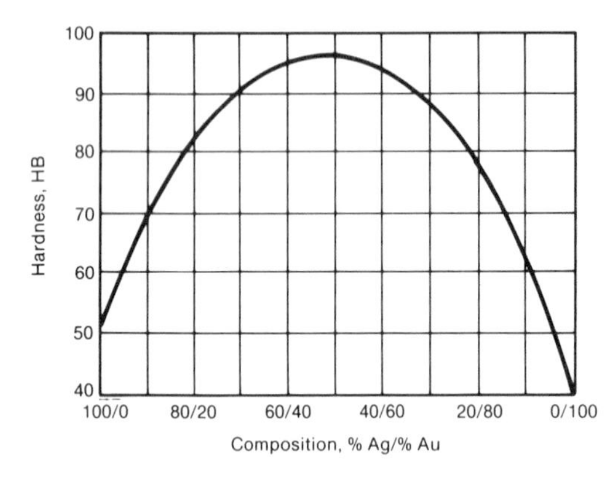

Figure 2P-6. Variation in hardness with proportions of silver and gold, in solid solution (Jeffries and Archer)

thickness, required a unit load of 157 tons per square inch.

In some backward extrusions, a shoulder is provided on the punch, and the metal flows until it meets this shoulder and completely fills the die. This might be described as closed-die extrusion, and there must be added to the punch load a pressure on the area of the punch shoulder exceeding the work-hardened yield point of the metal to square up the edge. This squaring up of a heavy wall may be desirable in preparation for forward extrusion to favor straightness of flow.

When hydraulic presses are employed for this type of work, it is best to use a pressure reversal of the ram to effect fill-out at the bottoming position, and to control workpiece dimensions by stops.

Figure 2P-5 shows a forward extrusion die for steel with high restriction to flow. The working area in this case is the cross-sectional area between the pilot on the punch and the outside diameter or shoulder.

It is interesting to note that the plastic ranges of most metals can be indicated by a straight-line relationship between reduction and stress (Figure 2P-7).

When a metal is reduced by tension, it moves internally along its slip planes until a point is reached where further movement becomes extremely difficult. At this point, fracture normally occurs and the straight-line relationship terminates.

In extrusion work, the material is in compression, and it follows the straight-line relationship between reduction and stress up to a point corresponding to that reached when the metal was in tension. When this point is reached, however, the metal "stiffens up," and the stress increases greatly with a small increase in reduction. Consequently, the curve takes a substantial upsweep and becomes what is known as the "secondary plastic range." The dashed portions of the curves in Figure 2P-7 indicate this range. Experimental data concerning this upper range is very limited at present, but it is known that this range can be reached only with internal work-heat and correct lubrication.

In regard to these high reductions, it might be noted here that reductions in area exceeding 70% are accomplished by "cold" rolling on 52100 spheroidized steel, and reductions over 90% can now be achieved on cold rolled and cold drawn lower-carbon steels with yield points far exceeding those shown in Table 2P-2.

Extrusion work is quite often performed in the upper range, especially in closed dies, and this accounts for the extra pressure required as the material meets high resistance to flow or when the die fills.

Correct lubrication in extrusion work is of paramount importance because it reduces considerably the pressures required and actually makes the cold extrusion of steel possible.

Ordinary lubrication is satisfactory under

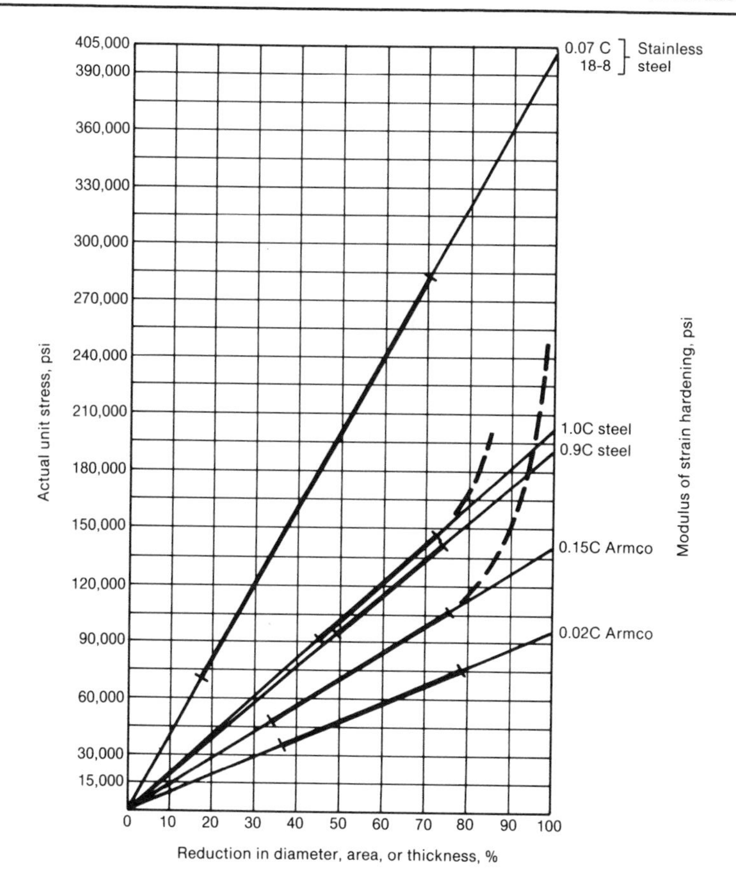

Figure 2P-7. Tentative curves showing commercial cold working ranges and rates of strain hardening for different steels (dashed portions indicate the higher range)

certain favorable conditions, but a breakdown in the lubricant results in surface heating and surface cracking. It has been found by experiment that a bonded steel-to-phosphate layer and a bonded posphate-to-lubricant layer are the most satisfactory. The phosphate layer serves as a parting layer to minimize metal-to-metal welding or pickup, and also as a host for the lubricant.

Reductions in cross-sectional area as high as 85% have been achieved in cold steel billets in a single operation. However, it seems better to limit the first reduction to approximately 60% and subsequent operations to about 40%.

For best results, the average hardness of steels before extrusion should not be over 60 HRB. It is possible to extrude harder steels, but the reductions obtained are considerably smaller.

2Q: Lubrication and Wear in Sheet Metalworking

Sheet metalworking processes present a wide range of demands on lubricants, from the almost negligible to the most severe. In these processes, as in other non-steady-state processes, whatever lubricant is introduced at the beginning must suffice for the entire course of deformation. Therefore, lubricant breakdown, tool pickup, and scoring of the workpiece are of major concern. For a given process geometry, conditions generally become more severe with increasing sheet thickness.

Stretching

At the start of deformation there is normal approach between punch and sheet, and, in the presence of a liquid lubricant, a squeeze film develops. Contact is made at the highest point; in stretching over a hemispherical punch, there is no sliding at this point (Figure 2Q-1). Away from it, sliding distance and velocity increase at a rate proportional to the distance, if friction is low. With high friction, sliding is arrested. Interface pressures are low, because the membrane stress is typically $4(t/d_p)$ UTS and is on the order of 0.01 to 0.1 UTS. Nevertheless, surface deformation is substantial.

A surface element of an originally smooth blank roughens prior to making contact with the punch, although details of the roughening process are affected by anisotropy and the strain ratio. Some sheet materials — notably, low-carbon steels and Al-Mg alloys — are prone to discontinuous yielding, and at low strains the surface is covered with highly visible, mutually intersecting families of slip bands. No analysis of the effects of such yield point phenomena on lubrication appears to have been made, and the remarks made here apply only to sheet that begins to deform plastically over the whole surface. The rate of roughening is lower in balanced biaxial tension than in plane strain, but roughness is proportional to the equivalent strain. Asperities deform and flatten, and the real (or boundary-lubricated) contact area increases rapidly to 60 to 70% of the total area. On a dry sheet, the *b* ratio increases in proportion to pressure, giving a constant μ. In the presence of a liquid lubricant, the *b* ratio is stabilized and μ decreases with interface pressure. The liquid film is squeezed out in a wedge which advances with the line of contact; the divergent gap is unfavorable for hydrodynamic lubrication. Nevertheless, some lubricant usually remains trapped in surface features and prevents metal-to-metal contact. Indeed, when the film is very thick (as it is when an oiled PE sheet is used), the surface roughens as in free stretching or hydraulic bulging.

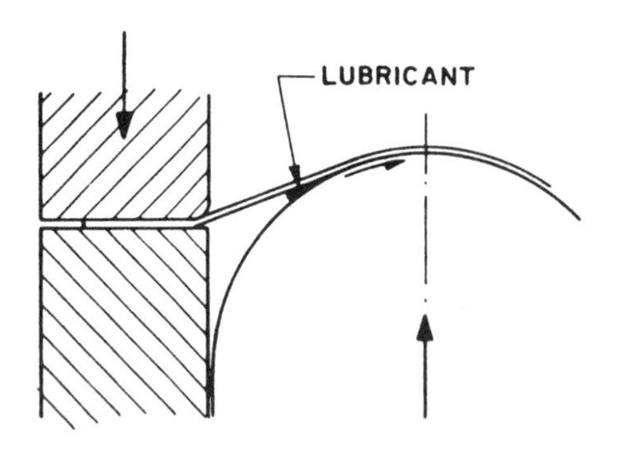

Figure 2Q-1. Negative wedge developing in stretch drawing

Deep Drawing

Conditions vary greatly at various points of a partially drawn cup (Figure 2Q-2):

a. Conditions on the punch bottom (zone 1) are the same as in stretching, except that friction around the radius (zone 2) limits the transmittal of stresses and interface pressures may drop close to zero. The sheet roughness is virtually unchanged. On the punch radius the stresses are higher, some relative sliding takes place, and the lubricant thins out.

b. Between the punch flank and the partially drawn cup (zone 3), the lubricant is trapped in a squeeze film, whereas it is drawn in under favorable conditions at the

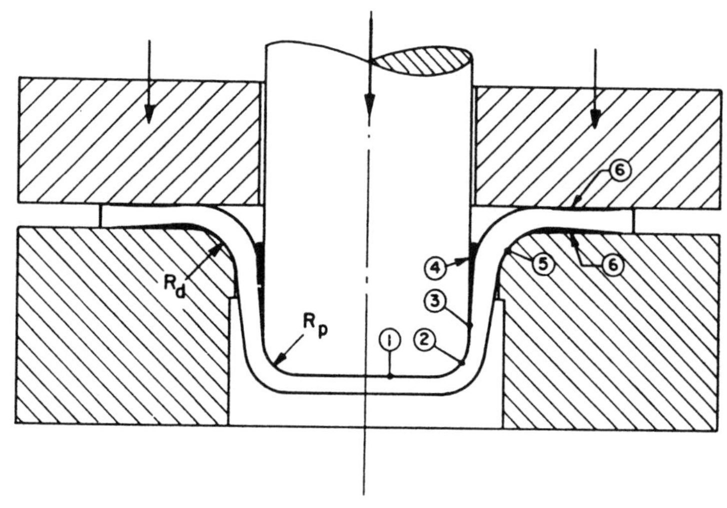

Figure 2Q-2. Zones of differing lubrication conditions in deep drawing (see text for identification)

point of transition (zone 4). Pressures are low and can easily drop to zero if elastic springback occurs, or if the clearance is large enough to allow the cup to separate from the punch in the early stages of drawing. The surface roughens as in free deformation, in combined tension/compression, and the presence of a lubricant makes little difference.

c. Around the die radius (zone 5), pressures are somewhat higher — by analogy to a frictionless pulley, on the order of (t/R_d) UTS. Since $R_d = 5t$ to $8t$, the pressure is around 0.2 UTS, which is still a quite small value. Sliding conditions are, however, severe. The speed is equal to the drawing speed and the surface, deformed by circumferential compression and bending and unbending, becomes progressively rougher. There is danger of exposing virgin surfaces, and tool pickup is observed only on the die radius, if at all.

d. In the blankholder zone (zone 6), pressures are low — typically 0.002 to 0.02 UTS. A liquid lubricant is trapped by a squeeze-film effect. Thickening of the flange creates a wedge which contributes to lubrication by hydrodynamic action. In materials of planar anisotropy there is more thickening in the low r-value direction, and thus the film is thicker in the high r-value direction. The surface roughens under the imposed tensile/compressive stresses at a higher rate than in pure tension, but roughness is still proportional to strain if the effects of anisotropy are considered.

e. When the clearance is such that ironing takes place at the rim, the surface becomes smoother and roughness depends on the thickness and nature of the lubricant film.

f. A sheet of directional surface finish or metallurgical structure exhibits visible parabolic markings, which are accentuated in the presence of a liquid lubricant.

Combined Stretching and Drawing

Conditions in combined stretching (on the punch) and drawing (around the die radius and under the blankholder), with the added demands set by the draw beads, have the effect of drawing over several die radii. A constant μ is suitable for modeling friction except when heavier sheet is drawn over tight radii; then μ increases and tool pickup can occur. If this is a major concern, the R/t ratio is increased.

Ironing

Problems in ironing are similar to those encountered in tube drawing on a bar, except that the part is shorter and the lubricant finds relatively easy access to the inside. Interface pressures are on the order of σ_f, and sliding velocity on the die is equal to the punch velocity. Sliding velocities on the punch are much lower.

LUBRICATION MECHANISMS

There is a vast variety of lubricants that are marketed for pressworking, but their modes of action can be fitted into the general framework of lubrication mechanisms.

Full-Fluid-Film Lubrication

Under favorable conditions it is possible to maintain a full fluid film, as evidenced by the increase of draw forces with increasing speed when highly viscous substances are used.

a. Upon impacting of the sheet surface by the punch, the initial film thickness h_0 is governed by speed and viscosity. Thereafter, the film thickness decays. The remaining oil film is thinner if the time between application of blankholder pressure and commencement of drawing is longer and if the punch speed is lower.

b. The situation around the die radius (zone 5 in Figure 2Q-2) is more complicated. Initially, the sheet is wrapped around the die radius, creating a squeeze-film effect which is partly counterbalanced by the negative wedge effect at the die exit. Thus a fluid film is more difficult to maintain. It has also been suggested that, at least in steady state, this zone could be regarded as a foil bearing, for which the film thickness h should increase with increasing die radius R_d, increasing ηv, and decreasing tension T:

$$h = 1.405\, R_d \left(\frac{\eta v}{T}\right)^{2/3} \qquad \text{(Eq 1)}$$

For the entry to the punch/flank interface (zone 4, Figure 2Q-2), the relevant tension is circumferential and is difficult to determine.

c. Application and removal of the highly viscous lubricant required for generation of a full fluid film can present difficulties. A lighter oil can maintain a film only if externally pressurized. In a practical application, the lubricant is supplied to orifices inserted into the die or punch at points where the lubricant would thin out most, and where good lubrication is desired to ensure free sliding and stretching. Thus, in pressing of a heat exchanger panel formed over a series of hemispherical punches, orifices were placed in each punch and in the adjacent die and blankholder surfaces. Capillary flow restrictors regulate the oil film. Oil pressure depends on the prevailing local pressures and ranges from 4 to 40 MPa in forming of 0.75-mm-thick sheets of aluminum and stainless steel. The required pressures and flow rates can be calculated by finite-element techniques.

d. In a version of phase-change lubrication, the surface of a stretch-forming punch was coated with a 2-mm-thick layer of ice kept at a temperature of −3.3 °C. Full-fluid-film lubrication was ensured when the sheet was kept at 15 °C. Lower-temperature operation is possible with solid CO_2 or solid krypton.

Mixed-Film Lubrication

Most liquid lubricants operate in the mixed-film regime. The coefficient of friction diminishes with increasing ηv, no doubt due to trapping of a thicker film.

a. In stretch forming this leads to the movement of the fracture site toward the apex, a larger pole strain, and a higher forming limit. Viscosity alone is not sufficient. For the same viscosity, silicones were much less effective than polyglycols. Tackiness is also needed and is imparted by additives such as polybutenes, polyisobutylenes, bentone clays, silica, and carbon black.

b. In deep drawing with a flat-nose punch, lower friction in the blankholder zone and around the die radius results in a lower draw force and a higher LDR (irrespective of whether lower friction is obtained through higher speed or higher viscosity; see Figures 2Q-3 and 2Q-4). A wide viscosity range was traversed by varying the sugar content in water. Draw force first decreased but, at high concentrations, increased again, presumably because full-

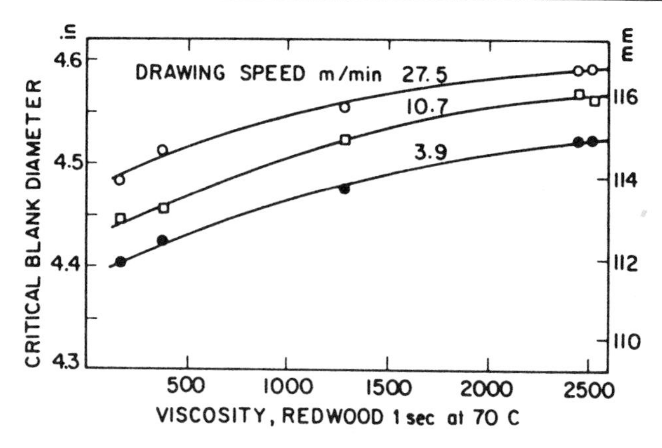

Figure 2Q-3. Effects of viscosity and speed on the maximum blank diameter that could be drawn from 1-mm-thick rimming steel sheet into 50.8-mm-diam cups

fluid-film lubrication had been attained. A similar trend can be seen in deep drawing of stainless steel with various oils. The mechanism does not remain unchanged over the entire draw; initially the lubricant film builds up and blankholder friction drops, only to rise again as more boundary contact develops. With increasing viscosity, the optimum blankholder pressure rises, partly because of the need for preventing wrinkling, which would then result in higher draw forces (Figure 2Q-4). The reason for the drop in LDR at intermediate blankholder forces in drawing of brass with low-viscosity lubricants is obscure (Figure 2Q-5). The draw force drops with increasing speed also in redrawing.

c. With a hemispherical punch the reduced support given by the punch leads to a lower LDR. The mechanism shifts toward the boundary end of the regime around the draw radius, although the ηv effect is still noticeable.

d. Increased ηv results in reduced ironing loads.

e. The ηv effect is observed also in simu-

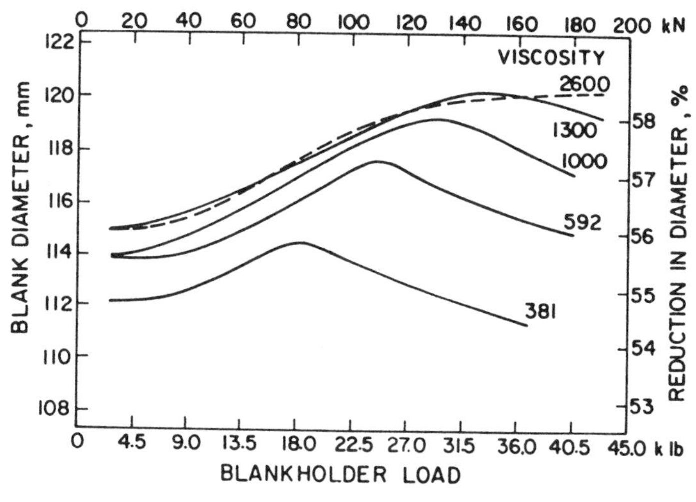

Figure 2Q-4. Effects of lubricant viscosity (measured in Redwood 1 seconds) and blankholder load on maximum blank diameter (0.9-mm-thick mild steel sheet; 50.8-mm-diam cup)

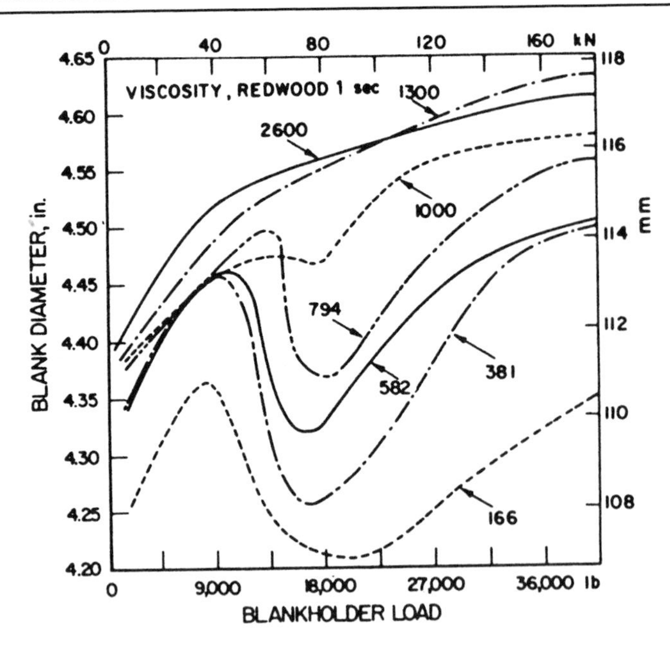

Figure 2Q-5. Effects of lubricant viscosity on maximum blank diameter and required blankholder force (viscosity in Redwood 1 seconds; 70/30 annealed brass blank 0.96 mm thick; 50.8-mm-diam cup)

lation tests. It was noted in drawing over a quadrant die and in the strip drawing test.

f. In bulk deformation processes, increased reduction and/or pressure shift the mechanism toward the boundary end, with a concurrent rise in friction. In sheet metalworking at low interface pressures, μ drops with pressure because yielding of the substrate in a tensile stress state allows the real contact area to increase and real contact pressure to decrease rapidly. Only when the pressure reaches the compressive flow stress of the material will μ become constant or even increase. The simplicity of the strip drawing test lends itself to an exploration of pressure effects, but a direct relation to pressworking is difficult to establish.

g. Some of the surface is in boundary contact, as evidenced by the improved performance obtained with compounded oils. The magnitude of the boundary-lubricated zone depends only on ηv, and not on composition. However, in the *b* zone even the base oil has an effect. For example, the aromatic constituents found in naphthenic oils gave lower values of blankholder friction than paraffinic oils of the same viscosity. However, the aromatic constituents also interfered with the functions of boundary and E.P. additives. Very light oils (below 20 cSt at 20 °C) dissolved large quantities of oxygen and provided much lower values of friction than could be expected on the basis of viscosity alone. In agreement with general principles, additives were more effective in highly refined, low-viscosity oils. The widely used emulsions and soap solutions also operate in the mixed-film regime, as shown by the reduced draw forces observed at higher concentrations. Except for ironing, heat generation is low, around 30 to 80 °C, but it must be higher on asperities, because otherwise the undeniable benefit of E.P. additives could not be explained. E.P. compounds do not necessarily reduce friction but limit pickup and scoring. Temperature is a powerful factor; drawing at elevated temperatures reduces viscosity and thus performance, whereas performance is improved by activation of E.P. additives. Unfortunately, tests conducted with commercial lubricants cannot separate these effects, even if infrared analysis is performed. The mixed-film mechanism is evident also from the reduced draw force measured in drawing of several blanks without cleaning the die. This improvement on aluminum was attributed to a black film identified as an oxide, although a friction polymer may also have formed with the mineral oil lubricant. Even greater improvement was obtained by dry polishing the blanks, presumably because the oxide debris thus generated acted as a parting agent. Much lesser improvements were noted with steel and copper, the natural oxides of which are much better parting agents than that of aluminum.

h. The deformed surface bears evidence of the mixed-film mechanism, with hydrostatic pockets and boundary-lubricated areas. Details of surface contact development can be followed by optical examination, particularly in the blankholder zone. The b ratio increases with increasing blankholder pressure and decreasing viscosity, corresponding to a shift toward a greater boundary contribution. Frictional force is directly proportional to b, indicating that friction in the hydrostatic pockets is negligible. Typical boundary contact ratios are shown in Figure 2Q-6, around the die radius and under the blankholder, for low and high ηv values, as a function of punch penetration. It will be noted that b under the bankholder increases monotonically during the stroke but changes around the die profile in a more complex manner. Initially, conformance is found only around the die entrance (zero position), and then conformance becomes more general around the whole radius, and finally a lower b ratio develops as the circumferentially compressed part is drawn through. Thickening under the blankholder is greater with high ηv. The b ratio decreased roughly linearly with the logarithm of the relative oil film thickness h/R_{max}, where R_{max} is the surface roughness developed in free stretching. The contact ratio and frictional force decreases with decreasing grain size, and the shear stress over the contact area can be described by a constant μ, the value of which is characteristic of the lubricant.

Solid-Film Lubrication

In most pressworking, relative sliding speeds are too low and the process geometry is not favorable enough to maintain a full fluid film, and lubrication with dry films is preferred so that the film thickness can be controlled by application rather than by process conditions alone. Such films also offer the possibility of differential or selective lubrication.

Dry Soaps. Even though dry soaps may be modeled as viscous substances and have speed-dependent actions, their fields of application place them in the dry-film category. For good adhesion, the surface must be clean, preferably freshly pickled. Soap-film

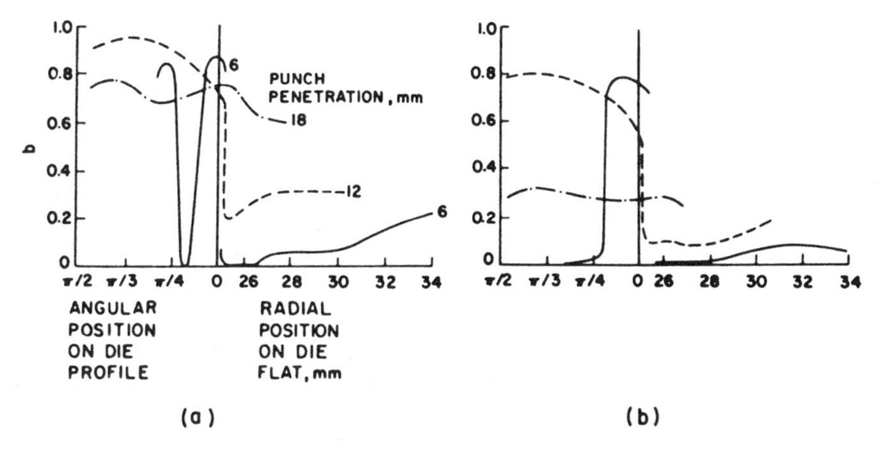

Figure 2Q-6. Proportion of boundary-lubricated contact area in drawing of steel sheet with (a) castor oil at 0.2 mm/s and (b) cylinder oil at 2 mm/s

thickness is controlled by the concentration and/or temperature of the aqueous solution. Subsequently, the water is driven off by heating in ovens, making the whole process most suitable for application in rolling mills. Coating thicknesses of 1.5 g/m^2 increasing to 3.0 g/m^2 are found on rough finished sheet. These coatings are related to the soaps used in wiredrawing and are based on mixed soaps of selected fatty acids. Borax is often added as a filler and presumably to improve attachment to the surface. Sensitivity to moisture can be reduced by replacing borax with a polymer. The melting point is adjusted according to the severity of operation, so as to avoid melting. Soaps have been most successful on steel. On stainless steel they are adequate only with coated tools. They are seldom used on nonferrous metals, because the slightly alkaline coating etches the metal and develops stiffer, higher-melting soaps on storage. The coating does not protect against corrosion or rough handling, but it has the advantage of water solubility and does not seriously interfere with welding. Proprietary formulations are claimed to be easier to apply.

Soaps applied to conversion-coated surfaces ensure a high lubricating power that is needed only in severe ironing or in drawing of stainless steel. Their use may be justified also when the part has to be phosphated anyway. The drawing and ironing loads decline slightly with speed, indicating mixed-film lubrication. However, when a reactive organic phosphate-oil system is applied, forces increase with speed, indicating that some breakdown must occur.

Waxes have excellent lubricating properties for all metals but have to be applied from and removed by organic solvents. Commercial water-soluble formulations exist. Friction is generally low; $\mu = 0.032$, virtually independent of speed.

Polymer Coatings. Loose films are practical only in prototype production or in the laboratory. After some early partial success with PE and polypropylene films, oiled PE became a standard for severe stretching and for stretchability and stretch-bend testing. PVC is also useful. The initial promise of PTFE was not fulfilled because of the limited stretching capacity of the film. An unusual product, suitable only for mild draws, is paper approximately 20 μm thick.

Most polymers are now used as coatings. Among the earliest were methacrylate copolymers deposited from trichloroethylene solutions or from aqueous dispersions. The value of PTFE, especially in the baked-on form, was discovered early on, and PTFE has potential application in drawing of aluminum hollowware. For better adhesion, these films are often deposited onto a ceramic substrate, fired onto the sheet surface. The pressworking of strip precoated in a special facility, often at the rolling mill, has received much attention. These coatings ensure clean shop conditions and protect the part during handling.

Polymer coatings impart greatly increased stretchability, approaching that obtained in hydraulic bulging. In deep drawing and incidental ironing the forces are substantially reduced, and the LDR is increased to 2.58 on steel. It has been observed that PTFE deposited on a ceramic substrate reduces draw forces and, in particular, the force peak associated with incidental ironing in drawing with a minimum or negative clearance. Force reduction was greater with a shot-blasted substrate. Surprisingly, the mean asperity height was also reduced relative to oil lubrication.

An important requirement is that the film should not wrinkle or stick to the dies. Coatings containing cross-linked polymers suffer damage through hairline cracking (crazing) under the influence of compressive stresses in the blankholder zone; the defect is similar to that found on the inside surface in a bend test. To avoid cracking and peeling in tensile regions, the polymer must have high elongation and good adhesion to the sheet. Coatings that soften at low temperatures may be damaged when worked on high-speed mechanical presses.

Under mild conditions, coated sheets can be formed dry. In more severe situations, additional lubrication is needed. Among mill-applied lubricants are esters 2 μm thick deposited on top of a 5-μm polymer film. Lubricants applied at the pressworking plant must be chosen for compatibility. For example, PVC is attacked by chlorine-containing lubricants. In general, water-base lubricants, mainly waxes, are preferred; often they can be left on the finished part. It is invariably necessary, however, to test these lubricants for possible delayed damage to the coating on the formed part.

Layer-Lattice Compounds. Until recently, layer-lattice compounds were used mostly in pastes, greases, and high-viscosity oils designed for use on heavy-gage material or in pressing at elevated temperatures. For demanding applications, particularly for metals that do not react readily with lubricants, dry films deposited from liquid or volatile carriers have been used. In agreement with experience in forging, the remnant of the carrier greatly affects friction and the response of the film to imposed pressure and sliding speed. In principle, a dry film shows no speed effect. Only when the liquid carrier remains, or the layer-lattice compound is added to an oil or grease, can there be a speed effect. In tests on deep drawing of steel, the optimum graphite content in oil, as judged from the level of the draw force maximum, increased with increased drawing speed. Proprietary or locally made lubricants often contain a variety of other ingredients, sometimes as tackifiers or parting agents and at other times with little clear justification.

Some bonded films can survive severe ironing. A lubricant containing 0.35-μm MoS_2 particles was reported to allow total reductions of 60%, typical of the three-stage ironing customary in two-piece beverage-can production. The ironing loads decreased linearly with increasing MoS_2 content to 20%, and with increasing coating weight to 5 g/m^2. Coating weights of 1 to 2 g/m^2 were found most cost effective. A proprietary system relying on MoS_2 in acrylic waxes for the outsides and thinner wax coatings for the insides of the cans has been claimed to be readily removable by standard cleaning techniques. In general, however, a common problem with layer-lattice compounds and nonorganic fillers is the relative difficulty of their removal.

Metal Coatings. Metal coatings find their most extensive use on sheet products. Most applications involve steel, but the principles are universal.

The dual function of coatings is evident from observations made on tinplate. Under mild conditions, tin alone can facilitate drawing; draw force and LDR were found to increase as coating thickness increased from 0.058 to 1.5 μm. In practice, a lubricant is invariably applied which then masks the effects of tinplate variables so that no improvement is found with coatings over 0.4 μm thick; blankholder friction keeps decreasing but so does the load-carrying capacity of the cup. The severe conditions of ironing bring out more subtle effects. A nonreflowed coating gives better lubricant entrapment. A thinner coating (say, 0.1 to 0.2 μm) on the punch side versus a 0.7-μm coating on the die side ensures differential lubrication and keeps punch friction high yet allows stripping of the can. Once a coating of minimum thickness (>0.2 μm) has been applied, no further drop in ironing force is observed. However, thicker coatings are needed to protect against localized breakdown at asperities. Not surprisingly, a superimposed layer of a polymer such as PTFE masks the effect of the tin coating. Chemical treatments applied to the tin coating do not appear to interfere with lubrication even though they may prevent wetting of the surface by water.

With increasing attention on the corrosion resistance of car bodies, galvanized (zinc-coated) sheet finds increasing use. LDR is improved in deep drawing, and higher blankholder forces are needed in press forming. Cracking or flaking of the film may occur, particularly in the compressive zone, and die pickup can be a problem. Polished, hard-chromium coated tools give best results.

Dry Pressworking

In light bending, stretching, and drawing it is quite possible to operate without an intentionally applied lubricant; surface oxides and incidental lubricant residues are sufficient, and often give lower friction than a light mineral oil which can dissolve surface films. Friction depends greatly on composition. For example, in a stretch-bend test friction was highest with 2036 (Al-Cu) alloy, lower with 6010 and 6009 (Al-Mg-Si), and lowest with 5182 (Al-Mg).

Application of Vibration

Vibration of the tool can have beneficial effects when the process geometry is not particularly favorable for film formation. Vibration can increase the film thickness and break incipient pickup points.

Benefits have been reported in sheet drawing, dimpling, deep drawing, and ironing, but improvements seldom justify the expense. There is no significant improvement on application of 20-Hz or 20-kHz vibration to a deep drawing or ironing punch when a good commercial lubricant is used. A substantial drop in friction was reported in drawing of strip and in deep drawing, but it is not clear what lubricant, if any, was used. Low-frequency vibrations in the range from 10 to 45 Hz have been applied by mechanical means such as cams. A vibrated blankholder reduced draw force and increased LDR in grease-lubricated deep drawing of aluminum blanks.

An important factor is heating due to vibration. Temperatures can reach 200 °C and impair lubrication by liquids or waxes, whereas layer-lattice compounds and polymers respond well. The application of ultrasound prevents rupture of films of limited stretchability, such as PTFE. High-amplitude vibration (in excess of 12 μm) induces galling, irrespective of whether the oscillation is axial, radial, or tangential (torsional).

EFFECTS OF SURFACE ROUGHNESS

As a result of the low pressures prevailing in pressworking, contact between die and workpiece is most often limited to asperities. Thus, details of the surface topography assume an even greater importance than in bulk deformation processes for two principal reasons:

First, many sheet products have large surface areas, and therefore roughness affects the appearance of surfaces, whether painted or unfinished. For a pleasing appearance, painted surfaces should usually have a controlled roughness, neither burnished nor excessively rough. Local variations in surface topography are to be avoided because of their high visibility.

Second, details of the surface topography determine the distribution of contact points between die and sheet, and the quantity and distribution of entrapped lubricants. Thus, topography affects press performance and the onset and consequences of lubricant breakdown. In general, rougher sheet is more tolerant: it accommodates slight unevenness such as that encountered with mechanical blankholders. Also, it is capable of entrapping more wear debris before surface damage occurs.

The problem of topography is most evident in automotive production but is also of importance in other applications.

Surface Characterization

A basic problem at the present state of development is that no truly relevant measure of surface quality has yet been found. The sheet surface carries the imprints of the last rolling passes. For good surface quality the sheet is cold rolled and the roll roughness, modified by the lubrication mechanism, is imprinted to give the finish of mill-hard or annealed-last sheet. Sheets subjected to temper rolling exhibit the surface finish of the temper rolls superimposed on the prior finish, to an extent determined by the temper reduction. Thus, a wide variety of finishes is encountered:

1. Easiest to characterize are finishes obtained with ground rolls. The height distribution is approximately Gaussian and R_a (CLA) or R_q (RMS) values are sufficient if taken in the rolling and transverse directions. When hydrodynamic pockets are numerous, the

Gaussian distribution is lost, and differences between the grinding (rolling) and transverse directions diminish.

2. In principle, shot-blasted surfaces should be easy to characterize if blasting took place with 100% coverage. Lesser coverage results in localized dimples in the as-ground surface of the roll. Of course, the strip carries the imprint (negative) of the roll surface profile. Therefore, the temper rolls produce generous asperity radii which, when superimposed on the roughness carried over from previous rolling, result in surfaces exhibiting plateaus and irregular valleys (Figure 2Q-7). Characterization has been attempted in various ways:

a. Early recommendations were based on statistical averages such as R_a or R_q. $R_a = 1.0$ μm has been suggested for general pressworking and 1.4 to 2 μm for automotive production.

b. It was soon recognized that statistical averages are quite inadequate in themselves. The surface can be visually characterized by stylus instrument recordings, but correlation with press performance is difficult.

c. Statistical averages combined with a count of the number of peaks per unit length give a better description of the surface. A difficulty is, however, that the number of peaks measured on a given surface depends on the cutoff wavelength, the height above the centerline at which the peaks are counted, and, ultimately, the definition of what constitutes a peak. Commercial instruments are available that give peak counts (see Table 2Q-1). Within these limitations, recommendations given for automotive sheet by various authors are combined in Figure 2Q-8. A criterion has been developed based on the mean peak (or valley) length on the centerline of the profile, ignoring peaks or valleys less than 5 μm wide (long). For the sake of uniformity, this has been converted into the equivalent number of peaks per millimetre in Figure 2Q-8, even though such conversion depends on the definition of a peak.

d. Surfaces of identical maximum roughness can have highly different real bearing areas. Sheets with peaked profiles allow relatively easy squashing of asperities. Contact points are closely spaced and well distributed, yet the real contact area is limited to a small fraction of the apparent area of contact. Therefore, sliding is well controlled without excessive frictional forces. The bearing area measured at 1.5 μm below the reference line gave the best correlation with press performance (Figure 2Q-9). A further development introduces a parameter p that combines peak density and bearing area:

$$p = \sum_{-3R_a}^{+3R_a} (N_{hT} \cdot A_{hT} \cdot N_{hL} \cdot N_{hL})\, \Delta h \qquad \text{(Eq 2)}$$

where N_h is the number of peaks at a given height, A_h is the bearing area at the

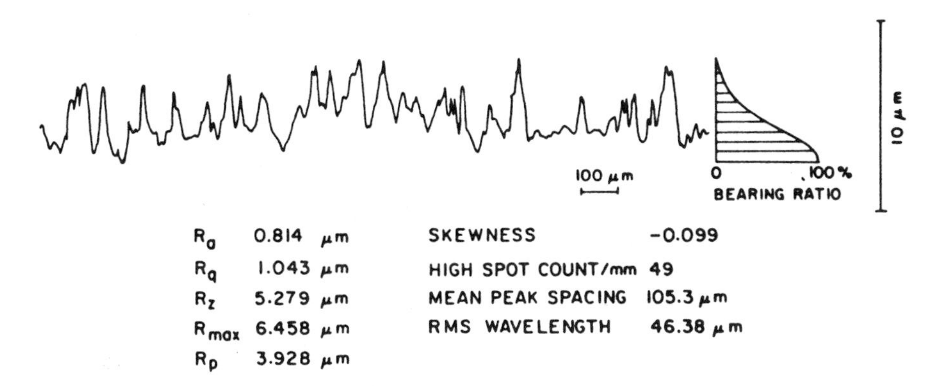

Figure 2Q-7. Surface roughness profile of typical steel sheet used in automotive body construction

Table 2Q-1. Effect of Measurement Method on Roughness Parameters

Sample	R_a	Cutoff (mm): Height(a)(μm):	0.75			2.5			Remarks
			0.25	1.25	6.25	0.25	1.25	6.25	
A:									
Top	0.95		197	137	10	187	118	13	Nongalling
Bottom	1.00		210	136	11	173	116	14	
B:									
Top	0.68		191	106	2	160	86	2	Galling
Bottom	0.55		192	87	0	156	68	0	

(a) Measured from centerline.

same height, Δh is the height increment, and the subscripts L and T refer to rolling direction and transverse direction, respectively.

e. Statistical measures such as the bearing area curve or, more likely, the skewness of the profile could give valuable information but do not appear to have been used extensively. Computer evaluation of surface topography (CEST) offers great promise. A lack of correlation has been noted between statistical measures and press performance. Analysis of the autocorrelation function yields little new information.

3. The roughness of the sheet is not constant even if the sheet was procured from a single source. With wear of the rolls, peaks are worn off, and both R_a (or R_t) and peak count diminish. The roughness changes also when surface treatments such as phosphating are applied.

4. Sheet roughness changes during deformation, and, if the sheet makes contact with the tool only after it has been stretched, the as-rolled roughness is modified by the straining history. The real contact area increases much more rapidly if the sheet is subjected to bulk deformation (as in the blankholder zone in deep drawing) than if it is only stretched over a punch or bent in U-forming.

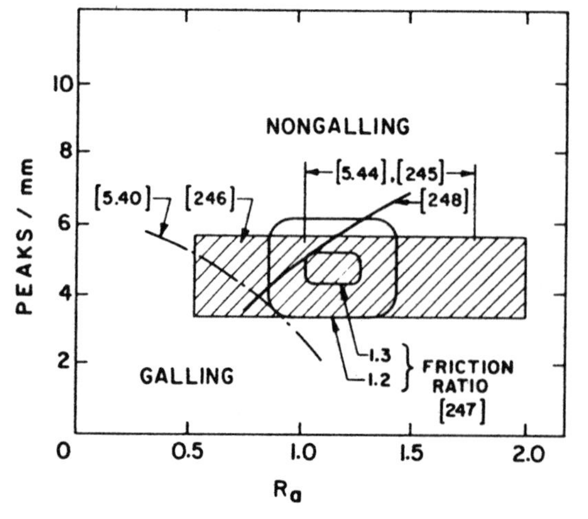

Figure 2Q-8. Recommended surface roughness parameters for automotive pressworking

Effect on Lubrication

The effect of roughness on lubrication is as would be expected from general principles. Increased roughness is beneficial in that more lubricant is entrained and wear debris is accommodated. The effect is more pronounced when the features of a directional surface finish are perpendicular to the sliding direction. A very rough surface can, however, be objectionable in appearance, and the lubricant may leak out in the direction of valleys. The desirability of roughness also depends on the process and, more directly, on interface pressure.

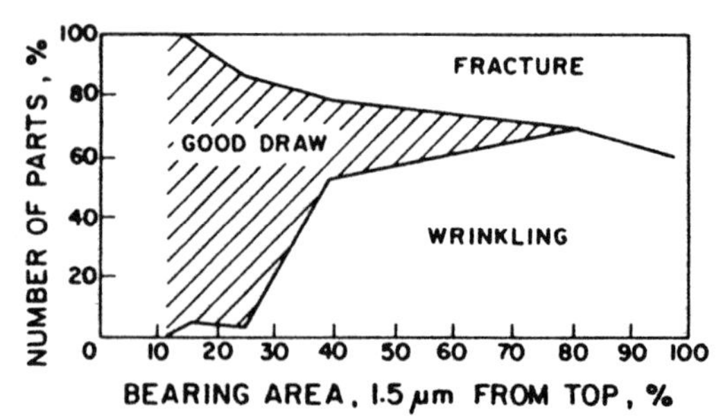

Figure 2Q-9. Success in drawing of an automotive part as a function of bearing area

a. At low interface pressures, increased roughness develops more asperity contact and μ increases. Because of the low pressures prevailing in the blankholder zone, a smooth sheet gives inadequate control, allows materials to run in, and may allow puckering to occur in stretch drawing.

b. At higher interface pressures, the effect of roughness depends on the lubricant. With a viscous fluid, more lubricant is entrained and μ drops, as observed in strip drawing at heavier reductions. The effect is more pronounced with higher viscosities and/or higher speeds. Because of the more severe conditions created at the draw radius, LDR generally increases and draw force decreases. This is true also of ironing. Little or no effect of roughness can be expected when the lubricant is solid or when a marginal lubricant such as a corrosion-protection oil is used. When the lubricant is of low viscosity or is aqueous, and does not fill out surface features, rougher sheet develops higher friction because of increased boundary contact.

c. Excess oil film thickness allows incipient wrinkling and thus gives higher draw forces; the effect is more marked on rougher sheet. In deep drawing with an emulsion, the optimum coating thickness corresponded to 1 g/m^2 of emulsified phase (dry weight) on top of 2 g/m^2 of corrosion-protection oil.

LUBRICANT BREAKDOWN AND GALLING

Even under processing conditions that appear mild in comparison with bulk deformation processes, lubricant breakdown is often encountered because the lubricant is wiped off in the course of sliding and asperities are exposed to direct contact with the die. Tool pickup then follows, which in turn leads to scoring of the workpiece surface. These phenomena are usually described by the term "galling;" even though the OECD glossary recommends that this term be avoided, it has gained such wide acceptance that we shall use it, too. Because galling is an event that follows local lubricant failure, it is greatly affected by workpiece and die topography and by adhesion between the contacting materials.

Incidence of Galling

Galling is typical of all processes in which the workpiece slides over the die. Of all galling problems observed, 40% are associated with draw beads, 22% with die radii, and less than 10% with ironing. Galling is responsible for some 70% of all rework (polishing) of automotive dies. Several points are to be noted:

a. Galling must still be judged visually, even though attempts at quantifying it have been made. The following grading based on SEM observations may be used:

Phenomenon	R_q (μm)
Smoothing (with no fragments on the deformed surface)	Original
Whitening	2 to 12
Scratching(a)	4 to 20
Light galling(a)	10 to 40
Severe galling(a)	20 and over

(a) Fragments visible.

More generally, a scratched surface for which $R_q > 10$ to 15 μm is regarded as galled, although the limit depends also on visibility.

b. As are all wear phenomena, galling is a function of pressure and total sliding distance. Pickup occurs after some length of sliding, and then proceeds in a manner that depends on the sheet/die material combination. The sliding distance increases with the depth of cup or pressing, and with the width of the blankholder or draw bead. In wedge drawing of strip, an initial pickup-free sliding with low, essentially boundary friction was observed. On further sliding of steel, increases in draw force and scoring set in. Draw force was proportional to the welded area of contact, which in turn increased with decreasing lubricant viscosity. Finally, a steady-state condition corresponding to dry lubrication was established. On aluminum workpieces, metal transfer was affected also by lubricant composition, with 5% sulfurized fatty oil or 5% stearic acid in paraffin giving the lowest friction and pickup. On both metals the shear strength of the welded zones was roughly equal to the shear strength of the strain-hardened workpiece material, indicating that adhesion had indeed taken place.

c. At low interface pressures, conformance of the sliding surfaces depends also on the elastic deformation of the sheet, and thus the effect of blankholder pressure can be normalized through:

$$BHP_{norm} = \left(\frac{t}{0.7}\right)^2 BHP_{actual} \qquad \text{(Eq 3)}$$

where t is sheet thickness in mm.

d. On sliding over a draw bead, galling is observed on the opposing surface, where reverse bending takes place over two radii. In deep drawing and ironing, galling is limited to the die side of the cup.

e. Systematic studies of galling in production are difficult, and simulation tests abound. All tests are based on sliding at low interface pressure and include sheet drawing tests, draw-bead simulators (Fig. 2Q-10), tests with combined drawing and bending, and tests in which breakaway friction is measured. Because galling is a function of sliding distance, several specimens are usually tested in sequence and, in addition to visual inspection, the trend in draw forces is noted. Tests based on breakaway friction have been quantified by a friction ratio which is the breakaway force divided by the running force. High values (over 1.3) are often accompanied by chattering and scoring of the sheet.

Mechanism of Galling

Galling is a result of localized adhesion and is, therefore, affected by a number of process variables:

a. The effects of asperity shape may be rationalized by analogy to bulge formation in ironing. An asperity of low slope deforms and is pushed into the bulk. As the asperity slope increases, a point is reached where a bulge forms, which is then removed by a shearing (cutting) action; the critical slope is less for higher friction. The shearing action is related to prow formation and to the cutting

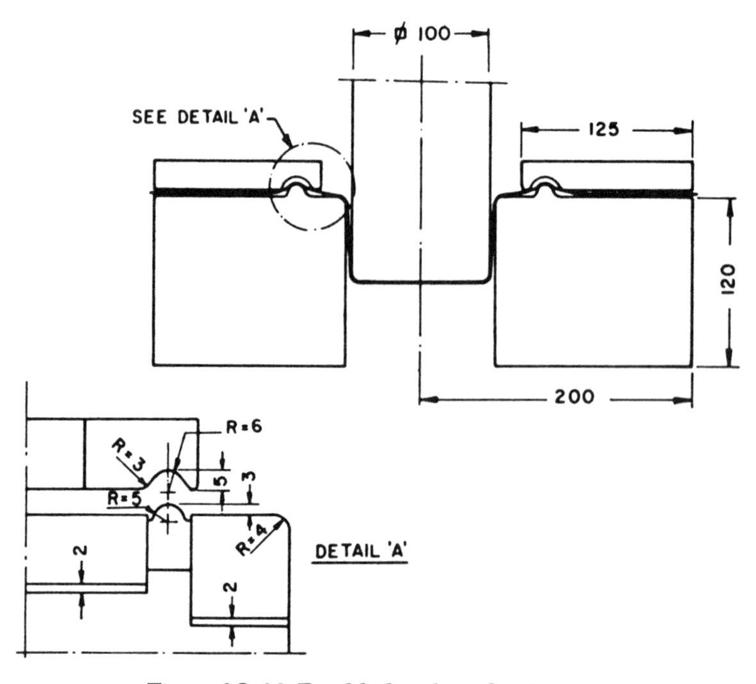

Figure 2Q-10. Double-bead test for drawing of steel strip

mode of wear. The sheared material is usually piled up at the ends of individual scratches.

b. Debris formed in the course of sliding may be trapped and rendered harmless in valleys of the sheet surface, may become embedded in other sheet asperities, or may lodge against die scratches. Repeated encounters at many points result in deformation, thinning, and lamination of particles; some of them may even roll up. If adhesion to the die is high, new large asperities form which lead to scratching, formation of more thick particles, and galling.

c. Galling is affected by the macroscopic tool geometry. It usually originates at transition points such as die radii, but its severity greatly increases if the already damaged sheet, with embedded particles, is dragged along a parallel surface such as a die land. These phenomena are classified as primary and secondary galling.

d. The incidence of galling is a function also of the stiffness of the system. Damage is localized in a soft (constant-force) system, whereas it is widespread and uniformly distributed in a stiff (constant-gap) setup.

Effects of Surface Roughness

Both workpiece and die roughness have significant influences on the incidence and severity of galling.

a. A smoother sheet surface presents a larger bearing area and is more prone to galling. For a given reduction in thickness, smoother sheet galls after shorter sliding or fewer repeat tests. This effect can be quite dramatic; a small change in roughness is sufficient to initiate galling in drawing of materials prone to pickup.

b. The size and distribution of asperities are most important. For a given average roughness, larger flat plateaus give longer contact (sliding) distance and more galling. For a given bearing area, a smoother sheet has smaller plateaus but a much greater peak density, and thus it presents more points for sliding contact and is more prone to galling.

c. The directionality of the sheet surface profile has a marked effect. Plateaus aligned in the direction of sliding present longer sliding distances and are more likely to allow embedment of the debris. Thus, they are more harmful than transverse ridges, which trap debris in valleys and may even act as rasps to remove incipient pickup from the tooling. In drawing aluminum, the weight of pickup increases with asperity width rather than length; however, this may simply represent the development of a thicker but no less stable surface coating.

d. The depth profile of the sheet surface is important because it determines the ability of the surface to entrap lubricants and accommodate wear debris.

e. There has been some success in defining what sheet surface is best, even though there is no general agreement. Recommendations are within the ranges shown in Figure 2Q-8. It is doubtful, however, that average roughness and peak count are sufficient descriptors of an antigalling profile. Almost certainly, the bearing area should also be specified. To exclude wild peaks, the baseline may be taken at a small distance from the highest peaks — e.g., by excluding the first 2% of the bearing area. Invariably, steel sheets with bearing areas of less than 20% at 1.5 μm from the surface prove to be nongalling, whereas those with larger bearing areas (and those whose bearing areas increase more rapidly with depth) are prone to galling. A large bearing area also yields a high friction ratio in the galling test. The bearing area may be calculated from flattened spots observed on sheet that has been slid between parallel jaws at different interface pressures; the bearing area increases more rapidly on sheet that is prone to galling. The mean peak length at the centerline, combined with R_a, has also been suggested as a measure. In view of the effect of asperity shape, a nongalling finish for low-carbon steel has been defined as one that has a mean plateau height of 10 to 25 μm, a plateau length of 60 to 120 μm, and a mean ratio of valley to plateau width of 3:2 to 5:2.

f. A rough die surface is desirable only on the punch because it then increases the LDR and the depth of draw in automotive pressworking. If the sheet material has to slide over die elements, roughness leads to higher friction, lubricant breakdown, and galling, although the effect depends also on die composition and hardness. The harder the tool, the more damaging its roughness. Thus a cast iron die or a nonhardened steel die can be rougher than a heat treated steel or tungsten carbide die.

Material Effects

Because of their importance in adhesion, material composition and hardness have decisive influences. Adhesion between die and workpiece governs whether die pickup will occur and, if it occurs, whether it will be cumulative. Details of the mechanism are still obscure, but the effects can be marked:

a. Workpiece material effects can be quite subtle. Thus, increasing the magnesium content of aluminum alloy 2036-T6 reduces adhesion. In steels, segregation of manganese, chromium, and other alloying elements has been suspected to reduce adhesion, perhaps by strengthening the surface layer. Steels precipitation hardened with niobium or titanium additions are more resistant to galling than solid-solution (Si, Mn) hardened steels for the same reason. These effects should be quite general. Martensitic transformation induced in austenitic stainless steel increases friction and draw forces but reduces adhesion.

b. Surface composition plays an important role. A sheet subjected to sliding contact galls more readily on subsequent contact. An untouched surface has surface layers which may or may not protect it. For example, steel pickled in HCl was found to be more prone to galling than that pickled in H_2SO_4. The oxides and carbonaceous deposits formed on annealing steel in a DX (exothermic) atmosphere prevented galling, whereas the cleaner surface produced in an HN atmosphere (mixture of N_2 and H_2) required good lubrication and more closely controlled surface topography. Galling tendency, as expressed by the friction ratio from the galling test decreases from 1.4 with a 10-nm-thick contaminant film to 1.1 or 1.2 with a 50-nm-thick film. The reactivity of a surface changes on cleaning and deformation, as shown by measurement of electron emission, and this too must affect adhesion.

c. The hardness of the workpiece material also plays a role. High-strength steels are not much more adhesive than low-carbon steels but generate higher pressures and are more prone to galling, especially on draw beads. Previously strain-hardened material undergoes less hardening during working, and this is beneficial because it reduces the thickness of wear debris. However, the sheet hardness should be not more than half the tool hardness; otherwise, strain hardening of the asperities makes them strong enough to cause abrasive wear of the die. Some researchers suggest that asperities should be softer than the bulk of the sheet to allow better conformance, while others claim that this leads to more galling. Perhaps no generalization can be made because of the influence of material variables.

d. The elastic properties of the die material affect conformance. This effect was investigated by removing partially drawn stainless steel cups at the point of reaching the force maximum, and by taking roughness measurements on the remaining flange and over the surface that contacted the die radius. A roughness parameter, formed by dividing the change in average roughness height by the original roughness, increased in direct proportion to maximum draw force (Figure 2Q-11). This, in turn, increased with the elastic modulus of the die material, indicating that a die that conforms less to the sheet generates a rougher surface and higher forces. A beryllium bronze was the only exception, because it wore by metal transfer.

e. The presence of a lubricant is important. If the film is thin and boundary contact dominates, the properties of the tool material influence draw force and galling. This is the case when only corrosion-protection oil is present. In simulation tests, a lubricant was applied only to the first strip and then the increase in force and the onset of galling in the double-bead tester were observed (Figure 2Q-10). As the film thickens through

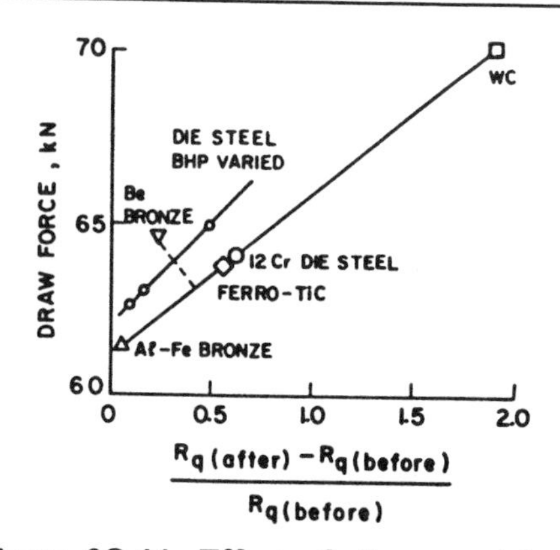

Figure 2Q-11. Effect of die material on smoothing or roughening of type 304 stainless steel cups drawn with a graphite/stearic acid mixture

higher ηv, galling diminishes also at any given pressure. On a very smooth surface, a liquid lubricant cannot prevent galling; a solid lubricant protects only when surface reaction is also possible. Thus, MoS_2 prevented galling on smooth steel and especially on copper-plated steel, but not on brass. Friction ratios of 1.2 to 1.3 call for controlled lubrication, higher ratios are indicative of galling with any lubricant, and lower ratios signify resistance to galling without regard to the type of lubricant. With a given steel, friction ratios of 1.02 to 1.05 with soap/borax, wax-base, and acrylic film lubricants were found. A synthetic ester/oil mixture and an emulsion gave ratios of 1.20 to 1.25, and rust-preventive oils and light paraffins allowed a ratio of 1.30, which is beyond the galling limit.

DIE MATERIALS

The high visibility of surface scratches dictates the choice of pressworking dies. Of course, wear is also of concern; however, if galling sets in, production has to be stopped long before wear could become a problem. The choice of die material depends greatly on workpiece material and on interactions with lubricants.

a. Die steels are still the most extensively used die materials. Graphitic steels combine adequate hardness with some built-in lubricity. Cast iron is advantageous for large dies and gives low friction, but cast iron dies allow more galling on low-carbon steel than do steel dies. To reduce the cost of dies, areas subject to wear (such as draw radii) can be made as inserts. Steel dies are adequate for stainless steel and other high-strength, high-adhesion materials only if the sheet is fully protected by an oxalate/soap or polymer film.

b. Of the nonferrous metals, zinc-base alloys are suitable for low-volume production. Of specific importance are bronzes typically containing 14% Al, 4% Fe, and 1% Ni, which are unique in that stainless steel can be drawn with them without pickup and scoring. Their high hardness keeps wear to a reasonable rate. Fats, soaps, and even water-soluble lubricants can be used with them. They are, however, brittle, and, apart from encasing them in steel, the problem of support can be solved by depositing them on a steel die by ion plating.

c. Tungsten carbide bonded with cobalt is the staple die material for large production runs in applications such as deep drawing and ironing — for example, in production of tin boxes or two-piece cans. In the presence of an aqueous lubricant, the tin coating forms a galvanic couple with the cobalt binder. It has been suggested that anodic dissolution of the binder is responsible for punch wear. Tool steel can be less damaging to a soft metal such as aluminum. TiC is reported to outperform WC in canmaking. Other carbides and ceramics have been explored experimentally. Their success is highly specific to the metal and lubricant, and results are sometimes contradictory.

d. Die coatings offer advantages for some of the difficult-to-draw metals. Most prominent of these is stainless steel, and several experiments have been aimed at finding coatings that would allow pressworking without oxalation. For austenitic steels, coatings of TiC appear to be best when used with waxes, soaps, or other well-adhering lubricants. For ferritic steels, TiC, TiN, and nitrided surfaces were found to be best, followed by WC-coated and boronized dies. Hard chromium was little better than D2 steel, perhaps because of the observed poor wetting of lubricants. For mild steel, no galling was experienced with carbide diffusion or vapor coating, but increasingly more galling with chromium-plated, nitrided, and sulfurized dies was found. Little information on industrial experience has been published for other metals.

e. Polymers such as epoxy resins are of relatively low strength and need metal inserts for tight radii. Polyurethane allows polished sheet to be formed without the danger of scratching.

LUBRICANT SELECTION, APPLICATION AND TESTING

Lubricant Selection

The major difficulty of a systematic selection of lubricants is the wide variety of conditions that is encountered. Classifications based on processes are not fully adequate because conditions may change greatly in, say, combined stretching/drawing, depending on the degree of drawing. Production rates affect equilibrium temperatures of the tooling and thus the potential breakdown or activation of lubricants. At this time, selection is best based on a knowledge of lubrication mechanisms, whereas those less familiar with the subject must rely on commercial recommendations (and on Table 2Q-2).

It is often economical to select a better-quality lubricant which then allows the use of a less-expensive grade of sheet metal. Beyond the lubricating function, compatibility with later treatment of the part is important. This may necessitate removal of residues or use of lubricants that do not interfere with processes such as welding, either by impairing weld quality or by generating excessive fumes.

Lubricant Application

Manual application with brushes, swabs, or rollers is still acceptable at low production rates. Drip application is the first step in mechanization and is suitable for lubricants of moderate viscosity. Roller coating is practiced mostly for coiled stock but is suitable also for precut parts. Mist application can be economical and versatile and is suitable for deposition of thin films of lubricants ranging from emulsions to waxes. Application to the die is also possible. Electrostatic deposition is attractive but more costly. Mill-applied coatings remove the burden from the press shop but may be too expensive. Some systems, such as a combined phosphating-oiling treatment with a lubricating oil that contains an organic phosphate, are suitable for press shop application.

Lubricants are often made in forms that allow several methods of application. Thus, polymeric additives endow the lubricant with thixotropic properties, which prevent dripping. Pastes and synthetic lubricants are often made to be water-extendable so that they can have consistencies ranging from semisolid to liquid.

We have seen that deeper draws are often possible when friction is low in the blankholder zone but high on the punch. This can be accomplished by differential lubrication of the blank. It is difficult to keep a liquid lubricant away from the punch, and roughening of the punch is then more practicable (provided that stripping is still easy enough). Differential lubrication can be better controlled with dry or solid lubricants. Spray phosphating and electrochemical deposition allow lubrication in a pattern. In deep drawing the critical zone is close to, but does not extend to, the outer edge of the blank. Otherwise, good lubrication on the die side and poorer lubrication on the punch side can be remarkably effective (Figure 2Q-12). Similar benefits can be reaped in ironing.

Lubricant Evaluation

The major effort is usually directed toward evaluating lubricating qualities. Part of the problem is that laboratory tests fail to show long-term trends unless run in large enough numbers to stabilize surface conditions and temperatures. Die and lubricants should be heated to operating temperatures so that lubricant breakdown or activation becomes evident. Trends are often more im-

Table 2Q-2. Commonly Used Sheet Metalworking Lubricants and Typical μ Values
(Lubricants are listed according to increasing severity of conditions)

Material	Shearing, bending(a)	Pressworking(a)	μ(b)
Steels	Dry (pickle oil)	Dry (pickle oil)	0.2
	EM (M.O. + E.P.)	Soap solution	0.15
	M.O. + E.P.	EM (M.O. + fat) (+ E.P.)	MF
		M.O. + fat (+ E.P.)	MF
		Fat (tallow)	0.07
		Soap (water-soluble) film	0.05
		Wax (chlorinated)	0.05
		MoS_2 or GR in grease	0.05
		Polymer coating (+ MoS_2)	0.05
		Phosphate + soap	0.05
		Metal (Sn) + EM	0.05
Stainless steels and Ni alloys	EM (M.O. + Cl)	EM (M.O. + Cl) (or fat)	0.2
	M.O. + Cl additive	M.O. + Cl additive	MF
		Chlorinated wax	0.07
		Polymer coating	0.07
		Oxalate + soap	0.07
		Metal (Cu) + M.O.	0.05
Al and Mg alloys	EM (M.O. + fatty derivatives)	EM (M.O. + fatty derivatives)	0.15
	M.O. + fatty derivatives	M.O. + fatty derivatives	MF
		Soap or wax (lanolin) coating	0.05
		Polymer coating	0.05
		Warm: GR film	0.1
Cu and Cu alloys	Soap solution	Soap solution	0.1
	EM (M.O. + fat)	EM (M.O. + fat) (or fat)	0.1
	M.O. (+ fat) (+ E.P.)	Drawing paste (EM of fat)	0.1
		Fat (tallow)	0.07
		Pigmented tallow (MoS_2, etc.)	0.05
Ti alloys	M.O. + E.P.	Wax coat	0.07
		Oxide + soap coating	0.07
		Polymer (+ MoS_2)	0.05
		Fluoride-phosphate + soap	0.05
		Metal (Cu or Zn) + M.O. (or soap)	0.1
Refractory metals	M.O. + E.P.	Warm: MoS_2 or graphite	0.2
		Cold: Al-Fe-bronze dies with wax	0.07

(a) EM = emulsion of ingredients shown in parentheses. M.O. = mineral oil; of higher viscosity for more severe duties, limited by staining. E.P. = E.P. additive (S, Cl, and/or P; also sulfochlorinated fats). (b) MF = mixed-film lubrication: μ = 0.15 to 0.05.

portant than absolute forces. As discussed in conjunction with galling, this is particularly true when adhesion is a problem. Lubricant evaluation can take different approaches:

a. Scaled-down production tests are feasible, especially in deep drawing, provided that the d_0/t ratio is kept large enough (usually in excess of 100). If enough cups (in excess of 50) are drawn with any one lubricant, the maximum draw force, its change, and pickup on the die give reasonable indications. The same applies to ironing tests. Correlation with production experience is usually adequate, although it can be poor when temperatures or d_0/t ratios are too low.

b. The large number of simulation tests advocated indicates that none of them can give full evaluation. Indeed, it has been repeatedly shown that several tests are needed for a reasonable prediction of plant performance. In general, the relevance of tests diminishes as they become further removed from the process conditions they are meant to simulate. For deep drawing, the wedge-

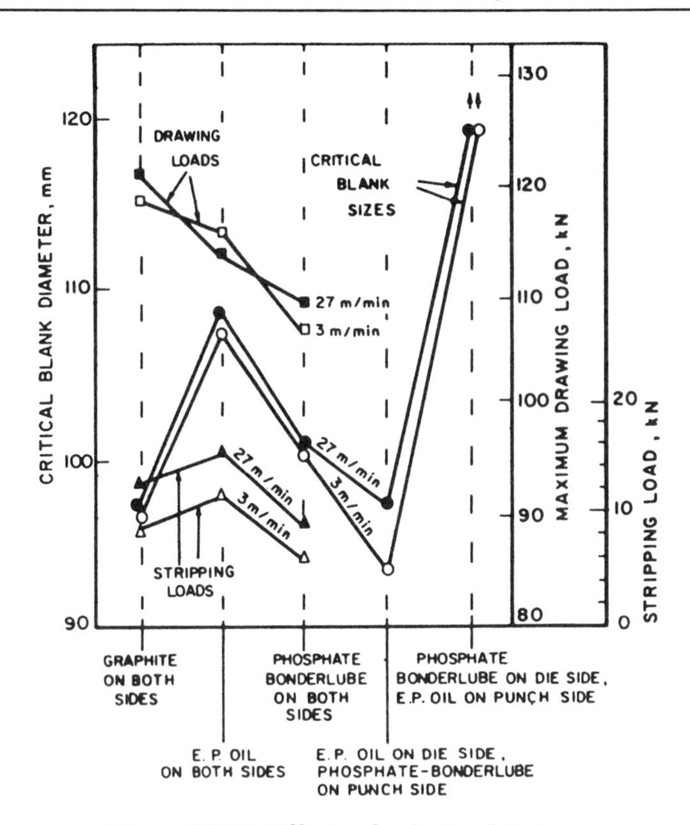

Figure 2Q-12. Effects of selective lubrication on limiting draw ratio, draw force, and stripping force in deep drawing of 1.6-mm-thick annealed low-carbon steel blanks into 50.8-mm-diam cups with simultaneous wall ironing

drawing test does not really simulate the blankholder zone from the material point of view but is adequate for lubricant exploration. This is less true of the strip drawing test, where lateral compression is absent. Because of the importance of friction around the draw radius, bend-draw tests are more relevant, although correlation with actual deep drawing can be poor. Draw bead tests can be conducted with variable penetration or with contact on the flat binder areas. Again, trends are more important than absolute values. The actual draw force versus drawing distance curve can be numerically represented by reading the force at the beginning of the draw and at, say, two points during the draw. By multiplying the three values by each other, a composite, more sensitive number is obtained.

c. Bench tests can show only limited correlation with some specific aspect of lubricant performance. The four-ball test gives adequate correlation with deep drawing when oil viscosity is dominant, but the correlation is nonexistent when boundary properties dominate with soap solutions. In general, poor correlation is found.

d. Functional tests for cleanability, corrosion due to the residual oil film, etc., are routinely performed.

2R: Lubricants for Forming Operations

In any forming operation there are interactions among three major components: work material, dies, and lubricant. Of these three, lurication is the least influential in correcting forming problems but, nonetheless, is a necessary component in any successful forming operation.

To better comprehend the role of lubrication, it is felt that some explanation of metal behavior during deformation would be helpful.

All forming begins with a blank from coiled or sheared sheet stock. Blanking is the process of roughly shaping a piece of metal to the required size on which finishing processes are carried out. Shearing is confined to a single straight-line blanking cut. Once a blank is cut, it is ready for forming.

Sheet-metal forming is a complex combination of many separate operations, including stretching, drawing, wrinkling, buckling, bending, unbending, shearing, piercing, etc. For practical purposes, however, a complex stamping can be classified primarily by the relative amounts of stretching and drawing required to form the part.

In pure stretch forming, the flange of a flat blank is securely clamped or held to the die while a punch (probably round) of a lesser diameter than the die opening is forced into the center of the blank. Deformation is restricted to the area initially within the die opening (Figure 2R-1). The material is elongated in one or more directions, and tensile stresses are applied in both directions. The stretching limit is the onset of material failure (Figure 2R-2). One soon realizes that the material behavior in stretching is similar to that in a tension test and therefore is limited to the tensile strength of the material, or the "n" value. An example of pure stretch forming would be forming of any hemispherical shape (Figure 2R-3).

Figure 2R-2. Splitting caused during stretch forming

Figure 2R-3. Hemispherical shape formed by pure stretch forming

At the other end of the forming scale is pure drawing. Here a parallel-wall cup is created from a flat blank. The blank is drawn into a die cavity by the action of a flat bottom punch (Figure 2R-4). Deformation is restricted to the flange area of the blank. No deformation occurs under the bottom of the punch — the area of the blank originally within the die opening. As the punch forms the cup (Figure 2R-5), the amount of material in the flange decreases. Other material in the flange is elongated while its width dimension is compressed. This is caused by the

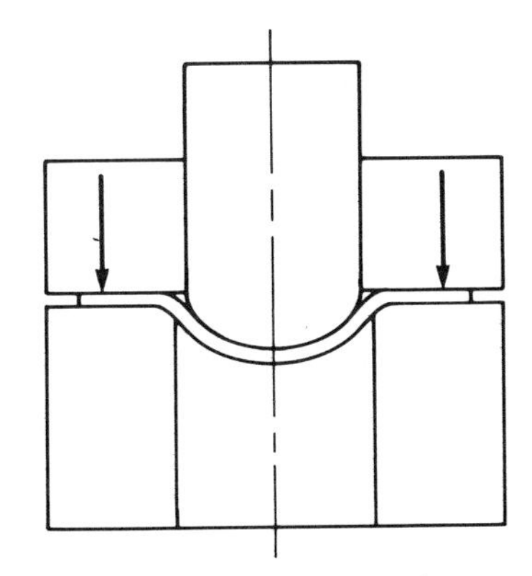

Figure 2R-1. Pure stretch forming

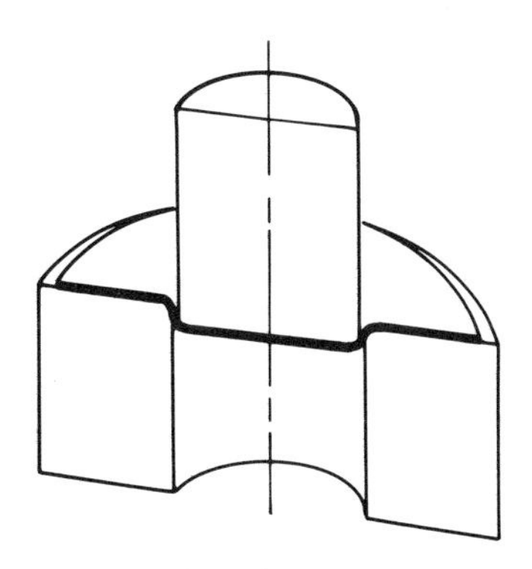

Figure 2R-4. Beginning of pure drawing

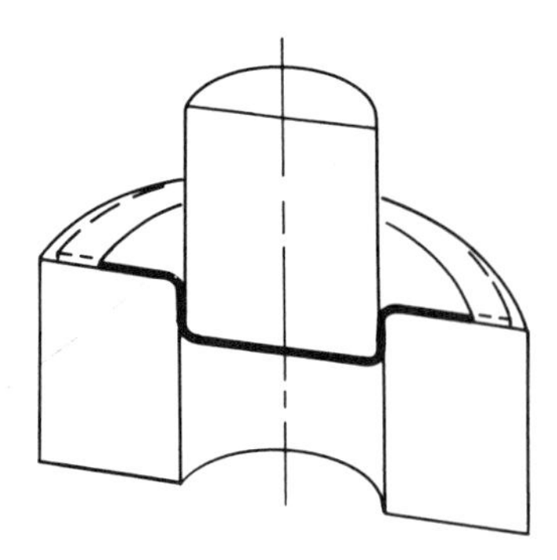

Figure 2R-5. Beginning of formation of cup in pure drawing

circumference of the blank continually being decreased as it is pulled in toward the die opening (Figure 2R-6).

Most manufacturing personnel agree on the basic metal deformation patterns which occur as flat blanks are drawn into cups. At first contact with the blank, the punch forces the sheet metal to conform (bend) to the radius of the die ring, as well as to its own (punch) radius. Further punch travel will cause tearing of the metal if the metal is not allowed to flow. In other words, the depth of contour alone is very limited. Greater depth is obtained by drawing or pulling in metal from the outer regions.

The metal previously bent to the die radius must now be straightened to create the cylindrical sidewall of the cup. Because this metal has already been hardened by bending, a much greater force is needed for this straightening action. Because metal from the outer blank is now sliding over the die surface and under a blankholder, the forces of static and dynamic friction have been overcome. The blank is being drawn toward the die radius and into an ever-decreasing

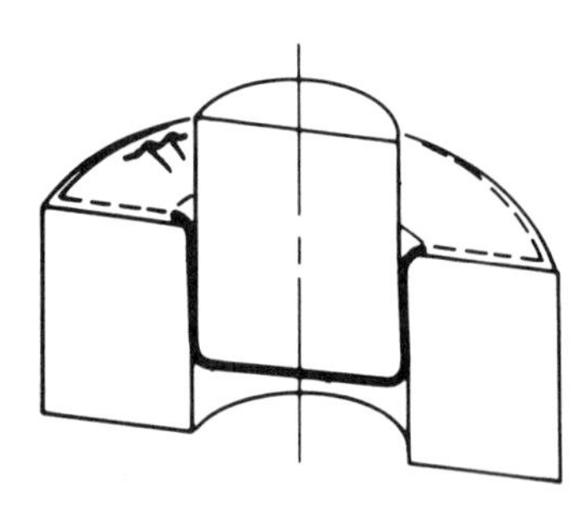

Figure 2R-6. Finished cup made by pure drawing

area. Very large compressive forces occur which tend to cause wrinkles, but firm pressure on the blank will prevent wrinkling. With the beginning of these actions, the maximum punch load is developed. This is the critical stage of drawing and occurs within the first 1/2 in. of punch travel.

As the punch continues its downward stroke, metal from the flange is drawn over the die radius to form a deeper cup and is, therefore, placed in a high state of tension. The sidewall, just above the punch radius, is stressed the most and becomes much thinner than the original blank. This is the area where tearing will occur (Figure 2R-7).

This pure draw, or measure of the resistance to thinning in the material, is referred to as the $\bar{r}$ value.

Some examples of items formed by pure drawing are oil pans, fire-extinguisher cylinders, and vegetable bins.

Most objects formed from sheet metal, however, are formed by combinations of stretching and drawing (Figure 2R-8). Herein lies the main problem of steel manufacturers. There are steels that stretch better and there are steels that draw better, but now a compromise must be made and a steel selected that can be subjected to both operations at the lowest cost.

The talents of the die designer must also be applied in order to smooth out a forming cycle within the safety factor of the metal for consistent performance with little breakage of parts formed.

It is necessary to develop a practical method of determining the strain curve and critical point for each formed part. The most practical method of determining the distribution of strain is with a circular grid pattern placed on the surface of the blank prior to forming. The blank is completely covered

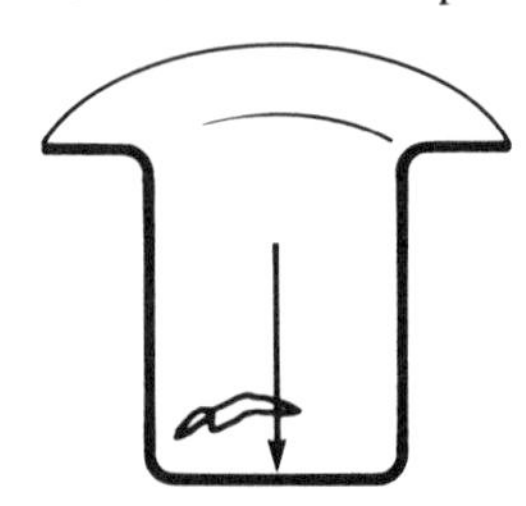

Figure 2R-7. Tearing in drawn cup

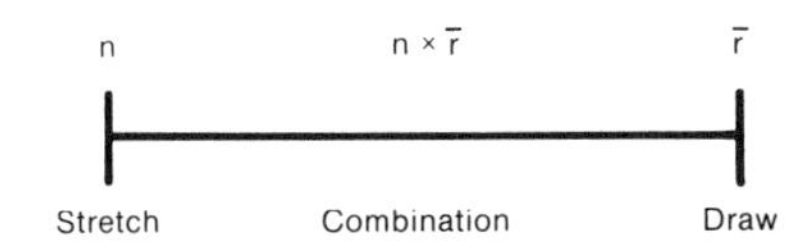

Figure 2R-8. Combined stretching and drawing

with small touching circles 0.1 to 0.2 in. in diameter which are applied by electrostenciling. Deformation to any area of the blank will alter the concentricity of the circle or circles in that area, and thus the amount of strain can be determined.

At this point, it would be beneficial to examine the effects of tooling and blank-surface variables on punch force. For instance, punch radii from two times the metal thickness on up to near the diameter of the cup have no effect on punch force once the cup is formed. On the other hand, changes in the die radius of the cup will produce significant changes in punch force. The general rule is that die radii from four to ten times the metal thickness give the best results. Larger radii may allow wrinkling, whereas smaller radii are more likely to cause tearing.

Increased punch velocity has no significant effect on punch force. High velocity does create vibration problems and necessitates increasing blankholding pressures up to three times in order to alleviate wrinkling. Another effect of high punch speed is that the rapid removal of the punch causes the cup bottom to be sucked in, thus making it concave. Venting corrects this problem.

Another parameter, related to the metal, that can affect drawability is surface roughness. A rough surface of 120 μ in. or more can cause breakage by limiting strain distribution in critical areas. On the other hand, an extremely smooth finish (15 μ in. or less) can cause oily lubricants to provide poor surface wetting. This promotes galling and fracturing. Finishes between 40 and 60 μ in. will provide the best all-around drawing results.

Die design is important, but the initial design and recommended alterations before final acceptance are left to the die designer.

Press-forming operations vary in degree of severity, from the light stamping of can lids and multistage progressive forming of small parts to the deep drawing of gas cylinders. Between these extremes are forming of appliance parts, kitchenware, automotive sections, and the like.

To make any of these operations perform smoothly, it is necessary to provide a lubricant film between the part being formed and the die so as to permit the desired shaping of the metal with a minimum of friction and wear.

In lubrication, "thick-film lubrication" is the ultimate in providing a low coefficient of friction (Figure 2R-9). Thick-film lubrication is complete separation of surfaces by a film of lubricant whose thickness is considerably greater than the height variation of asperities on the blank being formed and the die surface. Most press-forming processes, however, are considered to be operating in the boundary area of lubrication. In boundary or mixed-film lubrication there is at least some contact between the asperities of the blank and die surfaces (Figure 2R-10). Lubricants which are effective in boundary lubrication act to prevent or reduce contact between asperities. They also delay plastic flow of the peaks contacted, which can result in galling and increased levels of friction.

In light forming operations, this function can be accomplished with straight oils and emulsions. As increased pressures are required, polar materials such as fatty oils and

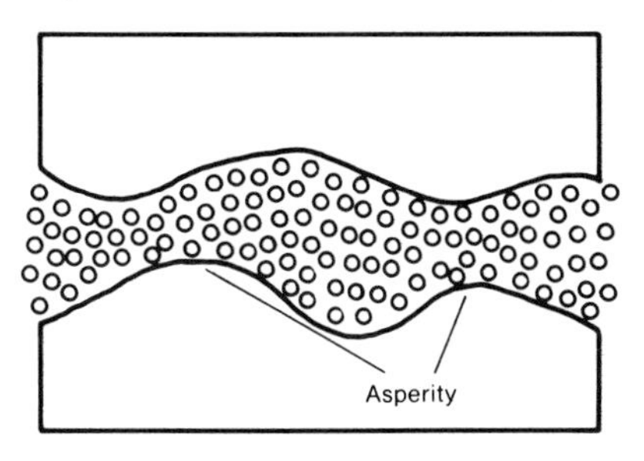

Figure 2R-9. Thick-film lubrication

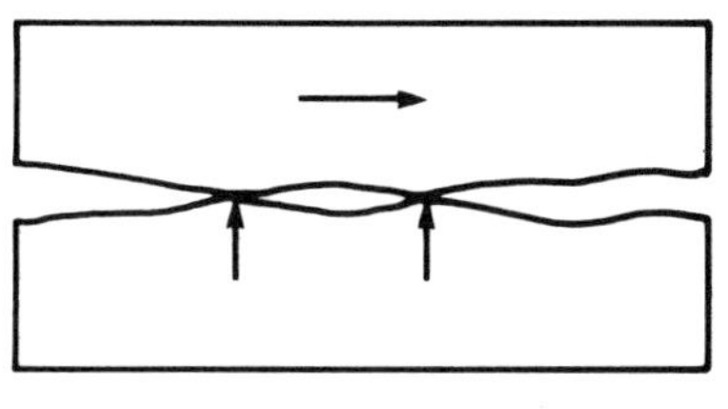

Figure 2R-10. Metal-to-metal contact (at vertical arrows)

soaps are incorporated into the lubricant. Such materials physically attach themselves to the metallic surface and increase the lubricant's ability to resist displacement. Under heavy pressure, extreme-pressure (E.P.) agents are required to provide boundary lubrication.

The lubricant must have a viscosity that is within a practical range and then be supplemented with polar or E.P. additives to provide metal separation.

In selection of lubricants (often called drawing compounds), consideration should be given to the work metal and its behavior as well as to die design. Additional factors that must be analyzed are:

1. The prime purpose of the lubricant
2. The method of lubricant application
3. The method of lubricant removal.

In slow-moving press work, the antiweld properties of the lubricant are primary, and lubricants fortified with additives should be selected. On eyelet machines and progressive forming machines, cooling is primary.

In progressive forming, the buildup of heat is directly proportional to operating speed. The heat must be controlled to a level compatible with the lubricant and the workpiece, and must be dissipated from the tools at a rate equal to its rate of input over the entire period of operation. The most effective means of controlling heat buildup is by use of a suitable lubricant.

The means by which a compound is to be applied helps dictate its selection. Press-forming lubricants are applied either at the press or in an area adjacent to the press. At the press, roller coating, dipping, swabbing, flow-on, and spray-on methods are used. Dry-film lubricants are applied to blanks in the adjacent area. In operations where blanking and forming are done in a continuous operation, lubricant is applied to the strip prior to any cold working and is generally expected to last through several stages. In eyelet equipment, where cooling is primary, lubricant is generally applied by recirculation or at several locations during forming, by dripping, flow-on, or spraying.

In selection of a lubricant, the readiness with which it can be removed at some later time must be considered. It may serve as an interoperational protectant and a rust inhibitor, but sooner or later it must be removed. In general, the longer the time interval between forming and cleaning of the part, the more difficult the removal will be. Another factor related to cleaning is the severity of the draw. Chemical changes brought about by heat and pressure during the drawing operation may make removal difficult.

Dried-on, pigmented compounds are difficult to remove also. Such compounds are removed more easily while still moist. The incorporation of some oil will keep the compound from drying out, because the oil will not evaporate as water does.

Removal of the residual films from drawn parts is done mainly by the use of two types of cleaning agents: solvent cleaners and alkaline cleaners. Solvent cleaners should be used only for removal of oil-type compounds. Emulsifiable and water-dilutable compounds are more efficiently removed by alkaline cleaners.

Solvent cleaners provide excellent removal of oil-type soils, but cannot remove inorganic residues. Alkaline cleaning, however, is the most widespread method for efficent removal of drawing films from formed parts. Soak tanks and power washers are the preferred types of equipment. Stubborn soils such as viscous oils and dried-on pigments are difficult to remove.

Today, there are two basic requirements of cleaners: efficient removal of oil-type compounds by solvent cleaners and safe handling of water-soluble alkaline cleaners. Both requirements often can be satisfied by emulsion cleaners which combine the good points of both cleaning techniques, but it must be remembered that there is no one all-purpose cleaner.

Having considered the obvious factors in selection of forming lubricants, we now turn to the several industrial requirements that must be met. These requirements are:

1. Satisfactory performance (first and foremost)
2. Toxicological safety
3. Freedom from corrosion.

4. Stability of product
5. Weldability
6. Economy.

A compound must perform with a reasonable degree of efficiency. The compound which gives the best performance is the one which permits the desired draw with a minimum of rejects and the smallest amount of tool wear. The selection of a compound based on its price alone can sometimes be costly.

The operator's health must always be considered. Toxic ingredients such as white lead and chlorinated solvents should be avoided if possible. When such components are necessary, adequate precautions must be taken, such as good ventilation and the use of suitable protective clothing. E.P.-treated compounds and high-pH water dispersions can irritate the operator's skin after continuous long contact. It is recommended that continuous contact be avoided and that protective creams be used. Often skin irritations are blamed on such contact; however, the more common cause of dermatitis is bacteria from "dirty hands and the like." The addition of biocides to lubricants has been helpful in minimizing such problems.

Corrosion or rusting is not a desirable condition on a finished or unfinished part. Any type of corrosion can be avoided if its source is understood and proper steps are taken to prevent its formation. Rusting on steel is most frequently encountered with water-dilutable compounds; prolonged surface contact with E.P.-treated and high-fatty-acid compounds will also cause corrosion. The same holds true for copper and aluminum. Effective means of combating corrosion from drawing compounds are:

1. Select a compound that offers good surface protection.
2. Remove residual films before corrosion develops and apply a suitable protective coating.
3. Use rust inhibitors in water dispersions.

Stability and uniformity of lubricants are mandatory. Bulk buying can require storage for several months, and stability and uniformity are expected unless otherwise stated. No chemical change or separation should take place. Pigments, such as whiting and graphite, must stay uniformly dispersed. With the wide range of climatic conditions in the United States, this can be difficult. Water-containing compounds can freeze en route, and oil-containing pastes and greases can bleed. Such compounds should be labeled. Problems can also arise during customer storage. The lubricant supplier must supply a good product, and the customer, in return, should handle and use it properly.

Some industries, such as those which manufacture kitchenware appliances, automotive parts, etc., conduct welding operations on formed parts prior to cleaning. Therefore, the press lubricant used should not interfere with good welding practices.

Economy is the watchword today, but this term can be overused and misapplied in relation to press lubricants. The initial cost of the lubricant can be too highly stressed: it is the total expense of the metalworking operation from start to finish that should be the only economical consideration. The efficiency of the lubricant from the over-all perspective is what really counts, and lubricants that materially reduce over-all production costs should be selected.

Proper selection of drawing compounds will be facilitated by a knowledge of the various components that these compounds contain. There are six principal classes of materials that may be present:

1. Polar agents
2. Extreme-pressure agents
3. Pigments
4. Emulsifiers
5. Diluents
6. Inhibitors.

Many operations, such as light stampings, operations in eyelet machines, and the like, can be done with lightly fortified straight oils, but as the pressure of deformation increases, so also does the accompanying temperature, which brings about the need to incorporate polar and E.P. agents in order to provide effective lubrication.

Polar agents are organic compounds

containing one or more polar groups with a strong affinity for metal surfaces. Typical polar groups are carboxyls, aldehydes, and hydroxyl-hydrocarbon combinations. These agents orient themselves over the metal surfaces and provide a thin molecular film between the workpiece and the die, which also aids in supplying "slip" between the metal surfaces. They also improve the viscosity index of the compound.

When pressures of deformation are exceedingly high, extreme-pressure agents are required in the drawing lubricant to prevent welding of the sliding surfaces. Chlorinated, sulfurized, and phosphated compounds are typical E.P. agents which under heat from pressure react with the metallic surface to form chlorides, sulfides, and phosphates, respectively. Such films have lower shear strengths than metal-to-metal welds and are constantly being broken and reformed so as to provide a continuous film separating the two surfaces.

The use of chlorinated, sulfurized, and phosphate-bearing materials has greatly extended our abilities in metal forming. It would appear that all drawing and forming compounds should contain E.P. additives in quantities as high as are economically feasible, but this is not true. E.P. agents should be considered as medicines and like any medicine, to have proper dosages beyond which there is no benefit. Such agents may even produce harm. Use of excessive amounts of E.P. agents, or of excessively active E.P. agents, can even cause die wear by chemical attrition.

E.P. agents are chemicals of varying degrees of activity. Their activity is triggered by a temperature rise, but their function is to prevent welding. The point to be born in mind is that their chemical activity will continue once started after temperatures subside. Corrosion often results even on stainless steels.

Pigments differ from E.P. agents in that, although they withstand extreme pressures, they do not react chemically with the die or the metal being formed. In effect, they are physical spacing agents which facilitate and yet control the flow of metal over the die. Common pigments now in use include whiting, lithopone, white lead, talc, mica, and graphite.

Pigments are relatively soft substances which are literally ground to a colloidal fineness as they separate the die from the work. In this respect they perform a beneficial polishing action, but the residual films are very difficult to remove in cleaning.

Pigments, in general, are noncompressible spacing agents, but supply little lubricity. Pigments are always mixed with oils or fatty materials, or both, which act as the sources of lubrication.

Most pigmented products are water-dispersible or emulsifiable. Most emulsifiable products do not contain pigments. Straight-oil-type drawing compounds, and particularly the polar- and E.P.-fortified types, are made emulsifiable. The emulifiers employed are quite varied and are too numerous to be given an all-inclusive treatment here. Soaps of animal and vegetable oils and greases have long been used, as have petroleum soaps and sulphonates.

Emulsifiable products may be in the form of pastes or viscous fluids. In many instances, the viscous fluids are not used in water dilutions; they are used undiluted. Their emulsifiability is utilized for purposes of ease of removal in subsequent cleaning operations. It should be additionally noted that many of the emulsifiers employed enhance the lubrication afforded.

Diluents include water, oils, and solvents. In "water-dispersible" compounds, diluents extend their use and also provide the important function of cooling. In paste-type compounds, there is always a percentage of water present which aids in the dispersibility of the product as well as keeping it in a soft enough state to be easily removed from the drum. Oils and solvents reduce viscosity and extend the use of oil-type compounds. Customers today do not like to mix their own and prefer to purchase oil-type compounds that are ready for use. Water-dispersible pastes and oils are made for dilution: customers do not desire to purchase, or pay freight on, water.

Inhibitors are necessary additions to

drawing compounds. Not only must inhibition against corrosion be considered, but also against foaming, rancidity, and oxidation, each of which requires its own small percentage of a controlling chemical.

From these classes of materials, all drawing compounds can be made, but so also can all industrial lubricants.

Now that we have described the components of forming lubricants and their functions, a discussion of finished products in terms of typical combinations of principal raw materials will be better understood. The two main classifications are "straight-oil" and "water-dispersible."

Under the straight-oil classification, lightly fortified oils of 45 to 300 SUS viscosity will provide good lubrication for many light stamping and progressive forming operations. As the severity of the draw increases, and greater pressures are required to form the part, polar agents are blended in. Such additions are made in the range of 5 to 50%. When pressures for deformation become extremely high, chemical E.P. agents are added. Chlorine can be present up to 50%, and active sulfur as high as 20%.

Let us now look at Figure 2R-11, which graphically illustrates the effects of temperature on combinations of straight-oil-type compounds. With mineral oil alone, the coefficient of friction increases with temperature. Mineral oil with an E.P. agent shows a marked decrease in the coefficient of friction when the E.P. agent is released by increased heat. Mineral oil with a polar additive shows an excellent low-friction curve right up to its melting point, and then the curve rises sharply.

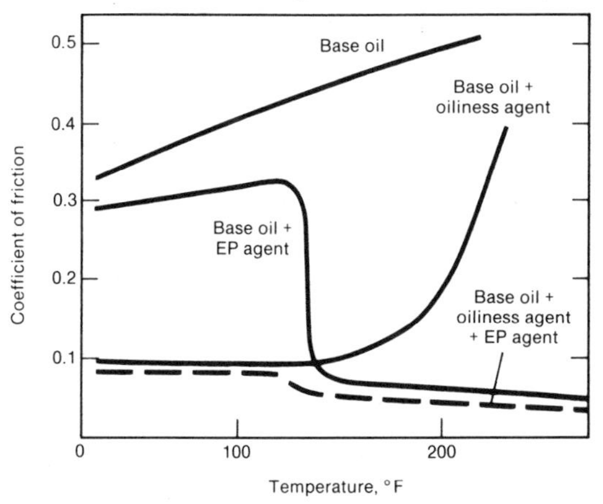

Figure 2R-11. Effects of temperature on behavior of lubricants

The combination of mineral oil, fatty oil, and an E.P. additive provides the lowest coefficient of friction through the widest temperature range; but even here there is the possibility of staining and corrosion, because the chemical action, once started, cannot be stopped. Fatty materials offer effective boundary lubrication up to 100 °C.

Straight oils find extensive use in light stamping and progressive forming of low-carbon steels, copper, and aluminum, particularly where surface protection after forming is required. As tensile strengths and metal thickness increase, polar agents are added. In working of stainless steels and high-carbon steels, E.P. additives are almost always incorporated. For light forming operations, the quantity may be as low as 1.0%, but even this amount is effective. Amounts up to 50% are used for deep drawing.

Sulfur is a very effective E.P. agent, but may be offensive in odor and has severe staining tendencies unless in odor and has severe staining tendencies unless controlled. Chlorinated agents are often preferred, because they are light in color, less offensive in odor, and less prone to staining. Phosphorus is the latest E.P. agent to arrive on the scene. It is an effective antiweld agent because it develops a light conversion coating on the metallic surface. Steels constitute the most effective area of application.

Inhibitors are incorporated to lower oxidation and to control bacteria. Inhibitors may include antirust agents.

Emulsifiers are added to straight-oil combinations to obtain effective cooling of water. The balance of ingredients now becomes more critical in production of combinations that are clear, stable, and uniform in the drum and that will disperse readily when mixed with water. Pigments, in general, replace chemical E.P. agents. The use of pigments brings up the problem of uniform dispersion in storage. Such products will almost always have a paste or grease consistency and use soap or another thickening agent to ensure uniform dispersion of the solids present.

Some press forming on steel is still done with conversion coatings, such as phosphate on carbon steels and oxalate on stainless steels. Their application is costly, but their efficiency in severe drawing operations frequently justifies their cost. Such coatings require supplementary applications of soap, wax, or oil.

Now that we have covered metal performance and touched on die design and lubricant composition, behavior, and parameters, there remain to be discussed the interactions of lubricants with the metal blank and die.

In pure stretch forming (see Figure 2R-1), lubrication is primarily necessary between the punch nose and the blank surface contacted. Increased pressures require lubricants which have greater resistance to displacement. Lubricants can aid wrinkling under the punch nose, but do not cause splits and cannot correct overstretching. Lubricants are used to smooth the metal flow under the punch nose, minimizing friction.

In pure drawing (see Figure 2R-5), lubrication is necessary primarily in the area of the die radius. Lubricant is applied to the blank areas passing over the die ring. Light lubrication aids in the movement of the flange under holddown pressures, but is not sufficient to cause wrinkling as the flange is compressed (see Figure 2R-6). In high-speed progressive forming, lubricant is necessary to allow smooth removal of the punch. Venting is the best method, but lubricant also helps.

Inasmuch as most forming is a combination of stretching and drawing, areas of lubricant application must be studied in the light of forming requirements. In forming of parts that require several stages, each stage should be studied to determine the areas necessary to lubricate in order to apply ample lubricant without being wasteful.

2S: Steel Rule Dies for Blanking and Forming of Sheet Metal and Nonmetallic Materials

A steel rule (also referred to as a metal-form die) consists of a strip steel section and a mating punch which may be mounted in a master die set, in a standard die set, or directly to the ram and bolster plate of the press (Figure 2S-1).

The steel rule type of tooling is quite simple, its cost of production is low compared with that of punch-and-die tooling, and it exhibits uniform reliability and performance when properly designed and constructed. The term "steel die" usually refers to a single-element tool with a knife edge that shears material — usually against a flat plate.

The steel rule die used for blanking and piercing is another type that uses flat-edge rules in conjunction with a shear template. The rule acts as the die or female member and the template as the punch. This type of tool must be aligned and have proper clearances similar to standard tooling (Table 2S-1).

Because the die section in this tooling is of comparatively thick stock called rule or flat ground stock, the edge which does the shearing against the template must be held in alignment so as to provide good shearing action and produce burr-free blanks. It is obvious that a narrow section or column will have a tendency to buckle under comparatively substantial shearing stresses. The shear coefficient of the metal to be worked in relation to the compressive strength of the rule steel section must be considered also. A safety factor of at least 2.0 should be used,

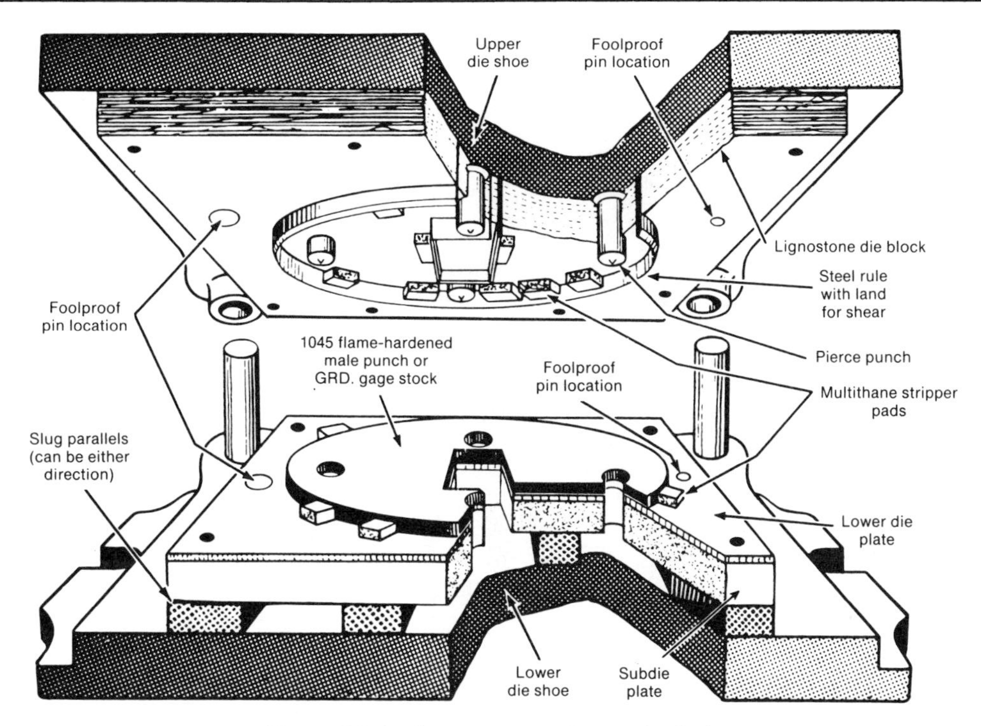

Figure 2S-1. Cross section of a steel rule die mounted in a die shoe

because the die will sustain impact loading and the internal forces will be considered as doubled under this stress. In shearing of 80,000-psi stainless steel, the die rule should be heat treated so as to afford a compressive strength in excess of 175,000 psi. Failure to provide a sufficient factor of safety will subject the die parts to critical stresses, and an increasing distortion and swelling of the die sections will occur under cyclical impact loading, such as in stamping operations. In forming the rule to the layout lines, prehardened rule should be used in the straight sections and dead soft rule to bend the curved section (Figure 2S-2). Small, intricate sections can be made from tool steel inserts. Keep checking the steel rule against the layout lines while bending, and leave the rule a little long (approximately 1/4 in.) before heat treating so that it may be fitted after the hardening process.

To heat treat the steel rule, the oven is preheated to 1500 to 1550 °F, and the steel rule is allowed to soak for approximately 20 minutes, removed from the oven, and immediately quenched in oil with *no time lag.* After quenching, wash in warm water. *Do*

Table 2S-1. Clearances for Steel Rule Dies

Stock thickness, in.	Clearance, in., for: Brass and soft steel	Medium rolled steel	Hard rolled steel
0.010	0.0005	0.0006	0.0007
0.020	0.001	0.0012	0.0014
0.030	0.0015	0.0018	0.0021
0.040	0.002	0.0024	0.0028
0.050	0.0025	0.003	0.0035
0.060	0.003	0.0036	0.0042
0.070	0.0035	0.0042	0.0049
0.080	0.004	0.0048	0.0056
0.090	0.0045	0.0054	0.0063
0.100	0.005	0.006	0.007
0.110	0.0055	0.0066	0.0077
0.120	0.006	0.0072	0.0084
0.130	0.0065	0.0078	0.0091
0.140	0.007	0.0084	0.0098
0.150	0.0075	0.009	0.0105
0.160	0.008	0.0096	0.0112
0.170	0.0085	0.0102	0.0119
0.180	0.009	0.0108	0.0126
0.190	0.0095	0.0114	0.0133
0.200	0.010	0.012	0.014
0.210	0.0105	0.0126	0.0147
0.220	0.011	0.0132	0.0154
0.230	0.0115	0.0138	0.0161
0.240	0.012	0.0144	0.0168
0.250	0.0125	0.015	0.0175

(a) Lay out die on a Ligniform die block using conventional vernier height gage. (b) Bend steel rule to scribe line using Multiform bender. (c) Place steel rule on scribed die block to check accuracy. (d) Cut steel rule to proper length using cutoff saw to effect dead square cuts. (e) Use optical scope to ensure location in boring die-block holes. (f) Cut steel die slots in Ligniform block using precision electromatic saw. (g) Insert steel rule into Ligniform die block using soft hammer to protect steel rule cutting edges. (h) Cut paper template from steel rule section for positioning punch steel on backer plate. (i) Strike punch plate with steel rule die section to transfer contours to punch section. (j) View of punch section after transfer, showing sharp, clear detail. (k) Cut punch plate section on band saw. (l) Punch section mounted on lower die plate and die set with rule to be sheared in. Grind flat land on top edge of rule section before shearing in. After shearing in, grind proper clearance on punch section. Heat treat punch section and check for clearance.

Figure 2S-2. Ligniform die process

not chill the part. After the rule has been washed, it is put in an oven and drawn at 475 °F for approximately 20 minutes and then removed from the oven and allowed to air cool.

Drill holes in the die for the bridges. The bridges are the means of holding the lignite section together in one piece. These bridge units are spaced approximately 6 in. apart where practical, or closer.

The step of jigsawing the die is where the accuracy of the steel rule die is obtained. It is imperative that a precision jigsaw unit be used to saw these dies, because the rule must be perfectly vertical in the board.

The mating punch section is normally made of 1045 flame-hardened material or oil-hardened tool steel, or any suitable material to produce the required run. The punch can be in one piece or in sections, depending on the shape of the part to be blanked.

Sharpening the punch can be done by any suitable grinder without removing the punch from the 1/4-in. adapter plate. After sharpening, urethane strippers are put on the steel rule die section and the punch section. The urethane should be about 1/32 in. higher than the steel rule die, and about the same height on the punch section.

To produce a steel rule die, the following procedures should be followed.

Using a predrilled lignite material, lay out the die by employing one of several methods. Apply blueing on one side of the lignite section using a height gage for scribing the layout on the lignite (Figure 2S-2).

One of the factors influencing the proper operation of steel rule dies is the design and condition of the press equipment. Another important factor in the use of rule-steel tooling in production is the use of a proper die set. For stripping in large areas, a 70 durometer neoprene may be used, whereas under confined or heavy stripping conditions, dense polyurethane is quite satisfactory.

The cost of steel rule dies ranges from 10 to 30% of the cost of regular dies, depending on the type of construction used.

Types of materials that can be stamped in production with steel rule dies include:

1. Hot rolled steel up to 3/8 in. thick
2. Cold rolled steel up to 3/8 in. thick
3. Copper, brass, and some types of aluminum and stainless steel thicker than 3/8 in. (in experimental quantities)
4. Titanium.

Major areas of application are:

1. Large parts which would require a large investment if blanked by conventional methods
2. Limited numbers of parts
3. Applications in which tools are needed on short lead time
4. Thick material (such as 1/2 in.)
5. Tolerances of ±0.005 in. are practical.

Table 2S-2 presents the tonnages required per linear inch when using steel rule dies.

The significant advantages of steel rule dies are as follows. When short-run production involves excessive tool costs, steel rule dies can reduce original costs by 70 to 90%. Personnel require only a short training period. Expensive machine tools are not required for fabrication. Low-cost, readily available materials can be used. The dies can be easily modified.

Steel rule dies can be used for production quantities of up to 100,000 parts with a die cost of only one-fourth that of corresponding conventional dies. Multiple dies may be made if the production requirements are high enough. Steel rule dies may be used as duplicate dies in case of failure of high-production dies.

Metals successfully blanked to date on rule dies include aluminum, magnesium, vanadium, and titanium, as well as low- and high-carbon stainless and alloy steels. Moderate forming of light sheet metal is also practical.

The equipment needed to manufacture steel rule dies (Figures 2S-3 and 2S-4) includes:

- Multiform electromatic jigsaws (available in throat sizes of 30, 50, and 74 in.)
- Multiform benders: #4HDC hand bender; #6 hydraulic power bender
- Multiform cutoff saw
- Magnifying lamp

Table 2S-2. Estimated Tonnages per Linear Inch for Steel Rule Dies

Length of cut	Tonnage per lineal inch for stock thickness of: 0.023	0.029	0.035	0.041	0.047	0.059	0.067	0.074	0.089	0.104	0.119	0.134	0.187	0.250
10	4.0	5.0	6.1	7.1	8.2	10.3	11.7	12.9	15.5	18.2	20.8	23.4	32.7	43.7
11	4.4	5.5	6.7	7.8	9.0	11.3	12.8	14.2	17.1	20.0	22.9	25.7	35.9	48.1
12	4.8	6.0	7.3	8.6	9.8	12.3	14.0	15.5	18.6	21.8	24.9	28.1	39.2	52.5
13	5.2	6.5	7.9	9.3	10.6	13.4	15.2	16.8	20.2	23.6	27.0	30.4	42.5	56.8
14	5.6	7.1	8.5	10.0	11.5	14.4	16.4	18.1	21.8	25.4	29.1	32.8	45.8	61.2
15	6.0	7.6	9.1	10.7	12.3	15.4	17.5	19.4	23.3	27.3	31.2	35.1	49.0	65.6
16	6.4	8.1	9.7	11.4	13.1	16.5	18.7	20.7	24.9	29.1	33.3	37.5	52.3	70.0
18	7.2	9.1	11.0	12.9	14.8	18.5	21.1	23.3	28.0	32.7	37.4	42.2	58.9	78.7
19	7.6	9.6	11.6	13.6	15.6	19.6	22.2	24.6	29.5	34.5	39.5	44.5	62.1	83.1
20	8.0	10.1	12.2	14.3	16.4	20.6	23.4	25.9	31.1	36.4	41.6	46.9	65.4	87.5
22	8.8	11.1	13.4	15.7	18.0	22.7	25.7	28.5	34.2	40.0	45.8	51.5	71.9	96.2
24	9.6	12.1	14.6	17.2	19.7	24.7	28.1	31.0	37.3	43.6	49.9	56.2	78.5	105.0

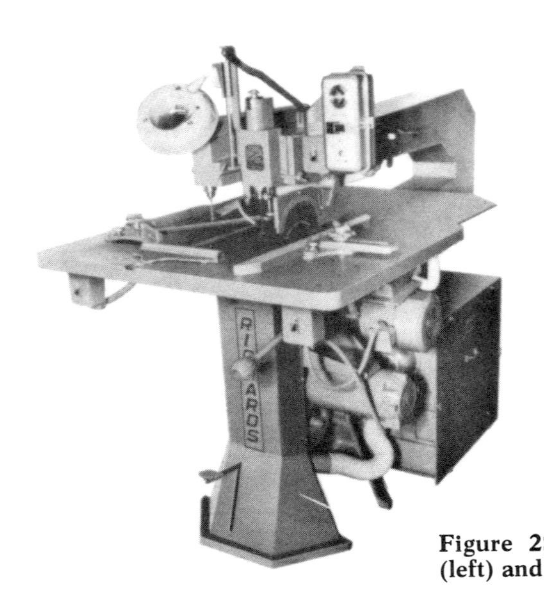

Figure 2S-3. Electromatic universal die saw (left) and electromatic jig and drill (right)

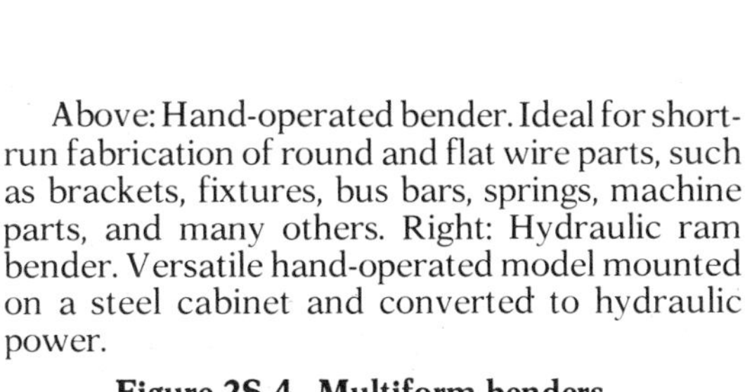

Above: Hand-operated bender. Ideal for short-run fabrication of round and flat wire parts, such as brackets, fixtures, bus bars, springs, machine parts, and many others. Right: Hydraulic ram bender. Versatile hand-operated model mounted on a steel cabinet and converted to hydraulic power.

Figure 2S-4. Multiform benders

- Master die sets
- Drill templates
- Electromatic furnace.

The following supplies are needed for making steel rule dies:

- Lignite
- Masonite SRD board
- Steel rule
- Punches
- Abrasive wheels
- Urethane squares, rounds, and sheets
- Epoxy
- 1045 flame-hardened die steel
- Tool steel
- 1/4-in. roller flat hot rolled steel
- 5/8-in. steel plate with parallels
- Master die sets.

2T: Lower Hole Cost With Proper Die Design

Many parts and components previously produced by costly machining operations are now produced much more economically in stamping presses. A well-designed and constructed die capable of long, uninterrupted runs is, of course, desirable.

The higher cost of a well-designed and constructed die is insignificant in the light of the high cost of the downtime it eliminates. A survey has also proven the fact that by far, the greatest percentage of downtime is caused by failure of what is perhaps the most inexpensive and most fragile of all the die's components — the punch or perforator.

It then is reasonable to assume that anything done in die design and die construction that would in turn lessen the susceptibility of the perforator to failure, thus entirely eliminating or substantially reducing costly downtime due to failure of this fragile component, would certainly be advantageous. Efforts directed along this course have led to the Durable System of Die Design.

There has evolved a unique item called an "intermeshing punch-support sleeve." The prime purpose of such a support sleeve is to give reinforcement to this most perishable and fragile of die components (Figure 2T-1).

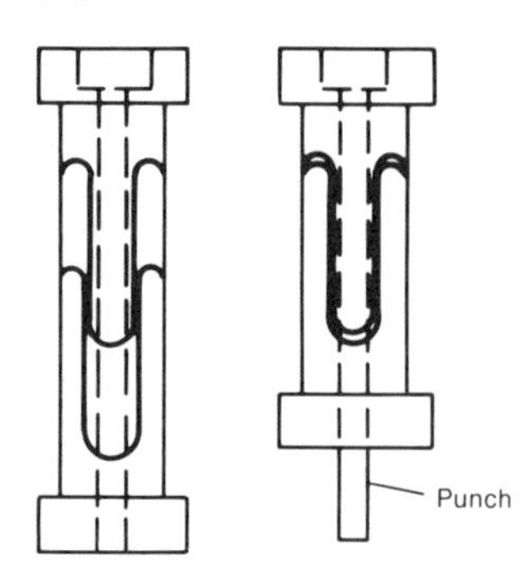

Figure 2T-1. Intermeshing punch-support sleeve

The punch-support sleeve, supporting as it does the punch over its entire length, will not permit the perforator to bend or break.

The intermeshing punch-support sleeve consists of two units: the upper section, which acts as a punch holder; and the lower section, which serves as a guide bushing. Each section is sufficiently long so that equally spaced slots may be milled in each section. The resultant splines intermesh, moving freely in and out of each other as the die opens and closes.

Fatigue from high-speed operation is substantially reduced due to the fact that the vibrations of the punch are throttled or "damped" due to enclosure of the perforator in the intermeshing sleeve.

The intermeshing sleeve, fully enclosing and supporting the perforator over its entire length, has broadened substantially the scope of practical metal piercing. It has removed and made obsolete a great restriction in the area of metal punching — the old and widely accepted fact that the limit of practical metal piercing was reached at a ratio of punch diameter to material thickness of 1 to 1.

Because the punch is supported over its entire length by the punch-support sleeve, which prevents any buckling, bending, or breaking, then it must accept the force applied to it and must penetrate the work material with which it comes in contact. It is most practical to pierce metal in thicknesses several times the punch diameter. A punch-diameter-to-material-thickness ratio of 1/3 to 1 is quite commonplace for mild steel.

Frequently, the requirement presents itself for a straight, well-sized hole. Such a requirement would naturally demand a drilling-and-reaming operation with its attendant higher costs, or, if a press application, at least piercing followed by shaving.

Lessening of punch-die clearances will produce a greater degree of hole straightness, minimizing substantially the amount of breakout. Such lessening of punch-die clearance also increases punching pressure. Because the punch is supported over its entire length by the intermeshing sleeve, clearances may be cut to an absolute minimum.

For example, one application involved a cluster of holes with relatively close center-to-center spacing. The hole-size requirement was 0.070 in., with a total permissible tolerance of 0.001 in. The material was aluminum with a thickness of 0.190 in. It was determined that drilling and reaming to secure close hole tolerances would be costly, and that punching would have substantial cost advantages.

The procedure that was used consisted of a very practical piercing operation. Intermeshing punch-support sleeves were employed, giving the punches the support necessary for use of a minimum total punch-die clearance of 0.0025 in. This resulted in straight, well-sized holes within the tolerance of 0.001 in., and there was no breakout whatsoever in the hole itself.

Figure 2T-2 illustrates the Durable System of Die Design. Subguide pins (1) are a distinct advantage in operation as well as initial construction. Two or four hardened pins uniformly spaced in the punch plate extending through the spring stripper and into die block ensure continuous alignment of punch plate, spring stripper, and die block. The pins are press fit in the punch plate and slip fit in the stripper and die block. Hardened insert bushings in the stripper and die block will resist wear.

Initial die construction is simplified if these subguide pins are first installed, followed by line boring of the three plates in one operation for acceptance of the intermeshing sleeve and die button. Mounting and alignment in the die set are simplified, and proper alignment is ensured. In operation, the subguide pins maintain continuous alignment of punch plate, stripper, and die block, preventing "cocking" of the stripper from any number of causes. The cost of subguide pins and their installation is a nominal premium to pay for protection against a costly die smash-up.

For a die in which the intermeshing punch-support sleeve is to be used, it is suggested that the punch plate and the stripper be of sufficient thickness so that when the support sleeve is in its closed position, there will be a clearance of approximately 1/32 to 1/16 in. between the punch plate and the stripper. Thus, the punch plate and stripper support the prongs of the intermeshing sleeve, which in turn supports the punch. The splines of the intermeshing sleeve have a "lead" or relief ground on them so that there is no interference when the splines enter the hole that retains the matching part. A travel-limit pin (4) is installed in the die to prevent overtravel, which would ruin the punch-support sleeve.

Stripper bolts (7), which limit the travel of the stripper, should be of proper length so that sleeve splines will never become disengaged. When the sleeve is in its open position (which is also the punch length), the sleeve prongs will still be engaged.

Stripper springs of sufficient strength should be properly positioned for good stripper balance.

A hardened backup plate, which could be a disk, an insert, or a complete filler plate between the punch plate and the upper die section, is definitely required.

An area of wide acceptance and use of the intermeshing punch-support sleeve is piercing of round stock or rods. This operation has always involved consistently severe ratios of punch diameter to material thickness.

Forming of a 0.076-in.-diam hole in a 0.218-in.-diam rod (punch-diameter-to-material-thickness ratio of approximately 1/3 to 1) has been and usually still is considered to be a drilling operation. However, the standard intermeshing sleeve can now be adapted to the piercing of round parts. Figure 2T-3 illustrates such a pierced part.

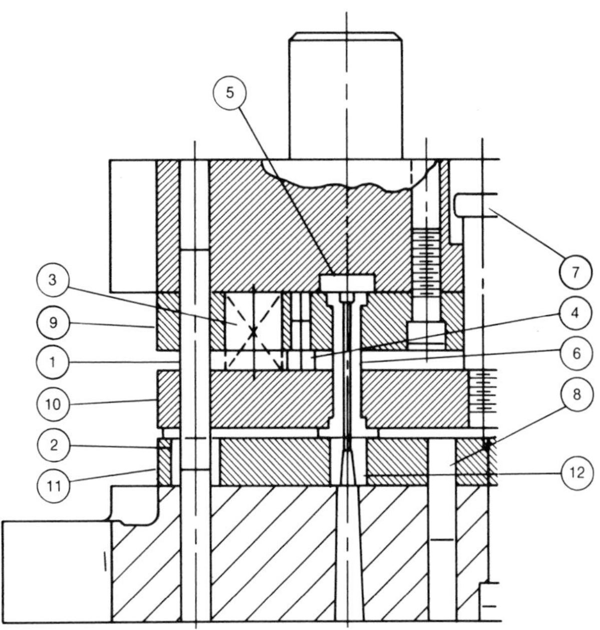

1. Subguide pin. 2. Bushing. 3. Stripper springs. 4. Travel-limit pin. 5. Hardened backing plate. 6. Intermeshing sleeve-type punch. 7. Stripper bolt. 8. Dowel pin. 9. Punch pad. 10. Stripper pad. 11. Die pad. 12. Die button.

Figure 2T-2. Illustration of the Durable System of Die Design

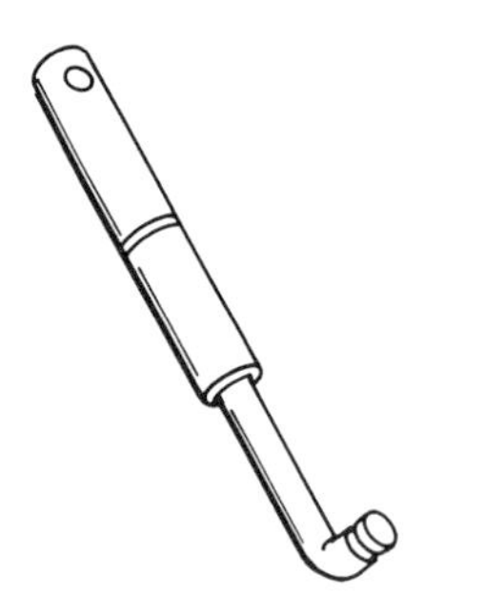

Figure 2T-3. Rod in which a hole was made in a piercing operation employing an intermeshing punch-support sleeve

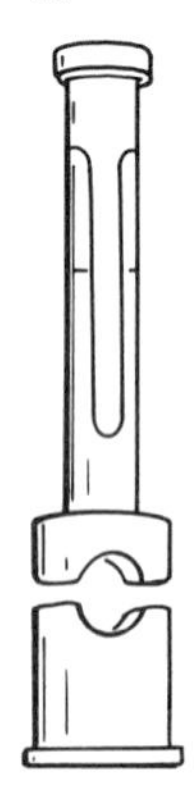

Figure 2T-4. Typical intermeshing punch-support sleeve and matching die button

Figure 2T-4 illustrates a typical rod-piercing punch-support sleeve and matching die button. It is necessary to tightly confine the bar diameter during piercing. Failure to do this will, due to the punching pressure required, cause the bar to swell in the area of the pierced hole, resulting in a swelled or barrel shape in the area of the hole. Naturally, such a condition would be unacceptable in most cases.

Such an undesirable condition is eliminated by grinding a radius matching that of the part to be pierced to a tolerance of +0.000, – 0.0005 in. in the lower sleeve section and die button. The depth of radius in each part is one-half the rod diameter minus 0.003 in. This ensures full bearing of the spring stripper on the part being pierced. Because the area of the hole being pierced is tightly confined, there is no swelling of the part in this area. Figure 2T-5 shows a typical die design for rod piercing. Die design should allow for application of plenty of stripper pressure. It is the stripper pressure brought to bear on the part during piercing that maintains its straightness. An insufficient amount of stripper pressure will allow the rod ends to move upward, precipitating a kink in the part. There have been cases wherein, to secure sufficient stripper pressure, urethane has been substituted for heavy-duty die springs.

Proper die design and die construction should never fail to give due consideration to a problem that is as old as metal punching itself — the tendency of the material to flow, stretch, or elongate due to the impact and pressure necessary to pierce it.

In the foregoing, the importance of tightly and securely holding the part in the die from the moment of die closure through completion of the press stroke has been mentioned. Upon closure of the die and initial penetration of the punch into the material, should the part move due to any cause, such as improper nesting, the punch point, either partially embedded in or completely through the part, has no choice but to flex or bend, going with the movement of the part or material flow. A high punch mortality rate along with soaring downtime and die maintenance costs result. Similarly, under the pressure of piercing, the metal has a tendency to flow along or follow the line of least resistance. The end result is no different from that of the part shifting or moving during piercing: the metal, in its flowing or stretching will carry the embedded punch point with it, resulting in angular holes and very short punch life.

It is suggested in piercing of rods that the centerline of the hole be a distance from the rod end equal to the rod diameter. This allows a sufficient sidewall thickness or "web" from the edge of the hole to the end of the part to resist the flow of material in that direction. Again, however, with proper die design and construction, this relatively restrictive hole-location factor can be circumvented.

In the operation illustrated in Figure 2T-5, simple modification of typical rod-piercing die design was made: a hardened tool steel insert was incorporated in the die block. Its location was such that, when the end of the

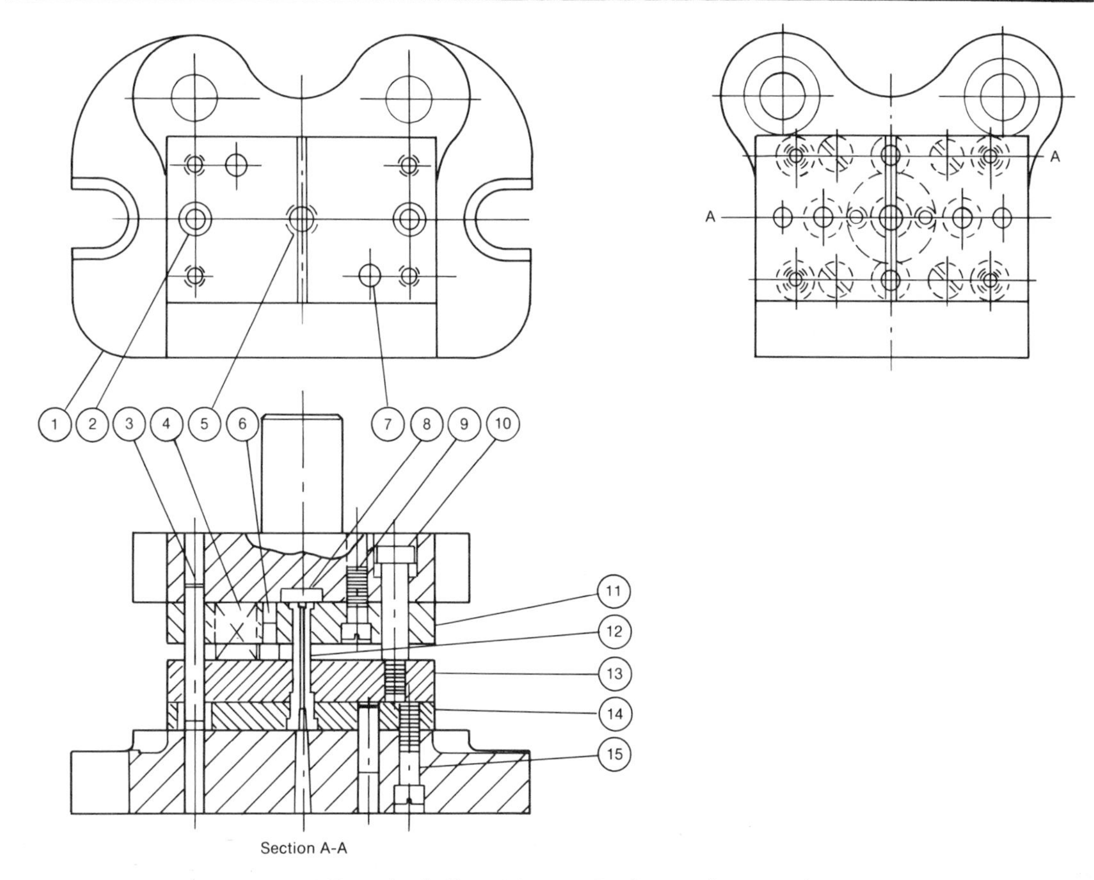

1. Baumbach die set, 4 × 5. 2. Bushings, 5/8 OD × 1/2 long. 3. Dowels, 3/8 × 2½. 4. Springs, 3/4 × 1½. 5. Die bushing, 7/16 OD. 6. Travel-limit pins, 3/4 diam. 7. Dowels, 3/8 × 1¼. 8. Backing plug, 3/4 diam × 1/4. 9. Screws, filister hd., 3/8 × 1. 10. Shoulder screws, 1/2 × 1¾. 11. Punch plate, 3/4 × 3 × 5. 12. Durable specification punch. 13. Stripper, 3/4 × 3 × 5. 14. Die, 1/2 × 3 × 5. 15. Screws, filister hd., 3/8 × 1⅛. (All dimensions are in inches.)

Figure 2T-5. Typical die design for piercing of rods

rod was against the "gate" or stop block, the hole would be pierced in the proper location with respect to the rod end. The insert extended sufficiently high above the die block. Pressure brought to bear against the opposite end of the rod, forcing and holding it tightly against the "gate" or stop block during the entire piercing cycle, accomplished the task. There was no flow of metal due to the closeness of the hole to the end of the rod, resulting in straight holes and long, trouble-free, productive runs.

Figure 2T-6 is another typical die design for confinement of the part during a severe piercing application. A rectangular business-machine part, 1¼ by 7/16 by 5/16 in. thick, required piercing of two 0.125-in.-diam holes. Under the pressure of piercing this severe ratio, the sides of the part were bulging. The part could be removed from the die only with difficulty, and holes often were angular and exceeded dimensional tolerances. This problem was solved by using the locking gage shown in Figure 2T-6.

The taper construction of the locking plate permitted positive locking regardless of stock variation. The backing plate was sufficiently strong to withstand the strain. For high-production runs, locking could be cam or air operated. After the part had been

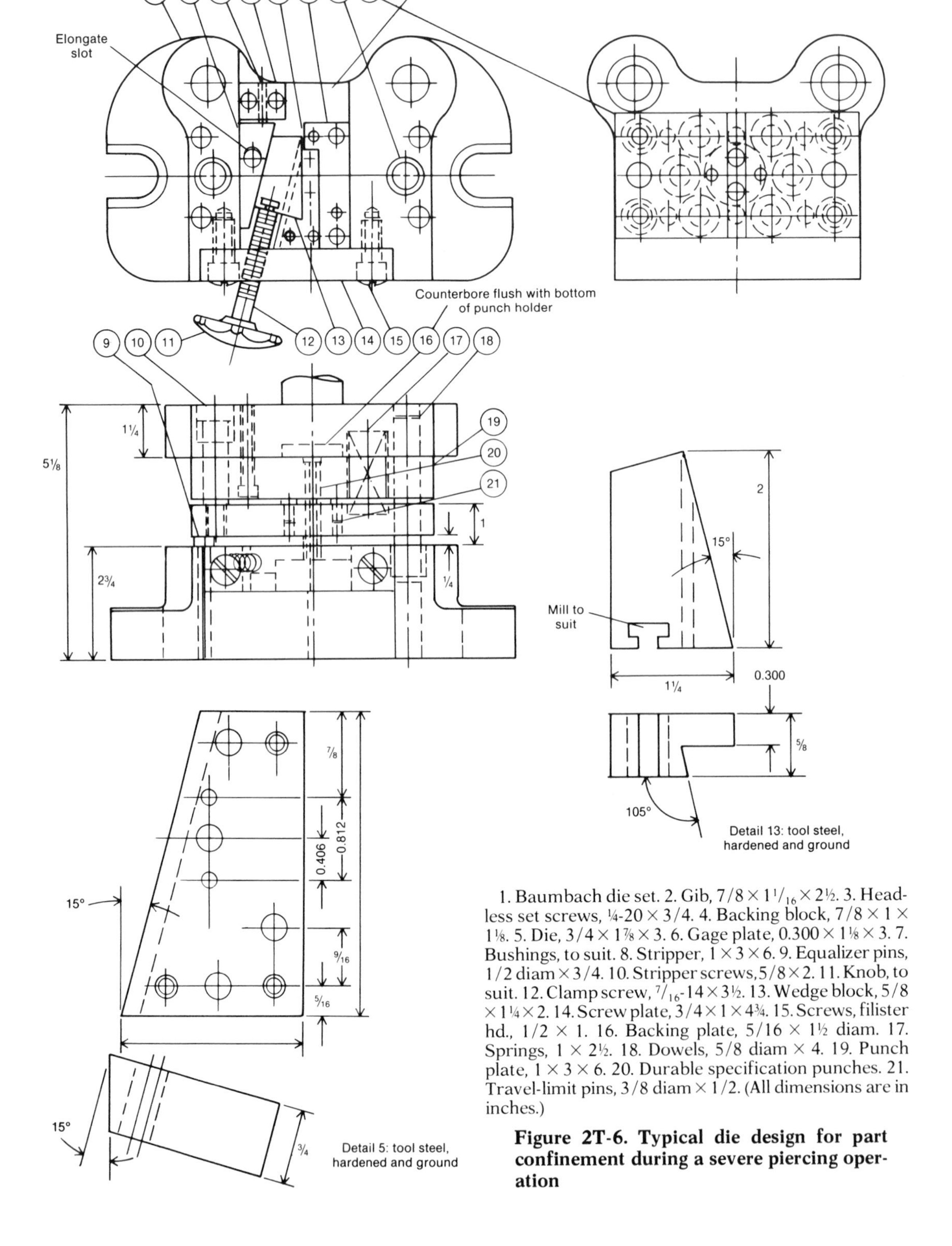

1. Baumbach die set. 2. Gib, 7/8 × 1 1/16 × 2½. 3. Headless set screws, ¼-20 × 3/4. 4. Backing block, 7/8 × 1 × 1⅛. 5. Die, 3/4 × 1⅞ × 3. 6. Gage plate, 0.300 × 1⅛ × 3. 7. Bushings, to suit. 8. Stripper, 1 × 3 × 6. 9. Equalizer pins, 1/2 diam × 3/4. 10. Stripper screws, 5/8 × 2. 11. Knob, to suit. 12. Clamp screw, 7/16-14 × 3½. 13. Wedge block, 5/8 × 1¼ × 2. 14. Screw plate, 3/4 × 1 × 4¾. 15. Screws, filister hd., 1/2 × 1. 16. Backing plate, 5/16 × 1½ diam. 17. Springs, 1 × 2½. 18. Dowels, 5/8 diam × 4. 19. Punch plate, 1 × 3 × 6. 20. Durable specification punches. 21. Travel-limit pins, 3/8 diam × 1/2. (All dimensions are in inches.)

Figure 2T-6. Typical die design for part confinement during a severe piercing operation

punched, the taper wedge was released and the part was freely removed manually or by air. When the parts were thus locked and confined in the die, the holes were straight and well within dimensional tolerances. Bulging of the part was eliminated.

Another area in which proper die design makes piercing both practical and economical is the area of printed circuit boards. Often the hole requirements of printed circuit boards constitute extemely severe ratios of punch diameter to material thickness. Support of the punch over its entire length by the intermeshing punch-support sleeve makes possible the piercing of thicknesses of this laminate which appreciably exceed the punch diameter.

A case in point is the piercing of 1/8-in.-thick epoxy glass, copper coated on one side, with 0.040-in.-diam holes. Total punch-die clearance is 0.003 in. Should the copper coating exist on both sides of the printed circuit board, it is necessary to have a very minimal total punch-die clearance for clean shearing of the copper cladding on the bottom side of the printed circuit board.

Total punch-die clearance when copper cladding exists on both sides of the material is from 0.001 to 0.0015 in. Such a minimal punch-die clearance is possible and practical when the punch-support sleeve is employed. Cleanly sheared, burr-free holes result.

The face of the lower sleeve shoulder is beveled slightly, leaving a flat surface approximately 1/32 in. wide around the inside diameter of the sleeve. This results in concentration of spring stripper pressure around hole periphery as the punch is withdrawn, eliminating the possibility of the copper coating being delaminated from the insulating material.

Concentration of stripper pressure around the periphery of the hole coupled with minimal punch-die clearance also minimizes appreciably the "crazing" that occurs around the hole.

To resist the wear to which punches are subjected during piercing of this abrasive material, it is advisable to secure punches made of high-alloy steel with a hardness of 65 to 67 HRC, rather than of steel with a conventional hardness of 60 to 62 HRC.

There is a consistent growth in the demand for miniaturized components, no doubt caused by the advances of present-day technology. Many such components demand very small holes that are consistent in size, often to very close tolerances. Although the thickness of the material in which such a small hole is required is not great by comparison, the ratio of hole size to material thickness is severe. A hole with a diameter of 0.004 ± 0.0001 in. through 0.005-in.-thick material constitutes an extreme ratio of punch diameter to material thickness. In addition, such a close tolerance is difficult to hold, and is almost impossible if the hole is made by drilling. Drilling in such thin material produces holes of irregular shape and inconsistent size, and usually produces obnoxious burrs as well.

The miniaturized intermeshing punch-support sleeve shown in Figure 2T-7 is a die component which, because it supports the very small-diameter perforator, makes punching of such holes both possible and practical.

Tolerances of die components as well as total punch-die clearances are on the order of a few ten-thousandths of an inch. Therefore, the importance of proper die design and construction in the area of miniaturized hole piercing cannot be overemphasized.

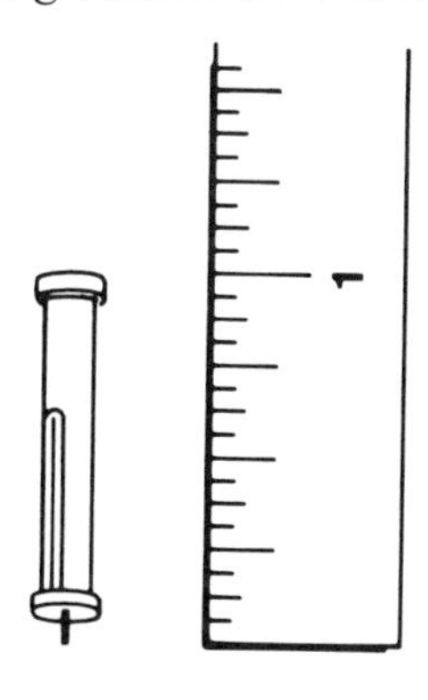

Figure 2T-7. Miniaturized intermeshing punch-support sleeve

2U: Unitized Tooling for Sheet Metal

Today, with rising costs squeezing profits, management looks hopefully to tooling and methods departments for new manufacturing techniques not only to offset these increases, but actually to lower manufacturing costs.

In the past, sheet-metal departments have been overlooked as an area for cost reduction and consequently have received very little consideration in the tooling budget due to their unique tooling requirements.

Generally summarized, the following are the requirements for tooling in a sheet-metal department:

1. Low-cost tooling, or tools that can be amortized over a large number and variety of parts
2. Versatile tools that can be adapted to light- and heavy-gage sheet-metal and structural steel parts
3. Flexible tools that can be easily handled and quickly changed from one setup to another
4. Simple tools that can be set up and maintained by average press-room operators without special skills
5. Compatible tools that can be adapted to existing press-room equipment
6. Reusable tools that are constantly "in work," eliminating dead, unproductive tools that must be stored, taking up valuable factory space.

The search for tooling that meets all of these requirements has led to the "unitized" tooling concept — a concept that has developed into today's most modern and economical system for fabrication of sheet-metal and structural steel parts.

The key to the entire system is the word "reusable." Punch and die units are reused over and over again. Setups are made in various patterns using interchangeable, low-cost punches and dies, thus offering tremendous savings when compared with the single-purpose, high-cost tooling set-ups that are still used throughout industry today.

Application of this unique tooling system presents an opportunity for the open-minded individual who is ready to throw off the shackles of "past practice" and be willing to accept a totally new and refreshing concept in reducing tooling costs.

THE DESIGN

Unitized tooling has been described, at various times, as self-contained tooling, unit tooling, subpress tooling, and C-frame tooling. Basically, however, a unitized hole-punching unit may be described more accurately as a quick-change punch and die held in alignment within a C-frame steel casting, as shown in Figure 2U-1. Manual alignment is a thing of the past. The C-frame holder is designed to accept, quickly, easily, and accurately, a wide range of punches and dies, both round and shaped.

When a unitized tooling unit is located in a press, nothing is attached to the press ram. The unit is completely independent of the ram, which provides the "hammer" or force to pierce the hole. This complete independence of the ram means that punch and die misalignment due to sloppy or worn ram gibs is eliminated. Much of the human-error

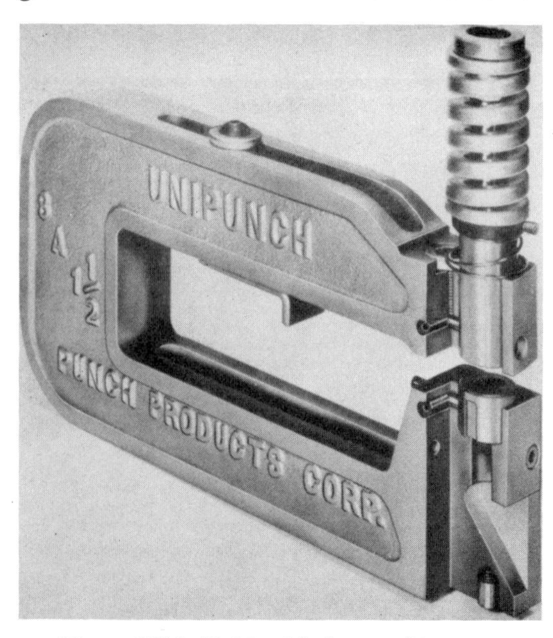

Figure 2U-1. Unitized hole-punching unit

factor is also removed with this design concept.

Located in the lower arm and concentric with the punch and die is a 3/8-in.-diam pilot pin. This unique design feature permits fast and accurate location of the units in relation to one another in any pattern.

Drilling and reaming of a 3/8-in.-diam hole in a steel template corresponding to the desired hole pattern of the part locates the unit. The templates, completely engineered with one or more piece-part hole patterns, are normally provided at nominal cost by the unitized tooling manufacturer. Due to their simple design, however, most shops with a minimum of equipment and semiskilled mechanics can, and do, provide their own templates.

The low-cost punches and dies are made of the highest-quality oil-hardened tool steel and are guaranteed to be 100% interchangeable between units. Fast changeover from one size to another, or from a round punch and die to a shaped punch and die, is ensured. The two-piece design of most of these punches offers the two-fold advantage of shimming to increase punch life and maintenance of all punches at the same length.

Each unit is complete with its own stripping mechanism consisting of a stripping guide and a stripping spring. As the ram of the press descends, pressure is applied to the part through the stripping guide before, during, and after punching of the hole. While this downward pressure is maintained through the guide, the stripping spring, being depressed during the downstroke, exerts an upward, or stripping, pressure on the punch to pull it free of the material it has pierced. These counteracting pressures maintain a flat, smooth surface around the hole and prevent, to a large degree, any part deformation.

The C-frame holder, in most cases, is a 100% steel casting, accurately machined to maintain punch and die alignment and provide long unit life. The width of the unit determines the maximum size of punch that may be used. Punches ranging in size from 0.030 to 4 in. are quickly and accurately changed in the holders.

Available throat depths (dimension from center line of punch back to inside edge of C-frame) are 4, 8, 12, and 18 in., and deeper in special applications.

The design of unitized tooling permits custom setups with standard interchangeable units. These units may be used and reused in innumerable setups, minimizing dead, unproductive storage. Because they are designed with shut heights compatible with most presses and press brakes, no special capital equipment is required for punching and notching of sheet metal, plates, angles, channels, or extrusions.

THE SYSTEM

This unique tooling system has developed from a single hole-punching unit into a complete fabrication system for all phases of sheet-metal fabrication. Simply by making a template for locating and mounting of units, a complete setup is ready to produce long- or short-run parts. A good mechanic can assemble these setups outside of the press in a matter of minutes, as shown in Figure 2U-2.

The units, being completely independent of the press ram, permit the setups to be made outside the press while another setup is running. Costly press downtime is thus held to a minimum. The press stops only for setup changes. New parts are produced with the first stroke of the press.

Because the units are located on the template with 3/8-in.-diam holes, it is a simple matter to incorporate quickly any necessary engineering changes, ensuring prompt and profitable delivery of the finished parts. Even changes in hole size or configuration become simple procedures with the quick-change punches and dies.

Many a single-purpose die has been made obsolete by a change in the material type or thickness to be punched. Not so with unitized tooling. Again, the easily available, completely interchangeable, quick-change die buttons add to the tremendous flexibility of this system.

Once the part has been run, only the template is stored. In most cases, these templates are 1/2 in. thick and are conveniently stored

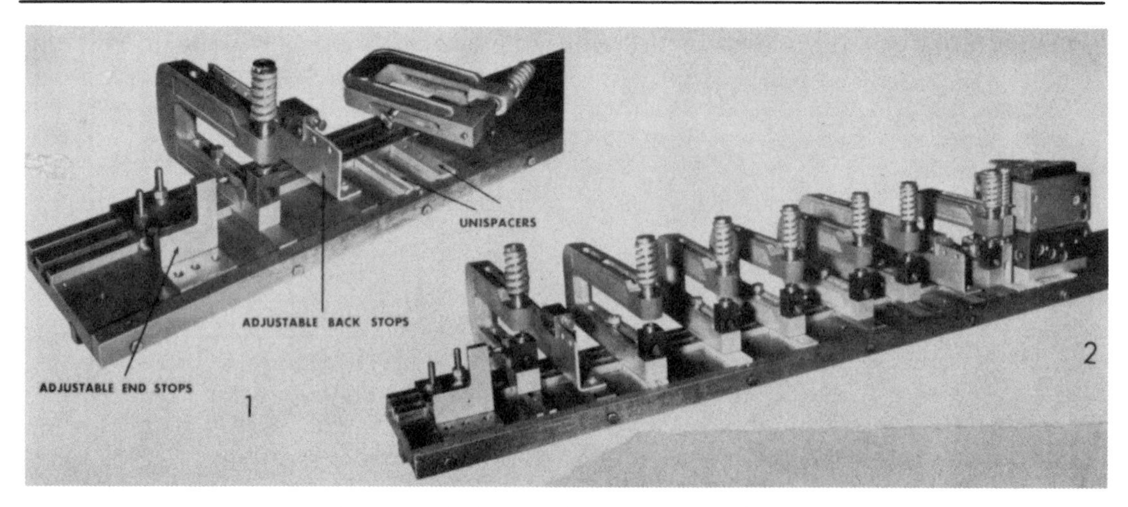

Figure 2U-2. Unitized tools mounted on bed rails. Setups can be adjusted for short-run production.

on edge, using a minimum amount of valuable factory space.

These templates permit accurate reruns of all parts. Regardless of whether the parts are run once a month or once a year, repeatability is ensured. Because the setups are so easily and quickly made without adding to press downtime, it is profitable to produce short-run lots of parts and thus eliminate costly inventory problems.

Simplicity in design allows the average press-room mechanic to make the templates and mount the unitized units, thus releasing skilled tool designers, tool-and die-makers, and highly skilled setup personnel to work on complicated dies that require constant maintenance.

Corner-notching units, using the same "unitized" designs as those of hole-punching units, are available to work in combination with them. Costly secondary operations are thus eliminated by combining the punching and notching operations. Corner notching, edge notching, vee notching, and coping of

Figure 2U-3. Universal mounting plate with thin-gage mounting template

angles are all operations that can and should be combined with hole punching.

Units are available for almost every conceivable sheet-metal operation. Lancing, forming, embossing, coining, marking, nibbling, blanking, and extrusion operations have all been designed into unitized units, proving their extreme versatility.

The bedrails that accept the templates and the ram plates are provided for all popular makes of press brakes. These two items are used to increase the ram and bed "working" area of a press brake, converting a single-purpose forming machine into a dual-purpose forming and punching machine. This "working" area may be further increased by adding angle supports to both the ram and the bed.

In some instances, when the cost of a drilled template cannot be justified for a "one-shot" job or an extremely short prototype run of a new part, the units may be mounted on a bedrail in the press brake. The units are adjustable from front to back, as well as from left to right, and are fastened to

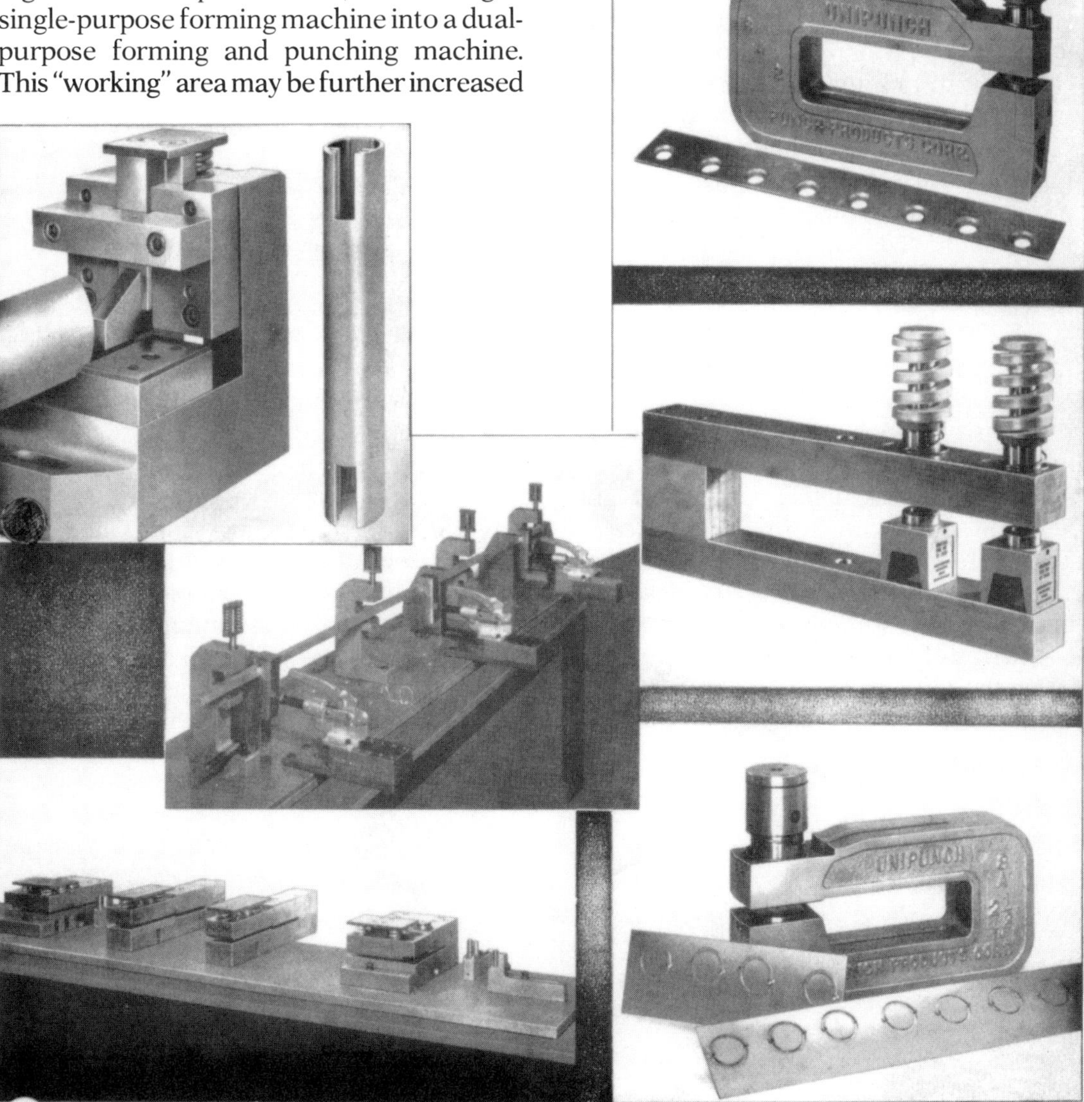

Figure 2U-4. Applications of unitized tooling

the rail by bolts in the T-slots of the bedrail.

Ram plates, T-slot bolster plates, and universal mounting plates are available for use in all popular styles of punch presses as well. A universal mounting plate is used with the "Thinplet" (thin-gage mounting template) system (Figure 2U-3).

Mounting accessories, such as T-bolts, adjustable stops, spring-loaded stops, pickup gages, feed rails with magnetic holddown, and even wrenches, help to round out the system.

For maximum efficiency, the punching and notching units, templates, and mounting accessories are stored adjacent to the press. The hole-punching units are stored on shelves, completely assembled (less the punch and die). Interchangeable punches and dies are assembled into the units as required when the setups are made. A cart adjusted to the bed height of the press is convenient for use in assembly and handling of setups.

APPLICATIONS

The actual applications of unitized tooling are as numerous and as varied as the entire range of sheet-metal operations (Figure 2U-4). Standard hole-punching units are available for punching any thickness of material from shim stock to 3/4-in.-thick boiler plate. Again, the quick-change die buttons allow many different die clearances to be used within the same unit.

Different types of materials, such as stainless steels, cold rolled steels, hot rolled steels, spring steel, brass, copper, titanium, aluminum, etc., are also punched or notched efficiently and profitably simply by varying the die clearance in the inexpensive, quick-change die buttons.

Punches are also manufactured from high speed steel for applications where the material being punched is very tough or abrasive, causing extreme punch wear. Stainless steel kitchen equipment is a typical example of such an application.

These same punches may also be manufactured from type S-7 shock-resistant steel for use on very tough, hard, tenacious material such as rerolled rail stock. This material is very popular in the bed and furniture industry.

The appliance, automobile, and farm-implement industries take tremendous advantage of the "reusable" feature of the unitized tooling system. Model changes from year to year demand flexible, versatile, low-cost, and readily available tooling. Tooling obsolescence, once considered to be part of the cost of a product, is now considered to be a factor in cost reduction.

The very nature of the unitized tooling design lends itself very readily and easily to "special" or "semistandard" units. Operations such as setting of electrical knockouts, extrusion of holes for tapping, piercing of holes and slots in steel or aluminum tubing, and notching and coping of steel and aluminum angles and channels are all easily performed with this highly versatile tooling.

Section 3
Plastics for Tooling

3A: Plastics for Tooling

Plastics today are no longer serving as substitute materials in the fields of tooling and prototype parts, but are being used as engineering materials in their own right, and in many applications they are doing a better job than any other material could do. Some authorities claim that the rapid advancement of plastics tooling would have been impossible without the versatility of the epoxy resins. These resins can be combined with various types of modifiers and fillers to create formulations having physical properties tailored for specific applications.

The major uses of plastics for tooling currently include prototype parts, mockups, model duplications, spotting racks, templates, fixtures, drill jigs, spray masks, foundry patterns, core boxes, and molds and dies for producing prototype plastic and sheet-metal parts.

Plastics tooling offers several advantages over conventional metal tooling. These include:

1. Lower cost, often less than half
2. Less time, generally a matter of hours or days rather than weeks or months
3. Light weight, easily handled and used
4. Easy modification for design changes and rework.

MATERIALS FOR PLASTICS TOOLING

Successful plastics tooling requires the use of proper resin formulations and reinforcing materials. Fiberglass cloth is one of the main reinforcements for resins used in plastic tools and prototypes. Nearly all tooling resins are modified with diluents, wetting agents, flow-control agents, and fillers. Liquid epoxy resins are by far the most popular materials for tooling formulations. The most common liquid epoxy resins are produced from bisphenol-A and epichlorohydrin. These resins are hardened by the addition of a curing agent which cross-links the molecules to form a rigid thermoset polymer. There are quite a few different types of curing agents, but the aliphatic amines are the most widely used for room-temperature applications.

Because an exact ratio of resin and hardener must be maintained for proper chemical reaction, the materials should be weighed on accurate scales. Diluents, flexibilizers, and other modifiers must also be added in exact proportions in order to obtain the proper results. For the more commonly used resins, modifiers, and hardeners, it may be desirable to have automatic metering and dispensing equipment to provide the correct ratio of materials. Machines are available which will handle as many as six different materials and dispense them singly or in any combination.

Various powdered fillers are used in resin formulations for plastics tooling applications. A typical filler is atomized aluminum powder. Aluminum oxide, silicon carbide, and iron powder are used in tooling formulations. The resin formulation must be mixed and blended thoroughly. There are plastics tooling materials companies which formulate epoxy compounds and supply them in premeasured pint and quart units along with the proper amount of hardener in small cans.

FABRICATION METHODS FOR PLASTICS TOOLING

Equally as important as the selection of the proper resin-and-hardener formulation is the method employed in the fabrication of the tool or part. Even the best of materials cannot produce good results without correct fabrication procedures. The two most common techniques for plastic tool fabrication are laminating (building up alternate layers of resin and fiberglass cloth) and casting (pouring the resin in a solid mass).

In the fabrication of a plastic tool or part, it is important to recognize that the end product is only as good as the original model. Surface details as minute as wood grain and fingerprints can be reproduced in plastics. Models can be made of wood, clay, plaster, metal, or any other material as long as the desired accuracy and surface finish are obtainable. A mold can be made from a model using plaster, plastic, or other suitable material. Here again, surface finish and dimensional accuracy are of prime importance. A plaster mold must be sealed, and nearly all molds should be polished with wax and sprayed with a release agent to prevent the plastic from adhering to the mold.

The first step in producing a plastic tool or part is to apply a specially formulated surface-coat or gel-coat resin to the prepared mold. This coating material should be flowed evenly over the surface, taking care to cover the mold uniformly without air bubbles. This thin layer of plastic should be allowed to harden for 30 or 40 minutes so that it will be firm enough to prevent the backup material from coming through to the surface. At this time, a suitable casting-resin formulation can be poured into the mold, or a fiberglass laminate can be made behind the surface coat.

If a fiberglass laminate is to be made, the cloth is precut to size and the appropriate laminating-resin formulation is mixed. A layer of resin is brushed onto the surface coat and is followed by a piece of fiberglass cloth. Alternate layers of resin and fiberglass are applied, taking care to wet the cloth thoroughly, until the desired thickness is attained. Twelve layers of cloth produce a laminate about 1/4 in. thick.

After the fiberglass laminate has been completed, it is usually reinforced or braced, such as in egg-crate fashion with fiberglass sheet stock or other suitable material. The braces are bonded into place with a resin-paste formulation. After the resin has hardened overnight at room temperature, the part can be removed from the mold, and the parting agent can be washed off. The model duplication has the accuracy of the original and has greater durability and dimensional stability in most cases.

When desired, fiberglass laminates can be filled with a casting-resin formulation to produce a solid tool. In many cases, an aggregate filler may be used to reduce shrinkage and the amount of resin required. By using the proper resin-and-hardener formulation with appropriate fillers, extremely large castings can be made successfully.

APPLICATIONS OF PLASTICS TOOLING

The many advantages of epoxy tools, such as ease of fabrication, durability, and accuracy, have promoted widespread use of these versatile materials in a great many applications. The uses of epoxy formulations in producing prototype parts for design and appearance evaluation are illustrated in Figures 3A-1 to 3A-6. The cast prototype nameplate emblems shown in Figure 3A-1 are representative of the ease with which plastics of

Figure 3A-1. Cast epoxy prototype emblems and mold

Figure 3A-2. Cast epoxy prototype trim ornament and mold

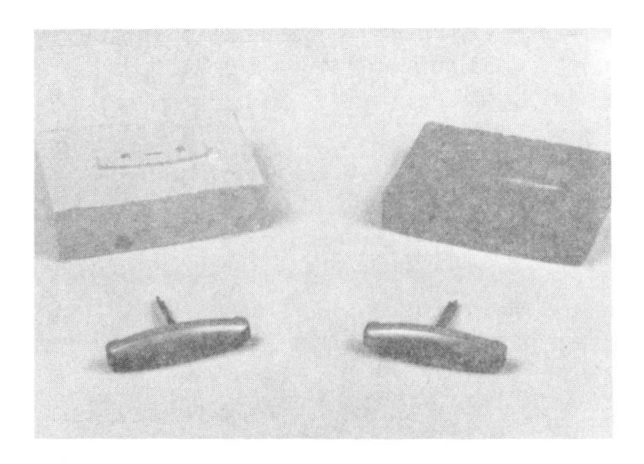

Figure 3A-3. Cast epoxy prototype upholstery buttons and mold

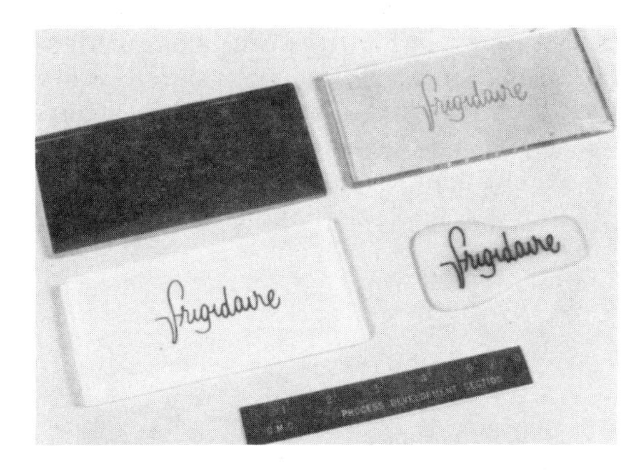

Figure 3A-4. Clear cast epoxy prototype emblems

Figure 3A-5. Painted epoxy-fiberglass laminated prototype panel

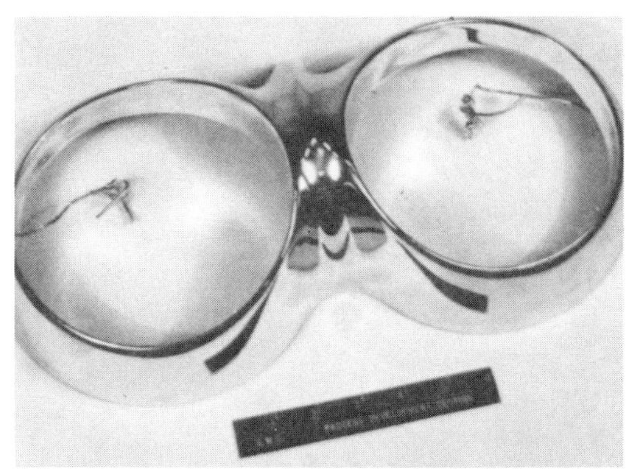

Figure 3A-6. Chromium-plated epoxy headlight bezel

various colors can be poured into a simple epoxy mold.

In producing trim ornaments (Figure 3A-2), two-piece epoxy molds often are required in order to duplicate the part completely. Combination rubber-and-plastic molds can be used to advantage in producing many types of parts, such as the upholstery buttons shown in Figure 3A-3. Metal inserts and clips can be incorporated easily into plastic prototypes by simply inserting them into the mold before the plastic is cast. A relatively new type of epoxy resin allows casting of parts which are water clear and simulate the appearance of injection molded acrylic. The prototype emblems shown in Figure 3A-4 were cast into a rubber mold and then decorated by painting and vacuum metalizing.

Large prototype panels that simulate sheet metal can be made by laminating fiberglass with epoxy resin. Figure 3A-5 shows an epoxy-fiberglass panel which has been painted to give it the appearance of a production part. Plastic parts can also be chromium plated by first coating the plastic surface with a silver paint to make it conductive. The part can then be run through the regular copper-nickel-chromium plating process. The epoxy headlight bezel shown in Figure 3A-6 was chromium plated to give it the appearance of an actual metal part.

Epoxy-fiberglass laminates also are used successfully for model duplications for use in producing spotting racks, fixtures, gages, dies, and other tooling. Figure 3A-7 shows model duplications made by the laminating technique with appropriate reinforcement. These models are accurate duplications of

Figure 3A-7. Epoxy-fiberglass model duplications

Figure 3A-8. Plastic-faced plaster model duplication

the original models and allow the originals to be kept safely. As many model duplications as necessary can be fabricated for tooling purposes. When lower-cost model duplications are required, a plastic-faced plaster fabrication can be made, as illustrated in Figure 3A-8. The use of an epoxy face coat combines the durability of plastic with the low cost of a plaster backup.

Plastics are also a valuable means of producing templates and fixtures of various types. Figure 3A-9 shows an epoxy-fiberglass overlay template made from a plaster model. The small assembly fixture shown in Figure 3A-10 was made by pouring an epoxy formulation into a rubber mold of a wood model. A steel locator became an integral part of the plastic fixture because it was placed into the mold prior to casting. Figure 3A-11 shows a cast epoxy fixture which is used to hold a part for buffing and polishing. Another cast epoxy holding fixture, used in stud driving, is shown in Figure 3A-12.

The ease of duplication in plastics lends itself to foundry applications. The cast epoxy pattern insert shown in Figure 3A-13 is one of several that are mounted in a standard jolt machine to produce sand molds. Pattern inserts are made by laminating and/or casting into a mold made from a master pattern. Figure 3A-14 shows an epoxy-fiberglass foundry pattern made in a laminated mold taken from an aluminum pattern. Epoxy resin formulations are also suitable for core boxes and core-box inserts. An epoxy core box is shown in Figure 3A-15, along with a plastic core master. Cast epoxy core-box inserts are shown in Figure 3A-16. Here again, the use of plastics eliminates considerable machining in producing multiple duplicates.

Epoxy resin formulations also can be

Figure 3A-9. Epoxy-fiberglass overlay template

Figure 3A-10. Epoxy assembly fixture with model and mold

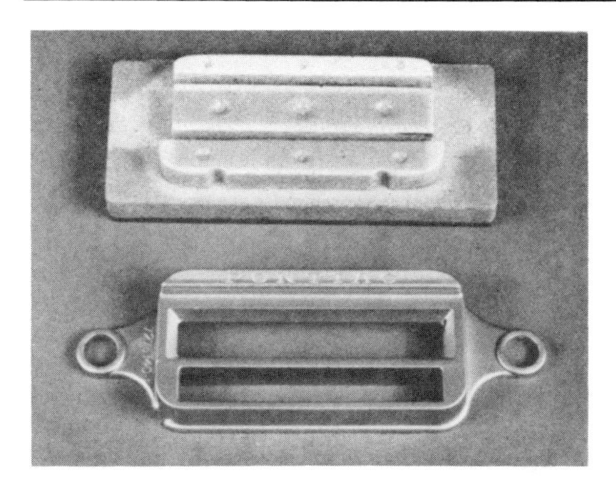

Figure 3A-11. Cast epoxy buffing fixture with part

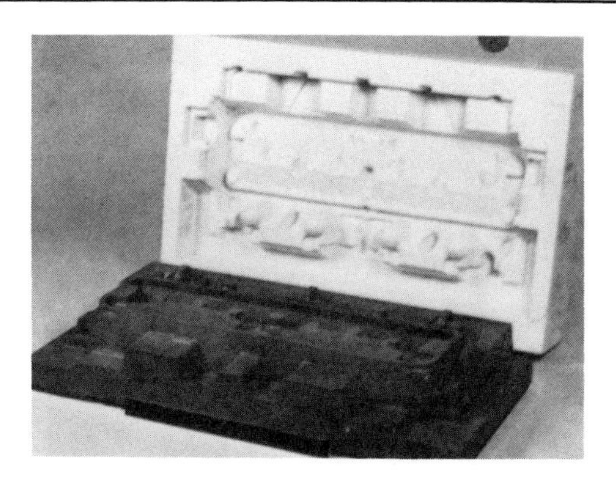

Figure 3A-14. Epoxy foundry pattern and mold

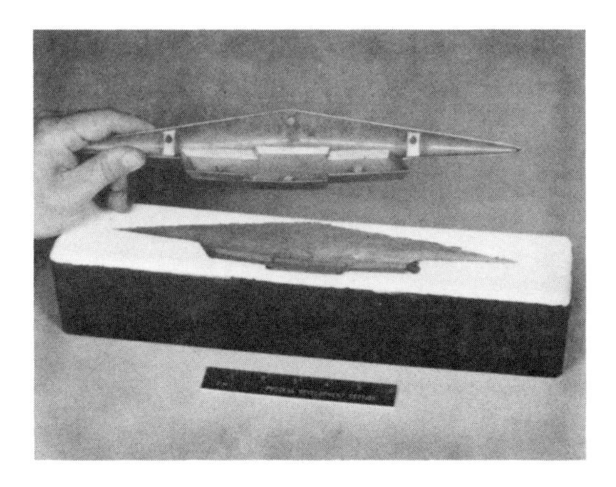

Figure 3A-12. Cast epoxy holding fixture with part

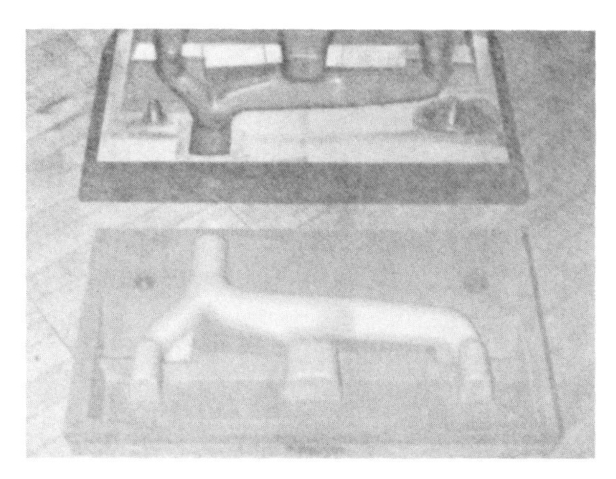

Figure 3A-15. Epoxy core box with core master

Figure 3A-13. Cast epoxy foundry pattern insert

Figure 3A-16. Cast epoxy core-box inserts

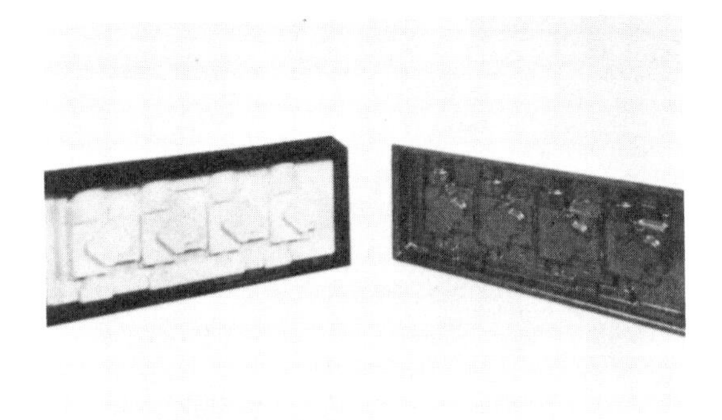

Figure 3A-17. Epoxy mold for vacuum forming of thermoplastics

Figure 3A-19. Epoxy punches for hydroforming of sheet metal

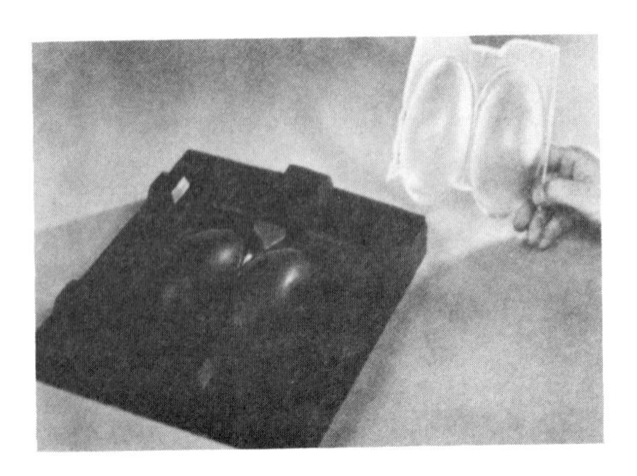

Figure 3A-18. Epoxy mold for forming of acrylic sheet

Figure 3A-20. Epoxy punch and die for forming of sheet metal

used to make molds and dies for forming prototype parts from thermoplastics and sheet metal. An epoxy mold for vacuum forming of thermoplastic sheet is shown in Figure 3A-17, along with a part formed in it. Figure 3A-18 shows a cast epoxy mold which was used to form a prototype part from acrylic sheet. Epoxy punches can be used for hydroforming of sheet-metal prototype parts, as shown in Figure 3A-19. Punches and dies for producing sheet-metal prototypes can be fabricated from epoxy formulations, as shown in Figure 3A-20.

SUMMARY

Plastics are already well established as tooling materials for many applications. As new materials and new applications are developed, the uses of plastics tooling will steadily increase. The continuing advancement in this fascinating field offers a challenge to development-conscious plastics engineers and tool engineers. The future growth of plastics tooling depends on the ability of engineers to make proper use of the materials and to avoid misapplications.

3B: Types of Plastics Tools and Methods of Fabrication

LAMINATES IN TOOLING

The most widely used method of plastic tool fabrication involves fiberglass reinforcement (in the form of cloth, matting, or chopped strands) which is impregnated with a resin system.

Most toolmakers prefer to have a smooth face on the finished tool, and this requires a gel or surface coat. Gel-coat epoxy resins are usually of paste consistency and are applied to the prepared pattern with a short-bristle brush or squeegee. Care must be exercised when applying the gel coat to avoid entrapment of air bubbles.

The application of the fiberglass reinforcement may begin as soon as the gel coat has become "tack free." This is the point when the gel coat may be lightly touched without having any of the resin stick to the finger. Once the gel coat is tack free, brush a coat of the laminating resin over the entire surface. If a layer of glass cloth were to be applied directly to the gel coat, it would, in all probability, stick to the gel coat and tear it loose. A coat of laminating resin will allow the toolmaker to move the glass cloth without ruining the gel coat.

Each application of a layer of glass cloth should be preceded by brushing on a coat of laminating resin. By applying dry cloth over the coat of laminating resin, a much more air-free laminate will be produced, because the glass cloth will act as a wick; the resin will push all air ahead of it as it wicks its way up through the glass cloth. Continue this procedure until the desired thickness is reached.

Larger laminated tools will require exterior reinforcement of some type to help retain configuration. Reinforcement selection is determined by tool use, economics, and the temperature ranges to which the tool will be subjected.

Tools for use at room temperature and at slightly elevated temperatures (up to 350 °F) are usually treated just as any normal laminated tool; however, tools which will be operating at or above 400 °F should be vacuum bagged during curing to remove all excess resin. Vacuum bagging will produce a tool with more uniform distribution of resin and glass cloth, and hence a tool more capable of withstanding thermal cycling.

CAST TOOLS

Because of cost and general brittleness, tools which are made entirely of a cast epoxy resin are usually small in size. This method of tool construction is obviously the easist and fastest way to build a tool. Simply mixing the resin and hardener and pouring the mixture into the prepared pattern can produce a tool if one or two basic principles are remembered.

Follow the resin manufacturer's recommendation for the maximum thickness to which the resin system may be cast; exceeding this maximum thickness will only cause excessive shrinkage and warpage of the tool. There are resin systems which are designed to be cast in all thicknesses. Cast resins shrink during curing, and this shrinkage occurs predominantly from the open side of the pattern. Small metal-forming dies, hydropress forming dies, chucking jaws, and holding fixtures have been successfully fabricated from cast epoxy resins.

Another widely accepted use of cast resins in tooling is in production of prototype parts. Small, intricate shapes are easily reproduced by casting. One camera manufacturer builds each prototype camera entirely of cast epoxy resins. After the prototype has been approved for production, the unit is disassembled and the parts are used for master patterns.

SURFACE CAST TOOLS

Where a solid cast tool is too expensive, or when the stress in the tool will surpass the normal properties of the epoxy resin, only the working face of the tool is made of cast epoxy.

A core is fabricated with the face 1/2 to 1 in. back from the desired contour. The prepared pattern is then spaced from this core, all edges are sealed off (except for pouring and venting spouts), and the epoxy is cast between the core and pattern to form the face over the core. Cores in the aerospace industry are usually constructed of kirksite or aluminum.

AGGREGATES

Tools which have high compressive strength and, in some cases, high heat-conduction properties are constructed with a core of epoxy resin blended with various aggregates such as aluminum, crushed rock, or other inorganic materials. A gel coat is used to impart a smooth surface to the tool, and after the gel coat has become tack free, the core material is prepared and packed behind the gel coat.

When great compressive strength is required of the tool, the ratio of resin to aggregate may be as low as two parts aggregate to one part epoxy resin. Prototype injection dies and limited-run compression mold dies are built using this ratio of epoxy resin to aluminum particles. This same type of construction has been used for years to build molds for vacuum forming of thermoplastic sheet materials. For vacuum-forming molds, an aggregate-to-resin ratio of 2 to 1 is used. This ratio will produce a core which is porous. The drilling of vacuum breathing holes is then limited to drilling through the gel coat, because the entire core acts as a vacuum chamber.

TOOLING WITH A SPLINE

Although building of master patterns with templates and splined contours is not a new practice, it has taken the epoxy resin formulators many years to develop resin systems which will have the "feel" and "drag" of plaster. Building of master patterns with the contours splined with epoxy resins offers many advantages, the greatest of which are the permanence and resistance to weathering and mechanical damage of the surface. Techniques employed in splining with epoxy resins are similar to those used for splining with plaster materials.

Epoxy pastes have been used for years to patch and repair thousands of items, from water valves to concrete blocks, but some of the newer uses of epoxy pastes are great assets to the tool builder who needs duplication of contours in limited areas.

Jig pads and contour boards formerly required many hours of hand grinding to obtain an accurate duplication of a compound surface. Use of epoxy pastes in this application involves fabricating the basic jig pad or contour board to the approximate contour and then spacing it away from the contour master. Epoxy paste resins are then forced into the resulting gap. Because it is of a paste consistency, the epoxy will not run out, and will exactly duplicate the contour.

RELEASE AGENTS

Because of the excellent adhesive qualities of epoxy resins, it is necessary to employ release agents on patterns used in plastic tooling. Of the wide variety of materials used for patterns, nearly all require either different release agents or methods:

1. **Plaster** must be thoroughly dried (8h at 150 °F) and sealed. Clear lacquer cut 50 – 50 with thinner is the best sealer. Do not use shellac. After sealing, apply a generous coat of paste wax and wipe off any excess. After the remaining wax is dry, it should be lightly buffed. Follow this with a spray coat of PVA or PVC to achieve the desired gloss.
2. **Plastic** and **metal** patterns may be treated in the same manner and usually require only a coat of wax. Silicone release or TFE-type release may also be used.
3. **Wood** is the most difficult material because of grain and gassing tendencies. Generally follow the procedure outlined for plaster patterns. If in doubt, contact your resin supplier.

MIXING OF RESINS

Misunderstanding of the difference between epoxy and polyester resins has caused many tool failures. Polyesters start to cure the day they are manufactured. Adding a catalyst only speeds up that which is going to happen anyway. Epoxies, on the other hand, never solidify until they are mixed with a hardener. Epoxies rely on a chemical cross-link with a hardener in a precise ratio. If the mix ratio for an epoxy is listed at 100 to 10, each molecule of hardener will cross-link with ten molecules of resin. Putting more hardener into the mix will only result in an excess of molecules with nothing to react with, and the excess hardener will remain liquid.

Thorough blending of resin and hardener is very important. Generally the density of epoxy resin is higher than that of a resin-and-hardener mixture. Because of this, any unmixed resin will settle to the bottom of the casting and cause soft spots on the tool face.

The following procedures should be used to achieve thorough blending:

1. **Hand mixing**: Using a sturdy paddle, mix for 5 min, stopping at least twice to scrape sides, bottom, and stirring paddle.
2. **Power mixing**: Mix for at least 2 min at 1800 rpm with a propeller-type mixing blade. Stop during mixing to scrape sides and bottom.

DE-AIRING OF RESINS

Some castable resin systems — notably polyurethanes — require degassing or de-airing in order to achieve high-quality castings. After blending, but before gellation, the resin system should be subjected to repeated short exposures to vacuum. Three or four cycles of rapid vacuum buildup and collapse will usually de-air any resin system.

TRIMMING AND FINISHING PROCEDURES

Trimming of fiberglass laminate tools is usually accomplished through the use of a diamond rotary saw or a band saw. If a band saw is used, follow the feed and speed rates recommended for aluminum. Carbide blades are recommended. Final dressing to finished trim is usually done by rotary disk sanding followed by hand filing. It has been found that an increased clearance rake angle will aid in cooling the drill bit. Because of the highly abrasive nature of fiberglass, use of carbide and diamond tools is suggested wherever practicable.

3C: Urethane Tools for Sheet-Metal Fabrication

Urethane has become a necessary and useful material in the manufacture of tools for sheet-metal fabrication. Urethane is now listed among such other tooling materials as flat ground tool steel, decarb free tool steel, carbide, and the like. Urethane offers many advantages that other forming-die construction materials cannot. In order to realize the full benefit of these advantages, the fabricator must learn the proper techniques of using urethane, as well as its proper applications. Of equal importance is the understanding of the limitations of urethane materials. There are many formulations of urethane, each having different characteristics that react quite differently in tooling.

Urethane is a thermosetting elastomer that combines in one basic material the flexibility of rubber and the hardness of structural plastics. Urethane is a plastic, although it looks like, and may feel similar to, rubber. In general, urethane offers high load-bearing capacity, high impact resistance, low compression set, high tear strength and cut resistance, and good abrasion resistance. Urethane is impervious to oil and ozone.

A good description of urethane as used in forming applications is "oil in suspension" or "a solid hydraulic." At the pressures used in forming, urethanes are incompressible yet deflectable solids. By controlling movement and displacement, and keeping within the established capabilities of the specific formulation of urethane, long life can be maintained.

URETHANE WORK

Work to urethane in metal forming should be considered as the total movement required and the time taken. In the tooling industry, the best way to describe this would be a forming die which deflected the urethane 10% at a rate of 30 cycles per minute.

From an engineering standpoint, urethane should be classified by maximum allowable deflection. This is one major factor in design criteria. Durometer hardness should be used for identification purposes only.

The most common grades of urethane presently used have allowable deflection ratings of 5, 15, 25, and 35%. These ratings indicate percentage deflections allowed for maximum life of the urethane. In a 1-in.-thick piece, 5% urethane can be deflected 0.050 in., 15% urethane can be deflected 0.150 in., 25% urethane can be deflected 0.250 in., and 35% urethane can be deflected 0.350 in. In designing with urethane, the proper amount and grade (formulation) of urethane now can be determined. Because of this design approach, urethane now appears more often as a standard tooling material.

ADVANTAGES OF URETHANE

The general advantages of urethane are as follows:

1. Nonmarking characteristic eliminates tool marks, resulting poor appearance, and need for costly subsequent polishing operations.
2. Automatic ability to compensate for metal-thickness variations. Urethane self-adjusts for thickness changes caused by commercial mill tolerances, as well as changes in gage of material.
3. Eliminates heat treating of steel components. Many tools can be built with mild steel because of the reduction in material cost and tool fabrication time. Substantial reduction in abrasive wear.
4. Compensates for other variations. Automatic adjustments for various misalignments, such as ram and bolster alignment, punch and die alignment, etc. Because part of tool construction is a flexible, self-adjusting material, less time is spent in assembly.
5. Eliminates air bending. Complete pressure forming in some tools. Piece part gripped with a blankholding force prior to initial bending. Accurate and consistent flange lengths result. Reduced deformation or fluting in pressure forming throughout the stroke.
6. Reduces press or die damage. Urethane has the ability to absorb operator and press errors when press is of sufficient tonnage.

USING URETHANE ADVANTAGEOUSLY IN METAL FORMING

The use of urethane in metal-forming operations can bring the tool engineer three distinct advantages: cost savings, improved quality, and decreased lead time. More specifically, there are five areas in which urethane may be used advantageously. In terms of how the material is used, these are:

1. Where a urethane pad deflects to become a female die
2. Where urethane is used as a male punch
3. Where urethane is used as wiping blocks
4. Where urethane springs and pads are used to provide pressures
5. Where urethane is used as wear pads, clamping jaws, and fixtures.

DESIGN CONSIDERATIONS

It should be remembered that with urethane, as with any other engineering material, there are certain design considerations that must be borne in mind. These are:

1. How pressure is controlled
2. How deflection is controlled
3. That undue strain on the urethane is avoided
4. That cutting of the urethane is avoided.

TOOLING ADVANTAGES

Some of the major advantages of urethane tooling are noncompressibility, resiliency, and ability to form and hold metal without marring its surface finish. Urethane behaves similarly to a solid fluid with a memory: under force it will change its shape, but its volume will remain constant. Urethane will produce high, uniform, and continuous counterpressure under load and, when the load is removed, quickly return to its original shape.

In addition, urethane provides a unique combination of properties, including high load-bearing capacity, toughness, hardness, abrasion resistance, and oil and ozone resistance.

A SECRET OF GOOD DIE DESIGN

One of the secrets of good die design is to provide enough pressure: blankholding pressure on the downstroke and stripping or shedding pressure on the upstroke.

It is frequently impossible with conventional steel springs to generate enough pressure to positively strip, shed, or hold blanks. In blanking and piercing dies, metal holdup or jamming due to double slugs is evidence of insufficient stripping or shedding pressure. In forming dies, a crown next to the bend line, or blank slippage, is evidence of insufficient blankholding pressure.

Another problem encountered in die design is fitting the required number of steel springs into a limited area, as for example in long, narrow press-brake dies.

Urethane springs, strippers, shedders, and pressure pads provide more pressure per unit area than conventional springs — an ideal situation for the narrow confines in dies and fixtures.

Urethane springs, strippers, and shedders consist of cylindrical tubes which can be used independently or to which one or two caps can be fitted. This unique combination offers:

- High pressures
- Close center distances
- Positive stripping
- Punch vibration damping
- Simple installation
- Non marring
- No shrapnelizing.

Springs or cylinders are made of a tough, resilient, high-modulus grade of urethane that will withstand more than 100,000 cycles when properly applied. They are quickly installed or removed, requiring neither set screws nor stripper bolts. When fractured under load, these springs will not shrapnelize (fly into pieces), thus averting damage to the die and possible injury to the worker.

ENGINEERING PRESSURE/ DEFLECTION DATA

In selection of urethane springs, strippers, and pressure pads, major consideration should be given to two factors: the modulus of elasticity and the shape factor.

The shape factor accounts for urethane blocks or cylinders bulging at their sides when under a compressive load. Increasing the area that is free to bulge permits greater vertical displacement, or the same displacement with less force. Conversely, decreasing the area that is free to bulge decreases vertical displacement, or for the same displacement requires greater force.

The concept of shape factor is numerically defined as the area of one loaded surface divided by the total area of the unloaded surfaces that are free to bulge.

Dimensionally, this may be written as:

$$S.F. = \frac{lw}{2t\,(l + w)} \text{ for rectangular blocks}$$

or

$$S.F. = \frac{d}{4h} \text{ for solid disks and cylinders}$$

where l is length, w is width, t is thickness, d is diameter, and h is height.

These equations are limited to:

1. Pieces which have parallel loading faces
2. Pieces whose thickness is not more than twice the smallest lateral dimension.

The modulus of elasticity, E, is defined as the force per unit area (stress) divided by the percentage of the change in height (strain), or:

$$E = \frac{F/A}{\Delta H/H_T}$$

where F is force, A is area, ΔH is change in height, and H_T is total height.

For many of the common engineering materials, such as steels, E is a specific value that remains constant within the elastic range of the material. With urethane, however, the E value changes as the shape factor changes. Further, its value also changes with each specific compound (see curves in Figure 3C-1).

The test data for these curves was determined by Kaufmann Tool and Engineering Corp. during a two-year period. Testing was conducted under controlled conditions, and the data reflects the variation of E with shape factor for three K-Prenes with dry and lubricated surfaces. The curves represent a statistical average of the test results, and are offered as a guide to help the engineer predict forces, size and grade of urethane required, or per cent deflection.

OTHER ENGINEERING CONSIDERATIONS

Heat buildup due to internal friction (hysteresis effect) is the most common cause of premature failure of urethane. The amount of heat generated is a direct function of effective strokes per hour and/or degree of deflection per stroke. Thus, in selecting springs or pressure pads, minimize the percentage of deflection for longer life — particularly when exceeding 700 strokes per hour.

For high-speed applications (12,000 strokes per hour), select the size that will provide the least deflection, or no greater than 15% of total height. For intermittent operation (700 effective strokes per hour), it is safe to select a size with as much as 25% deflection.

For short runs or low-speed operations (200 effective strokes per hour) where long life is not important, the maximum recommended deflection of 25% can be exceeded. There is no bottom position as in steel coil springs.

Urethane can withstand temperatures up to 250 °F, although it may soften before that point and lose some load-bearing capacity. However, upon cooling, it will return to its original physical characteristics. Urethane will also withstand temperatures as low as −70 °F. If stored cold, the urethane should ideally be brought up to room temperature before it is used.

Whether the load-bearing area is lubricated or in a dry condition is another factor that affects the stress-strain relationship. For urethane compressed between parallel plates, there is a tendency for the surfaces to spread laterally. Whereas a clean dry-loaded surface offers some resistance to this lateral movement, a lubricated surface will offer essentially none. If extremely high pressures are required, lateral movement can be prevented by bonding the urethane to metal with double-faced tape or some other type of adhesive.

The cut resistance of urethane is very

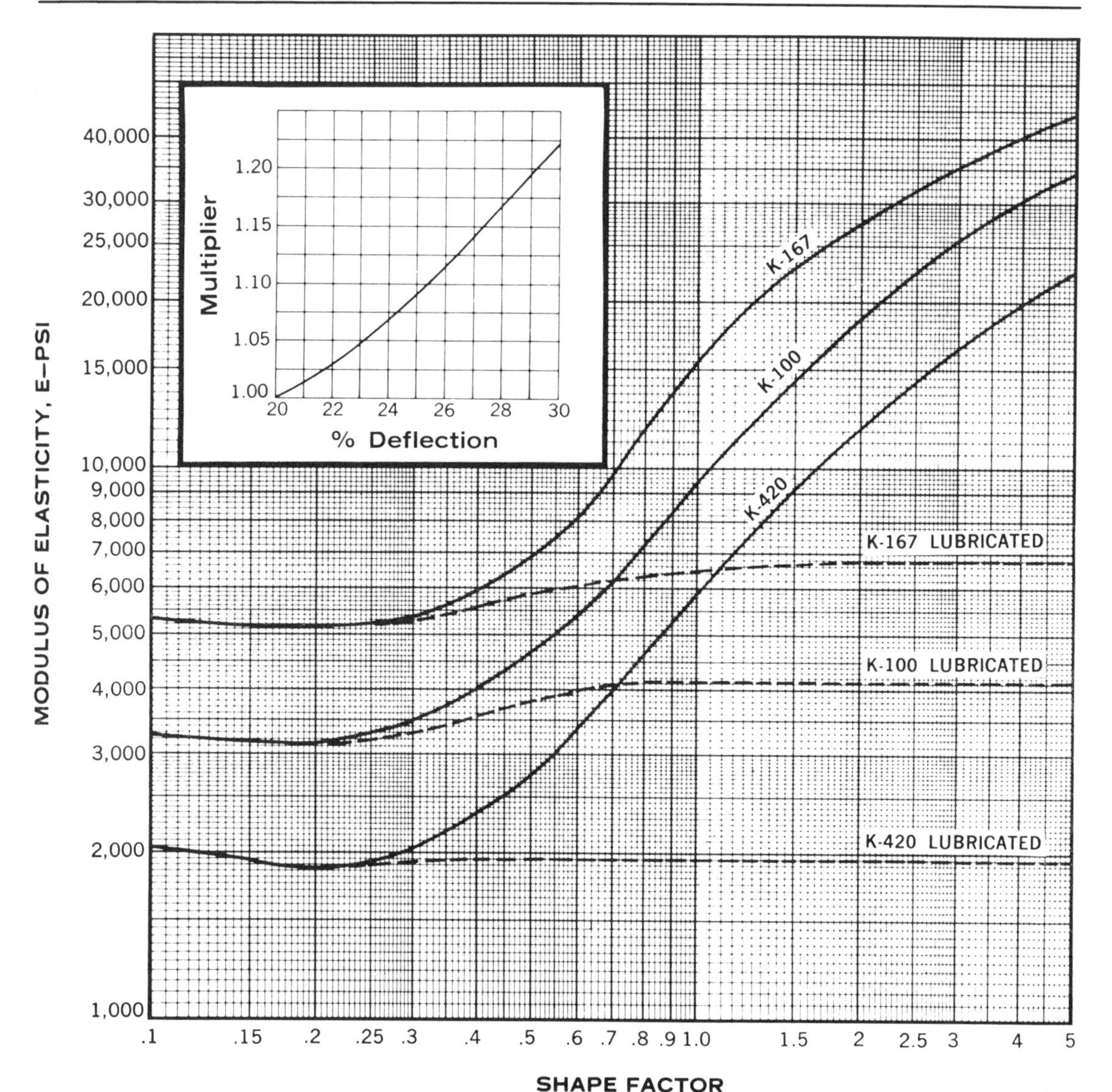

Figure 3C-1. Variation of modulus of elasticity with shape factor (inset curve is for deflections greater than 20%)

high. However, placing it near a sharp metal edge or permitting it to bulge over a sharp edge should be avoided. With the forces involved, the urethane may fail due to cutting or fracturing.

A final factor that the engineer should consider is permanent set, which, like heat buildup, is a function of the percentage of deflection. If a small amount of permanent set does occur, the engineer can compensate for it by shimming the urethane.

HOW TO USE THE CURVES IN FIGURE 3C-1

In solving problems, the engineer should first determine the shape factor using the formula previously introduced, and assume a particular size of urethane pressure pad.

Having once determined the shape factor, the engineer can then select the appropriate grade of urethane, and come horizontally across on the chart in Figure 3C-1 to find E.

With the modulus of elasticity known, the basic formula for E can then be applied to solve for either force or per cent deflection. That is:

$$F = \frac{\Delta H}{H_T} AE^* = (\%\ \text{Deflection})\ AE^*$$

$$\%\ \text{Deflection} = \frac{\Delta H}{H_T} = \frac{F}{AE^*}$$

Example, Problem A

A forming job on a 15-ton press requires a pressure pad which will exert no less than six tons at the bottom of the stroke. It cannot exert more than 11 tons of force, because four tons will be used for the actual forming. Die design has fixed the height of the pressure pad as 2 in., the amount of deflection as 3/8 in., and the space available for the pad as 4 by 7 in. For blankholding near bend lines, force must be maintained near the periphery of the pressure plate (same size as die opening). Thus, the pressure pad can be no smaller than 2 by 4 in.

Determine the size and grade of the urethane pressure pad.

Assuming that the largest pad that will fit into the given die opening, allowing space for bulging, is 3 by 6 in., we solve for the shape factor:

$$\text{S.F.} = \frac{\ell w}{2(t)\,(\ell + w)} = \frac{(3)(6)}{2(2)(3 + 6)}$$

$$= \frac{18}{36} = 0.5$$

With F_{max} = 11 tons = 22,000 lb, then:

$$E = \frac{F/A}{\Delta H/H_T} = \frac{22{,}000/18}{0.375/2} = 6{,}520 \text{ psi}$$

On the graph we find that the junction point of E and S.F. falls between the curves for K-167 dry and K-100 dry. Therefore, we must select K-100 and check that the pressure does not fall below the six-ton minimum. Note that if we used K-167, we would exceed the 11-ton limit. For K-100 a S.F. = 0.5 and E = 4,750 psi:

$$F = \frac{\Delta H}{H_T} EA = \frac{0.375}{2} (4{,}750)(18)$$

$$= 16{,}000 \text{ lb}$$

The tonnage at the bottom of the stroke, therefore, is about eight tons, which is satisfactory.

Example, Problem B

If a change in die design permitted a pad 1½ in. thick, what grade of urethane should be used?

$$\text{S.F.} = \frac{(3)(6)}{2(1 - 1/2)(3 + 6)} = \frac{18}{27} = 0.667$$

$$E = \frac{F/A}{\Delta H/H_T} = \frac{22{,}000/18}{0.375/1.5} = 4{,}900 \text{ psi}$$

Because the deflection in this case is 25%, we find the multiplier for E from the curve in the upper left-hand corner of Figure 3C-1. For 25%, this is 1.09. Therefore, E = (4,900) (1.09) = 5,340 psi.

On the graph we find that the new junction point for E and S.F. falls between K-100 dry and K-420 dry. Thus, we would choose K-420. Again checking the bottoming force:

$$F = \frac{\Delta H}{H_T} EA = \frac{0.375}{1.5} (1.09)(3{,}900)(18$$

$$= 19{,}000 \text{ lb}$$

We find that our bottoming force has now increased to almost 10 tons, which is still between the six-ton and 11-ton limits.

BASIC URETHANE TOOL DESIGNS

There are six basic designs for urethane tools. In addition, urethane is now being

* For deflections greater than 20%, E must be modified by the multiplier derived from the small graph in the upper left-hand corner of Figure 3C-1. For deflections equal to or smaller than 20% this multiplier is 1.

used in the industry as a shock absorber and noise reducer. This use is a built-in advantage in all urethane dies.

Female Die Pad

This is the first and most widely used concept in urethane dies. The urethane is used as a nonshaped female (Figure 3C-2). Placing the urethane in a box provides a control on pressure. The air space provides a place for the urethane to "flow." The urethane moves into the air space, reducing deflection. This effectively reduces the strain in the urethane, which increases its service life. Although this system proves that urethane is a useful tool and die material, it is the one concept that is most abused. The work requirement and cutting of the urethane under stretching strain can be very high. Required tonnage is from one to three times that of conventional air bending. Still, this system is widely used because of the advantages it offers, even considering the possible abuse and attendant replacement of urethane. More than one male punch can be used with this pad. Blank slippage is virtually eliminated. Use of a shaped urethane female die pad reduces the strain in the urethane and gives far longer service life. The tonnage required drops to about that of conventional dies. However, the shaped pad no longer represents a universal type of female pad, and

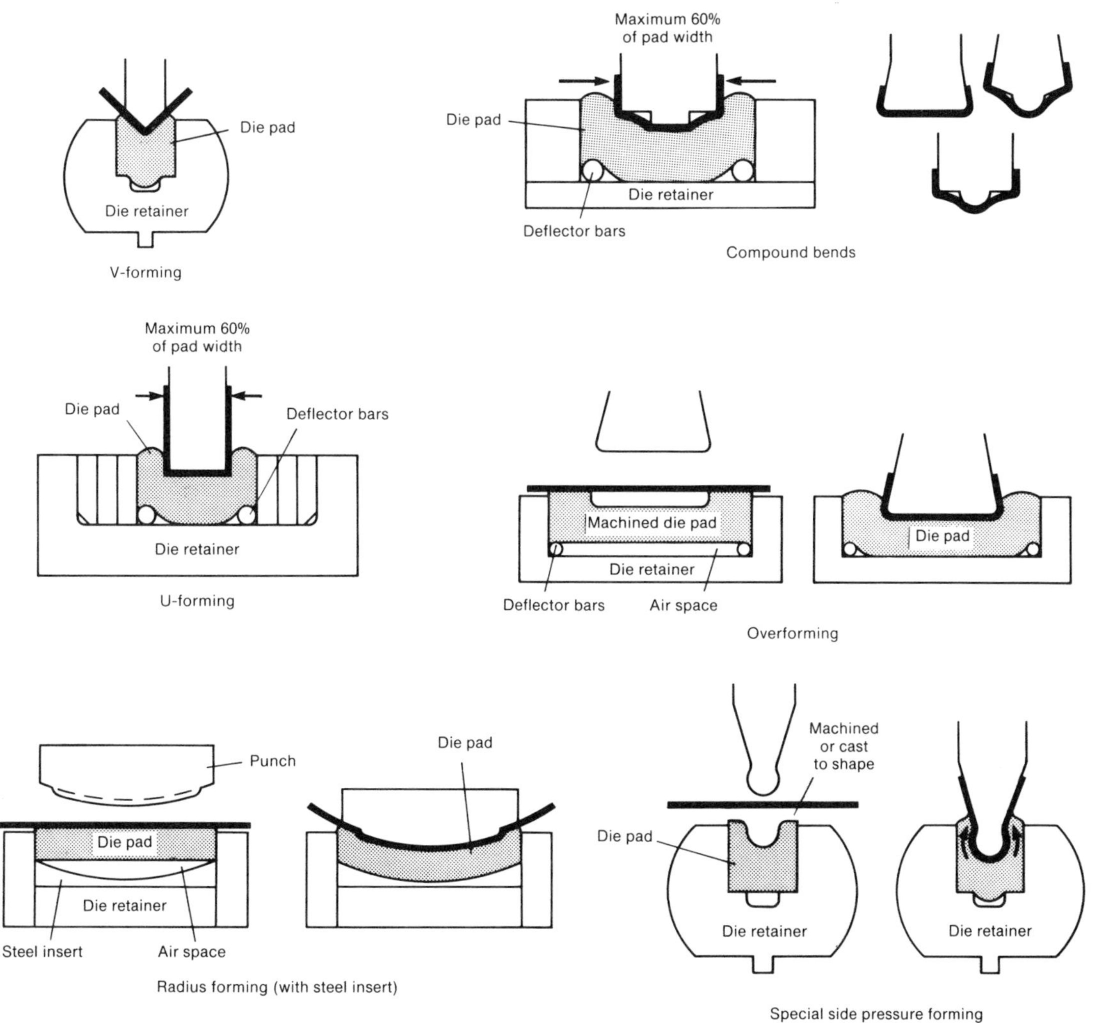

Figure 3C-2. Female dies formed by deflection of urethane pads

the type of forming and the production requirements will determine the value of the shaped pad.

Male Punch

The development of urethane for use as male punches has opened up many new applications for which urethane previously could not be employed. The male molded or machined urethane punch (Figure 3C-3) is usually bonded to a steel plate. This bonding retards uncontrolled side movement of the urethane. The shaped male punch has enough rigidity to form metal into steel or metal-working epoxy cavities and, at the bottom of the stroke, to produce "bulge forming." The urethane, normally 5 to 15% deflectable, has to deflect about 1 to 5%. The work acting on the urethane is so slow that long life results. The male punch may not always be totally shaped to the female die. For bulging of a tubular shape, the urethane is round with no other shape. The piece part is placed over the urethane. A plunger depresses and displaces the urethane until the bulging of the piece part into the shaped female die is completed. Retraction of the plunger releases the urethane to its original shape, ready for the next

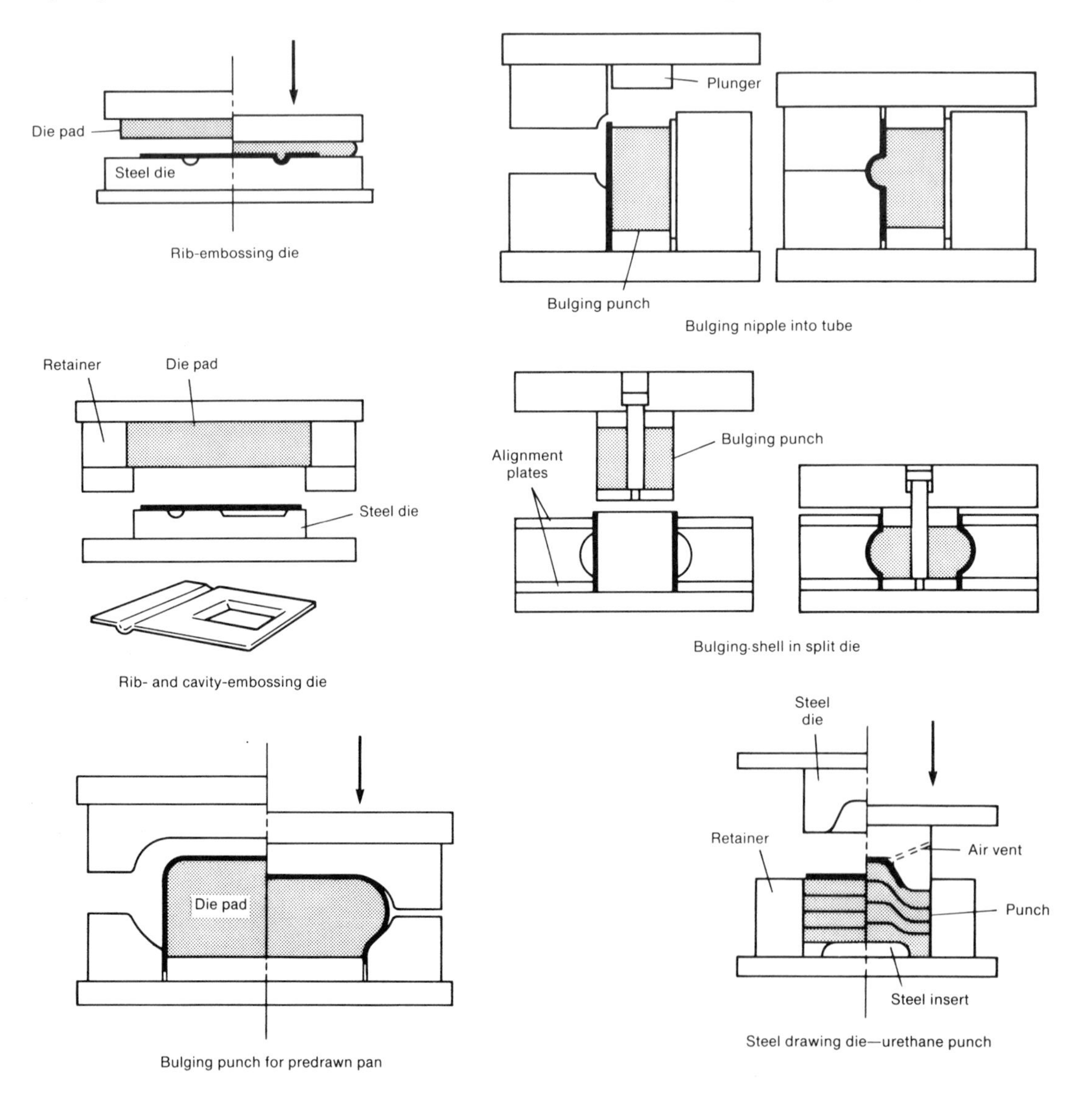

Figure 3C-3. Urethane male punches

cycle. "Leakage" of urethane under strain normally causes cutting or erosion at the top edge nearest the plunger and at the bottom edge of the die. By layering less-deflectable grades in these two areas, cutting and erosion can be minimized. The metal thickness and size of the part determine the amount (thickness) and grade (per cent deflection) of urethane to use.

Wipe Dies

The harder grades of urethane, 5 to 15% deflectable, are used as wipe edges of forming dies. Back tapering of the punch can be done to allow overflowing to compensate for springback of material. A wipe die using this principle will produce nonmarked, dimensionally accurate, consistant piece parts. Material up to 14 gage can be formed by this method. In forming of 16-gage and 14-gage material, the inside radius should be a minimum of 1½ to 2½ times metal thickness; otherwise, premature failure of the urethane due to overdeflection will occur. For forming 16-gage, 14-gage, and heavier materials with small inside radii, a combination wipe die should be used. In this die, tool steel or aluminum bronze wipe edges are backed up by a thin urethane section. The metal wipe edges can withstand the abrasive and high-load-bearing requirements, while the urethane provides adjustable die clearance. The deflection required of 15% deflectable urethane is approximately 5%, whereas for 5% deflectable urethane it would be 1 to 1½%. This approach has been used in high-production progressive dies.

Springs and Pressure Pads

Urethane rounds, tubes, bars, blocks, sheets, or slabs offer the ability to produce from one to five times the pressure provided by conventional die springs in the same space. The necessity of spring pocket holes is usually eliminated. Long life is achieved as long as maximum deflection limits and speeds are observed. Urethane springs and pressure pads will not shrapnelize and cause damage to tools, press, or operator. Stripper or pressure plates in many cases can be eliminated, because the urethane strips directly against the piece part.

Figures 3C-4 and 3C-5 illustrate the use of urethane for forming blocks and pressure pads.

Clamping Jaws and Locator Blocks

The harder grades usually 5 to 15% deflectable, are used as chunking jaws, expanding collets, and fixture-locating blocks. The urethane is rigid enough to securely grip the various parts for machining or locating, but flexible enough to accept part variations. A molded urethane part is usually made from an actual production sample, rather than being machined from a urethane section. With an irregular part, such as a casting or forging, projections would cause mislocation in solid jaws or fixtures. With urethane tools the irregularities can be absorbed into the urethane by displacement, allowing for more of the surface area of the piece part to be held.

Filler Mandrels

In tube bending, urethane has been successfully used to prevent internal collapse during forming. Channels are rolled without collapsing when urethane is used as the internal support. Removal of the urethane is possible due to its flexibility. The gage of the material, the size of the piece part, the configuration, and the amount of forming to be done will determine whether or not urethane is practical.

Shock Absorber and Noise Reducer

Urethane can absorb up to 60% of the shock applied against it. When urethane is used primarily for noise and shock absorption, urethane design criteria should be considered closely.

In a press that has considerable wear, urethane could be used as a built-in preload to reduce the shock during production runs. Tools requiring high tonnage in a short stroke would cause quick removal of all the play in the various press components, such as gibs, cranks, gears, bearings, etc. In modules, urethane could make contact with the ram 1/2 in. before the required tool tonnage is reached. This slow buildup of the "hydrau-

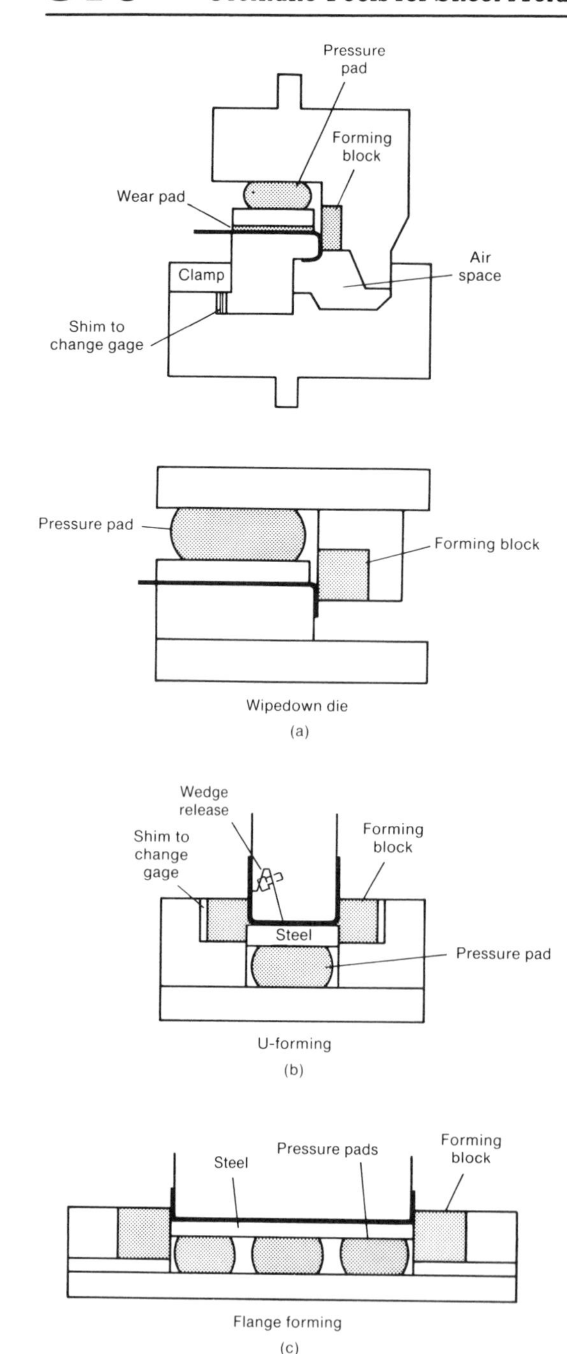

(a) Press-brake application. Urethane functions as a precision forming punch, with a different grade used as a pressure pad to clamp the piece part. Illustration shows part after second stroke. (b) Punch-press application. Simple setup showing one-stroke wiping action. (A die set also can be used.) (c) Two forming blocks are used. The steel bar on the pressure pads maintains bottom flatness of the piece part. Same setup can be used for a wide range of metal gages by shimming behind forming blocks. Wedge release is optional.

Figure 3C-4. Urethane forming blocks

lic" pressure from the urethane would load all the components of the press, so that the play would be removed prior to the tool tonnage requirement. In piercing, cutting, and blanking tools, there would be a backup, reducing the "break-through" noise by the ram releasing its play downward. On the upstroke, the pressure from the urethane against the ram would be diminished in the longer part of the stroke.

Urethane can be used in place of solid designed materials in assembly machines operating at high cycling rates to reduce noise levels up to 40% and to help reduce maintenance costs.

Urethane gage blocks, part chute linings, and guides will reduce shock and noise levels along with piece part damage. Each application should be examined thoroughly to fully utilize the advantage of urethane.

COMMON URETHANE FORMING PROBLEMS AND THEIR SOLUTIONS

Cutting, caused by sharp edges or burrs on the piece part, is the most common problem. Because these sharp features cannot be eliminated, they must be accounted for in design. Wear pads consisting of thin urethane sheets can be placed over the main urethane die pad to take the cutting. Such a sheet is easily replaceable at a cost far lower than that of the main die pad. Segmenting the urethane die pad will keep a cut that has started in an isolated area. Although this one segment may have to be replaced periodically, the need to replace the whole section is greatly reduced. Sharp edges of the tool itself should be rounded to eliminate cutting caused when the urethane is under high strain in stretching. It should be noted that urethane will react in much the same way as a rubber band being cut while under full extension or a balloon being popped by a pin at full inflation.

Variations in the desired angle of bend or the radius of a piece part are caused by two factors: changes in temper of the material, and inconsistent bending as a result of fixed

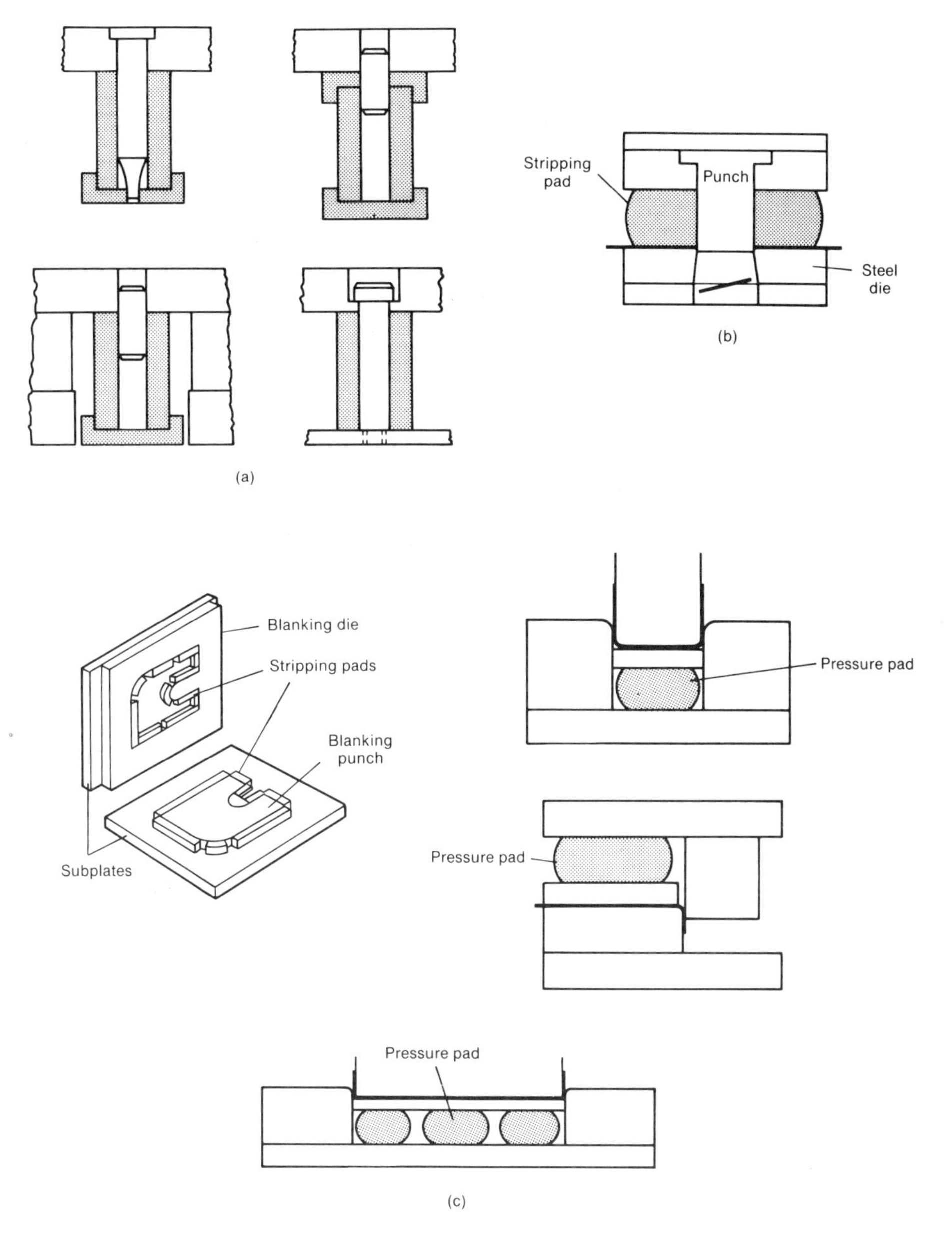

(a) Springs. (b) Stripping pad for piercing dies. A Standard urethane pressure pad machined to fit around the punch. This is an easy, economical method of constructing a durable, oil-resistant, high-pressure stripping pad. (c) Stripping pads for blanking dies. Urethane strips are fastened (with adhesive or double-faced clamping tape) around the punch and in the die cavity to function as reliable strippers. For use with template, steel rule, and conventional blanking dies. (Die illustrated is usually used with master die set.)

Figure 3C-5. Urethane pressure pads

die clearances which do not account for changes in stock thickness. Temper variations should not be considered a major problem. The springback caused by temper changes can be adjusted for by making changes in the angle or size of the male punch. Once this adjustment has been made, the springback will be consistent.

Metal-thickness variation is the greatest cause of angle-of-bend errors. The clearance between the forming punch and the urethane should be set for the minimum possible stock thickness. The die will then automatically adjust, if the material thickness increases, by additional deflection of the urethane. This will create additional side force. Calling will not occur, as it does in conventional dies. As much as 80% of the error in angle of bend can be caused by metal-thickness changes.

3D: Ceramic-Filled Epoxy Tooling

Plastics are currently used in a wide variety of tooling applications (dies, master models, etc.). Plastics have high-strength flexibility, and are easily machined. A prominent disadvantage of plastics, however, is that their normal usage is restricted to relatively low temperatures.

Ceramic materials offer high-temperature applications and high compressive strengths, but are generally brittle and porous in nature.

Ceramic additions to epoxy systems are used primarily because they are cheap fillers, they provide a simple means of obtaining a desired color, and they provide some additional thermal or electrical properties. Recent work has been done to correlate the properties of the individual ceramic materials with their effects on plastic-ceramic systems, resulting in new ceramic-filled epoxy systems with wide tooling applications. These new systems have better thermal-stability and thermal-shock characteristics; high compressive strength; and lower curing shrinkage, thermal expansion, and thermal conductivity.

A comparison of two ceramic fills on the properties of a commercial epoxy system is shown in Table 3D-1. Equal fill concentrations and grain-size distributions are assumed.

Table 3D-1. Properties of Two Commercial Ceramic-Filled Epoxy Systems

Property	Unfilled epoxy	Alumina (Al_2O_3) fill	Magnesia (MgO) fill
Compressive strength, psi	23,475	29,130	22,596
Curing shrinkage, %	0.67	0.17	0.12
Thermal expansion, in./in. · °F	42.3×10^{-6}	17.8×10^{-6}	18.9×10^{-6}
Flexural strength at 70 °F, psi	18,497	9185	7125
Flexural strength at 350 °F, psi	75	170	152

TOOLING APPLICATIONS

Hot Forming of Reinforced Plastic Pieces

Ceramic-filled epoxy tooling is currently used in the automotive and aerospace industries for hot forming of both resin-matt and "preimpregnated" products. Two approaches are used with the hot forming technique: heating of molding dies and pressurized diaphragm forming.

Heated molding dies (matched punch and die) are used to form both types of reinforced pieces.

In forming of fiberglass resin-matt pieces,

the trimmed matt is placed in the die cavity, and resin is poured along the periphery and into the cavity. On closure of the tool, the resin is squeezed homogeneously through the matt.

A telescoping die effect may be used to control flow of excess resin, or the excess may be forced into a catching trough. Curing time generally varies from 3 to 7 min depending on piece thickness.

Products of the preimpregnated-cloth type are generally formed by simply closing the tool on the trimmed material. A preforming operation may be required to prevent wrinkling and "starved" areas in complex or deep-drawn pieces. Curing of reinforced polyester pieces (with the tool temperature at about 300 °F) usually requires 1 to 2 min per layer of thickness (8 mils per layer). Reinforced epoxy pieces are normally cured at 375 to 400 °F, and require curing times of 8 to 12 min per layer of thickness.

Forming of resin-matt and preimpregnated pieces is shown, respectively, in Figures 3D-1 and 3D-2.

The pressure-diaphragm method is adapted only to forming pieces of the preimpregnated type. With this method only one tool side is used, usually a male section. The piece is formed by the squeezing action of a heated, flexible diaphragm, which is lowered onto the tool. The forming action may be facilitated by applying a vacuum from beneath the preimpregnated cloth. Curing rates are essentially the same as those of the molding-die process.

Use of the pressure-diaphragm method is illustrated in Figure 3D-3. Both molding dies and pressure-diaphragm tooling may be constructed by casting against a variety of reference materials (wood, metal, plaster, etc.), although plaster patterns are generally used. A generalized casting sequence for building matched molding dies is shown in Figure 3D-4. Reference points for control of tool alignment and piece thickness may be controlled to within ±0.010 in. Pressurized-diaphragm tooling employs only steps A and B in Figure 3D-4.

Figure 3D-2. Forming of a preimpregnated product

Figure 3D-1. Forming of a resin-matt product

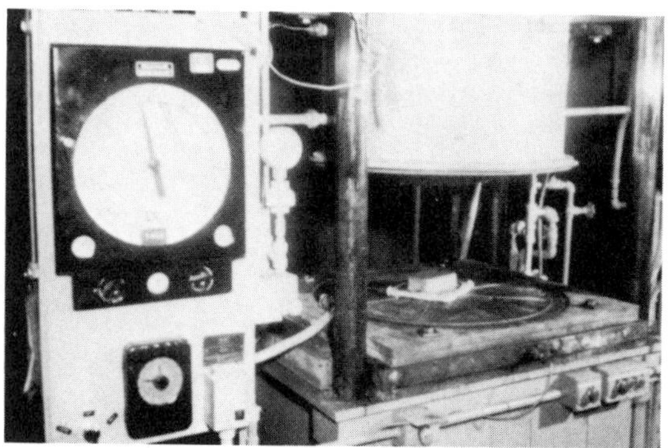

Figure 3D-3. Pressure-diaphragm method of hot forming reinforced plastic

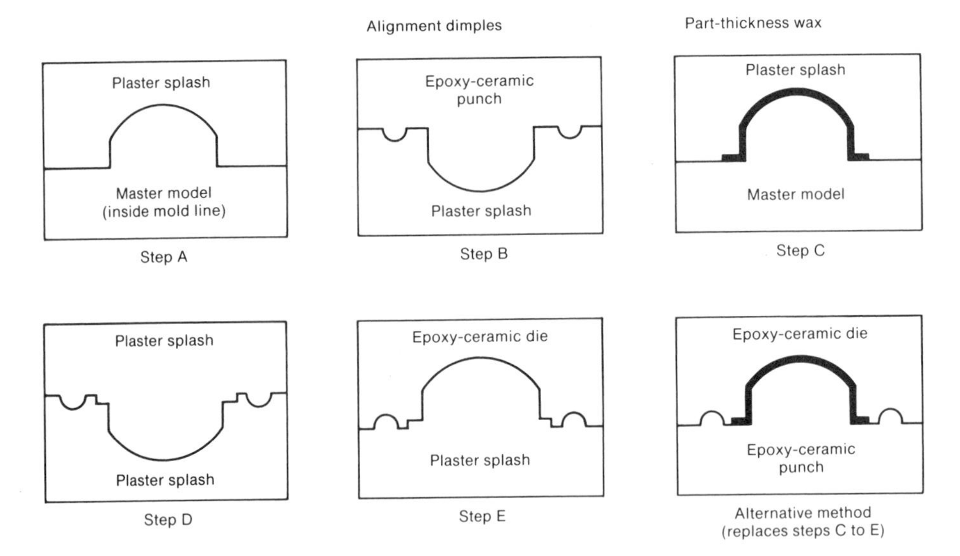

Figure 3D-4. Casting sequence for building matched molding dies

In cases where heavy loading of the tool is anticipated, additional reinforcement may be required. Molding dies are commonly reinforced with steel rods. Rod-alignment patterns may vary with model configuration. In the reinforcement pattern shown in Figure 3D-5, 5/8-in.-diam. concrete reinforcing rods are placed parallel to one another. The rods are secured by driving nails through the plywood Dyke boards into holes drilled in the rod ends. Provision for additional tooling details (embedded heaters, lifting lugs, etc.) may be made at this time.

The epoxy-ceramic mixtures may be blended by standard techniques where it is necessary to fill the resin before usage (as opposed to purchasing a system already filled). It is generally better to add the fill to the resin for better mixing. In cases of high fill content, it may be necessary to preheat both the resin and the grain to lower the mix viscosity. In preheating the resin, care should be taken to prevent excessive loss of volatile constituents. It is almost never wise to preheat the hardener, because volatization of hardener constituents is highly detrimental to cured properties of the system.

Figure 3D-5. Molding die reinforced with steel rods

Standard casting techniques may be employed with most hot forming tooling. The alternative casting sequence shown in Figure 3D-4 is not recommended where high exotherms are encountered, because the additional heat may cause softening and/or flow of the wax, resulting in changes in die clearance or piece thickness. When casting mixes are extremely viscous, as encountered with heavily filled systems, bubble-free surfaces are more readily obtained by vibrating the mix while pouring. The models in the casting operation shown in Figure 3D-6 are mounted on a vibrating table. Here the epoxy-ceramic mixture is poured into one corner of the mixed mold and allowed to "work" its way across the mold face. The vibration employed here is sinusoidal; the amplitude may be incorporated at this time, al-

Figure 3D-6. Casting operation in which models are mounted on a vibrating table to provide vibration of mix during pouring

though the tool bases are generally made flat and parallel after curing (either by grinding or by additional casting). A completed molding die is shown in Figure 3D-7.

Cost comparison figures based on 82 molding dies indicate the cost of building epoxy-ceramic tooling to be 25.7% that of Kirksite tooling for the same application. This cost saving represents labor and finished costs only. Tool repair and replacement costs are 1.0% of the "tool make" costs.

Hydropress Metal Forming

The compressive strength and wear resistance of epoxy-ceramic systems make their usage ideal for hydropress forming of metal sheet. In hydropress forming, the pressure exerted by a rubber press head forms the metal sheet to the configuration of the tool. This forming operation, although similar to the pressure-diaphragm method previously described, is normally conducted at room temperature. A production "hydroblock" and a sample part formed on it (0.025-in.-thick titanium) are shown in Figure 3D-8. "Hydroblocks" are normally constructed in the same manner as the pressure-diaphragm tooling.

Figure 3D-7. Completed molding die

Figure 3D-8. Production hydroblock and sample titanium part

Drop-Hammer Metal Forming

Sheet metal is readily formed by the stamping action of heavy drop hammers. Most drop-hammer tooling today is cast metal (usually Kirksite or steel). Cast epoxy-ceramic drop hammers, however, have shown that they can successfully form aluminum, titanium, and steel with little or no tool wear. The tool-construction cost for the epoxy-ceramic hammer die, exclusive of materials, is about one-fourth that of Kirksite. Data on tool life and repair costs for this tooling is currently incomplete.

Epoxy-ceramic hammer dies are generally constructed in the manner outlined for casting of molding dies (Figure 3D-4). The impact strength of such tooling is enhanced considerably if the casting is encased in a steel shell. An operational hammer die and an aluminum part formed on it are shown in Figure 3D-9.

Electrical-Discharge Metal Forming

Recent advances in tooling for electrical-discharge forming have incorporated the use of epoxy-ceramic materials. A liner of

Figure 3D-9. Operational drop-hammer die and aluminum part

epoxy-ceramic is molded to part configuration and encased in a steel shell. Metal pieces are formed to the linear configuration by the shock wave produced by a high-energy electrical discharge. Piece configuration normally is limited only by the metal ductility.

The liner used to form tubular pieces is usually referenced on a polished aluminum mandrel. Nontubular tools may be referenced to models of plaster, metal, or wood.

A steel shell mounted in preparation for pouring the liner is shown in Figure 3D-10 (top view). Provision has been made to incorporate vacuum seals, ports, and coolant lines. Liner casting mixes are normally highly viscous and require the mixing and casting techniques described for viscous molding-die casting. A completed electrical-discharge forming die is shown in Figure 3D-11.

Figure 3D-10. Steel shell mounted in preparation for pouring of an epoxy-ceramic liner used in electrical-discharge forming

Because this application is a relatively recent one, cost data are very limited. A review of the data available, however, indicates epoxy-ceramic fabrication costs (exclusive of materials) to be about one-fourth those of similar steel tooling.

Figure 3D-11. Completed electrical-discharge forming die

SUMMARY

Ceramic-filled epoxy systems may be used to good advantage as tooling materials for both metal and reinforced-plastic forming operations.

Epoxy-ceramic tooling may be cast readily from a variety of reference materials using standard techniques. Fabrication costs are noticeably lower than those of metal tooling for similar applications.

3E: Plastics Molds for Thermoforming

THERMOFORMING

Thermoforming generally refers to low-pressure forming methods for thermoplastic sheet materials. The sheet is made to conform to the shape of the mold by various combinations of heat, vacuum, air pressure, and positive tool pressure. In thermoforming, a thermoplastic sheet is clamped in a frame and is heated until it becomes pliable and formable. Vacuum or air pressure is then used to shape the sheet to the contours of the tool, and the sheet is cooled while being held against the tool to retain this shape.

There are numerous variations of the techniques used in thermoforming, but the following are probably the most important.

Vacuum Forming Into a Female Mold (Figure 3E-1)

The hot plastic sheet material is drawn into a female mold by vacuum. The tool contains numerous holes which connect the mold cavity to a vacuum source. Female molds are used where it is necessary to control accurately the outside dimensions of the part or where maximum strength is desired in the flange area.

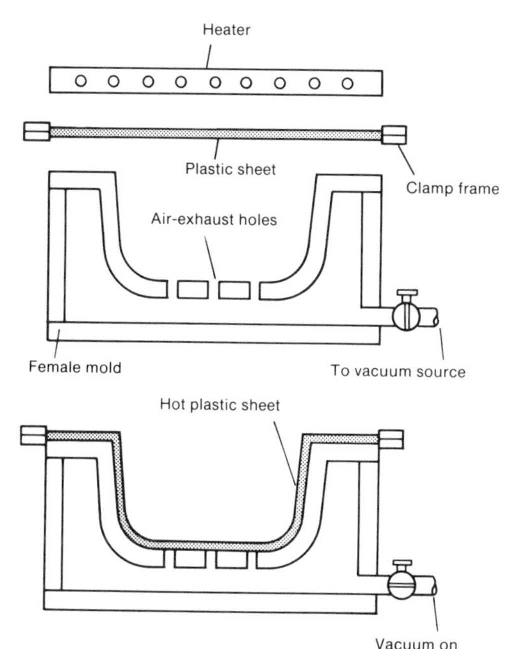

Figure 3E-1. Vacuum forming into a female mold

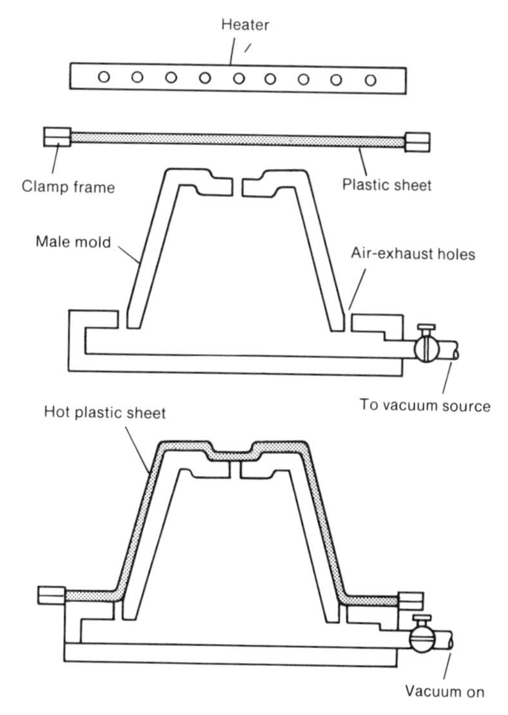

Figure 3E-2. Vacuum forming over a male mold

Vacuum Forming Over a Male Mold (Figure 3E-2)

In this technique, the hot sheet is drawn over a male mold. Fewer vacuum holes are needed than for a female mold because there is less closed cavity volume. Male molds produce parts with maximum strength in the top area and are used where the inside dimensions of the part are the most critical.

Female Vacuum Forming With Plug Assist (Figure 3E-3)

This technique is similar to vacuum forming into a female mold, except that a plug having dimensions smaller than those of the female mold is used to push the hot sheet into the mold. The plug prestretches the sheet and thins it down more uniformly as it assumes the shape of the female mold. In many cases, warm air is blown between the

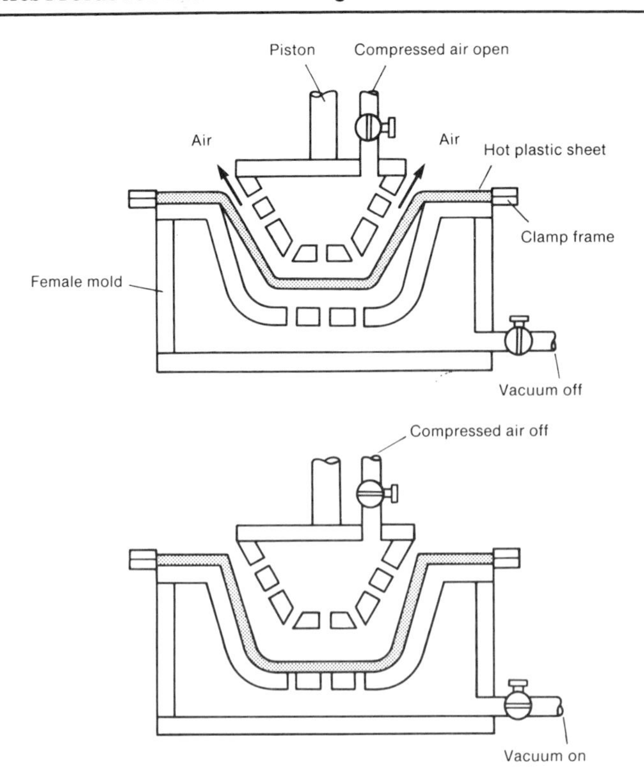

Figure 3E-3. Female vacuum forming with plug assist

plug and the sheet to keep the plug from actually contacting the hot sheet. Generally, the plug will need holes for vacuum or air pressure.

Pressure Forming Into a Female Mold (Figure 3E-4)

In this technique, a combination of air pressure and vacuum is used to draw the hot sheet into the female mold. The mold is capped with a chamber through which air can be introduced under pressure to force the sheet against the mold. To prevent trapping of air, the mold is evacuated at the same time. A similar technique can be used with male tooling.

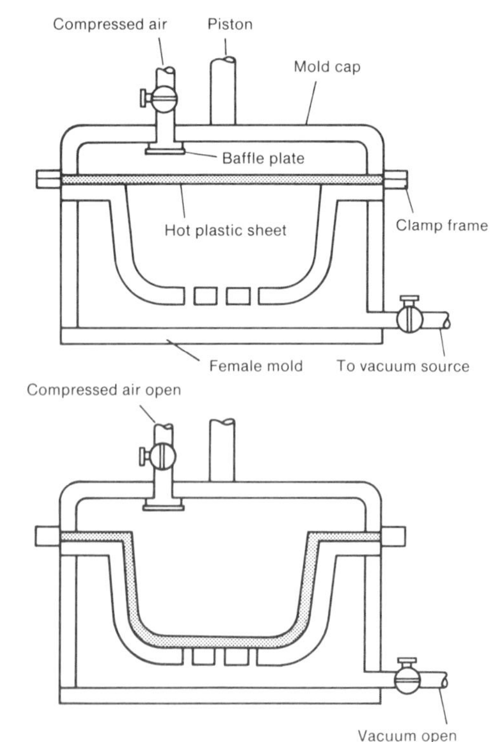

Figure 3E-4. Pressure forming into a female mold

TOOLING REQUIREMENTS

The brief descriptions of thermoforming techniques given above indicate that the tooling for this process requires consideration of the following:

1. The tool must be porous or contain holes to allow evacuation of the air be-

tween the sheet and the mold surface. In the case of a plug, the holes are used to pressurize the space between the sheet and the plug.

2. The mold surface must withstand relatively high temperatures, depending on the plastic sheet material being formed. These temperatures can range from 250 to 450 °F.
3. The tooling will have to withstand pressures ranging from 14 to 150 psi, depending on whether pressure or vacuum forming is used.
4. The mold should have sufficient rigidity to resist deformation under the temperatures and pressures to which it will be subjected.
5. Tooling should be capable of being cooled to a temperature below the deformation temperature of the plastic. Part-removal temperatures range from 120 to 170 °F. Simple air cooling is usually sufficient; however, where the cycle time is critical, it may be necessary to incorporate cooling coils or other mechanical means to rapidly lower the surface temperature.

The actual requirements for tooling will, of course, depend on the plastic being formed, the pressures involved, and the economics of the part design.

TOOLING MATERIALS

The materials selected for tooling will depend on product requirements, such as prototype or production, and, if the latter, the number of parts to be made, the economics involved, and the surface requirements of the part. Obviously, for prototype applications, molds with the lowest cost are used. For production applications, an economic balance has to be struck among mold cost, service life, and processing costs, which are, in part, determined by cycle times.

The low pressures required for thermoforming allow the use of a variety of materials. Some of the principal materials are discussed below.

Wood

Wood is inexpensive and is excellent for prototypes or possibly very short production runs. It should be used for parts with smooth contours that do not require intricate carving. The wood selected should be a hardwood with a close grain, such as birch, cherry, or mahogany. Hot plastic sheet will adhere to wood, and a heat-resistant coating, such as an epoxy paint, should be used to cover the wood. Wood has a limited life due to warping and cracking.

Plaster or Gypsum

Plaster is also low in cost, is very easy to work, and sets at room temperature. However, it abrades and cracks easily and its use is generally limited to the making of a few parts. Both wood and plaster have very poor heat conduction, which requires long cooling cycles.

Plastic

Plastic molds made from epoxies, polyesters, or phenolics have become very important in the thermoforming industry, with the epoxies being used most predominantly. Usually, the molds are either cast from a metal-filled epoxy or are laid up with glass fabrics and epoxy laminating resins. Plastic materials are noted for their excellent dimensional stability and high abrasion resistance, and they provide relatively smooth and nonporous surfaces that require little or no additional finishing or polishing. The high-heat-distortion materials, such as the epoxies, adequately resist the temperatures encountered in thermoforming. Epoxy molds are easily repaired and are much less expensive than plaster or wood. Epoxies are principally used for prototype or limited production runs because of their disadvantages of low thermal conductivity and short service life compared with metal.

Sprayed Metal

Sprayed metal molds have also increased in popularity and are being utilized for production runs in many applications. A metal, such as aluminum, copper, or nickel, is sprayed on a form to produce a shell which is then reinforced with an epoxy resin or with a low-melting metal, such as zinc. These molds are more expensive than plastic molds but are less costly than cast or machined aluminum molds, and they are used where the durability and thermal conductivity of aluminum are not required.

Aluminum

Aluminum molds are used extensively in the thermoforming industry for long production runs where fast cycling is required. They have the advantages of long service life and excellent thermal conductivity. However, their initial cost is high and they are not economical for prototype or short production runs.

EPOXY TOOLS

Epoxy tools are used for both vacuum and pressure forming. Very careful design and fabrication of the tool are required for pressure service. Our experience has been with vacuum forming, and the remainder of this paper will be confined to the use of epoxy molds for this purpose.

Most thermoforming epoxy tools are designed for vacuum service. The tool generally consists of a solid shell covering a porous chamber of some type. Holes are drilled through the shell into the porous chamber, which is enclosed and attached to a vacuum source.

Shell Construction

Fabrication of tools from epoxy materials and glass fabrics has been very adequately covered in numerous papers and will not be discussed here. Thermoforming tooling, in most cases, requires the use of high-temperature epoxy materials, and these materials are available from many sources.

The shell can be made by either casting or laminating techniques.

Casting. We use the casting system to make small molds, because it is a fast technique and provides adequate strength. We use aluminum-filled high-temperature casting epoxy systems in nearly all cases because of their high strength and good heat conduction. Casting is a slow process, and we generally cast the material in 1/2-in.-deep layers. Thicker sections, in our experience, generally blister due to the exothermic heat that develops. It is possible to cast depths of 3/4 to 1 in. for flat configurations, but varying depth contours are involved in most cases, and to prevent localized blistering, we limit depth to 1/2 in. There is generally a waiting period of 6 to 7 h after casting, to allow for curing and the dissipation of exothermic heat. This limits the casting to two 1/2-in. layers in an 8-h day, with the second layer curing overnight. Normally, a thickness of 3/4 to 1 in. is sufficient in thermoforming.

Laminating. For larger molds, which constitute the bulk of our work, we use laminating techniques. Normally, the model is covered with a thin surface epoxy coat, and after this coating reaches the tacky stage, a fine woven glass fabric is laminated to it to produce a smooth outside surface. Layers of heavier glass fabric are then added, using laminating resin. Usually, three to five layers of heavier cloth are used to provide the necessary strength. Laminating can be a slow process, because only two or three plies of fabric can be applied at a time, particularly on a vertical surface, and sufficient time for curing and cooling must be allowed before the next application to prevent the formation of blisters. To conserve resin, only enough resin should be prepared for the application at hand because of the short pot life involved. Vacuum bagging should be used to produce void-free structures where the highest strengths and smoothest surfaces are required.

Porous Chamber

The shell has to be reinforced with a porous chamber of some type to withstand the thermoforming pressures. It is not desirable to make the mold solid or impervious to air throughout, because this requires drilling of the air-exhaust holes completely through the tool. The best approach is to drill fine-diameter holes through the shell into a porous chamber which is enclosed and connected to a vacuum source.

Selection of the number and location of the holes needed in the shell to provide proper vacuum release is an art in itself. The number of holes must be sufficient to provide for rapid release of air from the cavity. This allows the hot plastic sheet to reach the mold surface while it is still at a deforming temperature. The locations of the holes depend on the contours of the mold. For a female mold, many holes are placed at the bottom of the cavity but only a few holes along the top flange area. We find that in the cavities, holes must be approximately 3/4 in. apart, whereas on plateau areas, a hole spacing of 3 to 6 in. is sufficient. Male molds require few holes, unless there are depressions in which air can be trapped. Holes spaced 2 to 3 in. apart along the base of the male tool are usually adequate.

Hole diameter is important because large-diameter holes will form marks on the surface of the thermoformed part. The size of the hole will vary with the plastic sheet material. For instance, polyethylene, being a more fluid material, will pick up holes of smaller diameter than ABS or polystyrene. We use holes 21 to 13½ mils in diameter (#75 to #80 drills).

The porous chamber can be fabricated by filling the shell with a paste made from a granulated filler and an epoxy binder. The filler can be aluminum chips or beads, commercial soft stones, or ground walnut shells. Aluminum chips are the best for heat-

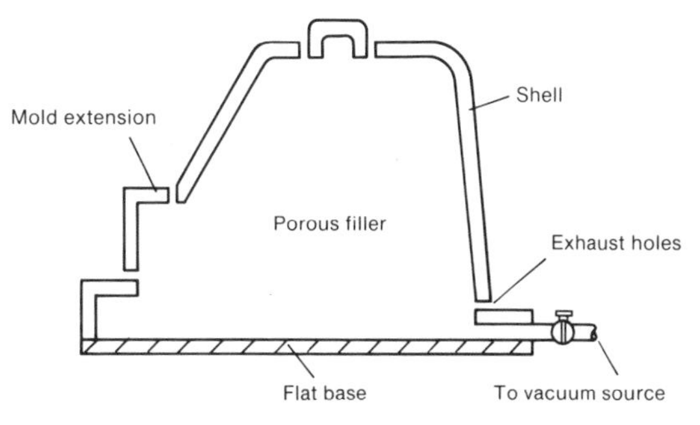

Figure 3E-5. Male mold, showing shell surface and porous chamber

conduction purposes, while soft stones are the lowest in cost. Walnut shells have the advantages of light weight and softness. The exhaust holes through the shell have to be drilled with very fine drills, and we find that stones and aluminum chips break these drills fairly frequently. For this reason, we prefer the softer walnut shells. We use the porous paste materials for small or medium-size molds.

Normally, the shell of a male mold is extended to form a flat base (Figure 3E-5), and the shell is filled to the base with the porous paste. A wood panel can be used to seal the base. Female molds require framing of the shell with wood or epoxy-glass fabric, which is also extended to form a planar base (Figure 3E-6). The space between the epoxy and the frame is filled with the porous paste, and the base is sealed with a wood panel.

For larger molds, where weight is important, paste materials are not very satisfactory and we use a sandwich construction to reduce weight. The sandwich consists of an inner and outer shell held together with

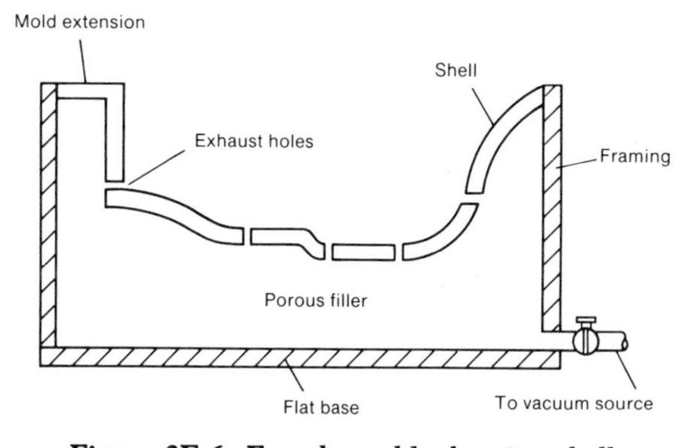

Figure 3E-6. Female mold, showing shell surface and porous chamber

small wood cubes. The inner shell is fabricated by the laminating technique. An adhesive layer of epoxy is applied, and small wood cubes approximately 1/2 to 1 in. on a side are placed in this adhesive layer approximately 1/4 to 1/2 in. apart. Metal screening is stapled to the wood cubes, and the outer shell of epoxy-glass fabric is laid up on top of this screen (Figure 3E-7). Both shells should have the same strength, because the vacuum force is applied to both surfaces. This technique produes a relatively low-weight mold that is portable. However, the placing of the individual blocks is slow and tedious. The exhaust holes either have to be predrilled with the cubes and then spaced around them, or have to be drilled from the back bewteen the cubes before the back layer of epoxy-glass is applied.

Other techniques can be used to connect the holes in the surface to a vacuum source. Copper tubing can be cast against the surface shell in a predetermined pattern and the vacuum holes drilled into the tubing. The tubing is connected to the vacuum source. We have found that this is a very difficult system to use in practice, because it requires that the locations of the vacuum holes be predetermiend and bcause it is very difficult to add new holes if needed.

Another technique that can be used with small molds is to make the mold solid and drill the vacuum holes completely through the mold (Figure 3E-8). In this case, a large-diameter hole is usually drilled from the back, stopping just short of the surface, and

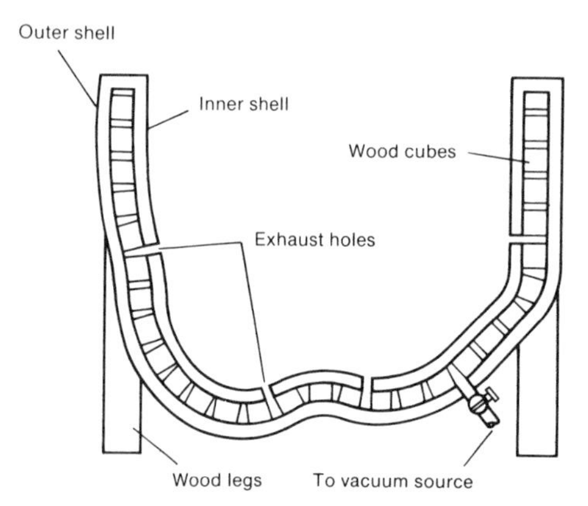

Figure 3E-7. Female mold of wood-cube sandwich construction

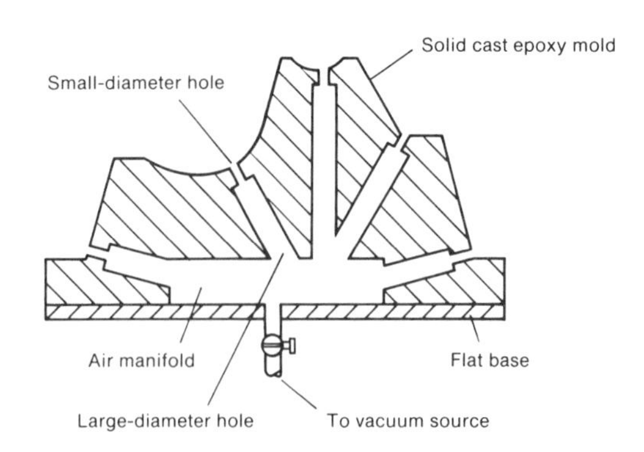

Figure 3E-8. Solid cast epoxy male mold with drilled holes

then a small hole is drilled through the surface. The large hole reduces the pressure drop of the air passing through the holes at a much faster rate. However, considerable skill is required to drill the holes properly without weakening the mold or ruining the surface, and we use this technique only for very small molds.

MOLD COOLING

The ability to adjust the mold temperature is very important for production tooling, but it is not very significant for prototype work. In production, the mold surface should be as warm as possible to reduce the strains formed in the plastic material by rapid cooling. However, the warm sheet has to be cooled below the deformation temperature before it can be removed from the tool, and the time required for this cooling should be as short as possible. Thus, there is a balance that must be achieved in the temperature of the mold to obtain a stable part with a short forming cycle.

Metal-filled epoxy compounds have the best thermal conductivity and, where mold cooling is important, should be used for the shell of the tool. These compounds are usually very viscous and are more difficult to work with than are normal surface casting compounds. For the best heat transfer, the shell should be kept as thin as possible.

To adjust mold temperature more rapidly, it is possible to cast in cooling coils of copper. The coils can be embedded in the porous paste material in direct contact with the shell material. The coils should be as close to the surface of the mold as possible and should be intimately bonded to the epoxy material to obtain the best heat transfer. A coolant, such as water, can be pumped through the coils to maintain the desired surface temperature. Aluminum-chip- or metal-fiber-filled epoxies provide the best heat transfer for the porous chamber.

Metal-fiber- and aluminum-chip-filled epoxies have thermal conductivities approximately 5 to 10 times higher than those of conventional epoxy compounds. However, cast or machined aluminum is still 100 to 1000 times as conductive as metal-filled epoxies and is the preferred choice where rapid heat transfer is required.

SHRINKAGE CONSIDERATIONS

Plastic sheet materials will shrink after thermoforming due to thermal contraction and the changes in density or molecular structure on cooling. Depending on the plastic sheet to be formed, mold-shrinkage allowances can vary from 0.002 to 0.008 in./in. A good average to use is 0.005 in./in. If close control of part dimensions is necessary, the supplier of the plastic material should be consulted to obtain more exact shrinkage values. Most of the plastic shrinkage takes place as the part cools from the forming temperature to room temperature. However, shrinkage will continue for 15 to 30 min after the part has reached room temperature, and it may be necessary to transfer the part to an auxiliary jig for this period in order to hold critical dimensions.

The shape of the mold will affect the amount of shrinkage that occurs. Parts formed in female molds will shrink away from the mold freely, and the maximum shrinkage will occur. Parts formed over male molds will be restricted by the mold itself and will shrink to a lesser degree, although the degree of taper on the mold will determine the amount of movement in the part along the mold surface. Unfortunately, many molds are combinations of male and female contours, and shrinkage factors will vary considerably throughout parts formed in them.

The degree of shrinkage in various areas of a part will depend on the stresses created in these areas during forming, and thus shrinkage will vary with the contour, the depth of draw, and the uniformity of heating. This makes determination of shrinkage factors more of an art than a science, and where very close control is needed, prototypes will probably have to be made from which more exact data can be determined.

In addition to the shrinkage of the plastic material itself, the shrinkage of the mold during fabrication may have to be taken into account. However, this is a factor only in the most critical cases, because epoxies shrink very little on cooling. Material suppliers can provide exact shrinkage factors for their products.

DRAFTS AND RADII

For female molds, it is possible to use drafts of as little as 1/2°, since the material tends to shrink away from the mold and may be removed easily. For male molds, however, minimum drafts between 1 and 2° are required, because the material tends to cling to the mold as it shrinks. However, to be on the safe side, many vacuum formers specify a minimum draft of 2 to 3° on all parts in which the shape or contour varies considerably.

All the radii on both male and female molds should be as generous as possible to provide maximum strength in the finished part. We use a minimum radius of 1/8 in.

SPACING OF CAVITIES

Multiple-cavity molds are used when many small parts are formed from the same plastic sheet. Multiple female molds can be spaced as close to each other as desired, limited only by the spacing required to trim the parts after molding (Figure 3E-9). In the case of drape or male molds, however, it is important to allow approximately the height of the mold between each mold in order to avoid webbing of the plastic sheet (Figure 3E-10).

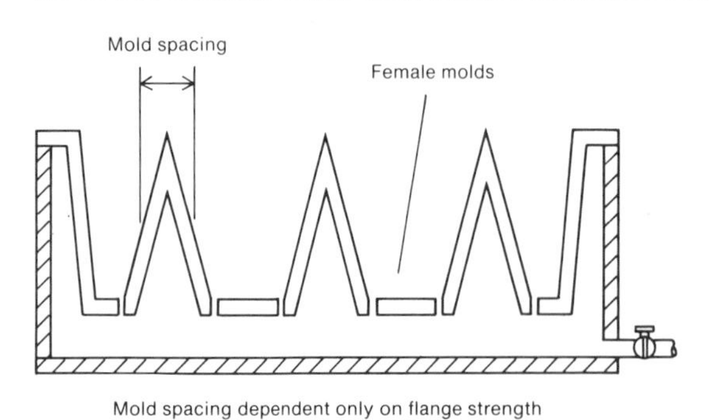

Figure 3E-9. Cavity spacing in female molds

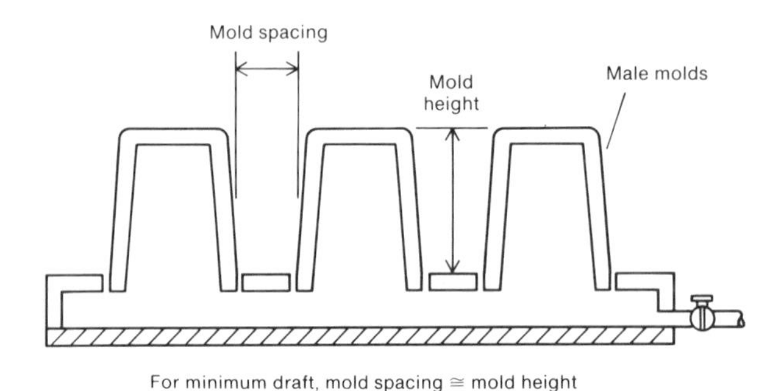

Figure 3E-10. Cavity spacing in male molds

DEPTH OF DRAW

The plastic material will thin out in proportion to the degree of stretching that occurs, and this is related to the depth of draw. Whenever possible, we try to follow the rule that for male molds, the maximum height should be no greater than the narrowest section at the base. For female molds, the average depth of the cavity should be no more than one-half the width of the narrowest part of the cavity.

CONCLUSIONS

Epoxy tools have an advantageous balance of physical properties and initial cost for prototype and short-run thermoforming operations. For long production runs or where fast cycling is important, the advantages of epoxies are not as obvious, and they should be considered along with metals and other tool materials.

We have found that for vacuum forming, the best epoxy tool design consists of a solid shell covering a porous chamber, with drilled holes connecting the exterior surface and the chamber. For medium-size tools, we prefer a walnut-shell-filled epoxy paste, and for larger tools, the wood-cube construction.

3F: Plastics Processing With Silicones

Silicone rubber may be characterized as the engineering-grade member of the synthetic rubber family. Their abilities to withstand extremes of heat and cold distinguish silicones from other elastomeric materials. The various available product types range from liquid RTV systems that cure at room temperature to semisolids requiring heat curing. RTV systems are either the familiar one-component adhesive-sealants, finding broad acceptance in industry and home, or two-component compounds primarily used for potting electrical components and for making flexible molds.

Flexible molds made from two-component liquid silicone rubber provide:

1. Accurate detail with a shrinkage factor as low as 0.2%
2. High self-release properties, which often eliminate the need for parting agents

3. High-temperature heat stability that will withstand exothermic heat generated by certain types of casting resins. There are also applications where the casting resin requires elevated-temperature curing.
4. A rapid means of making a production tool. This is especially true of the high-strength RTV-630, which will quickly develop maximum physical properties when exposed to an elevated-temperature curing cycle.

CASE HISTORY

Before describing production applications, we offer the following case history* as being indicative of advances being made in flexible-mold technology. In this particular application (Figure 3F-1), the master to be reproduced was a coil magazine machined from brass.

The brass master contained 100 slots with a width of 3/64 in., a depth to 3/4 in., and a length of 3 in. Each slot led into a cavity 11/32 in. in diameter and 3/8 in. deep.

Approximately 100 coil magazines were to be produced. The magazines were required for automation of both machine loading and quality-control testing of coils used in a grid-relay assembly.

Step 1: Cleaning of Brass Master

Normal procedure is first to clean with a solvent such as MEK, acetone, xylene, etc., followed by washing with hot water and Dreft. An alternative or additional operation is use of a barrier coating (i.e., SS-4101). Barrier coatings provide an inert continuous film between potential contaminant sites and silicone rubber.

Cure inhibition is described as a condition that occurs at the interface between the RTV and the master. RTV subject to contamination will exhibit a gummy surface to a depth of approximately 20 mils.

Figure 3F-1. Master coil magazine machined from brass

*This case history is provided through the generosity of Bill Lipsky of the Specialty Control Department, Waynesboro, VA.

Step 2: Application of Release Coating

Silicone rubber is self-releasing from most surfaces. In certain situations where there are deep undercoats or draws, it is expedient to apply a release coating. In this particular case, the master was dipped with Marblette #5. Excess material was removed by blasts of compressed air.

Step 3: Preparation of Silicone RTV

RTV-630 was supplied as a two-component system of blue RTV-630A and off-white RTV-630B. The blend ratio was ten parts A to one part B. Once uniform blending of A and B components had been achieved, the mixture was deaerated at vacuums to 10 mm of mercury. In container selection, allowance should be made for a volume expansion of four times the quantity being deaerated. In addition, the vacuum should be removed two or three times during deaeration to break bubble formation. For a 1-lb. blend, deaeration time is 30 to 40 min.

Step 4: Dispensing and Curing of Silicone RTV

The deaerated blend was loaded into an air cartridge and gun-injected into various holes and slots in the brass master (Figure 3F-2). After the cavities had been filled, the remainder of the blend was poured into the mold and the RTV allowed to cure for 48 h at room temperature. To obtain a maximum physical-property profile, postcuring is required. Suggested post curing cycles are 1/2 h at 350 °F or 2 h at 212 °F. Oven residence time is usually governed by the mass of RTV required for the application.

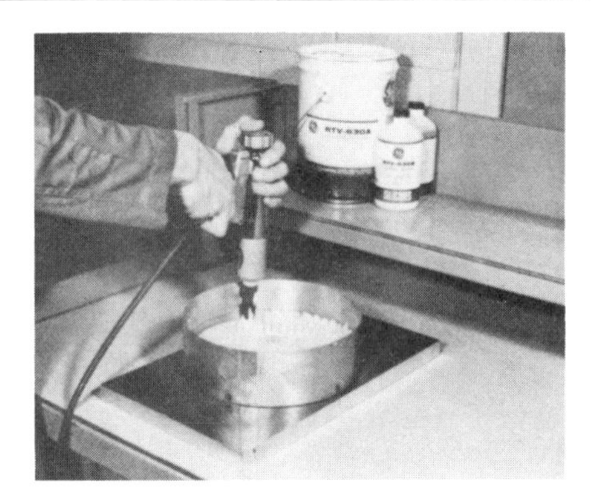

Figure 3F-2. Injection of deaerated RTV blend into holes and slots in brass master

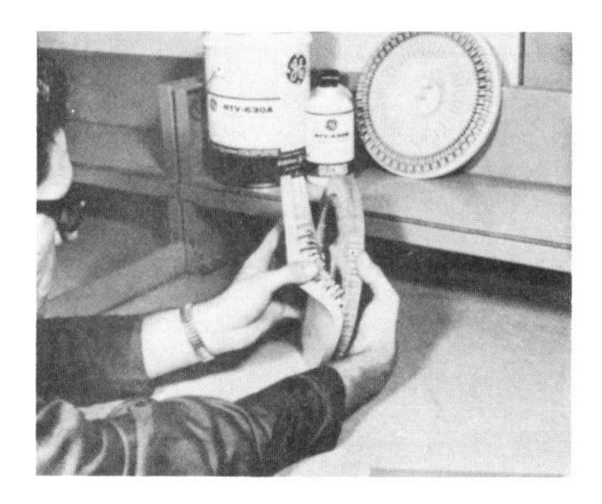

Figure 3F-3. Gradual removal of RTV-630 casting from master mold

Physical-property development can be quickly determined by a durometer scale. The optimum level for RTV-630 is in the 55 to 65 Shore A range.

Step 5: Mold Separation and Application of Release Coating

Air ports were provided in the brass master; consequently, when air pressure was applied, the natural seal that had developed between the RTV and the metal was broken. Due to the number of thin vanes comprising a major portion of the mold surface, it was still necessary to gradually remove the RTV-630 from the master mold (Figure 3F-3).

A Class H high-temperature epoxy (Scotch cast XR-5064) was selected for the casting resin because the coil magazine had to withstand a 400-°F degassing operation routinely performed on the coils. In view of the curing cycle required for the XR-5064 (8 h at 200 °F), the RTV-630 mold was immersed in Marblette #5 mold release (Figure 3F-4). Epoxy compounds that require elevated-temperature curing have a tendency to develop considerable adhesion to mold surfaces. Selection of a release agent is governed by the extent of postcuring operations performed on the cast article — e.g., painting or glueing. Under such circumstances, the release coating could plate out onto the casting with a resultant deleterious effect on subsequent bond development. On occasion, a solvent wash is employed to remove release agents picked up on the casting; however, this is expensive and often hazardous to surface detail. In this particular case, no post-cleaning operations were needed.

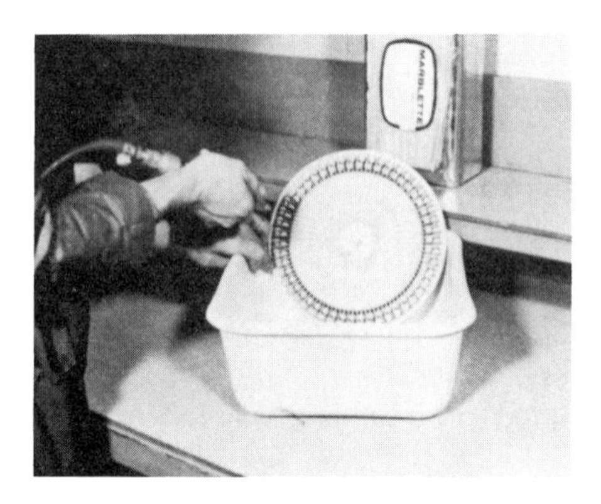

Figure 3F-4. Immersion of RTV-630 mold in Marblette #5 mold release

Step 6: Preparation and Pouring of XR-5064

The XR-5064 was supplied as a one-to-one mixture of A and B components. To facilitate mixing and deaeration, XR-5064B was heated to 95 °C and then blended with an equal quantity of XR-5064A; the mixture was heated to 60 °C and deaerated similar to the RTV-630. The deaerated mixture was then further heated to 70 °C and then syringed and brushed between projections and inner webs of the RTV-630 mold (Figures 3F-5 to 3F-7). The casting was then placed in the oven (making sure that the mold rested level) and cured for 8 h at 200 °F. After being cooled to room temperature, the casting was slowly stripped from the mold, producing an exact reproduction of the original brass

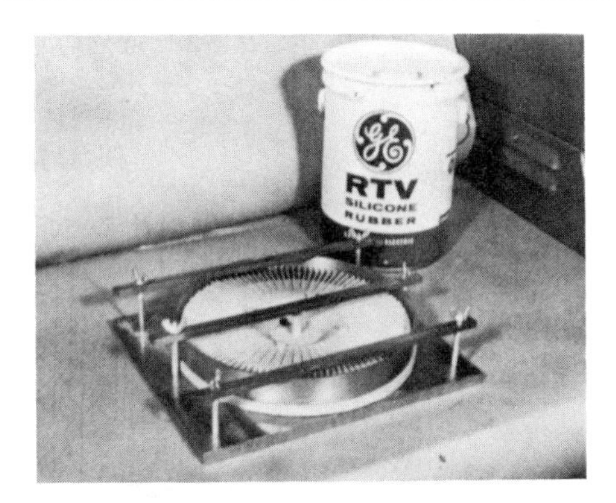

Figure 3F-5. RTV-630 mold

Figure 3F-6. Syringing of deaerated XR-5064 mixture between projections and inner webs of RTV-630 mold

Figure 3F-7. Brushing of deaerated XR-5064 mixture between projections and inner webs of RTV-630 mold

master (Figure 3F-8).

The estimated cost of producing these magazines by machining was $480 each in units of ten or more. If a zinc or aluminum die casting method was employed, tooling cost would be $1700 plus $10 per casting, re-

Figure 3F-8. Stripping of RTV-630 casting from master mold, providing exact reproduction of master

sulting in a cost of $27 per unit. Initial efforts with conventional RTV failed to produce more than two castings per mold due to excessive tearing in the silicone during extraction of the epoxy casting. Replacement of conventional RTV with high-strength RTV-630 resulted in an average of 15 castings per mold, giving a final cost of less than $10 per unit.

To review, the costs of producing 100 coil magazines by the three methods described are as follows:

Machining	$48,000 (highly impractical)
Zinc or aluminum die casting	$ 2,700
High-strength TRV-630	$ 1,000

APPLICATIONS OF RTV SYSTEMS

The skill and developmental activity being put forth on RTV in the model shop is now providing the foundation for much broader application in production processing of plastics. The incidences of making production castings of epoxies, urethanes, phenolics, and polyesters in flexible RTV molds is becoming more prevalent every day. Typical examples of production casting operations are:

1. Electronic components
2. Instrument knobs and buttons
3. Mannequins
4. Household interior decorations — e.g.,

Figure 3F-9. Glass-fiber-reinforced phenolic strips being laid up on steel die in preparation for vacuforming

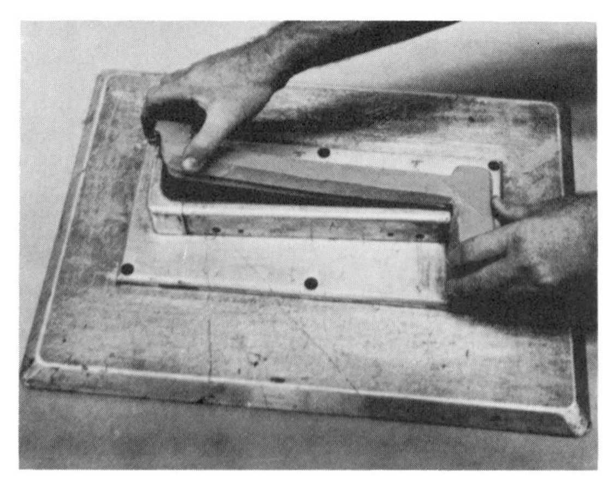

Figure 3F-10. RTV-630 mold being set into die over phenolic lay-up

picture frames, wall plaques, lamp bases, and figurines.

In the aerospace industry, RTV has been fabricated into male punch dies and pressure bags for thermoforming of reinforced plastics. A variation of this technique is the vacuforming operation illustrated in Figures 3F-9 to 3F-11. Figure 3F-9 shows the lay-up of glass-fiber-reinforced phenolic strips on a steel die. Figure 3F-10 shows the RTV-630 mold being set into the die over the phenolic lay-up, after which the total assembly is subjected to 500 psi and 350 °F in an autoclave. Figure 3F-11 illustrates a finished vacuformed laminated plastic structural part.

Implementing this transition to the use of RTV systems has been the introduction of high-strength RTV-630, which has the toughness and durability for utilization as a production tool.

Some recent examples of production tooling applications are matched die molds and embossing rolls.

Matched Die Molding

Precision reinforced plastic moldings of structural parts for aerospace and missile components have been made using high-strength RTV-630 as a male punch die. Continual design change plus the time delay associated with retooling were the prime reasons for selection of a synthetic tool in this application.

Prior to replacement with high-strength

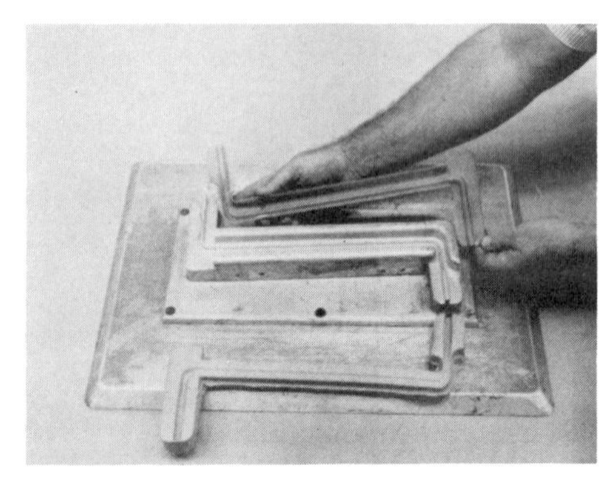

Figure 3F-11. Finished vacuformed laminated plastic structural part

RTV-630, conventional RTV could not withstand continual exposure to sealed processing conditions of 250 °F and 500 psi without reverting, often within the first ten cycles. In addition, molded parts formed with conventional RTV required machining to maintain tolerances. The punch die made from high-strength RTV-630 illustrated in Figure 3F-12 outperformed conventional silicone rubber by more than six to one with no visible signs of fatigue or surface deterioration.

Embossing Rolls

Use of RTV-630 as the impression surface on embossing rolls has received an enthusiastic welcome by style-conscious vinyl converters. The major advantage offered by silicone rubber is rapid and precise reproduction of a given pattern, be it a woven fabric (tapestry), a natural fiber (woven

Figure 3F-12. Punch die made from high-strength RTV-630

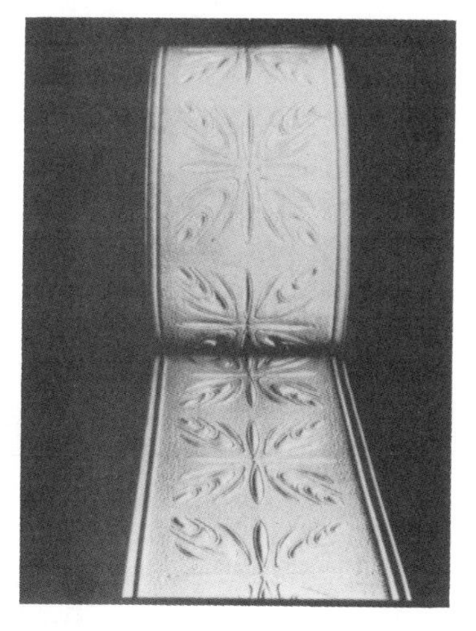

Figure 3F-13. Silicone reproduction of carved fleur-de-lis pattern

straw), or, as shown in Figure 3F-13, a carved fleur-de-lis pattern. The surface reproduction capacity of silicone enables transference of even the most minute detail.

Developmental effort is being made in this application because of the following circumstances:

1. Depending on intricacy of design, an engraved steel embossing roll usually requires 90 days or more for delivery.
2. A steel roll once subjected to surface damage is impractical to repair, and complete roll replacement often is necessary.
3. Industry desires to develop rapid style changeover techniques.
4. Releasability features of silicone.
5. Intricacy of design capabilities.
6. Potential cost reduction on embossing roll investment.

Immediate acceptance of silicone in this application is being hampered by a scarcity of available fabrication techniques. Some typical examples of fabrication problems are:

1. Hand tooling required at the seam area
2. Contour design of the roll
3. Adhesion development between the silicone surface and the mandrel.

Vinyl converters considering the purchase of silicone embossing rolls will have to strike a balance between urgency for a particular pattern and long-term demand. As more companies enter into production of embossing rolls with silicone rubber surfaces, it may become more and more worthwhile to buy replacement silicone rolls rather than make an initial investment in engraved steel rolls.

In addition to embossing rolls, there is considerable interest in cast silicone print rolls for valley printing and also for use as rubber backup rolls.

CONCLUSION

When the tool engineer is faced with a situation where the potential is unknown, the need urgent, or the shape cumbersome, then silicone rubber should be considered.

The immediate function of high-strength RTV-630 is to improve acceptance of silicone where conventional RTVs are marginal. Its long-range function is to further define opportunities for second- and third-generation product developments.

High-strength RTV-630 has exhibited a capacity to create new opportunities in plastics processing. Become acquainted with its qualifications, exploit its attributes, and gain a measure of the potential it represents.

3G: Plastics Tools for Forming of Prototype Sheet-Metal Parts

Prototype sheet-metal parts can be fabricated by the application of plastics tooling. This type of tooling can provide an economical and speedy method of making a desired part. Two types of plastics tools are commonly used for producing low-volume parts: hammer forms and die sets.

HAMMER FORMS

A hammer form is a die over which a sheet of metal is fastened down and, by the use of hand mallets, shaped to the contours of the form. There are three types of hammer forms: wood, zinc alloy or kirksite, and plastic-faced hammer forms.

Wood Hammer Forms

Hardwood hammer forms can be used where the desired part has a smooth, flowing contour and a limited number of parts are to be made. Metal inserts (Figure 3G-1), can be added in areas of detail or sharp corners to help withstand the pounding. This type of tool is made from a blueprint and is worked with templates to obtain the surface desired. Wood hammer forms are used when there is no model or pattern available for obtaining the required surface.

Zinc Alloy or Kirksite Hammer Forms

Zinc alloy or kirksite hammer forms are used where greater numbers of parts are required, or where the detail is such that wood forms will not stand up. A standard-size die model, or expanded wood pattern, is required to make this tool.

If a die model is used for the form, an expansion plaster should be applied in order to allow for metal shrinkage. An expansion plaster is a plaster that grows while being cured. This growth can be controlled to a certain extent. From this expansion plaster, a hard plaster pattern is formed, and then either the wood or plaster pattern is rammed up in sand (Figure 3G-2).

The zinc alloy or kirksite is poured into the open sand mold after the pattern has been removed (Figure 3G-3). When removed from the sand, the casting will be very close to size.

Plastic-Faced Hammer Forms

Typical applications for plastic hammer forms are those where the part is large and kirksite will not hold dimensionally, where there is more detail or a greater number of pieces than wood forms will make, or where there are no facilities for pouring kirksite.

Plastic hammer forms can be made in many ways, including plastic faces on wood cores and plastic faces over slurries of plastic and stone, aluminum shot, or any of several other fillers. The fillers will help prevent or dissipate the exothermic heat that all thermosetting resins produce.

A proper aid is required to make any plastic hammer form. Such an aid might be a standard-size die model, plaster, or plastic made the opposite of the hammer form.

Wood Core With Plastic Face. To make a plastic-faced wood form (Figure 3G-4), the hardwood core is fitted to the surface of the available aid. A gap of approximately

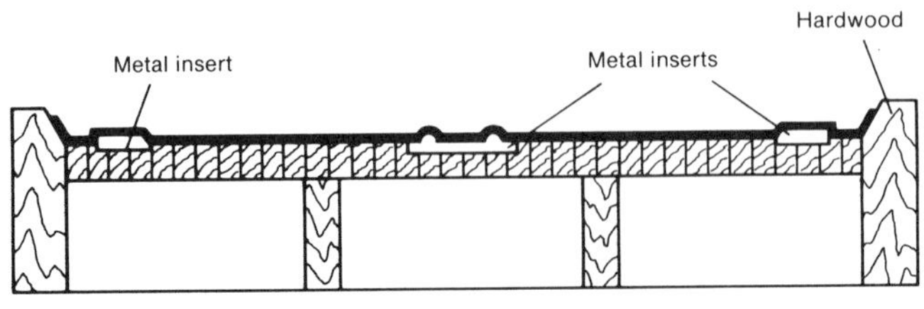

Figure 3G-1. Hardwood hammer form

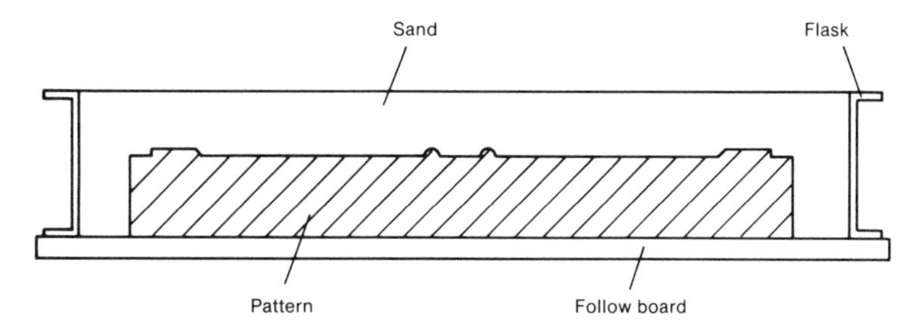

Figure 3G-2. Expanded wood or plaster pattern in sand

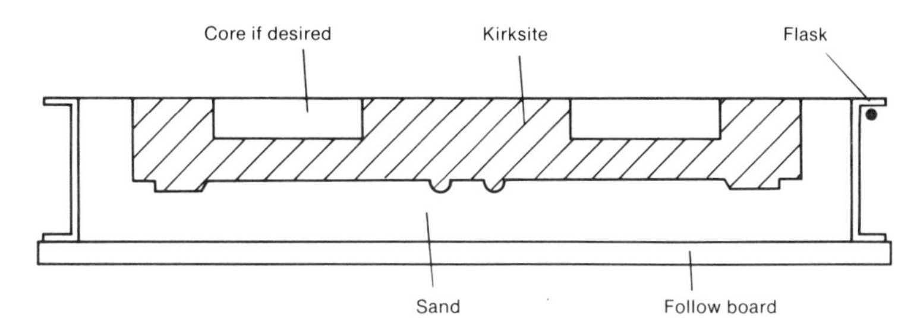

Figure 3G-3. Poured kirksite in open sand mold

1/4 in. is provided to allow for pouring of the plastic face. When the core has been placed, it is suspended in the cavity and sealed off, leaving sprues and air vents, and now the plastic can be poured. After being cured, the aid and the hammer form are separated. The surface should be checked for air voids and patched as required.

Kirksite Core. A plastic-faced hammer form with a kirksite core is made in basically the same way as a plastic-faced wood form (see Figure 3G-5). The difference is that, instead of wood being fitted to the aid surface, a plaster pattern is taken from it. This pattern is rammed in the sand. The casting should shrink so that it will allow a gap for pouring of the plastic. The core should be blasted or roughed up, and any oily film removed, to ensure a good bond between plastic and kirksite.

Epoxy Resin and Filler Core. In making a plastic-faced form with an epoxy resin and filler core (see Figure 3G-6), first paint a gel coat of resin on the surface of the

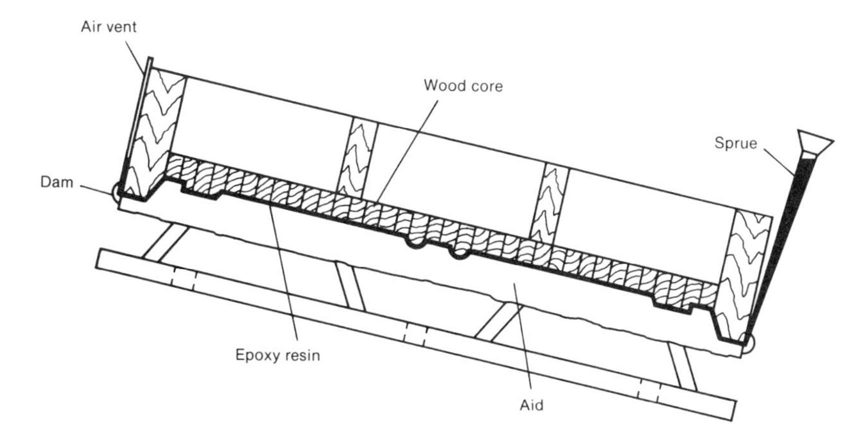

Figure 3G-4. Plastic face with wood core

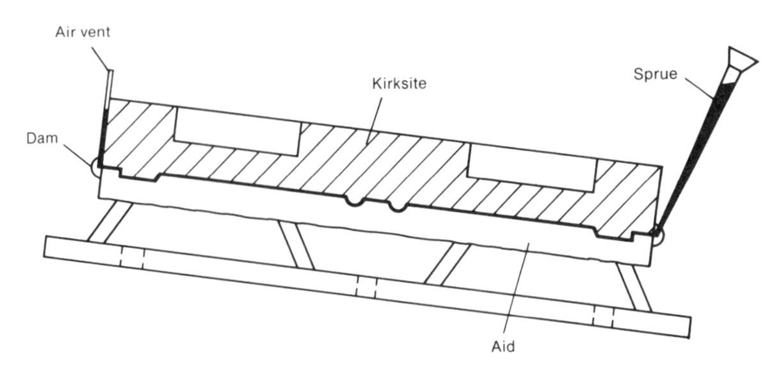

Figure 3G-5. Plastic face with kirksite core

aid. When the gel coat of resin becomes tacky, laminate a few layers of fiberglass cloth over it. Then mix a mass casting epoxy resin with the filler and pour it into the cavity. Following hardening or curing, carry out the same procedure as that used for other hammer forms.

Phenolic Resin and Phenolic Foam Core. This type of hammer form, which is recommended for general use, consists of an aluminum-filled epoxy resin face with a phenolic foam core that is encased in approximately 1 to 1½ in. of phenolic resin (see Figure 3G-7). This type of form is made by using a standard-size aid, and no shrinking is required. This aid should have a box built around it and anchored securely. Into this box a core of phenolic foam is fit very roughly. The foam can be worked easily with a saw or even a jackknife. This foam should be fit 1¼ to 1¾ in. away from the finished surface. This finished surface of the aid must now be caulked approximately 1/4 in. to allow for final surface pouring. The caulking can be accomplished by using plywood, clay, wax, or any other type of material that is suitable. Suspend the foam core in the cavity and fill the gap between the core and the aid with phenolic resin. Allow a curing time of 24 h and remove the phenolic-covered foam from the aid. Then remove the caulking, clean the surface of the aid, rough up the core, suspend the core back in the cavity, and make the final pour with aluminum-filled epoxy. After curing and removal from the aid, follow the same procedure as for other types.

The phenolic resin and phenolic foam core hammer form is favored for three reasons: cost, production time, and weight. The cost is less than that of using a kirksite core because more labor is required to make patterns and pour castings than to fit foam and pour phenolic. The same thing holds true for wood cores. Production time is also less. A phenolic-core hammer form can be made in five days compared with seven days for the same tool with a kirksite core. This is because the phenolic core is made at the same time a standard pattern is made. Also, plastic is lighter than kirksite, making it easier to move around and work with. If it becomes necessary to make an engineering change in the form, it is much easier to cut away plastic than kirksite.

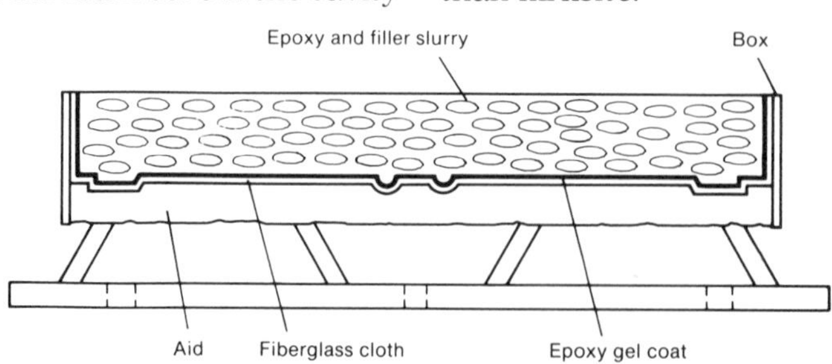

Figure 3G-6. Epoxy face with epoxy resin and filler core

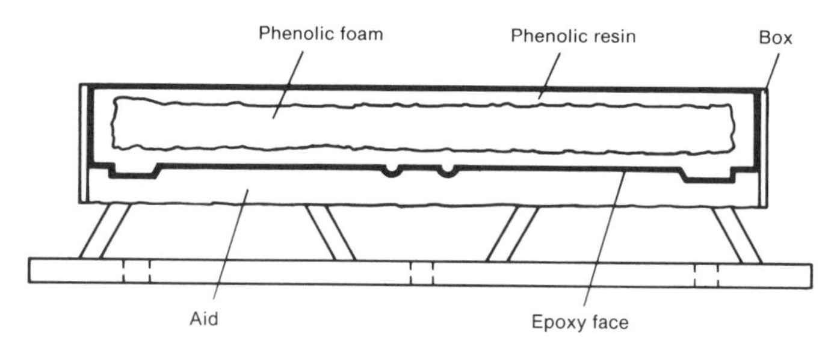

Figure 3G-7. Epoxy face with phenolic foam and phenolic resin core

DIE SETS

One of the high-priced methods of making prototype sheet-metal parts is by use of die sets. The cost of making a die set is considerably higher than the cost of making a hammer form. The reasons that justify spending the extra money for die sets are as follows. The part will more closely represent the production stamping if made with a die set than if hammer formed. Use of a die set also makes it much easier to maintain proper metal thickness. In addition, the production quantity has to be taken into consideration. The higher the number of pieces required, the cheaper a die set becomes, because pieces can be stamped faster than they can be hammer formed. Therefore, the part to be produced must be analyzed thoroughly to decide the most economical method that will achieve the desired results.

The same rule on size holds true as that for hammer forms. For smaller jobs, die sets can be made of all kirksite. The die set normally consists of three pieces mounted on larger panels. The three pieces are the die, punch, and ring. To make these tools, the desired aid is placed inside the metal female cast-off. From this female, a hard male plaster is made. After adding runoff and the required die surface on this plaster, a new female plaster is taken (see Figure 3G-8). The next step is to develop a plaster in the female to use for a punch pattern. Also, another plaster is developed around this punch for a ring pattern. A small amount of revising is necessary to allow for shrinkage on detail only. The castings will come out smaller, and this will allow a gap for pouring of the plastic face. Using these patterns, the cores can be made.

Next, take the original female plaster with the die buildup on and make a male plaster with the desired die surface (see Figure 3G-9). Then add sheet wax to the surface of the male plaster to allow for metal thickness, and after cleaning and roughing the surface of the kirksite core, match them together. Leave a gap of approximately 1/4 in. between them; seal off the joint, leaving sprues and air vents; and pour the resin (see Figure 3G-10). Cure overnight, and disassemble. The sheet wax should remain with the die if possible. Suspend the punch core and repeat the whole operation (see Figure 3G-11). After curing, disassemble, clean flash

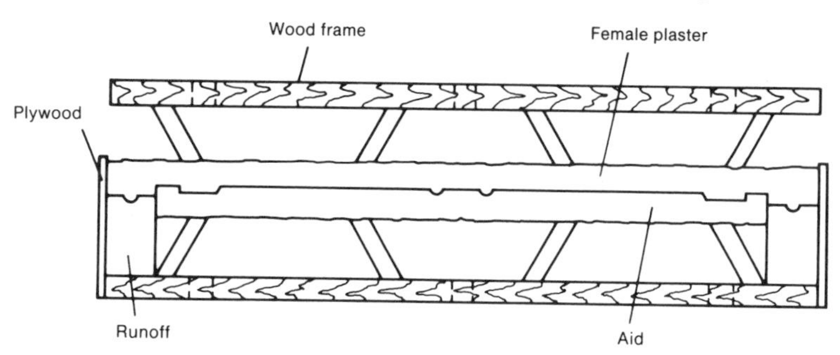

Figure 3G-8. Method of making female plaster in production of kirksite die set

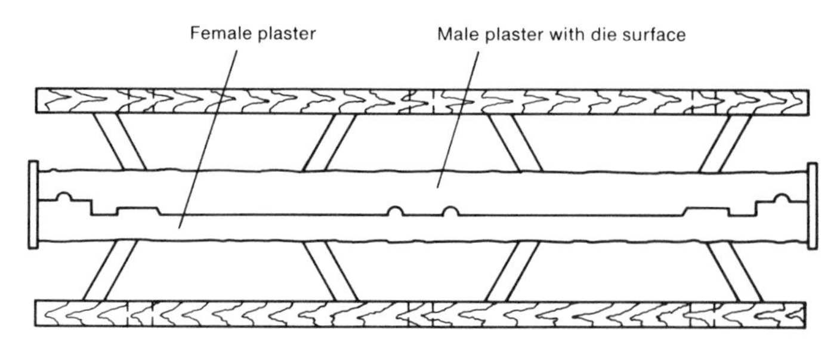

Figure 3G-9. Method of making male plaster for kirksite die

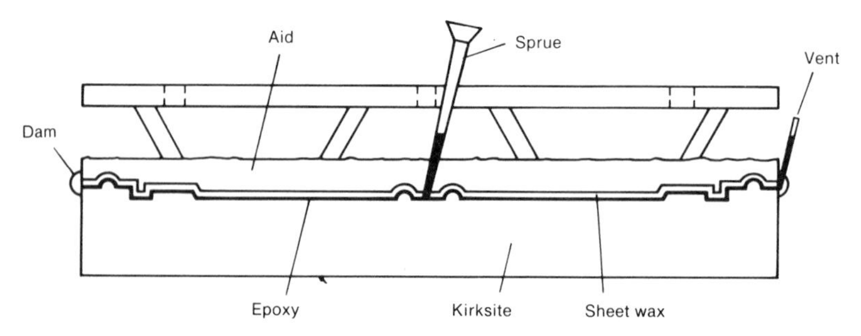

Figure 3G-10. Pouring of epoxy face on kirksite die

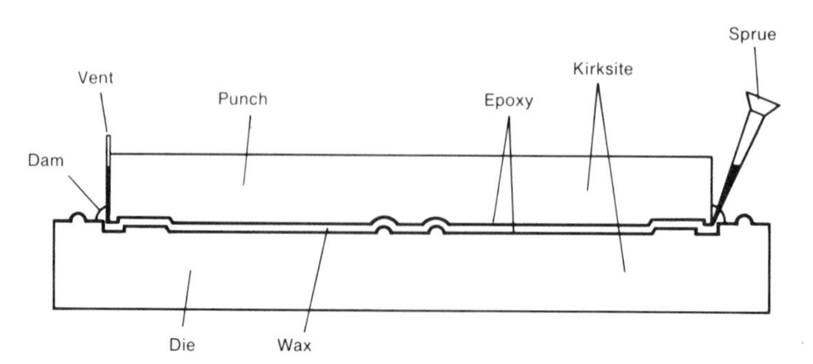

Figure 3G-11. Pouring of epoxy face on kirksite punch

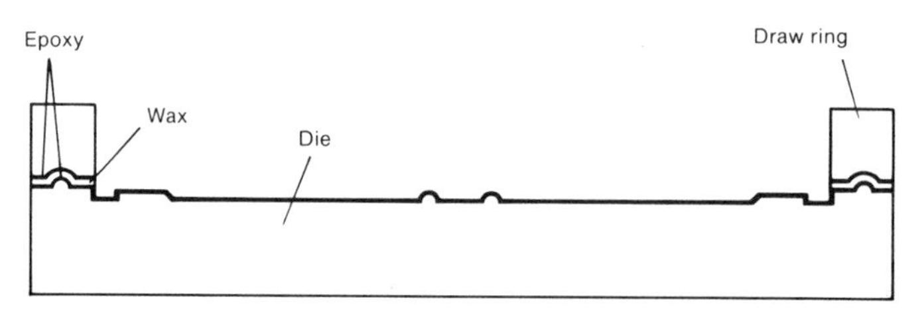

Figure 3G-12. Squashing of epoxy face on ring

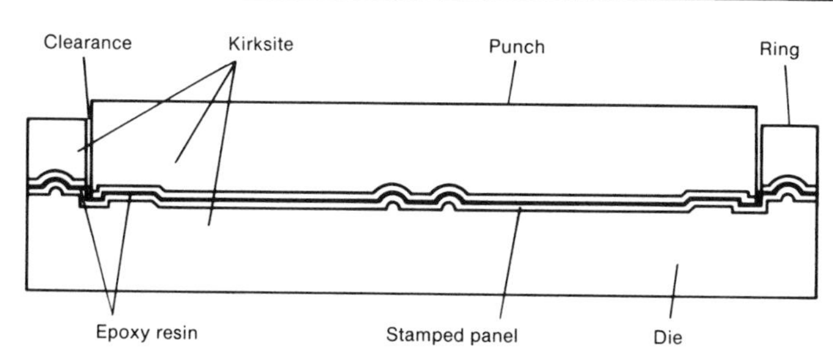

Figure 3G-13. Three-piece plastic-faced kirksite die set

off the punch, and place the punch back in the die. Locate the ring core around the punch in a manner that allows it to be relocated, and then remove the ring core and punch (see Figure 3G-12). Now put the ring core back in the die after adding plastic between them to be squashed. At this time, the bottom of all three castings should be either machined or squashed flat and parallel with the plastic.

Clean the surfaces of all three castings and check for air voids. Place in the press and proceed the same as for any prototype die (see Figure 3G-13). In many cases, these die sets will not make a complete part due to conditions that will not allow a one-hit operation. In such cases, use partial hammer forms to complete the flanging, etc.

Timing is always important in prototype tooling, and so is the cost factor. For comparison purposes, let us consider a complete hood assembly. If sufficient manpower and facilities are available to work all components simultaneously, hammer forms can be made in approximately two to three weeks, and parts can be assembled and delivered in five weeks. If the die-set method is used, tools can be made in four to five weeks, assemblies can be delivered in approximately two to three weeks, and parts can be assembled and delivered in five weeks.

The costs of these tools are difficult to compare because of the various ways in which they can be made. But in general, a complete set of hammer forms will cost around 20 to 30% of what a complete set of dies will cost. This difference in cost may be justified by the number of pieces involved or the quality of the part required. In either case, the cost of the prototype tooling is very small compared with that of production tooling.

3H: Basic Principles of Mold Design Used in Casting of Plastics

Some of the basic materials and methods used in making models and molds have been known for many centuries. New materials are constantly being discovered or developed and are often combined with established methods to produce relatively new processes. The result of this kind of application research is a much greater diversity of techniques from which a choice can be made according to the necessities of the work to be done. The past 20 years have witnessed the addition of many new plastic materials to the techniques employed in making models, molds, and cast plastics prototypes. The primary consideration of which mold material to use is determined by the intricacy of the model and the material used in its construction. The other important factors are the number of castings required and the properties of the casting material. The basic mold design is based on the conditions established by the foregoing considerations. For the sake of convenience, molds for casting of plastics can be categorized as being either rigid or flexible.

RIGID MOLDS

The fundamental design principle of a rigid mold is the same for all materials. Rigid molds can be constructed in plaster, cast or laminated plastic, electroformed metal, and cast low-melting-point metal alloys.

Plaster Molds

Plaster molds are used when it is necessary to cast only a few pieces in the most economical way. When properly sealed with lacquer and thoroughly waxed in order to ensure the release of the mold from the cast plaster part, a plaster mold will give excellent service. If it is used to cast or laminate polyester resins or high-temperature epoxy resin systems, it is advisable to use a film-forming PVA parting agent.

The simplest type of plaster mold is one which is made in one piece with an open face. This is possible when the baseline of the model is straight and all draft angles permit the draw to be in one direction (Figure 3H-1).

A cored-out casting is made by suspending a core inside of the mold cavity. The space between the core and mold face is determined by the amount of wall thickness desired in the casting. The core is maintained in the proper position by a series of bridges whose ends are keyed into the mold. The casting resin is poured through the open space between the core and the cavity (Figure 3H-2).

If, however, the draft of the model is in two directions and a solid casting is to be

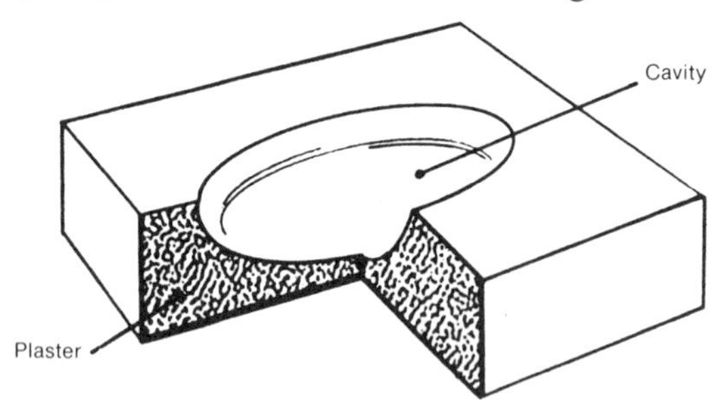

Figure 3H-1. Open-face plaster mold

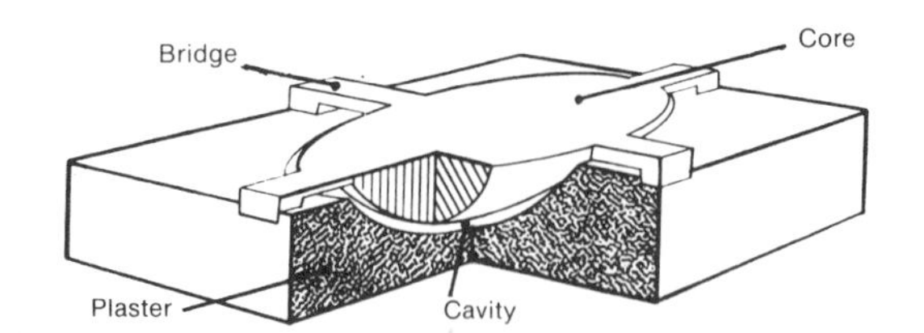

Figure 3H-2. Plaster mold with open core

made, the mold is made in two mating halves. Each half mold section will then be removable in the direction of the draw. The parting line will be a point on the model which is tangent to a line perpendicular to a common base plane. The parting line may be in a straight plane or, according to the configuration of the model, may be an irregular line. The two mold halves are properly registered by the keys on the mating faces of the mold (Figure 3H-3).

Due to the fact that the average casting resin system exhibits a degree of exothermic reaction during the curing cycle, it is necessary to cast models with a wall thickness of not more than 1/4 in. Therefore, in addition to the mold pieces that form the cavity, a core must be constructed which will not only determine the wall thickness but also register the internal webs and bosses in the casting. This arrangement is shown in Figure 3H-4.

Reinforced Plastic Molds

Reinforced plastic molds are used when many parts are needed. The surface finish

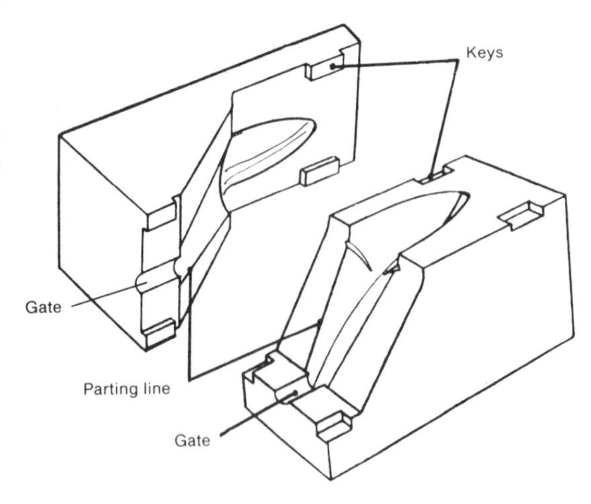

Figure 3H-3. Two-piece plaster mold

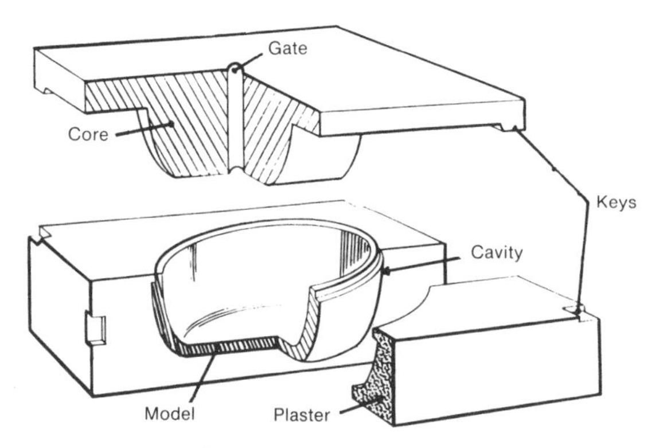

Figure 3H-4. Two-piece plaster mold with closed core

achieved in the finished parts, the methods used in casting them, and the durability of the mold are the determining factors in the use of this material. Reinforced plastic molds are constructed by the same procedure used in making any reinforced plastic tool. A surface coat of resin is applied to the model, fiberglass cloth reinforcement is laminated with resin onto the face coat, and a backup of mass casting resin is cast over the laminate. The finished mold can be made in whatever thickness is necessary according to the size and application of the mold. The same mold designs shown in Figures 3H-1 through 3H-3 can be used in constructing a reinforced plastic mold. Small molds can be made by using the mass casting method. Sometimes the mold design is influenced by the internal design of the part. The location of webs and bosses will establish where the gate will be placed and the manner in which the mold will be parted to permit proper venting of entrapped air in the mold. The removal of the core pieces is often facilitated by the use of a draw bar. Figure 3H-5 illustrates this design principle.

In order to ensure bubble-free castings, several methods are used. One method is vacuuming of the resin before pouring and of the mold during the pouring process. Placing the filled mold assembly in a pressure tank at 50 to 80 psi during the curing cycle is another, simpler method. Excellent results are achieved by exerting pressure on the casting resin by means of a plunger in the pouring gate. The gate is located on the parting line of the mold so that half of the gate is in each half mold section. Filling of the mold is accomplished first by gravity feed. The core of the mold is partially opened to permit venting until the mold cavity is filled. The core is then firmly clamped and the plunger is inserted into the gate. Pressure is applied after the gel time has begun and is continued until the resin ceases to flow. This mold design is shown in Figure 3H-6.

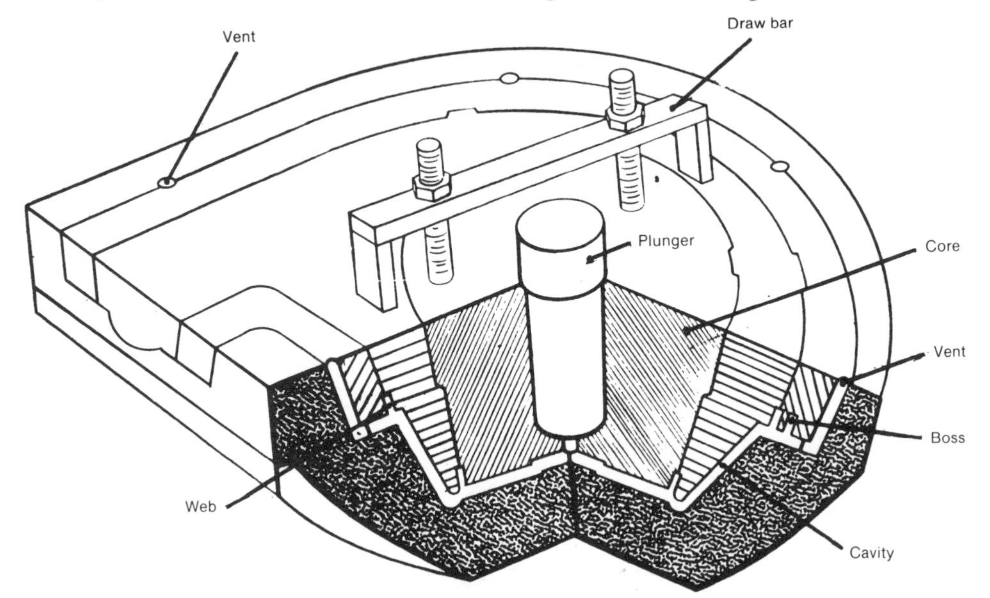

Figure 3H-5. Reinforced plastic mold, showing web, boss, air vents, and draw bar

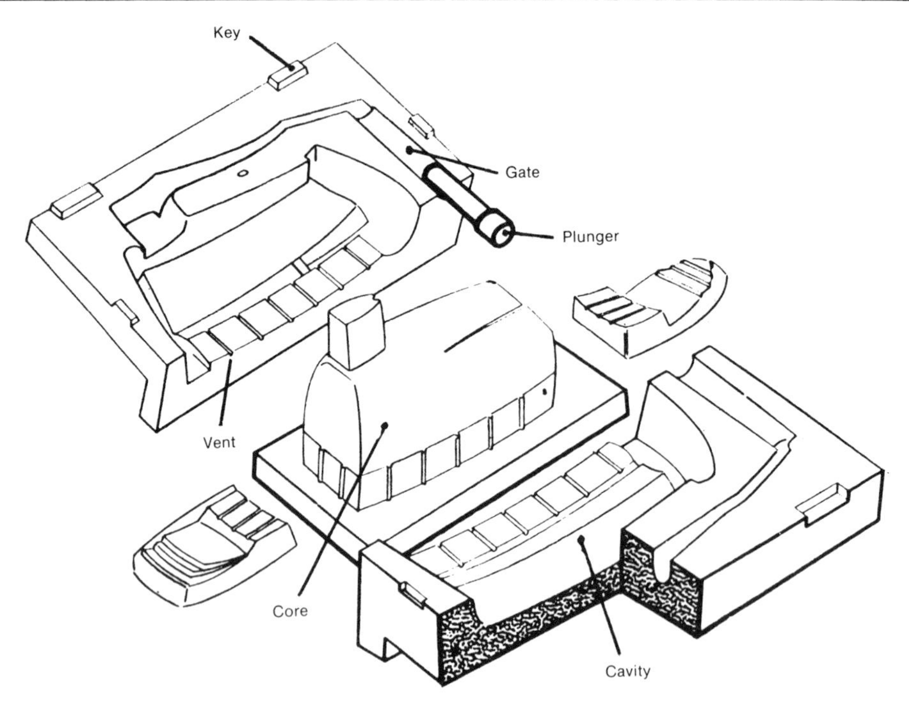

Figure 3H-6. Reinforced plastic mold, showing plunger used for exerting pressure on casting resin to prevent bubble formation

Metal Molds

Metal molds are used for casting clear cast epoxy resins and vinyl plastisols. The reason for this is that in order to achieve maximum clarity in clear cast parts, the surface of the casting must have a good polish. This is best achieved by having a polished surface in the mold. Vinyl plastisols cannot be polished in the finished casting and therefore they also require the finish to be in the mold. Two kinds of metal molds are used; electroformed molds and cast low-melting-point metal alloy molds.

Electroformed Molds. Electroformed molds must be made on a plastic

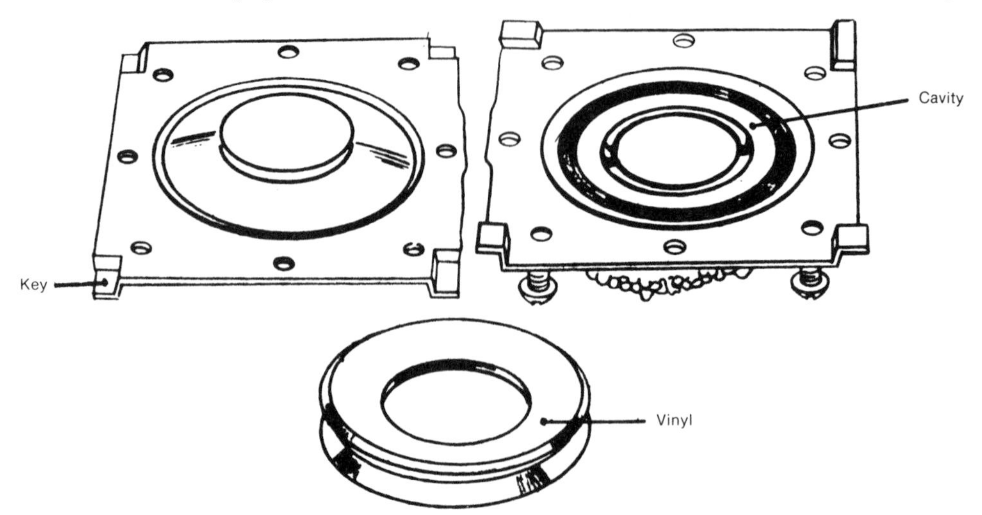

Figure 3H-7. Electroformed mold

model in order to avoid any reaction with the plating solution. They are made in whatever number of pieces may be required according to the design of the part. Each piece is electroformed individually on the model. Figure 3H-7 illustrates a two-part electroformed mold.

Cast Metal Alloy Molds. Cast metal alloy molds are made by casting the alloy in a plaster mold of each piece required for the mold. If necessary, the mold surface can be finished to a smooth, polished condition and nickel plated. Figure 3H-8 shows a two-part mold.

FLEXIBLE MOLDS

A flexible mold is used when a model which has intricate surface detail or contains deep draws and undercuts is to be reproduced in cast plastic. A flexible mold not only will register every surface detail, but also will release readily from any deep draws or undercuts. Two kinds of flexible mold materials are most commonly used: polysulfide compounds and room-temperature-vulcanizing (RTV) silicone rubber. Each one has a distinct area of usefulness according to the conditions set down by the model material and the casting material to be used.

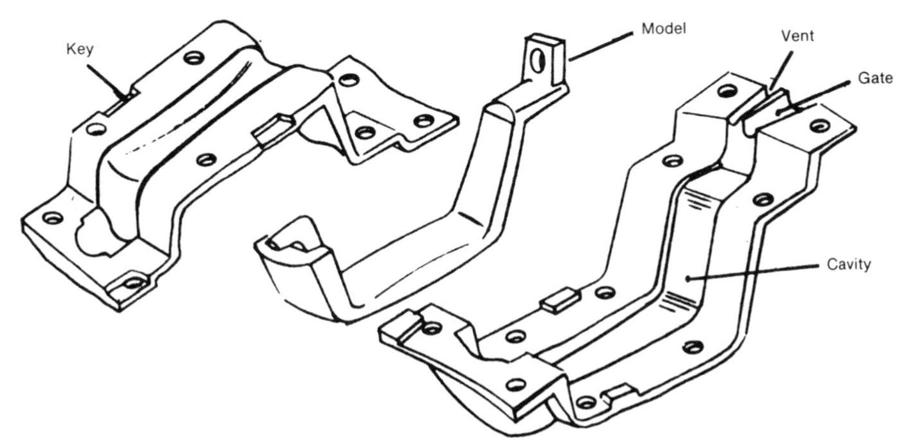

Figure 3H-8. Cast metal alloy mold

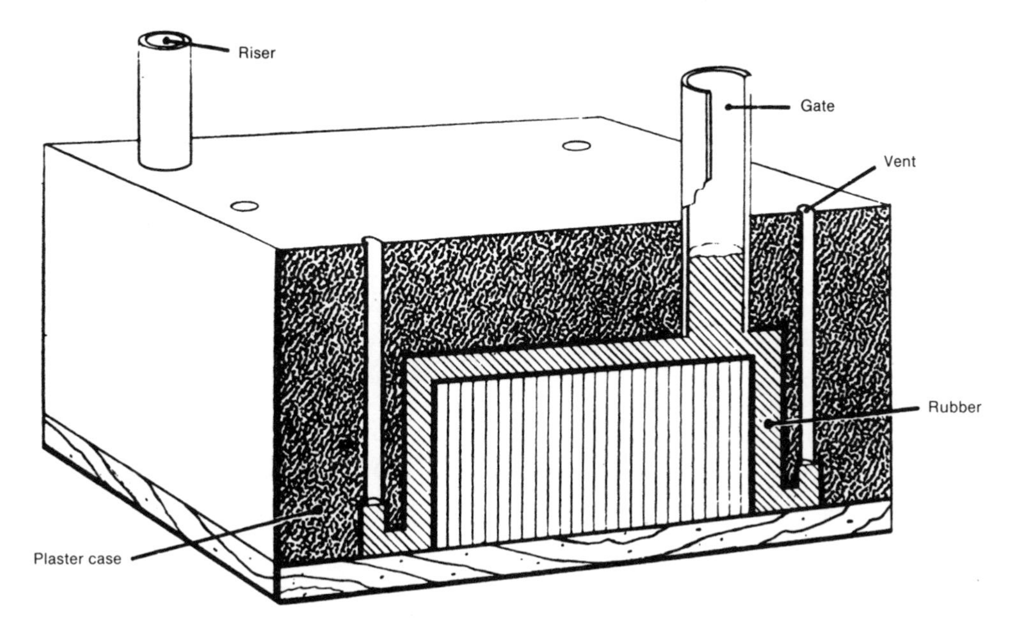

Figure 3H-9 Polysulfide rubber mold

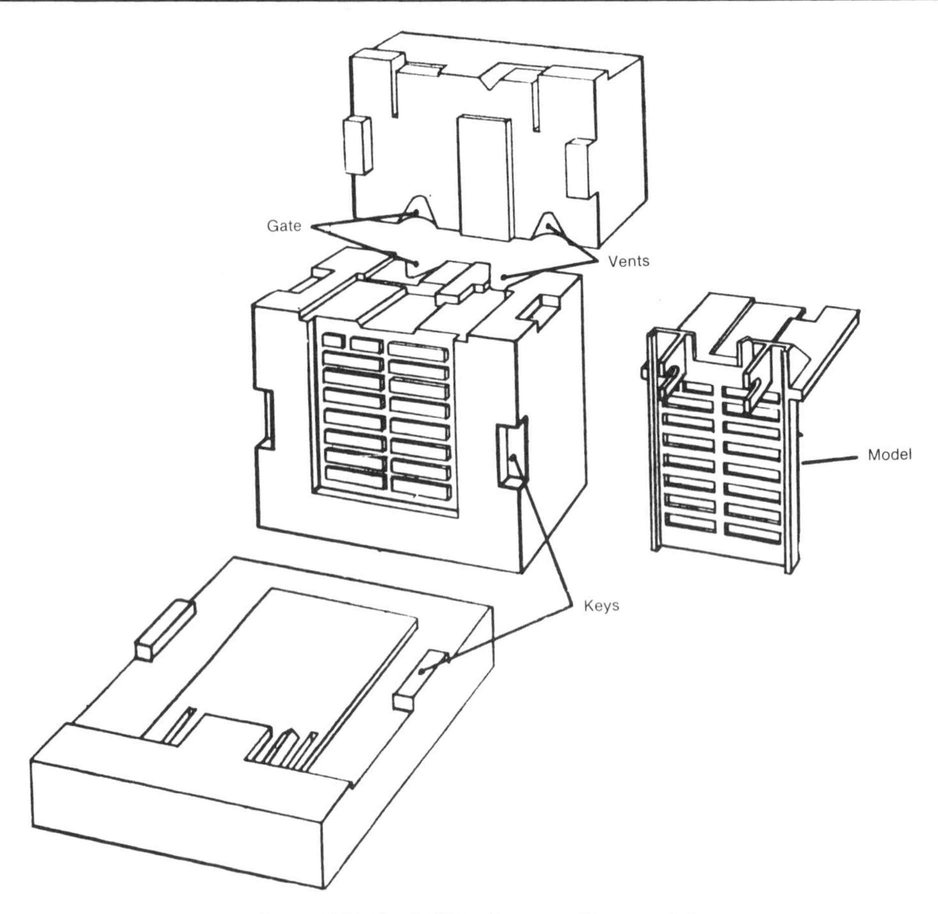

Figure 3H-10. RTV silicone rubber mold

Polysulfide Rubber Molds

Polysulfide rubber compounds are polymercaptan-base synthetic resins which are reacted with lead dioxide. They are used when the removal of the mold from the part requires good flexibility and excellent tear strength. They are formulated in viscosities ranging from a pourable liquid to a thixotropic paste. Again, the same principle illustrated in Figures 3H-1 to 3H-3 applies in the construction of a polysulfide rubber mold. If, due to the larger size of the mold, it is necessary to strengthen the mold, a plaster backup is constructed over the rubber mold before it is removed from the model. If the depth of draw is a factor, it is sometimes necessary to contruct the plaster matrix first over the model and then pour the polysulfide into it. Application of a PVA-film-forming mold-release agent will ensure longer mold life. Also, polysulfides must not be used beyond an operating temperature of 200 °F. Figure 3H-9 shows the design of this kind of mold.

RTV Silicone Rubber Molds

Room-temperature-vulcanizing (RTV) silicone rubber molds are used when many castings are to be made. The ability of this mold material to register the finest detail on the original model, plus low shrinkage and temperature resistance (600 °F), are factors that determine its use. A single- or multiple piece mold can be made of RTV silicone rubber according to the design of the model. Again, larger molds are normally backed up with a plaster matrix. Although no parting agent is necessary, mold life can be greatly increased by the use of a paintable silicone release agent. Figure 3H-10 shows a three-part silicone rubber mold.

3I: Evaluating Cast Urethanes for Plastics Tooling

The emergence of castable polyurethanes on the industrial scene has made available to fabricators of plastics tooling new materials of remarkable toughness. Some cast urethanes surpass vulcanized rubber in those physical qualities important to tool-manufacturing operations. More importantly, cast urethanes lend themselves to well-known processing methods for liquid plastics — techniques with which the tooling industry is very familiar. In analyzing the contributions of polyurethanes to plastics-tooling activities, the emphasis here will be on the characteristics which are important for tool-fabrication processes.

Foamed polyurethanes fit into tooling applications to a limited extent. Polyurethanes, prepared from polyisocyanates and polyols, represent many polymer structures, some of which are adaptable as coatings, others as foams or elastomeric derivatives. Polyurethanes selected for their casting qualities are prepared with a basically different objective: minimum foaming qualities. Casting demands ease of pourability, a useful working life, ready curing at convenient temperatures, and the development of physical properties useful to a tooling program.

The salient characteristics offered by cast urethanes are essentially rubberlike qualities and abrasion resistance of a high order of magnitude. To be more specific, Table 3I-1 compares typical physical properties of several castable plastics which have been developed for their rubberlike qualities, and which are capable of curing at room temperatures of 75 to 85 °F. Even heat-curable plasticized polyvinyl chlorides are inferior in properties to room-temperature-cured polyurethanes of compatible hardness.

It is clearly evident in Table 3I-1 that polyurethanes are a unique class of materials with "rubberlike" qualities that are superior to those of other grades of plastics. It should also be noted that these characteristics are attainable in readily pourable materials which can be handled in the shop, and that the property values are not selected from sheet stock produced for other purposes. Published values for some elastomeric polyurethanes are predicated on materials available only for heated, press-molded operations. The materials listed in Table 3I-1 are based on liquid systems which can be cast into tools of various sizes and complexities. Of course, flat sheets also are available in the various grades of polyurethanes shown in Table 3I-1. These flat polyurethane sheets are frequently employed in press forming of metals.

The net effect of these products is to lend new dimensions of toughness and resilience to plastics tooling, qualities woefully lacking in the past. High-impact tools, abrasion-resistant surfaces, durable molds, flexible

Table 3I-1. Properties of Room-Temperature-Cured Cast Elastomers

Product	Typical hardness range	Typical tensile strength, psi	Typical elongation, %
Epoxy-polysulfides	50 to 70 Shore D	1000 to 3000	25 to 90
Epoxy-polyamides	40 to 75 Shore D	500 to 4000	10 to 75
RTV silicones	50 to 80 Shore A	300 to 800	Up to 100
Polysulfides	20 to 60 Shore A	100 to 500	50 to 300
Polyurethanes:			
Hi-Extensible	30 to 50 Shore A	300 to 600	600 to 800
General purpose	60 to 80 Shore A	1500 to 3000	300 to 500
Maximum toughness	85 to 90 Shore A	4000 to 6000	300 to 400
Semirigid	60 to 70 Shore D	1000 to 2000	50 to 80

cores and snakes, shock-absorbing devices, gaskets, rollers, and mechanical goods of wide scope now lie within the experience and capabilities of plastic-tool fabricators.

DESIRABLE QUALITIES IN POLYURETHANE CASTING RESINS

Pourability

The quality of pourability is given special attention here because this characteristic usually has been obscured in earlier developments of cast urethanes. Earlier developments might have required heat to dissolve solid catalysts such as MOCA into liquid urethanes, or may have offered prefrozen kits. These measure represent procedures which we feel are obsolete for plastics tooling, because good liquid curing agents and resins are now available. When the resin prepolymer and curing agent are mixed together in the specified proportions, good pourability will be apparent and, for room-temperature curing systems, 20 to 30 min of working life may be expected. Longer working lives, up to 60 min, may be obtained from some systems, which may require heat to accelerate curing, although in many cases would completely cure at room temperature over a longer time (seven days) as contrasted with the more rapid setting materials.

The two components, a liquid resin and a liquid hardener, are weighed accurately in their required proportions and immediately poured. Viscosity-time curves for a typical general-purpose cast polyurethane system are presented in Figure 3I-1 for two different starting temperatures. The higher the initial temperature, the lower the initial viscosity to facilitate pouring. By the way of comparison, the viscosity-time curve for a more fluid cast urethane used for complex molds is reproduced in Figure 3I-2. Useful working life may be gleaned from viscosity-time curves. With restricted openings for pouring, 25,000 to 30,000 centipoises may represent top viscosity limits. However, for large pour vents, useful pourable viscosities up to 100,000 centipoises may be considered. Pourable

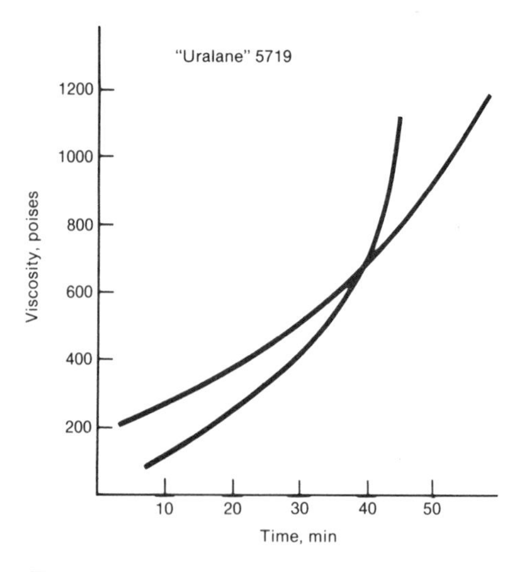

Figure 3I-1. Viscosity-time curves for a typical general-purpose cast polyurethane system

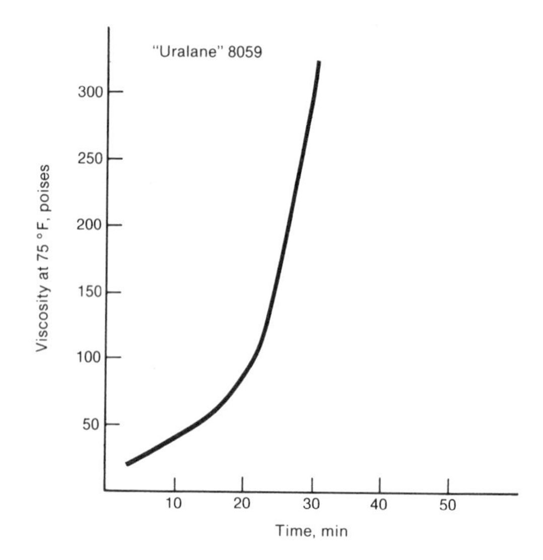

Figure 3I-2. Viscosity-time curve for a fluid cast urethane used for complex molds

urethanes under 8000 centipoises at 75 °F are available for tool fabrication. In general, the more rigid, tougher polyurethanes for tooling are more viscous, and some preheating before mixing will facilitate pouring.

The contents of opened containers will tend to thicken slowly if contact is made with moist air. Dessicant air filters providing entry to drums are suggested, as well as flushing of the drums with dry nitrogen. This precaution will preserve storage life, and dry nitrogen flushing is a packaging procedure practiced by more experienced polyurethane processors. In summation, the tool fabricator should carefully scrutinize viscosity specifications and polyurethane resins and curing agents before purchase is made.

Prepolymers

The resin prepolymer and its chemistry determine the properties of the final product. In general, prepolymers are prepared from polyols of selected molecular weight and reacted with polyisocyanates, such as toluene di-isocyanate. Among those qualities of good prepolymers desirable in plastic tooling, aside from an easy pourable viscosity, is a low NCO (isocyanate) content. Not only are the better elastomers prepared from prepolymers with less than 6.5% NCO, but they also are less susceptible to moisture contamination and less likely to present safety hazards. High toluene di-isocyanate (TDI) concentrations (which appear as high NCO contents) are to be avoided. In any event, good ventilation and skin protection are essential, although the problem is not as acute with excessive amounts of TDI.

Storage

Prepolymers and their curing agents should be stored in closed, airtight containers, preferably flushed out with dry nitrogen. Once containers have been opened, storage life is of necessity limited by contact with moist air. With proper sealing and packaging, storage lives of at least three to six months may be expected. Freezing weather may crystallize some curing agents, which will resolubilize on heating.

Mixing Proportions

The recommended mixing proportions for resin and curing agent should be carefully followed. This ratio is more critical than for epoxies, phenolics, or polyesters. In some instances, it may be more prudent to purchase materials in smaller units with contents preweighed and ready for blending. This is the more economical procedure, because premixing and freezing, as has been done by some manufacturers, increases the cost three or four fold.

Fillers

The selection of fillers for polyurethanes is limited. Traces of moisture or highly polar compounds will contribute to frothing and bubbles. Because thorough drying out is essential, it is recommended that this be performed by the manufacturer of the urethane prepolymer. In principle, the presence of moist fillers does not enhance the tensile strength or elongation of the cured elastomer.

Release Systems

In the processing of cast urethane elastomers, release from molds and patterns must be considered. Several proprietary releases are available to provide effective separation from other surfaces. Various waxes and silicone release systems have proven effective. In fact, some casting polyurethanes do not effectively bond to fully cured urethanes, suggesting, of course, that the urethanes themselves would make effective releases.

Adhesion

Plastics tool fabricators, well familiar with the exceptional bonding qualities of epoxy resins, will find that the resilient urethanes do not possess adhesive qualities to so marked a degree. In consequence, if an application requires casting of a resilient facing on a hard metal tool, it is suggested that a primer (such as Primer J) be used on the metal surfaces to enhance adhesion, and that mechanical means to be utilized to ensure permanent retention on metal faces. Lap shear strengths in excess of 2000 psi can be obtained on steel and aluminum surfaces when using semirigid polyurethanes and approximate primers.

Tensile Product

A useful criterion for evaluation of a resilient urethane resin is its tensile product, which provides some measure of its toughness. The tensile product is obtained by multiplying ultimate tensile strength by elongation. For the typical urethane casting resin, 2000 psi times 400% equals a tensile product of 800,000. This arbitrary unit gives an indication of the area under a stress-strain curve, which, after all, is an indication of toughness. Thus, for example, a castable urethane with a tensile strength of 500 psi and elongation of 300%, and thus a tensile product of 150,000, would be that much poorer in performance than the first example cited. Performance is thought of in terms of ability to resist abrasion and tearing and ability to absorb high impact loads without fracturing.

To put this evaluation on a more scientific basis, a comparison is made in Figure 3I-3 of the tensile stress – elongation curves for three materials, including a high-elongation urethane (UR 8118). These data are the fingerprints of materials of construction and underscore the qualities of castable urethanes being proposed for tooling applications.

The paramount criteria for the selection of castable urethanes are the qualities of tensile strength (psi) and ultimate elongation (%). For a given Shore durometer hardness, the product offering the best tensile product would be considered optimum.

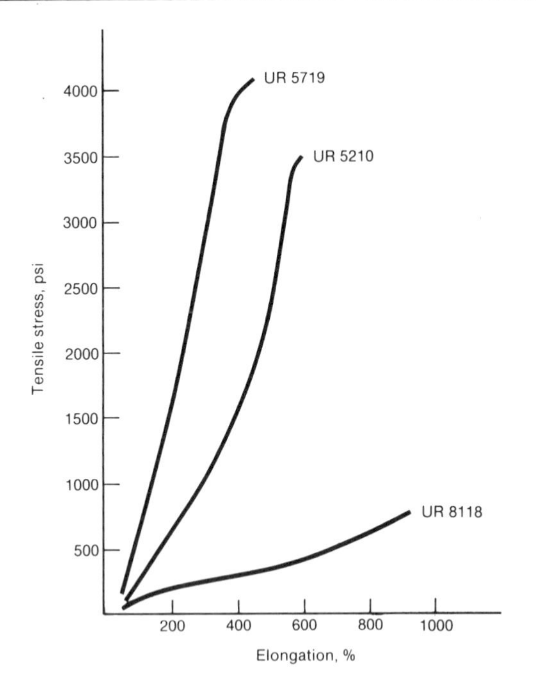

Figure 3I-3. Tensile strength - elongation curves for three urethanes

Hardness

There are considerable differences of opinion as to the optimum hardness for the forming and shaping of metal parts. In some instances where a great deal of elongation is necessary, a softer grade, such as 70 to 80 Shore A, would be desirable. On the other hand, harder grades (85 to 90 Shore A) might be more desirable for impact forming of metals. In any event, it is recommended that the real guides be the tensile strength and elongation of the materials, because these are, after all, measures of their toughness.

Tear Strength

A testing technique that is common to the rubber industry but not to plastics tooling is the ASTM D-624 cone-ring tear-strength test. An oddly angled tensile specimen is cut out of a sheet of urethane by a shopout die. This specimen is pulled in a testing machine, the breaking load is observed, and tear strength is reported in pounds per inch. This provides a useful method of comparing the tear strengths of rubberlike products. The better grades and stronger grades of cast urethane are surprisingly insensitive to notches. If problems are posed, there are adequate solutions in the use of reinforcing fiberglass inserted into the molds before the urethanes

Table 3I-2. Typical Tensile Products and Tear Strengths of Cast Urethanes

Grade of urethane	Hardness, Shore A	Elongation, %	Tensile product (tensile strength times elongation)	Tear strength, lb/in.
Hi-Extensible (8118)	50 to 60	800	48,000	100
Flexible mold (8059)	50 to 60	500	15,000	80
General-purpose (5715)	70 to 75	300	45,000	200
Maximum toughness (5719)	85 to 90	400	200,000	300

are poured. Typical tensile products and tear strengths for cast urethanes are given in Table 3I-2.

SUMMARY

The quality and usefulness of castible polyurethanes for plastics tooling depend on:

1. High tensile strength and elongation
2. Ease of pourability (mixed viscosity preferably less than 20,000 centipoises)
3. Minimum free isocyanate content (say, less than 6.5% free NCO)
4. Availability of good primers for adhesion to metal
5. Preparation for storage by dry N_2 flushing
6. Avoidance of haphazard filler addition.

3J: Composites

Composites are perhaps the newest group of engineering materials, following centuries of mankind's use of pure and alloyed metals and of many other materials found in nature. One could seek out applications such as the use of straw in brickmaking, laminated bows, or "Damascus" blades and barrels, but this discussion will be limited to substances based on the strength of intentionally included fibers.

Mathematicians dealing with solids usually make simplifying assumptions that these objects are both homogeneous and isotropic. Metallurgists recognize that this assumption often is invalid on a macroscopic scale, and always is invalid on a microscopic scale. However, flow patterns in forgings, crystal textures in sheet, and dual-phase steels use inhomogeneity and anisotropy to great advantage in controlling grain size, improving electrical or drawing properties, and otherwise obtaining specific properties where they are most useful.

The most omnipresent of fiber composites are automobile tires, in which elastomers are strengthened by many strands of steel, nylon, rayon or other cords incorporated in different numbers and orientations.

Perhaps the greatest advantage of composites over metals lies in the fact that they permit designers to obtain properties in exactly the locations and orientations desired. Prepregged fabrics, and their combinations, can be laid up with specific orientations so that their end products will have higher moduli or strengths in particular directions or can be bent more readily in one direction than another.

Prepregged tows and tapes offer essentially unidirectional strength, whereas prepregged cross-plied fabrics provide multidirectional properties. So-called quasi-isotropic structures can be laid up in multiples of four orientations at 0, +45, −45, and 90°.

Advanced carbon-fiber composites are most attractive where stiffness is at a premium or where sizable compressive forces exist. They may find application in some portions of aircraft landing gear, or they might provide yokes for some swing-wing aircraft. Their use would eliminate some difficulties now encountered in machining and forging operations.

During the last decade, design engineers have been using thermosetting carbon-fiber composites as replacements for metals in aircraft. However, recent advances in thermoplastic resin technology and the proposed cabin fire safety requirements have caused design engineers to take a closer look at these plastic materials. Their strength-to-weight ratios, toughness, corrosion resistance, stiffness-to-weight ratios, wear resistance, and producibility make them suitable contenders.

Engineers will have to be more precise when referring to graphite- or carbon-fiber composites. This term today can mean either thermosetting carbon fiber/resin composites or thermoplastic carbon fiber/resin composites. Both "composites" use a resin and reinforcing carbon fibers; however, there are basic differences, especially in regard to resins. A thermosetting resin is one which solidifies when heated under pressure and which cannot be remelted or remolded without destroying its original characteristics (epoxy resins). On the other hand, thermoplastic resins soften when heated and harden when cooled. This heating and cooling process can be repeated, and the material remolded at will. This basic difference is due to the molecular characteristic of the resins. When a thermosetting resin is heated under pressure, its molecular structure becomes cross-linked, whereas a thermoplastic resin's molecular structure remains linear before and after heating and cooling. The cross-linking of the thermosetting resins is responsible for the resulting brittle matrix, which has a high stiffness-to-weight ratio and high yield strength, but a low strain to failure.

These new advanced thermoplastic composites can be manufactured by three basic methods: molding, filament winding, and pultrusion. The prepreg, lay-up, and molding process is being considered for the production of large aircraft panel components. The plastics companies are developing films which will be 48 in. wide for this application. The prepreg is made up of laminates of resin film, and the fiber reinforcement can be a woven mat or unidirectional fibers (either carbon or glass). This laminate is heated in an infrared oven to the softening temperature, which is related to the resin, and is inserted into a mold. Pressure is applied to the mold for a few minutes, as opposed to hours in an autoclave for thermosetting resins, and the final part is removed, ready for use. This thermoforming process can be used to form a variety of components.

Spools of what is called "black thread" are the end result of one manufacturing process but the beginning of most advanced composite parts used in the aerospace industry. The "thread" on the spools is actually a group of untangled fibers of carbon referred to as a "tow." In fact, 90% of the fibers currently used in the manufacture of advanced composites are really carbon fibers, but the term has been generalized and these fibers are often referred to as "graphite" fibers. The term "composite" denotes that more than one component material is used (such as fiber and resin). Although a resin containing glass fibers can be termed a composite, the term "advanced composite" is generally used when referring to a high-performance part consisting of graphite (carbon), boron, or aramid fibers and a resin. Typical resins can be epoxies, polyesters, or polyimide or thermoplastic resins such as nylon and polysulfone. Engineering takes place not only in the development of the fiber and resin, but also in the manufacturer of the final product.

Filament winding (FW) has been called the sleeping giant of the plastics industry. The world's largest single application of fiberglass composites consists of a 28-ft-diam by 45-ft-long filament-wound chimney-stack liner using 1.5 million lb. of fiberglass. Feasibility studies are underway for 200-ft filament-wound fiberglass ship hulls. FW composites offer an enormous potential for production of structures that will be more efficient and less expensive than present metallic construction.

The FW process applies individual spools of fiber and prepreg tow or prepreg tape to a male mandrel. Current technology permits winding on mandrels of virtually any shape except those with concave surfaces.

Hybrid composite materials are simply made by mixing fiber spools in the delivery system. Material thickness can be easily varied by adding or subtracting spools during winding. Multiple plies can be wound on a mandrel to build up thick stacks of machine-deposited uncured material that can be cut and stacked in a mold to make thick panels with minimal manual cutting and lay-up.

Over the last three decades, new high-specific-strength fibers have been used on FW missile cases. Major improvements in fiber specific strength have occurred nearly every decade for the last half-century, from E glass to S glass to graphite to Kevlar and now to the IM-X series of graphite fibers that approach tensile strengths of 800 ksi in experimental lots. Some graphite-fiber suppliers predict that fibers with tensile strengths of 1 Msi will be available by the start of the next decade. Such fibers would have nearly the weight of today's composite pressure vessels.

Wet-wound epoxy resins used in FW exhibit performance and environmental resistance rivaling those of the common aerospace hot-melt epoxy prepreg tape resins. Moreover, breakthroughs in filament winding of polyimides and thermoplastics should yield tougher FW structures that can be used at higher temperatures.

No longer are plastics regarded as being inferior to steel. Composites — often called engineered plastics or fiberglass, but more commonly known as composites — have gained acceptance by consumers and industrial users. Whether thermoplastic (those that can be remolded by applying heat) or thermosetting (those that cannot be transformed back to a liquid state), plastics continue to work their way into applications in the automobile, transportation, structural, architectural, and consumer markets. These materials are replacing steel in many of its traditional uses because they offer excellent strength-to-weight ratios, corrosion resistance, dielectrics, design flexibility, dimensional stability, parts consolidation, moderate tooling costs, and lower fininshing costs. Together, these benefits add up to products that have value, are economical, and better serve the needs of the consumer.

There are numerous types of composites. The term "composite" is a generic term meaning an assembly of dissimilar materials intended to do a job that none of the individual materials can do by itself. For example, reinforced concrete is a composite. It is made of a concrete matrix reinforced with steel. In the plastics industry, the term "composite" can mean anything from a material that consists of reinforcing fibers in a resin matrix to a material in which a core material such as balsa wood, or honeycomb, is sandwiched between thin reinforced plastic skins.

Since 1980, over 500 steel fabricating companies have gone out of business. Although there are many reasons for this, one contributing factor has been the shift from steel to composites.

By 1996, it is estimated that there will be less than 50 lb of steel in the average American-made automobile.

RESINS

Resins are by-products of the petroleum industry. Although they are petroleum-base products, they currently account for only 2% of our national consumption of crude oil. However, this 2% is insignificant in comparison with the total fuel savings accounted for by the entire plastics industry, because plastics and plastic composites require less energy for production of finished products and for transportation of these products to market due to their light weight. These benefits make the use of composites attractive to today's energy conservationist.

There are two types of resin groups: thermosets and thermoplastics. Thermosets include epoxies, polyesters, and vinylesters, each offering different benefits to the end-user. For example, epoxies offer the greatest versatility and widest range of physical properties, polyesters are the least expensive but offer a limited range of physical characteristics, and vinylesters are both versatile and economical. The various characteristics of the available resins make the applications of these materials almost limitless.

Although the resin itself accounts for little in the way of structural strength, it is nonetheless important to select the appropriate resin for a given application. Failure to do so can result in deterioration of the resin, eventually weakening the end product by changing its physical properties.

REINFORCEMENTS

As with resins, there are numerous types of reinforcements to suit any possible design need. However, the glass fiber is the principal reinforcement material for strengthening of resins, because it provides significant cost-to-performance ratios. The most common is "E" fiberglass, which provides the composite with average strength. Another commonly used glass fiber is "S" glass.

"S" glass is an aircraft-quality glass reinforcement that provides about 33% more strength in the composite product than does "E" glass. Although it is more expensive than "E" glass, it is quickly gaining popularity in the boat industry because of the increased value it brings to the end product in reduced fabrication costs.

Less economical, but providing excellent strength-to-weight ratios, are the kevlar, graphite, carbon, and boron fibers. Generally, economics and the strength, stiffness, and weight requirements of the composite material govern the selection of the reinforcement.

CORE MATERIALS

Core materials most commonly include balsa wood, urethane foam, or honeycomb, with the selection depending on the application. Core materials add high bending strength to the composite without much additional weight. In the boat industry, where cost is a consideration, the most commonly used core materials are balsa and foam. In the aerospace industry, where light weight and fuel economy are the primary concerns rather than costs, honeycomb cores are widely used.

AUTOMOTIVE COMPOSITES

There is a great potential for use of composites in the automotive industry. I think we can all relate to the Pontiac Fiero as a volume production car which demonstrates the beginning of an aggressive, forward-looking approach for the future.

The Fiero not only demonstrates a new concept in design, manufacturing, and assembly through the use of composite body panels on a steel space frame, but it is also providing for production evaluation of various composite materials and processes: SMC/urethane materials for the hood, roof, deck lid and upper rear quarter panels; RIM and reinforced RIM urethane for bumpers, fascias, fenders, and door panels; and injection molded thermoplastics for the rocker panels and lower rear quarter panels (Figure 3J-1).

Figure 3J-1. Pontiac Fiero

Figure 3J-2. Pontiac Fiero GT

The Fiero GT has a blow molded nylon spoiler in addition to restyled body panels and bumper fascias. The Fiero is demonstrating the many advantages of composites for customer value and appeal, and this technology is being refined and supplemented for greater value in the future (Figure 3J-2).

An automotive study conducted by the University of Michigan included projected material trends which showed that the total weight of steel and iron in an average car is dropping whereas the use of plastics and composities continues to increase.

This same study projected an increase in the use of plastic and composite body panels from the current 5% level to a level of 70% by the year 2000. A considerable amount of development, planning, and simultaneous engineering must be done if an evolutionary transition of this magnitude is to be economically accomplished.

The interest in composite car bodies dates way back to Henry Ford's famous sledgehammer demonstration of the impact resistance and durability of a trunk lid made of a soybean-base composite.

In 1953, the Chevrolet Corvette launched the all-plastic body as a production reality. The Corvette has been a continuing test bed for composites applications.

Since 1953 the Corvette has introduced new composite materials and processes including advanced mold coated SMC, front and rear soft bumper fascias, composite bumper beams, seat shells, and fiberglass leaf springs. Experimental composite components, such as drive shafts and wheels, have also been evaluated on Corvettes.

The Corvette Indy is a futuristic two-seat mid-engine research vehicle. It is the basis for a running prototype featuring a total wheel-control system that incorporates all-wheel drive, all-wheel steering, antilock braking, traction control, and active suspension (Figure 3J-3).

The Ford Probe V is another futuristic concept car which illustrates the potential for composites in design and manufacturing.

An earlier Ford project was an experimental car for demonstration and evaluation of graphite-fiber-reinforced composites. Not only was the body made from composite materials, but structural and functional components were used as well.

Many automotive companies have built concept cars which demonstrate composite applications. The Pontiac "Phoenix" concept car was 700 lb lighter than the original Ventura from which it was converted. This car was displayed showing the composite body and structural components along with the conventional steel components (Figure 3J-4).

The "Phoenix" weight-reduction concept car illustrated one of the fastest-growing areas of plastic body panels: full soft bumper fascias. This car also demonstrated flex fenders, which is another area with potential for rapid growth (Figure 3J-5), and the use of structural fiberglass-reinforced plastic for fender supports, radiator supports, bumper beams, and wheels (Figure 3J-6).

A step beyond the Phoenix was the Chev-

Figure 3J-3. Corvette Indy

Figure 3J-4. Pontiac Phoenix

Figure 3J-5. Pontiac Phoenix

Figure 3J-6. Pontiac Phoenix

Figure 3J-7. Chevrolet XP-898

rolet XP-898 concept car, which integrated composites into the body structure using large composite moldings for the underbody and the front and rear upper structures (Figures 3J-7 and 3J-8).

The XP-898 was a running concept vehicle which was based on a 3/8 scale model of a composite integral structure concept car which was fabricated and assembled from large composite moldings. This concept model illustrated the feasibility and potential of using composites to replace steel for achieving weight reduction and design flexibility.

Matra, Lotus, and DeLorean cars incorporate this type of integral body structure.

Peugeot built a research vehicle called V.E.R.A., which was based on the Peugeot 305. The use of plastics in the 305, which already accounted for 8% of total weight, was more than doubled in the V.E.R.A. An additional 205 lb of plastics were added to bring the total plastics usage to 368 lb, or over 22% of the new weight. SMC doors are used on the V.E.R.A., saving about 8 lb each. The two-piece bonded assembly contains continuous glass fibers in the high-stress areas. Peugeot's "VERA PLUS" research vehicle is a further extension of weight reduction and aerodynamic styling through additional composites usage.

Audi has built a research car which demonstrates the potential for use of major plastic components. This research car has a lightweight composite roof, and the floor is also molded of lightweight composite material.

VW has built an Auto 2000 concept car based on the Rabbit. The front-end module consists of an SMC support panel that holds all of the components as well as providing the bumper structure behind the RIM urethane fascia. The SMC support panel for the rear-end module not only provides the bumper structure behind the fascia, but also provides the exterior body surface above the fascia. The wheels are molded of a composite material and weigh 40% less than standard steel wheels. This car also uses a fiberglass-reinforced plastic rear torque axle which has less than half the weight of a comparable steel assembly and saves 20 lb.

Fiat has built a VSS concept car. VSS is an Italian acronym for "experimental subsystem car." This car is based on the RITMO. The VSS uses a steel cage to which all of the mechanical components and preassembled subsystems are bolted. All of the body components are composites, such as the hood, which saves 29% in weight, and the doors, which save 32% in weight. The total plastics content of the VSS is 27% of vehicle weight. Only 1800 spot welds are used instead of 3000, and the required production time is 5 h shorter. Fiat plans to use a common "space frame" for various models in which only the hang-on body panels need to be changed.

Composite concept cars are being built and shown in Japan. The Nissan Mid-4 has a complete body made of fiberglass-reinforced plastic. The Mitsubishi MP 90X also has a fiberglass-reinforced plastic body, and the Toyota AXV has a plastic body as well as fiber-reinforced plastic suspension and chassis components.

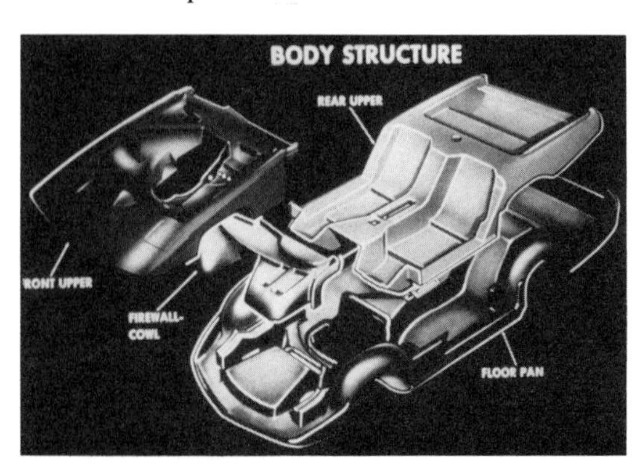

Figure 3J-8. Chevrolet XP-898

The Buick Wildcat is a driveable, four-wheel-drive, antilock-braking, two-passenger concept car that is a total integration of design, structure, electronics, powertrain, and driver and passenger. The structure of Wildcat is glass-and-carbon-fiber composite with front and rear steel cross-members bolted to the underbody. Access to the driver and passenger compartment is via a raised canopy made of cast gray acrylic-and-fiber-reinforced polyester resin (Figure 3J-9).

The Pontiac Trans Sport is a new multipurpose concept vehicle showing ideas which could be functional and producible in the future, and it exemplifies the design freedom possible through the use of composite body panels (Figure 3J-10).

The Saab 900 Turbo 16 experimental vehicle weighs only about 2000 lb. Composites were used in an innovative way to reduce weight and improve aerodynamics. The intrusion beams in the doors are made of glass-and-graphite-fiber-reinforced plastics. The bumpers in front and rear have been replaced by front and rear impact panels which are integrated into the body. These panels are aramid-reinforced composite materials which absorb the energy in a collision, but are flexible enough to deform elastically in a minor collision and then return to original shape.

Other composite concept cars have been proposed or built, and still others are on the drawing boards and in the styling studios, but the materials and processes must be further developed before high-volume production of these concepts can be achieved.

Figure 3J-9. Buick Wildcat

Figure 3J-10. Pontiac Trans Sport

In the area of reaction injection molding, fast-reacting polyurea and prepolymer chemicals are being developed which shorten the mold-closed time. Some of them react in 10 s or less and are self-releasing from the mold. New equipment has been developed that provides precise control of temperatures, ratios, and injection of filled and reinforced materials. New injection systems provide accuracy and high productivity, and new mix heads accommodate multicomponent systems with a variety of fillers. Fillers such as glass flake, mica, and calcium carbonate provide stiffness, thermal stability, and cost effectiveness.

The 1981 Sport Omega was the world's first car to use reinforced RIM fenders as well as a full front-end RIM fascia. The Porsche Carrera GT used nine major RIM body panels, including fascias, fenders, and spoilers.

RIM technology is being expanded into structural areas through its potential to bring high productivity and reduced costs to the resin transfer molding process wherein fiber preforms are placed in a mold prior to resin injection (Figure 3J-11).

Ford engineers have demonstrated this combination of RIM and preform molding in a front quarter apron assembly which weighs 45 lb compared with the 65-lb steel assembly consisting of 44 individual parts. Lotus uses this type of structural preform RIM in molding the body shells for their cars such as the Series 3 Esprit and the Turbo Esprit, which has a 0-to-60 performance of 5.5 s. Lotus also used their vacuum-assisted resin injection technology to produce the composite mono-

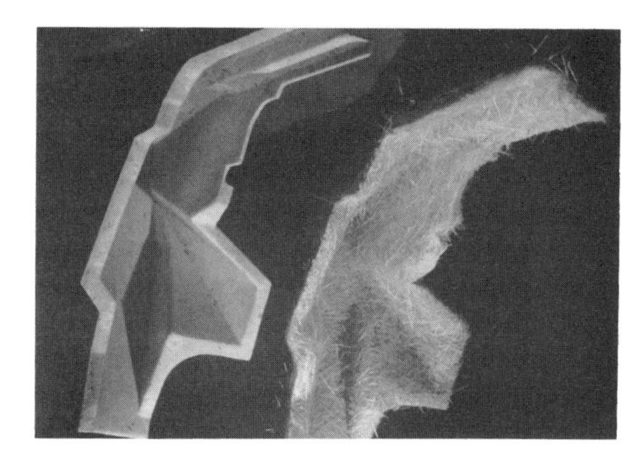

Figure 3J-11. Parts made by RIM technology

coque "clamshell" body for the ETNA "high technology" concept show car which received a six-page review in *Road and Track* magazine.

In the area of SMC, new material formulations have been developed and experimentally molded which exhibit greatly improved surface appearance and greatly reduced curing time. Unique sheetmaking equipment has been developed to produce SMC with improved consistency and uniformity. Materials are accurately weighed, mixed, metered, dispensed, and compounded.

A new generation of fast dual-acting presses, with accurately controlled parallelism, closing speed, and pressing force, are currently ready for implementation into production. Molding-cycle times with this type of press can be reduced to less than 1 min because the curing reaction can be controlled, analyzed, and optimized while the mechanical portion of the process is very fast and always under control.

Molded coating is used to fill surface porosity and uniformly coat the SMC part to improve subsequent painting operations for exterior body panels. The coating material is a thermosetting polyester which is filled with a black, conductive filler to allow electrostatic painting.

Robots have been demonstrated that can remove a molded part from a mold while loading the SMC charge for the next molding (Figure 3J-12).

The adhesive bonding of the outer panel to the inner panel creates box sections and provides stiffness and strength to the assembly. Bonding is a critical operation in the manufacturing process because it affects structural integrity and may in some cases affect surface appearance.

Robots and mix heads have been developed for applying adhesive to the SMC panels. The panels are held in a heated bonding fixture while the adhesive cures. New adhesives and processes are currently being developed to reduce adhesive curing time to less than 1 min.

SMC has been used for the hoods of many cars such as the Chevy X-11. SMC permits unique styling features and a weight reduction of 30 to 50% depending on design and material formulation.

Two-piece SMC tailgates have been used on Oldsmobile and Buick station wagons to replace seven-piece steel assemblies (Figure 3J-13).

Ford has demonstrated the feasibility and advantages of an SMC tailgate for pickup

Figure 3J-12. Robots in molding operations

Figure 3J-13. Two-piece SMC tailgate used on Oldsmobile and Buick station wagons

trucks. The SMC showed considerably better impact and damage resistance than steel during testing.

Owens-Corning Fiberglas has built a basic transportation vehicle to demonstrate the potential of an all-fiberglass-reinforced plastic pickup truck body and box.

SMC hoods and fenders have long been used on heavy-duty trucks. Truck doors are now in use with three pieces of SMC replacing seven pieces of steel. SMC truck cabs have been built in which the number of parts was reduced by over 80%. The Mack Ultraliner, with annual production of around 10,000, uses SMC for all of the cab panels.

In Europe, Mercedes is developing an SMC rear deck lid for their 190, and Citroen is using an SMC hood on their BX. The Citroen BX also has a fiberglass-reinforced plastic rear hatch. The BX rear hatch is injection molded from an SMC-like material called ZMC. Several new machines have been developed for injection molding of fiberglass-reinforced polyester compounds. Large vertical machines are being installed in Japan for injection molding. Mazda has a 2000-ton machine that has a dual injection system — one for thermosets and one for thermoplastics — capable of injecting over 20 lb each. Honda has several 2500-ton vertical thermoplastic injection molding machines with 360-oz injection capacity to produce parts for their CRX model. These machines have a quick mold-change capability of approximately 6 min. A robot is used to remove the parts from the molding machine.

The Honda CRX fenders are molded from a polycarbonate/ABS alloy. Such fenders have approximately half the weight of steel. The same material is used for the lower door panel in which an extruded insert is used to cover the attachment fasteners. The combination rocker panel and lower quarter panel is injection molded of the same alloy. The front-end header panel and headlamp covers are also molded from the polycarbonate/ABS alloy.

A thermoplastic olefin blend is injection molded for the front and rear bumper covers on the Honda CRX.

A 4000-ton hydromechanical horizontal clamp injection molding machine has recently been installed by DuPont in their new development center in Troy, Michigan, for molding prototypes for evaluation, such as door panels and other body panels.

General Electric has been using a 5000-ton machine for development of polycarbonate alloy bumpers and other body panels in their Louisville facility. GE is installing a 3000-ton machine in Livonia, Michigan, for increased development work. Even larger injection molding machines have been designed and are being built, such as Crauss-Maffei's 6000-ton machine.

Not all of the excitement in automotive composites for the future is in body panels. Much aerospace and aircraft composites technology is coming down to earth for automotive structures.

Bell Helicopter's Advanced Composites Airframe Program demonstrates the design, analysis, and productionizing of an optimized composites structure. Nearly every component has been designed for maximum parts consolidation and efficiency with a clear understanding of functional performance requirements.

The Beech Starship One is but one example of the advances being made in graphite-epoxy composite structural design providing increased accuracy and reduced tooling.

The fiberglass-epoxy leaf spring introduced on the 1981 Corvette set a whole new standard. The durability bogie for a fiberglass spring is five times that of a steel spring, and there have been no failures. Fiberglass springs are now also used on vans and on the Seville, Eldorado, Toronado, and Riviera (Figure 3J-14).

Ford has introduced a composite drive shaft on its Econoline van. The single shaft replaces a two-piece metal assembly which required a center yoke.

An automated filament-winding machine for high-volume production of multiple driveshafts has been developed. Three shafts are wound at one time in about 3 min on removable shuttling mandrels.

Automated fiber-placement technology for complex shapes is also being developed using multiple motions. Composite trans-

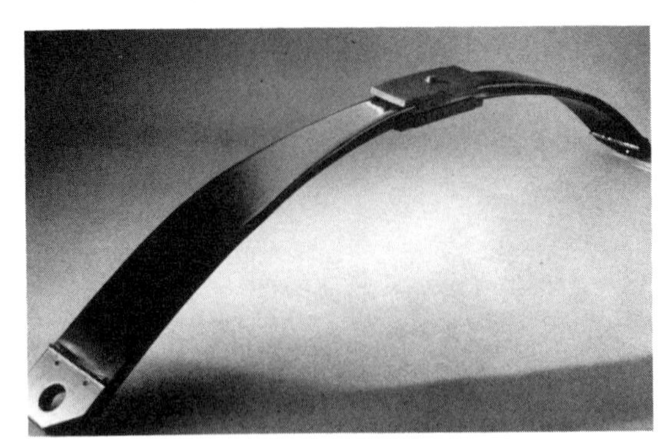

Figure 3J-14. Fiberglass-epoxy leaf spring

mission supports have been molded and evaluated which show that the use of 4% graphite fiber with glass fiber can increase stiffness by 20%. In cases where space or weight constraints prevail, graphite fibers can provide the additional structural requirements if necessary.

Composite bumper beams are in use on the Corvette and provide parts consolidation and energy management.

Many under-the-hood items have potential for composites in the near or distant future. Ford has been developing an intake manifold in Europe which is molded over a core made of a tin-bismuth alloy which can be melted out and reused.

Composite oil pans have been evaluated as well as valve covers where complex shapes are required. Amoco Chemical is evaluating composites for engine components such as connecting rods, wrist pins, and piston skirts. The reduction of inertial forces could be significant.

It may be quite a while before the Polimotor composites engine moves from race cars into high-volume production, but along the way we should be able to continue to increase our knowledge of the unique capabilities of composite materials for a wide variety of applications.

There are two production vans which are further examples of the potential of automotive composites in the next decade. The Renault Espace has a production volume of around 20,000 per year. All exterior panels are composites. The door panels, the front and rear bumpers, and the front spoiler are SMC. Side panels, the rear lift gate, the hood, and the roof are polyester resin injection molded with the hood and roof being stiffened with molded-in-place foam sheets. The roof and one-piece side panels are bonded to the sheet-metal chassis structure.

The Ford Aerostar, at an annual volume of around 170,000, also demonstrates the production feasibility of many composite materials and processes including SMC hood and rear lift gate, injection molded polycarbonate/PBT bumpers, and a blow molded polyethylene fuel tank.

Based on these concepts, development, and production realities, our challenge is to change the way we look at automotive design and manufacture.

The automobile industry is currently seeing a phenomenal growth in the use of reinforced plastics as the need for lightweight, low-maintenace, high-strength materials becomes more acute.

Section 4
Structural Shapes

4A: Structural and Bar Shearing

In structural and bar shearing, the cost of cutoff is a small percentage of the over-all cost. Material handling and setup, along with die changes, are most important from a cost standpoint.

The cost of cutting off a bar of steel (Figure 4A-1), regardless of how it is cut, is determined by the following formula:

$$C = \frac{S + T}{N} + R$$

where C is total cost per part, S is setup cost, T is tool cost, N is number of pieces to be cut, and R is running cost (feed, cut, discharge).

ALTERNATIVE CUTOFF METHODS

Sawing with a hacksaw, a radial saw, or a band saw is the most widely used cutting method, but it is also the most costly. Sawing must be used for cutting certain shapes and sizes for economic or quality reasons — for example, cutting of thick-wall round tubing. But the average plant does not have to saw everything, and consequently vast savings can be realized on parts that can be sheared. Table 4A-1 is a chart compiling the times required to cut various types of bars using a hacksaw, a band saw, a high speed steel band saw, and a hydraulic shear. The cost of deburring the sawed ends is not included, but in most cases this operation is required after shearing. Another cost not included is the blade cost per cut, which is much higher for sawing than for shearing.

Shearing, to some people, is associated with crude cuts and distorted parts — an attitude usually held by those who have seen parts cut only on inexpensive universal-type shears. However, with proper tooling and machines, nondistorted, burr-free cuts are possible, and this article will illustrate how this is achieved.

Material handling and gaging have presented problems with many of the commonly used universal shears. Figure 4A-2 illustrates a typical universal shear that will cut angle iron, flat stock, round stock, square stock, channel iron, and "T" iron. Note the different die locations and feed and stop heights. It's very difficult to equip this shear

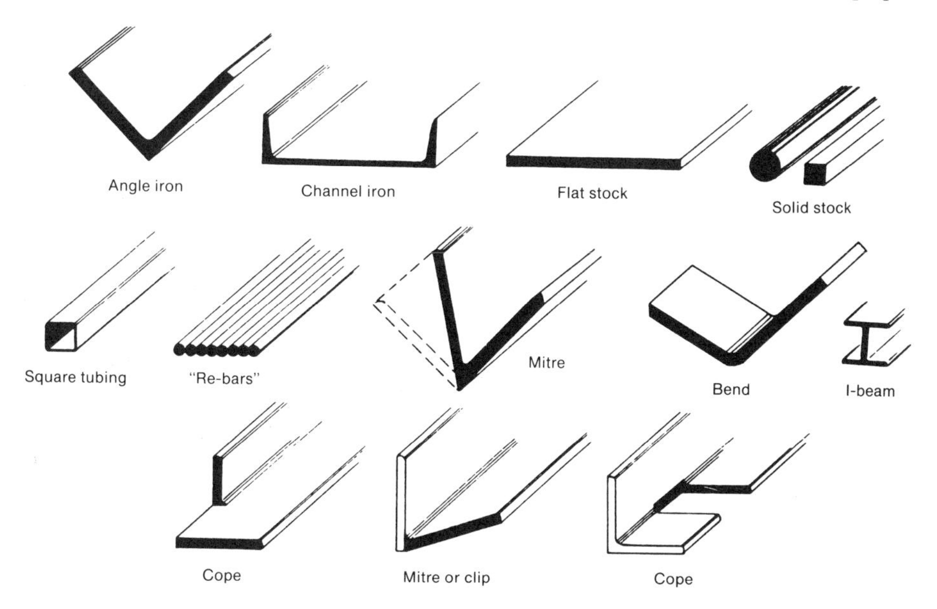

Figure 4A-1. Typical shapes that can be sheared

Table 4A-1. Time Comparison Between Sawing and Hydraulic Shearing

Size, in.	Material or shape	Time required for cutting by: Hacksawing	Band sawing	DoALL band sawing	Shearing	Time savings, %, realized by shearing compared with: Hacksawing	Band sawing	DoALL band sawing
1¼	C-1017 C.F.	30 s	1 min, 2 s	7 s	2 s	93	97	71
3/16×5	C-1018 H.R.	27 s	1 min, 15 s	5 s	3 s	89	96	40
1/2×9	C-1018	3 min, 54 s	6 min, 15 s	18 s	3 s	99	99	83
6	Channel	45 s	2 min, 54 s	23 s	6 s	89	96	74
1/2×4×4	Angle iron	54 s	3 min, 38 s	21 s	6 s	88	97	71
1/2×6×6	Angle iron	1 min, 59 s	5 min, 50 s	34 s	6 s	94	98	82
Mitre 3/16×3×3	Angle iron	—	1 min, 10 s	12 s	6 s	—	91	50
Mitre 1/2×4×4	Angle iron	—	4 min	30 s	6 s	—	97	80

Figure 4A-2. Universal shear with various feed heights and locations

with a feed and stop system that will result in fast cutting to length.

Abrasive and friction cutting offer the advantages of cutting various shapes without changeover costs, production of flat nondistorted ends, and rapid cutting. Their main drawbacks are the high blade cost per cut and the cost of deburring both ends of the cut part. A very sharp razor edge is usually produced and is dangerous if not completely removed.

HYDRAULIC BAR SHEARS

To fill the need for a fast-cycling shear for cutting structural and bar-shape parts that would cut with little or no distortion or burring, a series of hydraulically powered shears were developed. Hydraulic power was chosen because high forces are required throughout a relatively long stroke, due to the rake angle of the shear blade. The reason for the rake angle will be seen later in this article. Hydraulic power is noted for its maintenance-free performance, shock-free operation, ability to control the movement and stroke of the ram, and ability to operate more than one machine from a single power source.

Figure 4A-3 illustrates one of the earlier

hydraulic shear models, an 18-ton shear for cutting angle iron in sizes up to 4 by 4 by 1/4 in., flats up to 6 by 5/16 in., rounds up to 7/8 in. in diameter, as well as other shapes. Different dies are used to cut the various shapes and do the different operations, and a time of 10 min is required to make a tooling change.

Tool changeover time was often too long when the two-post shear shown in Figure 4A-3 was used. Therefore, a new series of four-post quick-change shears was developed, as shown in Figure 4A-4. With this shear, changing from cutting of angle iron to cutting of some other shape, such as channel iron, for example, takes less than 45 s. The die assemblies are self-contained with their own blade-guide arrangement and merely have to be clamped in place by two cam locks. The blades engage the press ram through a "T" slot and holder design. Because all die assemblies, except the round stock which is spring loaded, have the same shut height, the dies are easily interchanged with the shear ram at the bottom of the stroke. A tool carrier (Figure 4A-5), located next to the machine on tracks, stores the inactive die assemblies. Changing dies is accomplished simply by sliding the old die out of the shear onto the tool carrier, moving the tool carrier to align the desired die assembly with the shear, and sliding the new die into the shear. One operator can easily make the change.

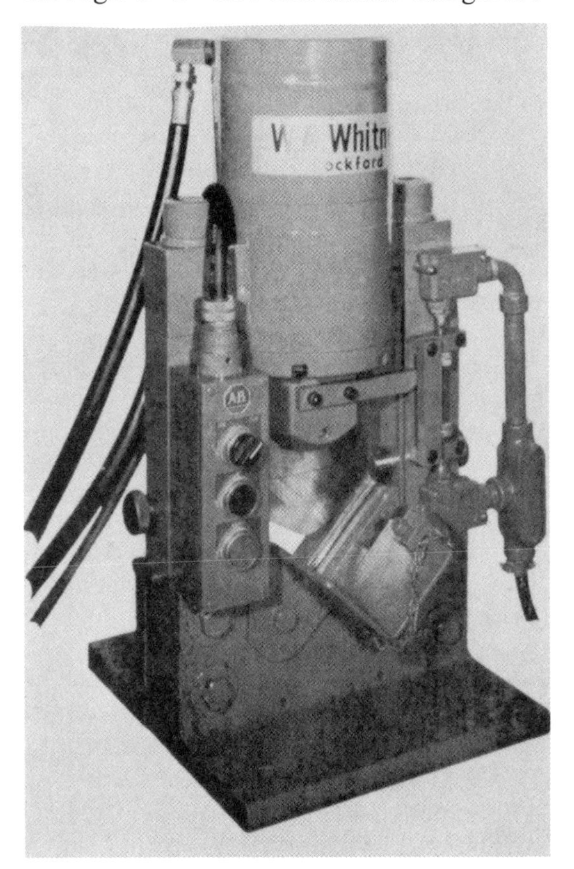

Figure 4A-3. Eighteen-ton two-post hydraulic shear

In a survey made by the American Metal Stamping Association, the quick-change shear provided 22% improved die materials, 34% improved tooling for short runs, and 44% faster die-changing methods.

Figure 4A-4. Quick-change 50- and 150-ton shears featuring "injector" tooling that can be changed in less than 45 s

Different stroke lengths are used to shear the various bars, and since the machine is a constant speed-force hydraulic unit, the cycle time is reduced by reducing the stroke. For example, the normal cycle time required for cutting 6-by-6-by-1/2-in. angle iron is 7 s, whereas for cutting 2-by-2-by-1/4-in. angle iron the stroke length is reduced by more than half and the cycle time is only 3 s. Round bars can be cut in 1 s. Six different stroke lengths are available and can be se-

Figure 4A.5. Die-change cart that stores dies near press for quick die changes

Preset dogs are mounted on a hex shaft. Each 1/6 turn reveals a new combination.

Figure 4A-6. Method for fast setting of press-stroke dogs

lected by the operator simply by turning a knob to the proper setting marked for the die being used. This is done in a few seconds, as shown in Figure 4A-6.

DOUBLE-CUT VS. SINGLE-CUT SHEARING

Two types of dies are used for shearing of structural and bar shapes: double-cut dies and single-cut dies.

Double-Cut Shearing

In shearing with double-cut dies, the shear blade acts as a punch and has a die on either side of it. A thin slug is sheared or punched out of the bar, its width depending on the size of the shear. Shears with capacities of 18, 50, and 150 tons have slug (blade) thicknesses of, respectively, 3/8, 1/2, and 3/4 in.

Double-cut shearing has three advantages. First, it produces distortion-free cuts on shapes that can be closely matched to the contour of the dies and where the surface in contact with the blade varies in shape. For example, angle iron and channel iron have square outside corners that can be closely matched by the dies, but the inside fillets vary with size. Angles measuring 2 by 2 by 1/4 in., 4 by 4 by 1/4 in., 6 by 6 by 1/2 in., and 8 by 8 by 3/4 in. have fillet radii of, respectively, 3/16, 3/8, 1/2, and 5/8 in. Both ends of a sheared angle or channel have the same square, burr-free quality after being double cut. Double cutting does not improve the quality of flat-stock, round-stock, or square-stock shearing.

Second, shearing forces are greatly reduced due to the rake angle on the blades. The angle-iron shear blade has a 120° included angle, which results in a 15° effective rake angle per side. Table 4A-2 gives a comparison of the forces required for double-cut and single-cut shearing. The rake angle on the blade badly distorts the slug, which is thrown away. Flat stock is double-cut sheared to increase the capacity of the shears.

Capacities of double-cut dies are determined primarily by stock thickness, but there is a limit on the bar width or flange size. With the rake angle, only a small area is being sheared at a time; therefore, the only way to increase that area is to increase the bar thickness. For example, an 18-ton shear will cut 4-by-4-by-1/4-in. angle iron but will not

Table 4A-2. Forces Required for Single-Cut and Double-Cut Shearing of Angle Iron

Size of angle iron(a), in.	Required force, tons Single cut	Double cut
4×4×1/4	60	15
6×6×1/2	175	40
8×8×3/4	345	140

(a) A-36; shear strength, 60,000 psi.

cut 2-by-2-by-3/8-in. angle iron even though its total area is smaller.

The third advantage of double-cut shearing is that no hold-down is required to keep the material from "kicking" up. The blade holds the material firmly against the dies, which creates a balanced condition. The material must span the two dies, and a minimum length equal to the stock thickness must be cut off to prevent damage to the blade or dies.

Single-Cut Shearing

In shearing with single-cut dies, the material is sheared by two opposed cutting surfaces that closely pass by one another — as in, for example, a pair of scissors or a typical mechanically driven universal shear. Nondistorted cuts are possible only where both the upper blade and lower die match the contour of the bar. For example, flat bars are cut free of distortion when the upper blade closely parallels the lower die. If a rake angle were applied to the blade, it would distort the part in contact with the blade; however, the other side of the cut in contact with the die would be free of distortion. Round and square bars are always single cut. Flat bars are usually single cut but may be double cut for increased size capacity. Angle iron must be double cut for best nondistorted cuts, but can be single cut if distortion is not a problem.

Single-cut shearing causes distortion of angle iron mainly because of the mismatch between the radius on the blade and the fillet size in the angle. The fillet size varies with the angle size, as noted above. Single-cut shearing requires a hold-down ahead of the die and, for best results, a support on the opposite side. Figure 4A-7 illustrates the importance of a hold-down and a support. The support is not required on single-cut angle-iron dies due to the shape of the bar, which prevents it from bending. A fixed support, as shown, is satisfactory, provided that the part being sheared is sufficiently long (approximately 6 in. on 50-ton shears). Shorter parts can be sheared if a spring-loaded support pad is used. The new style of round-stock dies used in four-post injector-type shears do not require a support due to their unique design, which will be explained later.

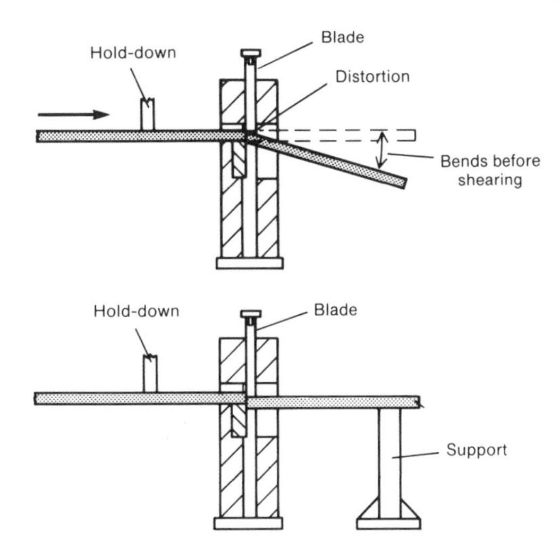

Figure 4A.7. Single-cut shearing of flat, round, and square bars, showing hold-down and support required for nondistorted cutting

NONDISTORTED, BURR-FREE CUTS

Figure 4A-8 illustrates what is meant by the phrase "nondistorted, burr-free cut." Normal deformation on the opposite side contacting the die still exists due to the inherent shearing characteristics of materials. The amount of deformation depends on the hardness of the material: the harder the material, the less severe the deformation. The amount of "break" depends on the thickness and type of material.

SHEARING OF ANGLE IRON

Angle iron can be sheared with double-cut or single-cut dies. Figure 4A-9 shows a

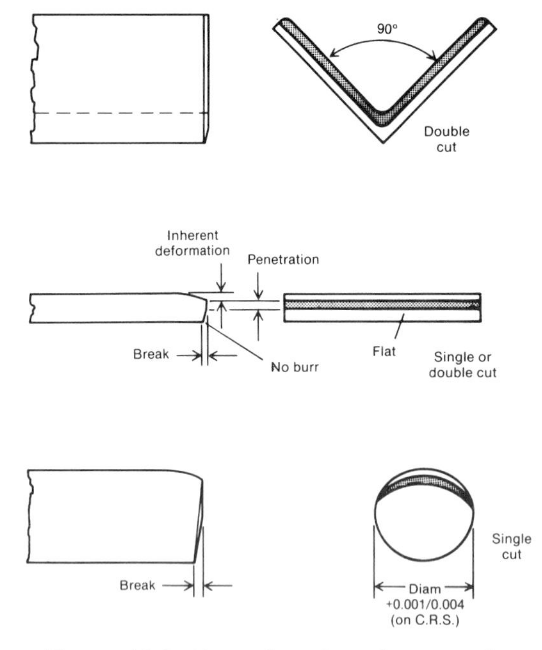

Figure 4A.8. Typical quality of cuts made by hydraulic shears

double-cut die and how it can be used to mitre angle iron up to 45° and cut square tubing and flat stock. This die will also cut small bar-size channels where flange height does not exceed 9/16 in. The versatility of this assembly, plus the fact that a hold-down is not required, makes it extremely popular.

Square tubing can be cut with little distortion provided that the wall thickness in proportion to the tube size is great enough to withstand the shearing force. The angle-iron blades normally have a large radius on the point, but in shearing of square tubing it is best to use a pointed blade. The type of material has a large effect on the quality of cutting.

Shearing of flat stock in a double-cut angle die creates a side thrust which is taken by blade guides. For this reason, it is best to cut two bars at once, one on each side, and in high production it is best to use a flat-stock die designed for that purpose.

The clearance between the blade and the dies must be adjusted to suit the angle-iron thickness. This is accomplished by inserting spacers behind the die inserts or holders. No spacers are required on 18-ton shears, one set on 45- and 50-ton models, and two sets on 150-ton models.

Single-cut angle dies are limited to angle-iron cutting. Figure 4A-10 shows an injector-type die with a manually adjusted hold-down on the feed side of the machine. The hold-down must be adjusted so that the material clears by only 1/16 in. An air-operated hold-down is used with feed systems that have a lift under the conveyor for lifting the material off the die during the feed cycle. The radius on the single-cut angle shear blade is of an intermediate size so that it can be used to cut a range of angle sizes with minimum distortion.

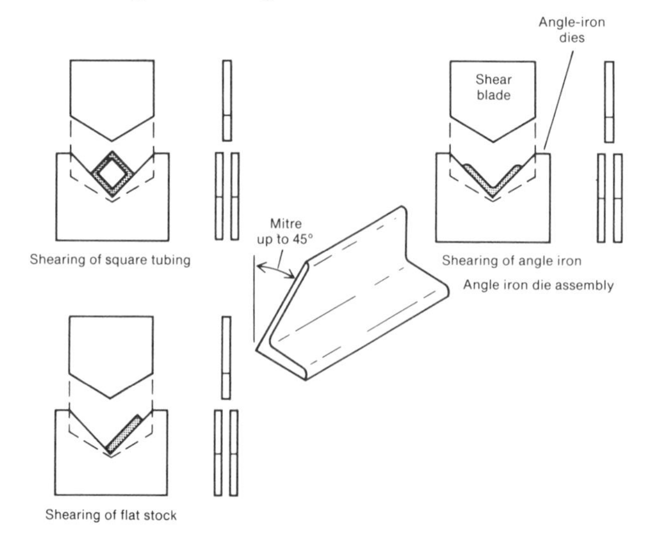

Figure 4A.9. Double-cut angle die for cutting various shapes without changeover

Figure 4A.10. Single-cut angle die with adjustable hold-down

SHEARING OF FLAT STOCK

Figure 4A-11 illustrates the double-cut flat-stock tooling assembly for a 150-ton shear. Note the concave rake angle on the leading face of the blade. Its concave shape balances the material through the center of the die.

The single-cut flat-stock die looks similar to the double-cut die except that the blade does not have a severe rake, but a mere 3/8 in. to the foot. This rake goes straight across

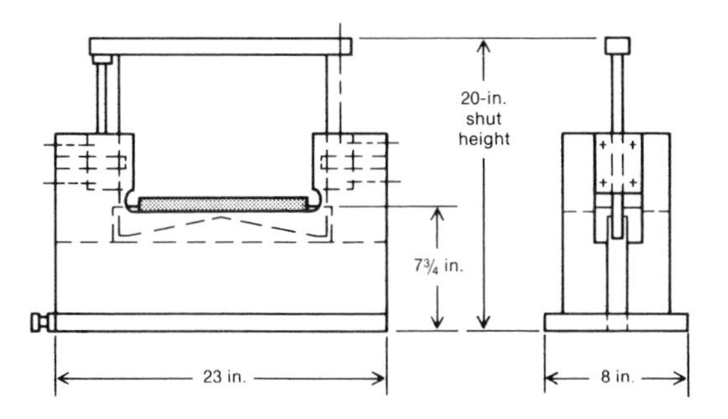

Figure 4A.11. Double-cut tooling assembly for cutting flat stock and bar stock

the blade and causes little or no distortion. If absolute flatness on the part is required, the blade should not have a rake angle. The 18- and 45-ton shear models have no rake.

SHEARING OF ROUND, HEXAGONAL, AND SQUARE BARS

Figure 4A-12 illustrates the type of die used in the 18-and 45-ton shear models. Five die cavities are provided — three for round bars, one for hex bars, and one for squares. The die cavity is made to fit the bars very closely for minimum distortion, and therefore this die is limited to shearing of five bar sizes in the shapes noted. The die cavity or hole for cold finished round bars is 0.002 to 0.004 in. larger than the bar. After the bar has been sheared, it will still freely slip back through the hole, which points out how little the bar is distorted. Hex and square bars can be sheared with the same accuracy. Smaller squares can be sheared with good success using a larger die cavity provided that a hold-down is added to the front of the machine. Normally, the die itself acts as a hold-down, because it completely confines the bar.

The single-cut blade shown in Figure 4A-12 has only a halfhole to fit over the bar and must be used with a rear support. This blade is simple in design and inexpensive. Figure 4A-13 shows a round-stock die for use in injector-type shears. This type of die features interchangeable bushings to suit the bar being sheared. Two identical bushings are used: one is for the fixed die and the other fits into a moveable slide, forming what is normally called the blade. Three bushing sizes (outside diameter) are used to cover a range of bar diameters up to 2½ in. in the 150-ton shear. The reason that there are three sizes is cost, because they are perishable. 7¾-in. die height (feed-conveyor height) is maintained only when the bar or hole size for each of the three bushings is at a maximum. Therefore, the height increases slightly when going to smaller bar.

The round-stock die in Figure 4A-13 has a spring-loaded slide which can be set in the open position to align very precisely the

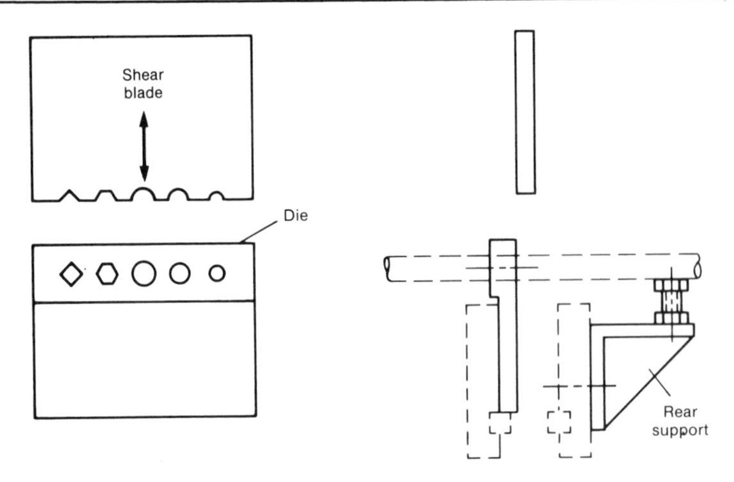

Figure 4A.12. Single-cut shearing die for cutting round, square, and hex stock in two-post shears

front and rear die bushings for feeding. Bars are sheared in less than a second as the ram forces the slide down, offsetting the bushings. Hexagonal and square bushings can be made for those shapes. A hold-down is used on the larger bars due to the high amount of torque involved.

As cited above, cold finished round bars are closely fit to the die hole within 0.002 to 0.004 in., but hot rolled bars need more clearance due to their oversize, out-of-round, scaly condition. Usually 1/32 in. is allowed for bars up to 1¼ in. in diameter, and 1/16 in. for bar diameters up to 2½ in.

SHEARING OF CHANNEL AND "I" SECTIONS

Channel iron and "I" sections have always been difficult to shear without distor-

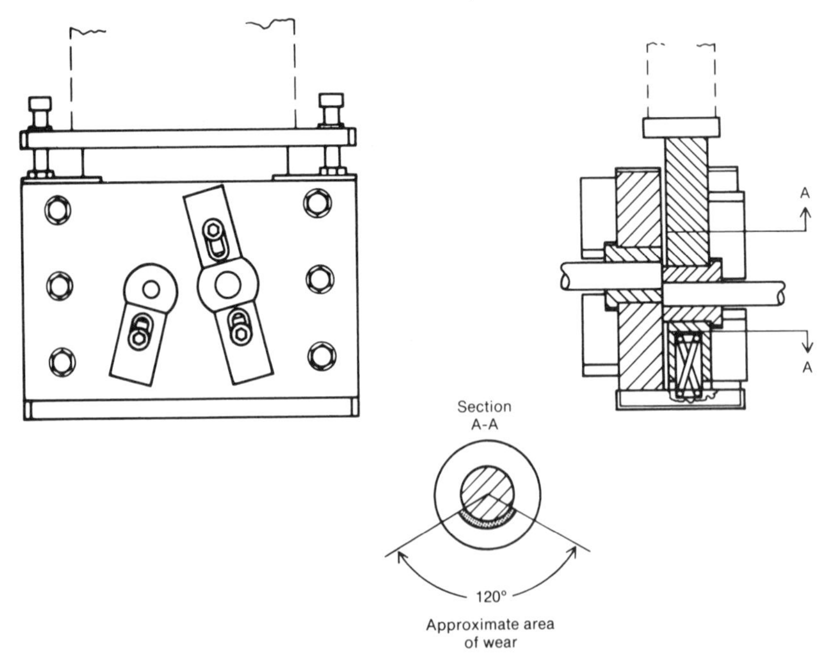

Figure 4A.13. Round-stock die assembly for four-post shear, featuring interchangeable bushings

tion, mainly due to the mill tolerance which allows their width to vary, depending on size, by as much as 3/16 in. But now there is a new type of die that overcomes this problem. Figure 4A-14 shows how a channel is cut in this double-cut die, which supports it on three sides. The "W"-shape blade shears the flanges outward against the inserts. This new die automatically, through hydraulics, causes the side inserts to move in and clamp the channel prior to contact by the blade. After the cut, when the blade starts to ascend, the inserts unclamp and move approximately 7/32 in. to allow the channel to be fed through the die.

One channel die assembly cuts a range of sizes. For example, a 150-ton shearing die cuts American Standard tapered flange channels and Ship and Car channels in sizes from 3 to 10 in. To adjust from one size to the next, the clamping cylinder located on the side of the shear (Figure 4A-15) is disengaged and a lever is turned to move the inserts in or out. Right- and left-hand screws are rotated simultaneously through a line shaft which keeps the inserts always centered about the blade.

Two shear blades are used to cut the 3-through-10-in. American Standard channel sizes. One cuts 3-to-5-in. channels and the other cuts 6-to-10-in. channels. Changing the blades takes only 30 s.

"I" beams are cut in the channel die by changing the lower die inserts and sometimes the blade. Figure 4A-14 illustrates how the inserts are made higher to fit up into the "I" for supporting the web sections. A set of two lower inserts is required for each size of "I" section, but in some cases one shear blade will cut more than one size.

Figure 4A.15. Shearing die with hydraulic cylinder that clamps channel for distortion-free shearing

The 18- and 45-ton shears use fixed dies for cutting channels, as shown in Figure 4A-16. These allow the flanges to spread, causing distortion, and the stock cannot be fed as easily as in the dies described above. Fixed channel dies are not recommended for high production shearing.

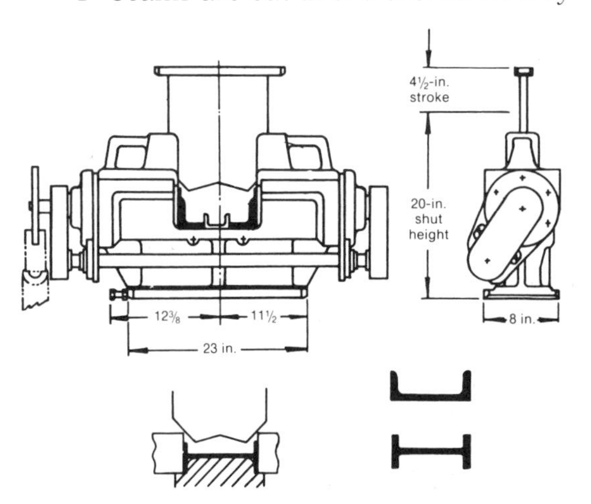

Figure 4A.14. Double-cut tooling assembly with adjustable die for nondistorted shearing of channels and "I" beams

SHEARING OF "T" BARS

"T" bars are sheared in double-cut angle-iron dies by using a special set of side inserts

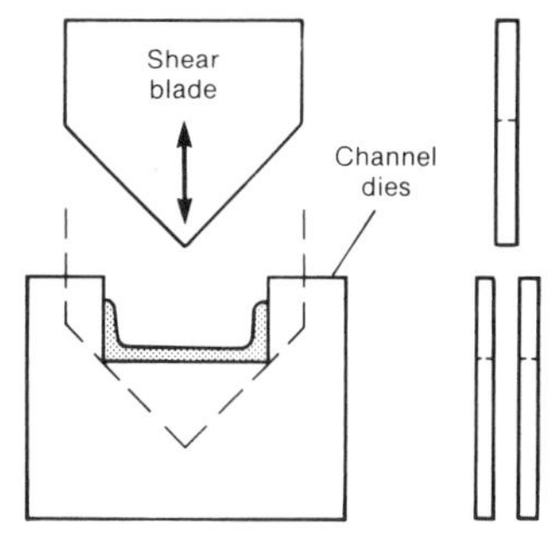

Figure 4A.16. Fixed channel die for two-post shears

as shown in Figure 4A-17. The angle-iron shear blade is used. Stroke is the limiting factor when determining shearing capacity. An 18-ton shear will cut "T" bars up to 1½ by 1½ by 3/16 in., a 45-ton shear will cut bars up to 2½ by 2½ by 1/4 in., a 50-ton shear will cut bars up to 3 by 3 by 3/8 in., and a 150-ton shear will cut bars up to 4 by 4 by 1/2 in. A slight amount of distortion can be expected in the web due to shearing.

SHEARING OF "Z" BARS

Structural "Z" bars are single cut with a special die designed for that purpose. The blade and insert must be made to match the "Z" bar's contour; however, due to the mill tolerance, some distortion can be expected. Only the 50- and 150-ton shears are adaptable for shearing of "Z" bars, and a special hold-down is required (Figure 4A-18).

Structural "Z" bars cannot be double cut due to their great thickness. Light-gage roll-formed "Z" purlins, used in the metal building trade, have to be double-cut sheared, as shown in Figure 4A-19. In this application, a 50-ton shear was mounted on a slide and used in a "flying shear" operation behind the roll forming machine.

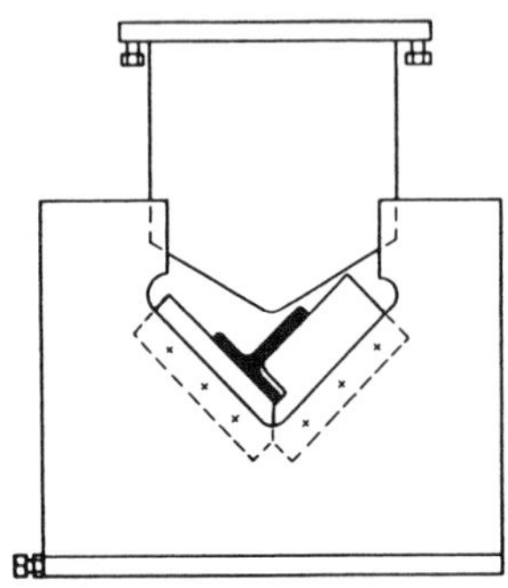

Figure 4A.17. Double-cut angle-iron die assembly with side inserts for shearing of "T" bars

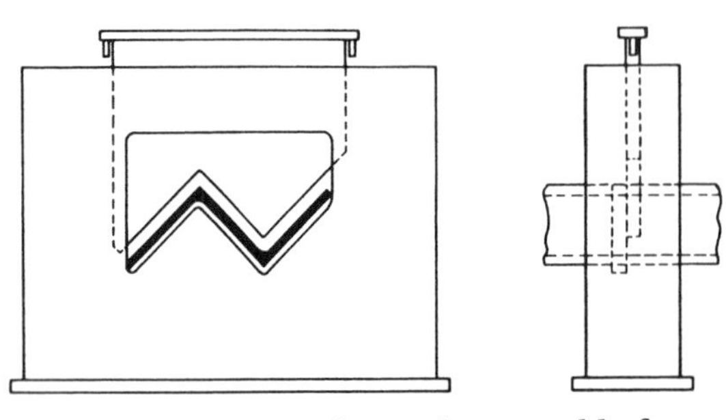

Figure 4A.18. Single-cut die assembly for shearing of "Z" bars

Figure 4A.19. Double-cut shear for cutting of "Z" purlins in 4.5 s

MATERIAL HANDLING MADE EASY

Otherwise inexpensive shear cutting can be very costly if feeding and gaging of the material are not done efficiently. Material handling on some shears constitutes 95% of over-all cost. With hydraulic bar shears, a common feed conveyor and stop that readily adapt to the various bar shapes reduce material-handling costs and the over-all shearing cost to a minimum.

Because four-post injector-type shears are normally used for production cutting, their feed systems will be discussed; however, much will be applicable to other models.

Two die heights are used on the four-post shears, one for angle iron and the other for flats, channels, rounds, and other shapes. Figure 4A-20 shows how a dual-purpose conveyor is made to adapt to the two die heights and provide positive alignment for the bars. "V" rolls are permanently mounted

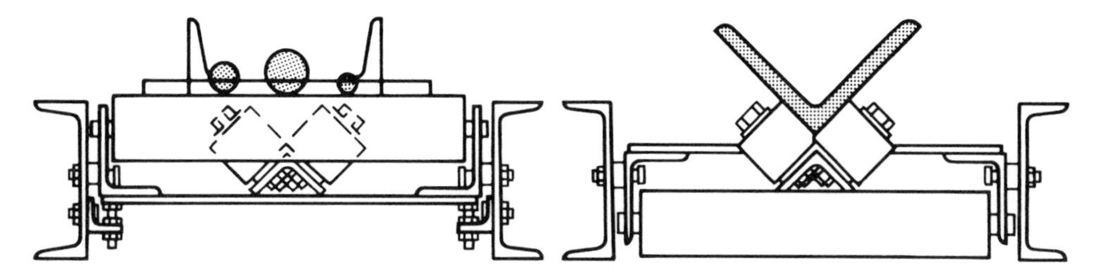

Figure 4A.20. Dual-height system with fixed "V" rolls and movable flat rolls

to the conveyor base for guiding of angle iron. The other dies, for flats and channels, are 1½ in. higher, which makes it convenient simply to raise the flat rolls to allow the material to pass over the "V" rolls, as shown at left in Figure 4A-20. The right-hand view shows the flat roll in the lowered position.

Figure 4A-21 shows a manual feed conveyor set for feeding of angle iron (top view) and then being converted for feeding of flats or channels (bottom view). In seconds, the operator can make the change simply by pulling a lever which lifts the entire flat roll conveyor section above the "V" rolls.

A cut can be no squarer than the guiding system. Thus, the "V" rolls for angles maintain positive alignment for both single- and double-cut dies. Adjustable side guides must be set for guiding the other shapes such as flats and channels. Figure 4A-21 (bottom) shows the side guides, which are provided with scales for setting them quickly and accurately.

Bars up to 60 ft long are cut on the shearing line shown in Figure 4A-22, which is equipped with a power-feed drive wheel and an adjustable back gage for gaging lengths up to 120 in. Angles, flats, and channels are power fed at 100 ft/min against the back gage, which is equipped with a shock absorber and a limit switch to trip the shear. The material is lifted off the dies, to prevent wear, during the feed cycle and automatically drops as the shear blade descends. The end of the in-feed conveyor next to the shear is raised and lowered by an air cylinder. This conveyor is 10 ft long. Additional 5-ft-long conveyor sections are spaced on 10-ft centers to accommodate whatever length of stock material might be chosen. Material handling is reduced by using greater stock

Figure 4A.21. Manual feed conveyor set for feeding of angle iron (top) and being converted for feeding of flats or channels (bottom)

lengths, such as 60 ft.

One person can operate a powered feed system, whereas a manual system normally requires two operators. The operator on the powered system normally stands by the end stop where he or she can stack the cut-off parts and, with the push-button station mounted on the stop, operate the powered

Figure 4A.22. Cut-to-length shearing system with power feed and retractable end gage

feed and shear. When the system is set to operate in the semiautomatic mode, a complete feed and shear cycle is completed by pressing a "start cycle" button.

Besides being provided with a shock absorber, the adjustable back gage retracts 4 in. to prevent the material from binding between the shear blade and the stop and to allow easy removal of the cut-off part. In shearing of short parts up to 3 ft long, the first 3 ft of the out-feed conveyor (Figure 4A-23) is dropped down to allow the parts to fall into a box. Short pieces of angle iron, normally used as clip angles for beam connections, can be single cut at a rate of 20 cuts per minute by one operator. The end stop can also be manually pivoted in cutting off long parts where it is desirable to roll the part off the end of the conveyor.

Figure 4A.23. Out-feed conveyor that drops down to allow short pieces to fall into a box

The conveyor system can be operated without a power drive, but it will not be as efficient and normally a second operator will be required. Whether powered or manually fed, the cut-to-length system greatly reduces shearing costs and increases production.

Hydraulic shears lend themselves to many nonstandard applications. One such case is illustrated in Figure 4A-24, which shows how an 18-ton shear for cutting angle iron up to 4 by 4 by 1/4 in. is used in line with a hydraulic modular gang punching machine. This approach to metal fabrication is a prime example of how operations can be grouped to reduce material handling.

Another example of a nonstandard application, which required only a special die assembly, is shown in Figure 4A-25. Here a 6-in.-wide flat bar is sheared, leaving one end with a large radius and the other end with a slot. Three operations were combined into one, producing significant savings in labor.

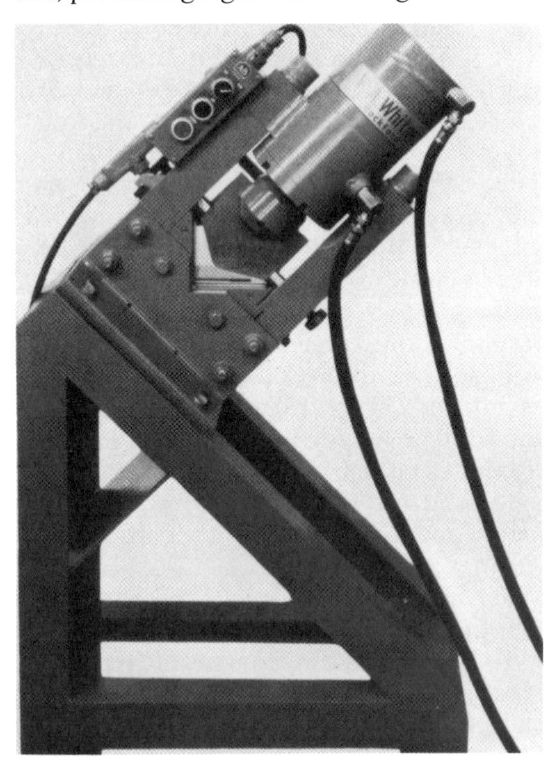

Figure 4A.24. Eighteen-ton shear mounted at 45° angle for use with gang punching machine

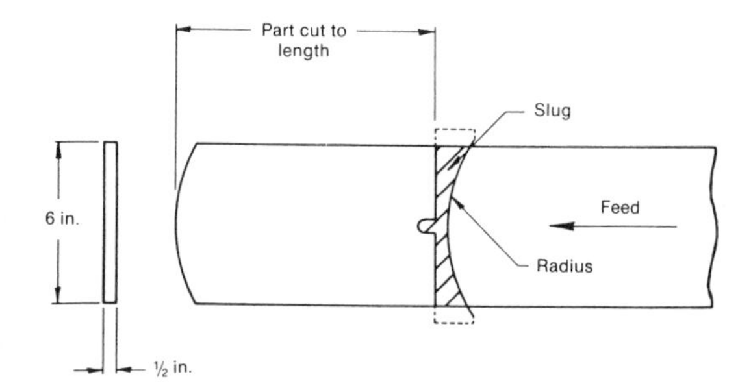

Figure 4A.25. Shaped blade and die that produce radiused and slotted ends with each stroke

These are only two of many applications that prove the versatility of hydraulic shears. These shears can also be used for forming, notching, punching, riveting, and many other pressworking operations. The four-post shear has a large tooling area and is adaptable to many different applications, with the main limitation being imagination.

The advantages of hydraulic bar and structural shears are as follows:

1. Reduced setup, tooling, and running costs: quick-change tooling for all shapes; long tool life — no shock; fast, efficient feeding; only one operator required
2. Ease of material handling: common feed line for all shapes; powered feed available; back gage and conveyor provide efficient takeaway
3. Improved quality of cut: little or no distortion; no burrs; square cuts gaged to close length tolerances
4. Secondary grinding operation to remove burrs eliminated
5. Reduced physical strain
6. Increased production
7. Increased versatility (shearing, forming, notching, punching, riveting, etc.).

4B: Systemized Beam Punching

Punching of holes is generally an inexpensive operation, but not when the holes are punched in structural steel. The problems of material handling, layout, and locating make this minor operation expensive. Layout involves clamping of a tape to the beam and using a scale, a square, chalk, a center punch, and a heavy hammer, after which the beam (or press, if portable) must be manhandled in order to locate the center of the punch in reference to the prick punch marks. If a stationary press (Figure 4B-1) is used, the beam must be swung around end to end and rotated several times to punch holes in the web and flanges. This is time-consuming, costly, and dangerous, and requires much floor space. Portable presses help greatly, but layout is still required.

One answer to the above problems is a machine commonly referred to as a Beamline (Figure 4B-2). Beamlines with spacing tables of the past have two conveyor and punching stations; one for the web section and the other for the flanges. A row of punches is set up for punching the rows of connection holes in the web, the number of holes varying with the beam size and size of connector required. Two punches, set to the proper gage dimensions, are set for the flange holes. A roller-mounted carriage on rails grips the end of the beam for positioning, first through the web punches and then through the flange punches, which sometimes requires several passes. The operator

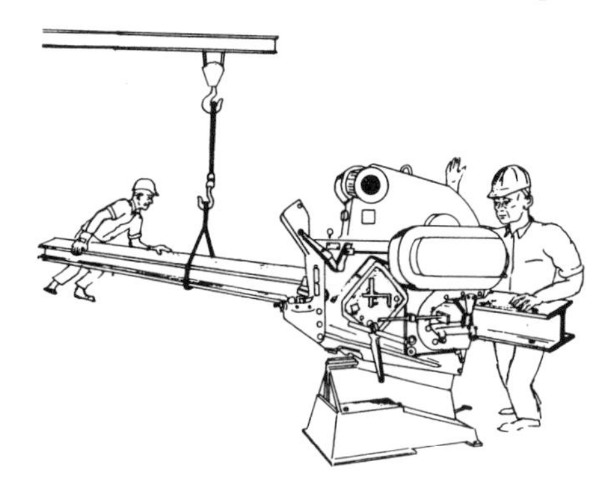

Figure 4B-1. Beam suspended for punching on a stationary press

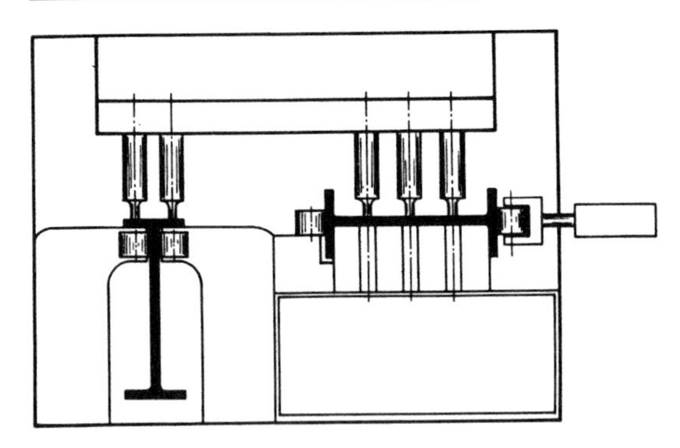

Figure 4B-2. Typical Beamline machine with two in-and-out punching stations

reads the linear dimensions from a steel tape and stops the carriage at the proper settings. Advantages of this type of Beamline are:

1. Layout is eliminated.
2. Two or more holes can be punched at once.
3. Close tolerances can be maintained.

The disadvantages are:

1. Setup time and cost are high. Setup consists of setting gage rolls and clamp cylinder on web punch; setting gage rolls and clamp cylinder on flange punch; setting punch holders, die holders, and strippers on web punch; setting punch holders, die holders, and strippers on flange punch; adjusting conveyor wheels in or out on infeed and carriage sides of press to suit beam size; and changing four sets of punches and dies, if necessary.
2. The beam is passed through presses three times to punch the web and flanges. It must be lifted to transfer it from the web to the flange line and then rotated 180° to punch both flanges.
3. Parts are entered and discharged at the same end of the machine, which increases material handling.
4. Machine cost is high.

Other methods of processing beams, such as using magnetic-base drills, radial drills, and Beamline drills, are available but are not generally considered due to their low speed or high cost.

NEW BEAMLINE

A new beam-fabricating line (Figure 4B-3) that has the primary advantages of the unit described above (layout eliminated and close tolerances maintained) has been developed to minimize material handling, setup time, and cost, and to offer versatility for various shapes. One-of-a-kind beams are more economically produced on this unit than by any other method, due to the fast setup features.

Material Handling Minimized

Beams up to 60 ft long are processed on conveyors and passed through a web press/flange press combination. With positive gaging, holes up to 1½ in. in diameter are accurately punched using three 100-ton hydraulic presses, completing many beams having holes in one set of flanges and the web in one pass. The web press has an 18-in. throat that will punch holes in beams up to 21 in. wide, and some 24-in.-wide flange beams, in one pass. Larger beams are rotated 180° and passed through a second time, which also allows punching of holes in the second set of flanges. Most beams have holes in only one set of flanges. If a range of beams from the smallest to the largest size within the punching capacity are to be punched, this can be accomplished by utilizing a larger web press

Figure 4B-3. New Beamline machine with three 100-ton presses

and two additional flange presses. The end result is a true one-pass machine.

One machine operator can efficiently operate the entire machine. A control lever at the control console enables the operator to traverse the beam through the presses and precisely stop it at any point.

"In-line processing" is achieved by entering the beam from one side of the presses and out the other side. Figure 4B-4 illustrates a typical installation. Raw material, which consists of random stock lengths or material either bought or cut to the exact length, is stored in an outside area.

Material from the raw-material area is loaded onto a raw stock feed rack where it first passed through a saw for cutting to the proper length. From there, the material is transferred onto a beam transfer, to await the punching operation. Material bought to the exact length is simply passed through the saw area and onto the beam transfer, or is placed directly onto the beam transfer.

The Beamline operator, from the operator's console, pulls a beam onto the powered infeed conveyor, which extends 60 ft* on the right side of the presses, and then starts processing the beam through the punches.

After being punched the beams are located on the exit powered conveyors and are ready to be unloaded by the Beamline operator to unload stations or horizontal work stations. All of the operations from the beam transfer through unloading of the beam is done from the Beamline control console by one operator.

Flat bars up to 36 in. wide and 1⅛ in. thick, angle iron up to 8 by 8 by 1⅛ in., American standard channels and "I" beams from 6 in. up, and wide flange beams from W8 × 17 through W36 × 170, can be handled. Wide flange beams with flanges thicker than 1⅛ in. cannot be punched on a standard Beamline due to their extreme thickness.

No Layout Required (Figure 4B-5)

Gaging or locating of holes on the new Beamline is simply and effectively achieved. A program sheet, prepared by the structural draftsman, gives the hole locations so that the beamline operator does not have to search through drawings. The program sheet (Figure 4B-6) instructs the cutoff saw operator on how long to cut the beam and how many to cut. It tells the Beamline operator what size holes to punch and where to punch them.

The lineal or lengthwise dimensions are gaged using a carriage with rack and pinion and with an encoder as a feedback for a visual digital display. Visual digital displays located in the Beamline control console give a direct lineal readout from the end of the beam to the centerlines of both punches. Figure 4B-7 shows the digital displays for each operation. The operator is able to stop and position the beam very accurately for both the flange and web holes. Upon completion of the punching operation, the carriage

* Lengths may vary depending on fabricator's requirements.

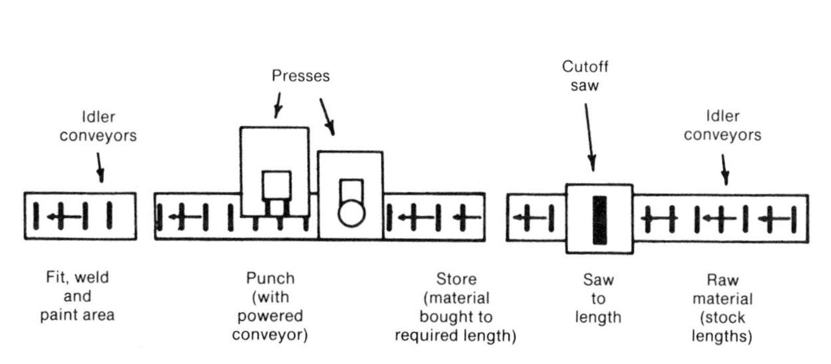

Figure 4B-4. Typical floor plan for in-line beam-punching operation

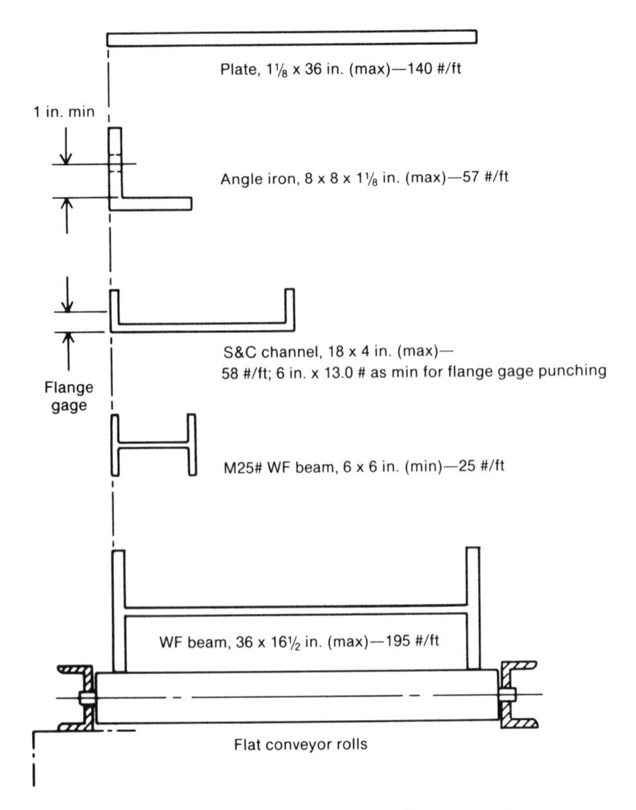

Figure 4B-5. Various shapes that can be punched on Beamline without layout

probe is retracted and the beam is powered down the line ready for unloading.

A gage roller inside the throat of the web press produces a direct readout from the outside of the beam to the centerline of the web punch. The roller is firmly held against the beam as it passes through the press. Figure 4B-8 shows how the web press is mounted on wheels and is moved across the beam by a servo-motor drive or free-wheeling hand wheel and screw. As the press moves across the beam, the distance from the gage roller to the punch changes. This distance is read by the operator at the front of the press (see Figure 4B-9). One hole at a time is punched at the rate of six to eight holes per minute when they are spaced on standard 3-in. centers.

Beam camber does not affect web-hole accuracy, because the gage roll follows the beam directly behind the punch. Flats, angles, and channels can also be gaged with the same roller.

The flange-gage distance is preset by the operator before the beam enters the presses. The flange-gage motorized adjustment control is located and operated from the control console. The gage-line setting is displayed by a direct readout in inches and fractions of an inch. The readout of the flange-gage setting is the distance between the two holes minus the thickness of the web. In the sample program sheet shown in Figure 4B-6, the desired flange gage is 5½ in. and the web is 7/16 in. thick, making the flange-gage setting $5^1/_{16}$ in. The flange-press gaging is accomplished by four gage screws, two on the upper press (Figure 4B-10) and two on the lower press. All four screws are turned simultaneously with the electric gear motor. The holes are gaged from the web of the beam, making sure that the holes are centered. The flange presses are mounted on a hydraulically counterbalanced carriage which the operator controls. One hole is punched at a time. The operator first selects the upper press, the lower press, or both presses, and then the selected holes are automatically punched when

S.O. NO. 6301	WEB HOLE 13/16 ϕ
SECTION 18" WF 70#	FLANGE HOLE 13/16 ϕ
QTY. 1	CUT LGTH. 22' 6-1/2"
MARK A423	DWG. NO. 423
	WEB TH'K. .438

PUNCH	LINEAL	GAGE	FLANGE GAGE SETTING
1	1-3/4	4½, 7½, 10½, 13½	
2-3	1-3/4	5-½	5-1/16
2-3	11'-3/4"	5-½	"
1	22' 4-3/4"	4½, 7½, 10½, 13½	
2-3	22' 4-3/4"	5½	"
		ROTATE	
2-3	1-3/4	5½	"

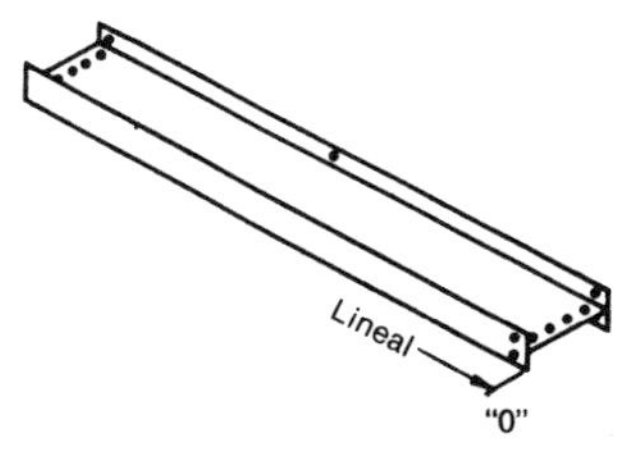

Figure 4B-6. Program sheet giving hole locations

Figure 4B-7. Tape giving direct readout from end of beam to centerline of presses

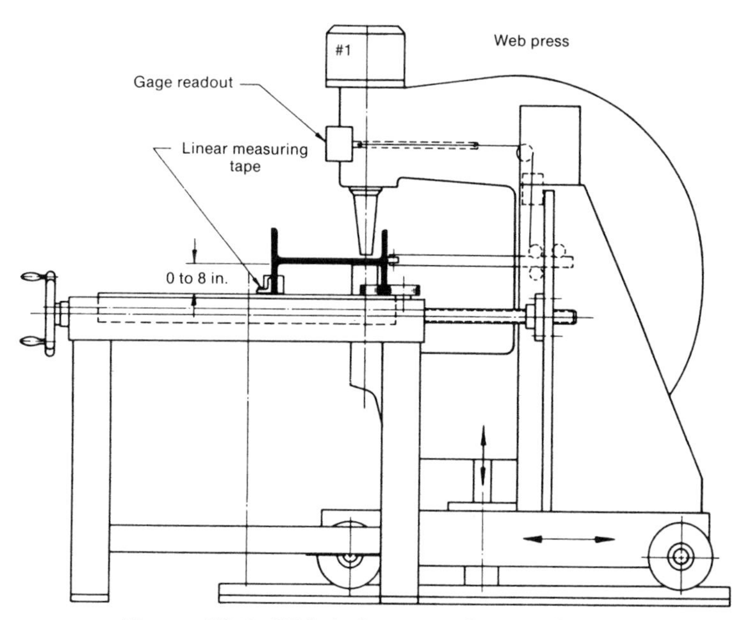

Figure 4B-8. Web hole gaging from roller arm to centerline of punch

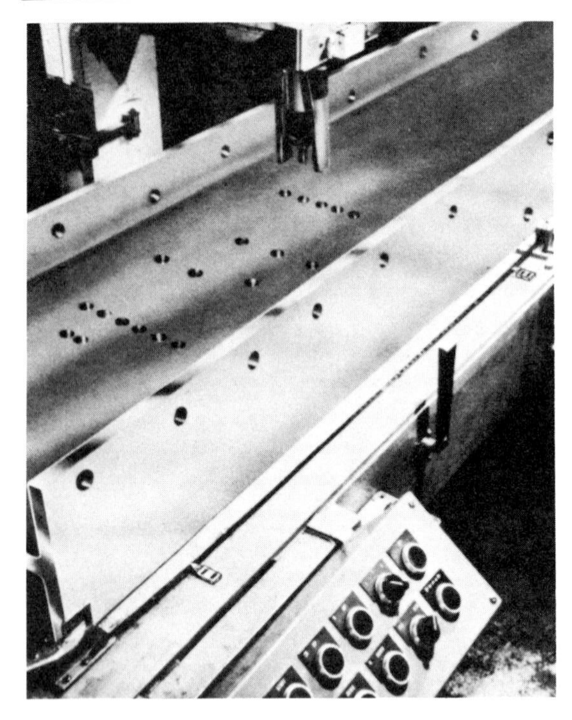

Figure 4B-9. Web-gage readout at front of press

the punch button is depressed. Example: When "Both" is selected and the punch button is depressed, the press assembly comes down and makes contact with the beam through the gage probe, a hole is produced, the press assembly automatically starts up until the bottom gage probe makes contact, producing a hole in the bottom flange, and then the press assembly returns to neutral or pass clearance.

Very Low Setup Cost

Changing from one beam size or shape to another takes less than a minute if the punches and dies do not have to be changed, or less than 5 min if they do. This feature eliminates the need for planning the punching schedule according to beam size or type, which is the case with other types of Beamlines.

Setup involves the following elements:

1. Check punch and die sizes. Change if required. The punches and dies are self-aligning; the shaped ones are keyed.
2. Adjust the die height of the web press relative to the web section. The operator uses control levers to hydraulically raise or lower the press in a matter of seconds. The web press is hydraulically counterbalanced, which allows it to float during the punch-and-strip cycle.
3. Adjust the flange-press gage setting as described earlier. The flange press has both angular and in-and-out float.

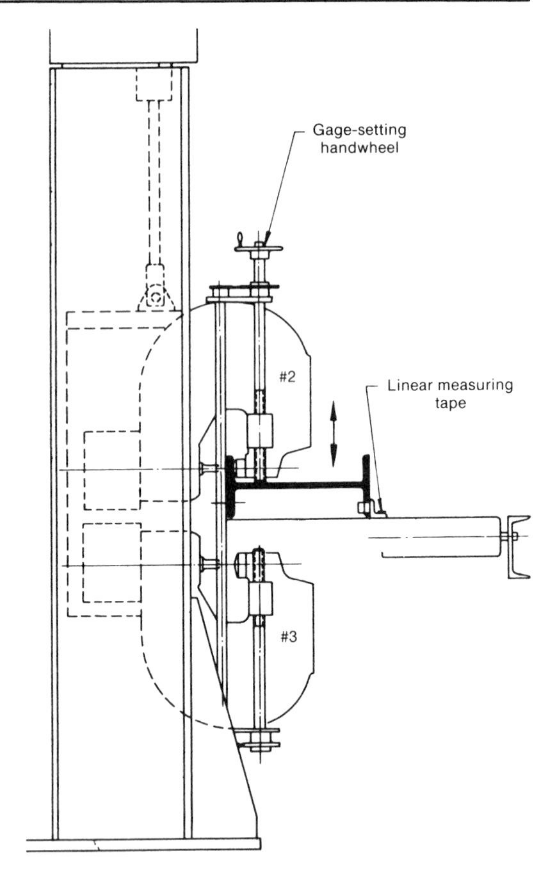

Figure 4B-10. Flange-press gaging from centerline of web

Ease of Operation

Once the setup steps (noted above) have been taken, the operator stands at the control console and does the following:

1. Traverses beam through web press and releases web-press gage roller, allowing it to follow beam.
2. Engages carriage probe to make contact with end of beam for zero reference point. Reference point is centerline of web press.
3. Moves beam to proper lineal dimen-

sion indicated by visual digital display at operator's console. Extends "Y" axis gage probe.

4. Turns hand wheel or motorized control to move web press to proper gage dimension, indicated by visual digital display or steel-tape readout.
5. Turns punch selector located on console to "Web" for punching hole in web. Depresses punch button.
6. Turns selector to "Flange." Moves beam to proper lineal dimension indicated by digitial display for flange presses.
7. Selects upper-both-lower and depresses punch button, causing desired operation to be performed automatically.
8. Repeats as required.
9. Retracts web-gage roller.
10. Retracts carriage probe to up position to allow beam to pass down line for unloading.

Central Hydraulic System

A 10-horsepower hydraulic power unit provides the power for all powered machine functions, which are:

1. Jog and semiautomatic cycling of all three 100-ton presses. Each press is independently controlled and has a two-speed action for minimum cycle time.
2. Elevating and counterbalancing of web press.
3. Positioning and counterbalancing of flange presses.

The power unit can also be equipped with another control valve for operating two additional flange presses that can be used to punch holes in the opposite flanges of the beams, which will eliminate some rotating.

Benefits

With the rising cost of labor, the structural steel plant must seek out new and better methods. The new Beamline concept provides a way of cutting labor costs and increasing output. It is designed to overcome the difficulty of producing one-of-a-kind beams and to make this type of work profitable and, needless to say, more competitive.

Accurate job estimating becomes possible due to the machine's routine operational characteristics.

The benefits of the new Beamline are:

1. Reduced material handling. This in-line system has an infeed speed of 80 ft/min and punches the web and one flange in a single pass. If both flanges are to be punched, an additional set of flange presses can be added.
2. Short setup and running times.
3. Adaptable to all shapes normally processed by steel fabricators.
4. No layout required.
5. No shock when punching. Will punch mild steel in thicknesses from 13/16 through 1⅛ in.
6. With 100-ton-capacity presses, will punch mild steel 1 1/16 to 1 in. thick.
7. Self-aligning, quick-change punches and dies, standard style.
8. Detail drawing not required.

4C: Tooling Techniques for Punching and Notching of Plate and Sheet Metal

In conjunction with the problems that exist in this area of sheet-metal and plate fabrication, the needs of the press owner are also many. Probably the greatest need is for a press with a tooling design which incorporates the following features:

1. Low initial cost for tool holders
2. Minimum cost for perishable tooling
3. Easy and quick tool change
4. Accurate alignment on shaped punches and dies
5. Wide variety of tooling available for standard and special applications.

Figure 4C-1 shows typical single-station-press punches and dies. The punches, 0.050 to 5.000 in. in diameter, both round and shaped, have 1-in.-diam shanks. The punch shank is concentric with the punch end within 0.001 in. total indicator reading. All punches are 3⅛ in. long except for special applications where lengths may vary up to 3½ in. All punches have center points, and punches with diameters greater than 1 15/32 in. have shear. Dies have been standardized and are available in outside diameters of 1¼, 2⅛, 2¾, 3¾, 4¾, and 5¾ in. Inside diameters of dies are concentric with outside diameters within 0.001 in. total indicator reading. Shaped punches and dies are dowel pinned and whistle notched so as to allow for perfect alignment and 90° indexing. Standard clearances are 0.006 and 0.012 in., but by selection of the next 1/64th or 1/32nd larger die, various other clearance combinations can be used for punching of thicker materials.

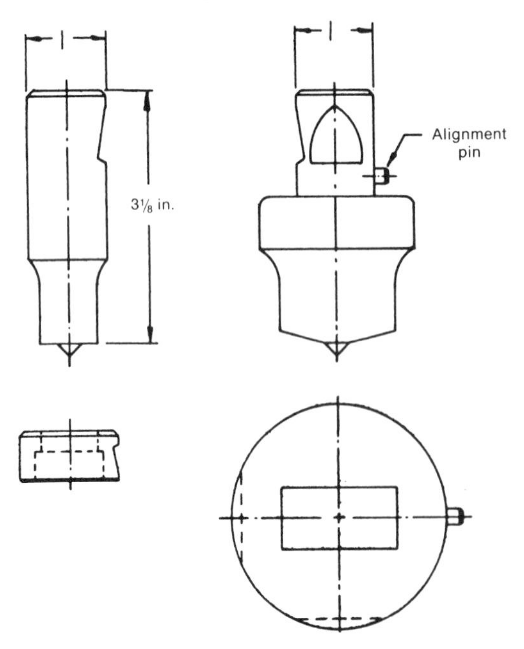

Figure 4C-1. Typical punches and dies for a single-station press

Center points are provided primarily for locating to layout marks. Because these tools are used for a variety of applications and presses, the center points can be removed from the punches when used on machines not requiring center-point locating, such as N/C machines, duplicators, and gaging presses. The centers will induce distortion in the blank and may also tend to distort the sheet. A large piece of sheet metal having many holes spaced on close centers will become "bowed" if punched with a punch having a center point, but will remain flat if the center point is removed.

The center point is also used in manufacturing the punch. It acts as a center for grinding of the shank and punch in order to maintain 0.001-in. total indicator reading concentricity.

Grinding of shear on the face of the punch or on the top of the die can be an effective way to reduce tonnage requirements in many punching applications. Figure 4C-2 shows "criss-cross" shear ground on the face of a punch.

This type of shear is used primarily on round punches. A small amount of distortion will occur at the beginnings and ends of the

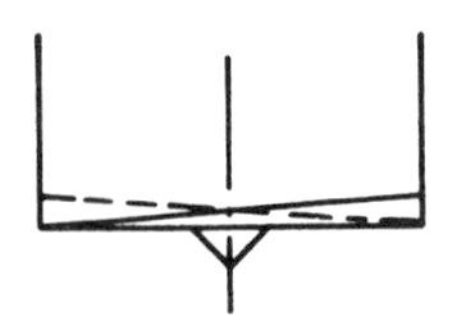

Figure 4C-2. Criss-cross shear on punch

shear points. This, however, is very seldom detrimental to the punching operation.

Figure 4C-3 illustrates "housetop" shear as ground on an irregularly shaped punch. This type of shear also causes a minimum amount of distortion at the last point of cutting.

Concave shear, as shown in Figure 4C-4, is a unique way to reduce punching tonnage and end up with burr-free, nondistorted holes. The one disadvantage of this punch is that there is no center point for punching to a layout line or center punch mark.

Figure 4C-5 illustrates "housetop" shear as ground on the top of a die; the die can be either round or irregular. This type of shear is used when the slug must remain flat and nondistorted. Because it is the slug that is desired, the material from which it is to be blanked should be cut into strip stock. These strips should be only slightly wider than the blanked piece part, because the scrap material will become distorted due to the shear on the face of the die.

Perhaps the most important and yet most often overlooked area of punching is the economics of tooling. Selection of tooling can often "make or break" the job. The use of tooling that is not intended for the application at hand is more often the rule than the exception. Too often, the order to the shop is to get the job done, now, immediately, no matter what the consequences may be. Invariably, the first thought is to save time and money by the use of tooling already available even though it might not be right for the job. This is particularly foolish in light of the fact that so much of today's single-station press tooling is so inexpensive and can be delivered so quickly.

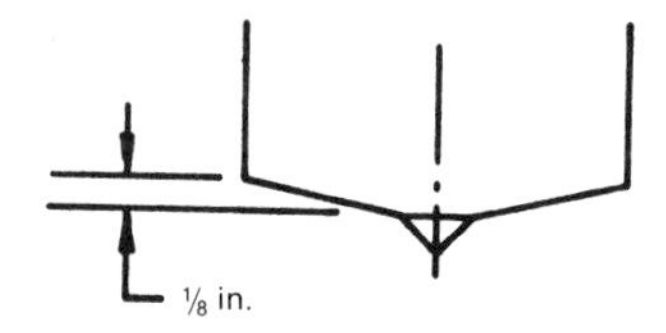

Figure 4C-3. Housetop shear on punch

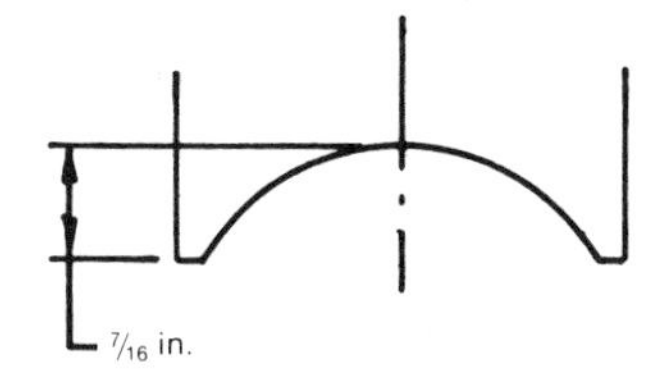

Figure 4C-4. Maximum concave shear on punch

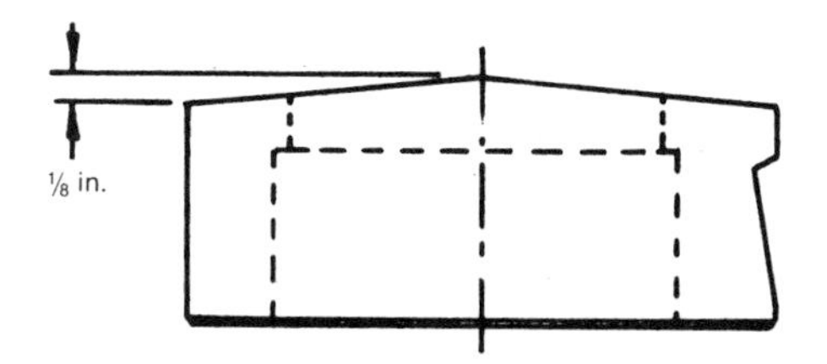

Figure 4C-5. Housetop shear on die

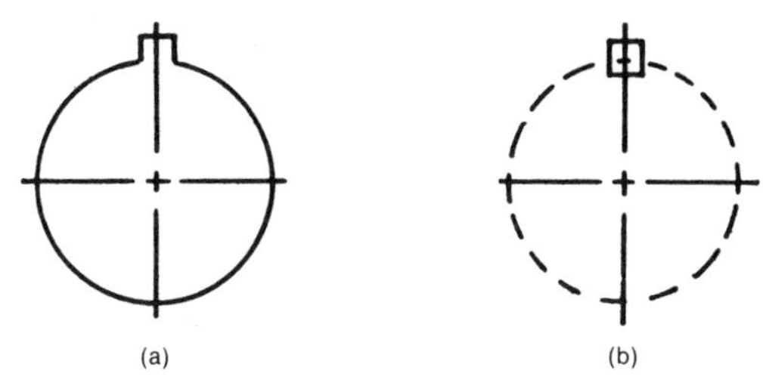

Figure 4C-6. Inexpensive method of punching an irregular hole with separate square and round punches and dies

Punching of special-shape holes in sheet metal and plate can be done in many ways. Figure 4C-6(a) shows a typical hole for an electrical pushbutton. A special punch-and-die set could be manufactured for this application at a cost of approximately \$225. However, this punching job can be done at a considerably lower tooling cost. Figure 4C-6(b) shows how. First punch the 1/8-in. square hole. Then punch the $1^{7}/_{32}$-in.-diam hole. Total cost for the required square and round punches and dies would be approximately \$60. Figure 4C-7(a) shows a similar

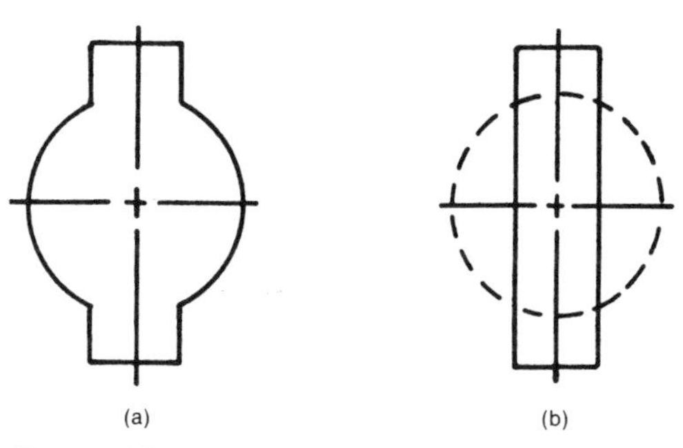

Figure 4C-7. Inexpensive method of punching an irregular hole with separate rectangular and round punches and dies

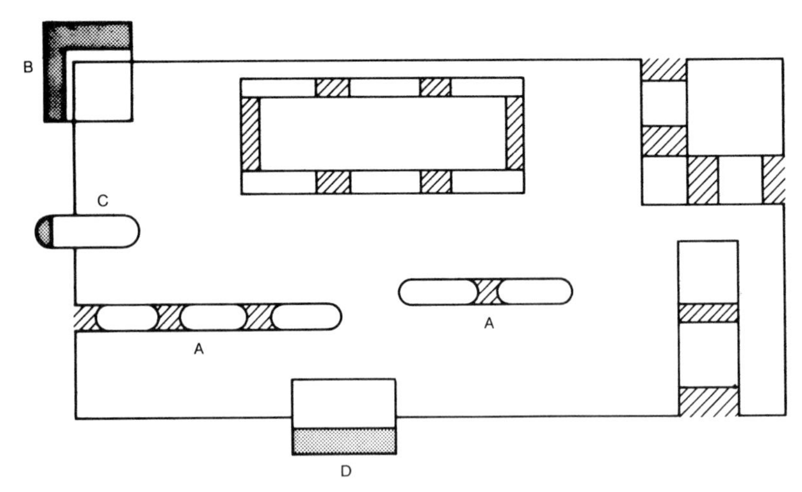

Figure 4C-8. Illustration of hole punching by the skip punch method (A) and by use of punches with heels (B, C, and D)

hole. The special punch and die would again cost about $225. Rectangular and round punches and dies could be used as shown in Figure 4C-7(b). Cost of the required tooling would be about $90.

It is a management decision as to whether special or standard tooling is to be used in any punching application. Areas to be considered should include:

1. Availability of tooling, both standard and special
2. Tool-change time
3. Number of holes to be punched
4. Cost of tooling

Many types of tooling are available, and there are many ways of cutting large holes in sheet metal and plate. The drawings labeled "A" in Figure 4C-8 show the "skip punch" method of punching. This method must be used to prevent damage to the punch and die when making three or more hits in cutting a large hole or notch. Due to side-thrust forces applied to the punch as it penetrates the material, the punch will deflect sideways when doing unbalanced punching. For example, if a standard square punch and die were used to make notches in blanks, cutting could be done on three sides, causing the punch to deflect to the point where the edge not cutting would "shear" itself into the die. This "shearing" action would destroy the cutting edges of the punch and die on that side. By use of "skip punching," cutting is done either on all four sides or on two sides of the punch, balancing the side-thrust forces.

Another way to counteract the side thrust is to use a punch with a heel, as illustrated by the shaded areas at points B, C, and D in Figure 4C-8. The heel extends down ahead of the cutting surfaces and thus enters the die before cutting begins, allowing the heel to transmit the side thrust into the die.

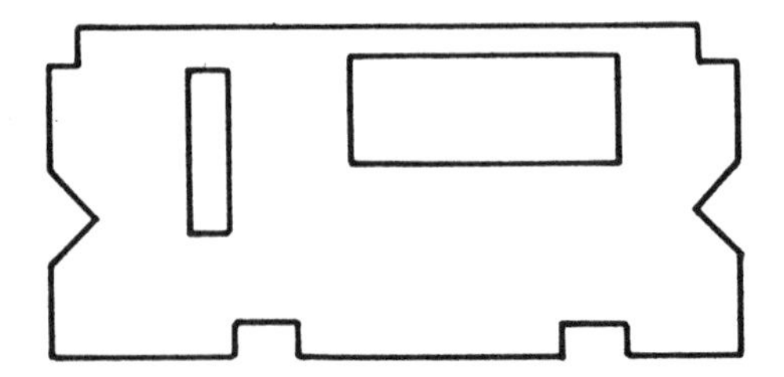

Figure 4C-9. Typical piece part punched with square shearproof punch

Figure 4C-10. Square shearproof punch and die

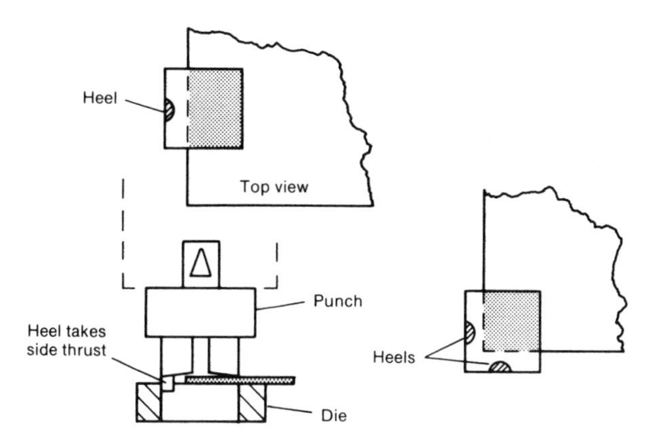

Figure 4C-11. Setup in which spring-loaded guide heels on punch face are used to eliminate shearing of punch and die during punching of partial holes

Sheet-metal parts and plate up through 10 gage, similar to the panel shown in Figure 4C-9, can be notched and pierced using a new versatile "shearproof" punch. The new punch (Figure 4C-10), which is used with standard square dies, pierces a square hole or any part of a square hole. For example, all the corner and edge notches as well as the large rectangular hole in the center of the part shown in Figure 4C-9 were cut with the "shearproof" punch by progressive notching and nibbling.

Spring-loaded guide heels protruding from the face of the punch eliminate the problem of shearing of the punch and die during punching of partial holes. The side thrust, caused by notching, is absorbed by one of the guide heels that has entered the die before the punch contacts the material, as shown in Figure 4C-11. All four heels retract upon punching of a full square hole in the center of the sheet.

One or two of the heels, depending on the type of notch, must be allowed to enter the die, which limits the depth of notch per hit. For example, a 2-in.-square "shearproof" punch will notch up to 1¾ in. per hit. Exceeding this amount will prevent the heel from entering the die and will result in damage to the punch and die.

Edge notching of a "V" is also done with the "shearproof" punch by indexing the punch and die 450°. Several hits might be required, depending on the size of the notch. Costly handling time can be eliminated by combining these notching operations with piercing operations through the use of "shearproof" punches.

Round "shearproof" punches and dies (Figure 4C-12) are ideal for making holes in a one-hit operation or for making large round holes (e.g., 5 in. and larger) while leaving a minimum scalloped effect.

Figure 4C-13 shows how a 7-in.-diam hole can be made with a 3-in. round "shearproof"

Figure 4C-12. Round shearproof punch and die

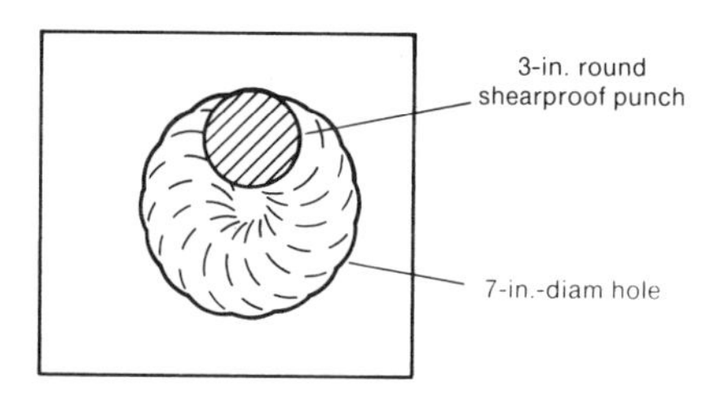

Figure 4C-13. Method of making a 7-in.-diam hole with a 3-in. round shearproof punch

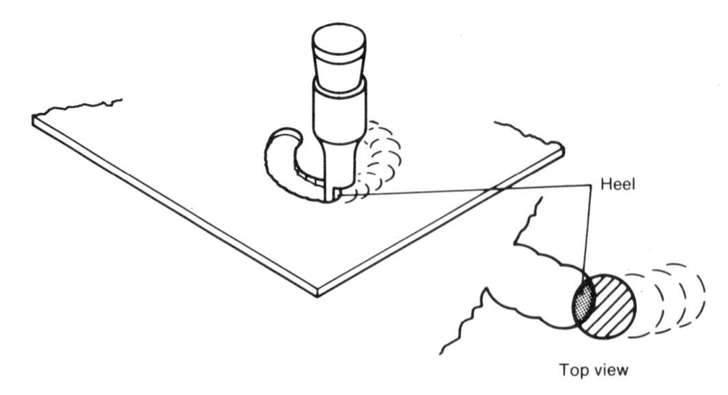

Figure 4C-14. Rectangular finished-edge nibbling punch

punch. This type of punch-and-die set is best used in a press with an adjustable stroke where a template or numerical control can be used.

Two types of nibbling punches and dies are in use: finished-edge and scalloped-edge.

Finished-edge nibbling punches (Figure 4C-14) are usually rectangular punches with heels and are used with standard rectangular dies. Long slots, notches, and large holes are easily produced, leaving a smooth edge condition. The heel has two functions:

1. It absorbs the side thrust caused by cutting three sides of the punch.
2. It controls the depth of notch per stroke, because the press stroke is adjusted to keep the heel in the die at all times.

The operator can index the punch and die 90° for making large cutouts or notches.

Scalloped-edge nibbling punches and dies (Figure 4C-15) are used to nibble large round holes and to make curved contours. The punch and die are round, usually ½ in. in diameter, and the punch has a heel which absorbs the side thrust and also serves to control the depth of cut. Circular cutouts or holes are possible, because the punch is free to rotate a full 360°. Without this capability, the punch heel would limit the cutting path.

A starting hole is required for both types of heeled nibbling punches for entering the workpiece. Inserting the heel in the starting hole is accomplished very easily on a hydraulic fabricating press due to the ability to jog the punch down and stop at any point. Once the heel is inserted into the work, the press is switched to a nibbling cycle which allows the ram to cycle continuously as long as the pushbutton or footswitch is held.

Special cluster punches and dies greatly reduce punching time, and are used where production justifies the increased tooling cost. Figure 4C-16 shows a typical set made for punching four rectangular holes. The cluster punch is a removable insert (shown at lower left) and is mounted in a punch

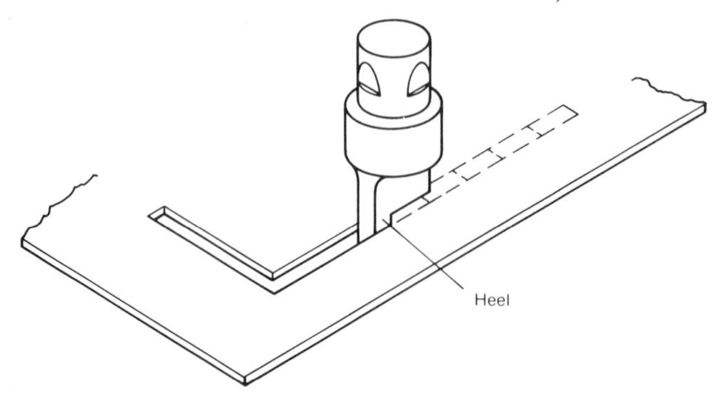

Figure 4C-15. Scalloped-edge nibbling punch and die

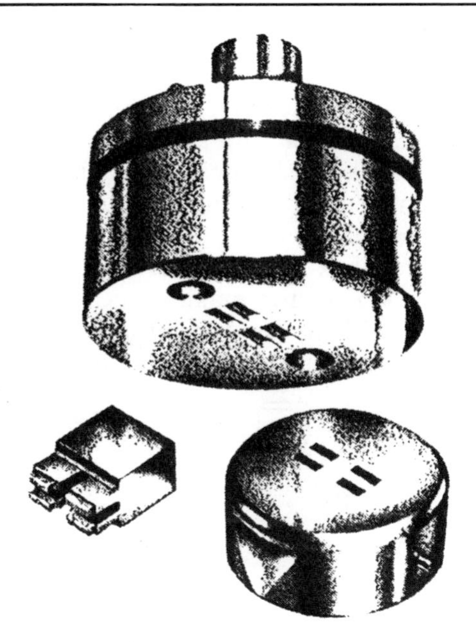

Figure 4C-16. Cluster punch and die

holder that has built-in spring stripping. Many similar tools have been made to produce holes for panel mounting of electrical instruments and at the same time punch three or four screw holes for holding the instrument in place. Not only does this save time, but it also ensures proper location of all related holes.

New electrical knockout tooling overcomes problems that sheet-metal fabricators and electrical-control-cabinet manufacturers have encountered when attempting to economically produce knockouts in low quantities.

Electrical knockouts are partially punched holes provided in sheet-metal wiring boxes and enclosures, as illustrated in Figure 4C-17.

The slug is retained in the hole only by one to four small tabs and can be removed by an electrician using hand tools such as a screwdriver and pliers. Single and double knockouts are the most common. The double knockout gives the electrician a choice of two hole sizes. If the inner slug is removed, the outer ring is still sufficiently strong to make a conduit connection.

The new tooling (Figure 4C-18) is used for making single and double knockouts. Three- and four-ring knockouts can be produced by first making a single or double knockout and then making a double knockout in the same location.

Automatic piece-part positioning on numerically controlled presses is not affected by the partially punched hole, because the slug is raised above the sheet. This is accomplished by inverting the punch and die so that the punch is on the bottom. If the slug were pushed down, it would catch on the lower die and cause the part to be pulled out of the work clamps.

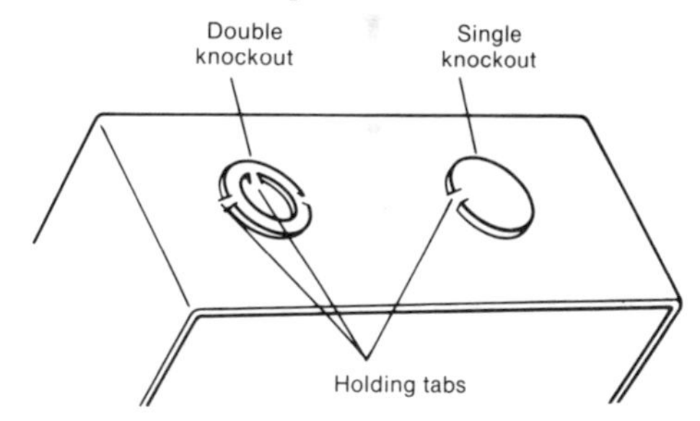

Figure 4C-17. Double and single electrical knockouts

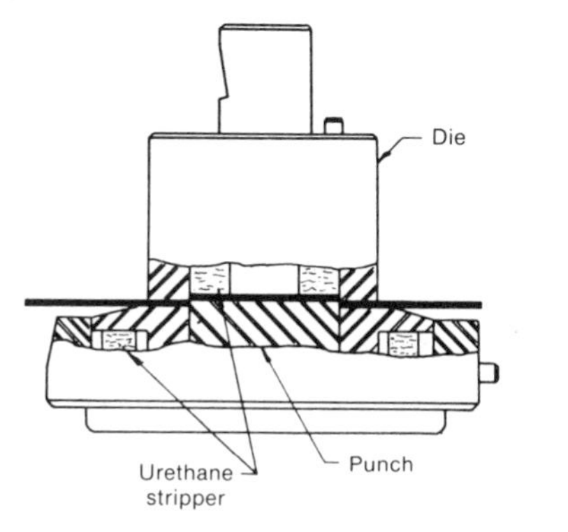

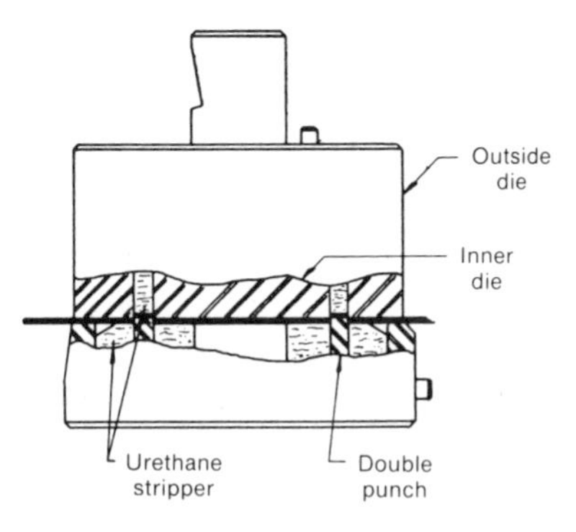

Figure 4C-18. Single-knockout (left) and double-knockout (right) punches and dies

The material is stripped free of the punch and die by built-in plastic strippers, allowing the material to be freely positioned over the tooling.

A built-in stop controls the amount that the punch penetrates the material, ensuring a perfect knockout on the first hit. If the punch were allowed to go too deep, the holding tabs would break, causing the slug to fall free. Shallow penetration would result in the slugs being held too securely, making it very difficult for the electrician to knock out the slug. The positive stop is factory set to provide optimum results in 16-gage (0.060-in.) and 14-gage (0.074-in.) material.

The shut height on the hydraulic press can be precisely set to reverse the press ram as the stop contacts the material. In addition, a pressure-switch-reversing circuit can be added to the press, which eliminates the need for accurately adjusting the shut height. The pressure switch senses when the stop on the punch and die has contacted the material and automatically reverses the press ram. The limit-switch shut-height adjustment is then set to satisfy the other punches and dies used on the same piece part. The pressure-switch feature is ideal where several knockout punches and dies are used in succession on a part. Install or changing the knockout tooling requires only 15 to 30 s.

The number of tabs and the tab width are varied with the size of the knockout to ensure best results. For example, on a double knockout, the outer ring must be kept intact while the inner slug is being removed. If the tabs holding the outer ring were too few or too small, the outer ring would break out. As the knockout size increases, additional holding tabs are required.

The knockout size is expressed in terms of pipe size. For example, the 1/2 knockout is actually 0.875 in. in diameter. The actual diameters on the new knockout tooling for all pipe sizes meets National Electrical Manufacturers Association (NEMA) and Underwriters Laboratory (UL) specifications.

The knockouts are usually pushed back into the sheet on the same press, leaving the sheet flat, dustproof, and in some cases watertight. A flattening punch and die quickly do the job.

Care must be taken in selecting the proper size of press for use with knockout tooling, because no rake or shear can be provided on the punch or die. Double-knockout tooling requires double tonnage.

Electrical-box manufacturers can now afford to set up for making one or two knockouts with the new tooling and hydraulic presses. Inventories of boxes can be re-

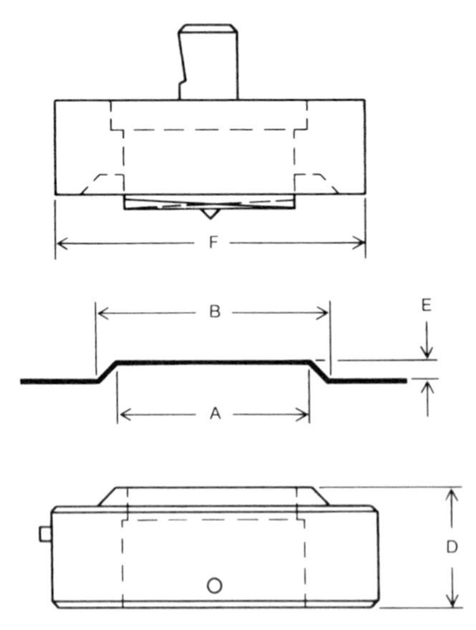

Figure 4C-19. Pierce-and-countersink (left) and form-and-pierce (right) punches and dies

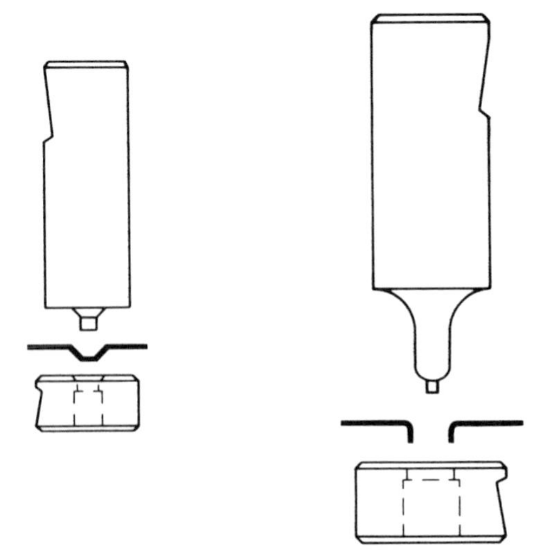

Figure 4C-20. Form-and-pierce tooling for use with numerically controlled presses

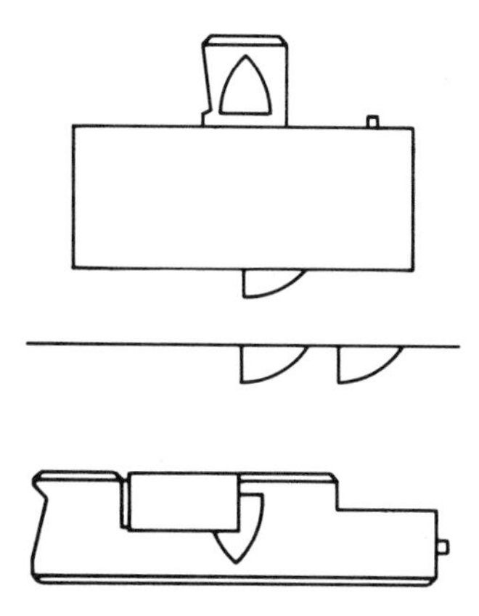

Figure 4C-21. Louver punch and die for downward forming

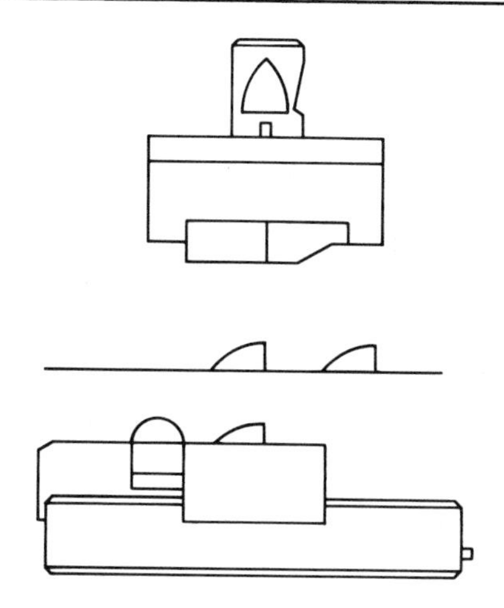

Figure 4C-22. Louver punch and die for upward forming

duced, and scrap should be virtually eliminated.

Two variations on one style of form-and-pierce tooling as used in a single-station punch press are illustrated in Figure 4C-19. The punch is of a compound design in that it begins to form the material upon first contacting it and then continues forming as it advances downward through the piece part. It punches a hole and completes downward forming of the material.

The inside diameter of the die is radiused so as to eliminate punching of a hole during forming.

This type of tooling is used mainly when two thicknesses of sheet metal are to be fastened together with sheet-metal screws.

Another style of form-and-pierce tooling is shown in Figure 4C-20.

This type of tooling is used mainly with numerically controlled presses where the material cannot be formed down into the die. The most common use of this punch and die is for forming and piercing of sink drains.

Louver punches and dies are also available in two basic designs. The tooling shown in Figure 4C-21 forms the louvers down into the die. The cutting insert in the die is removeable for sharpening and ultimate replacement. The step milled on the die serves as a gage for minimum distance between louvers.

The louver tooling illustrated in Figure 4C-22 forms louvers in an "up" position. This type of tooling was designed for numerical-control punching applications where projections cannot be formed down into the die.

In conclusion, it is easy to see that today's owners of single-station punch presses, whether hydraulic or mechanical, must be aware of circumstances in all phases of each punching application if they are going to make profits with this equipment. Correct clearances, the least expensive tooling, and punches and dies properly designed for each punching application are mandatory if work is to be completed correctly the first time it is attempted.

4D: Uncoilers

There are several different types of uncoiling units, and there are also several terms used to identify this procedure. Uncoiling is sometimes called "dereeling" or "decoiling," but all three terms mean the same thing.

Strip materials produced in final form in steel and aluminum mills must be rolled into coils for shipment to customers. When a coil has been received, the customer must unroll it into flat sheet or blank form so that it can be cut or sheared into the required sizes.

CRADLE-TYPE UNCOILERS

One of the oldest and most abused types of uncoilers known is the cradle-type uncoiler. For some purposes it is good, but for others it is absolutely useless. A cradle-type uncoiler (Figure 4D-1) can be either powered or nonpowered.

With a nonpowered uncoiler, the machine being fed by the uncoiler actually pulls the metal off the uncoiler, and sometimes this may necessitate a larger drive system. There are some advantages to this type of uncoiler. For example, coils quite often can be "manhandled" into position, provided that labor costs are low, eliminating the need for a large capital expenditure for a crane or fork-truck.

It also should be pointed out that although a fork-truck is mobile and thus can be used for other purposes, its use requires sufficient space for it to be turned around, and this must be considered when preplanning the floor space and surrounding area involved in setting up coil-handling facilities.

If marking or marring of the metal is not a problem, if "roll marks" do not constitute a reason for rejection of the finished blank, and if a lack of "jerking" is not a manifest requisite, the cradle-type uncoiler can often be used to good advantage.

Because the outside diameter (OD) of the coil changes as metal is uncoiled from it, the fixed spacing of the rolls and the radius of the arc formed by the rolls (Figure 4D-1) will cause the coil to be supported by less than the total number of rolls, which often means that the mere weight of the coil itself will mark the metal at each roll station or position.

A powered cradle-type uncoiler is often preferable, because motorized uncoiling eliminates "jerking."

If the coil ID is not greatly different from the coil OD, then a cradle-type uncoiler will probably be most satisfactory. However, such coils contain relatively short lengths of strip, which necessitates more frequent loading of new coils. A powered cradle-type uncoiler must be operated at a speed near that of the machine it feeds, which requires merely a dancer arm or similar type of

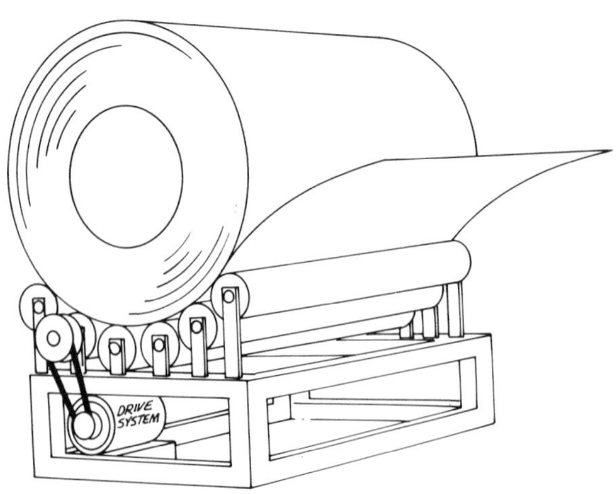

Figure 4D-1. Cradle-type uncoiler

switching device that is capable of activating the drive sytem of the uncoiler so that the metal is uncoiled only as it is needed.

One of the advantages of the cradle-type uncoiler is that it is extremely easy to load. Coils can readily be loaded into the cradle by means of either a fork-truck or a crane without any additional mechanical coil-handling equipment. No C-hook, sling, or steel belt is required, because all that needs to be done is to lower the coil into position.

DROP-IN UNCOILERS

The drop-in, or fixed-ID, uncoiler is a relative newcomer to the industry. It is only within the last 15 years that it has become well known and widely used.

The drop-in uncoiler (Figures 4D-2 and 4D-3) consists of a drum assembly and a frame to support the drum assembly when it is loaded with a coil. The frame can be made of either rectangular tubing or channel iron.

The drum assembly consists of a shaft, two side plates, and two 6-in.-wide drums attached to the side plates. The coil rests on the drums and is pulled down against them by gravity. One of the side plates is also a drive pulley. The drive motor turns this pulley, and thus the drum assembly and the coil, by means of two v-belts. The ends of the shaft rotate in self-aligning bearings which merely rest in bearing supports welded to the frame.

Figure 4D-2. Drop-in uncoiler

Figure 4D-3. Drop-in uncoiler with coil and side plate removed

The drum assembly must have a fixed OD that matches the ID of the coil placed in it, but often a variance of 1 to 1½ in. is allowed in order to facilitate loading of the coil. For instance, a 19-in.-OD drum is often used for 20-in.-ID coils, and a 23-in.-OD drum is often used for 24-in.-ID coils.

Several different types of switches can be used to activate the drive system. One of the most common is an insulated "slack bar" placed immediately below the coil. This slack bar, which can be powered with a 24-V power source, acts somewhat like an automatic switch to turn off the motorized drive system. The coil is no longer uncoiled until the machine being fed again calls for more metal, thus lifting the metal off of the slack bar, and again starting to uncoil more metal. This process is repeated over and over, and thus the motor used must be sized to accommodate this "start-stop" operational cycle.

This type of uncoiler is certainly limited in its capabilities. It is not a good type for use with very light-gage metals or with prefinished or prepainted metals, since marring of the coil could easily result due to the fact that there is no inside expansion capability holding the coil and thus keeping the various coil wraps from rubbing against one another. This type of uncoiler is good for galvanized or cold rolled materials, for materials where marring is not a problem, and also for medium-gage materials — say from 16 gage

Figure 4D-4. Multiple-uncoiler setup

(0.063 in.) through 26 gage (0.021 in.) or possibly 28 gage (0.018 in.).

The drop-in uncoiler is also quite valuable when it is desired to use only part of the coil but to keep the remainder readily available for immediate use. With this type of activity, quite often more than one uncoiler can be used for partial storage of various gages, and with this in mind, drop-in uncoilers are often sold in multiples of two, three, four, etc., up to a known maximum of eight in a row (Figure 4D-4).

A fork-truck or crane can be used in loading a drop-in uncoiler. An endless nylon sling can be placed around the coil in order to move the predrummed coil into position. A mansaver-type coil-handling unit can also be put to good advantage in both predrumming the coil and placing the predrummed coil onto the frame (Figure 4D-5) with the assistance of either a fork-truck or a crane.

CONE-TYPE UNCOILERS

Cone-type uncoilers are used in a number of applications for the specific purpose of eliminating the need for either fixed-typed drums or expanding mandrels to support the coils. A cone-type uncoiler can be used on a large variety of inside coil dimensions by inserting the cone into the coil to only the depth needed for support purposes.

Cone-type uncoilers are always used with a cone on both ends of the coil, since there is no inside coil center support, but basically a pressure-type application in the ends of the coil for support purposes (Figure 4D-6).

Figure 4D-5. Mansaver-type coil-handling unit

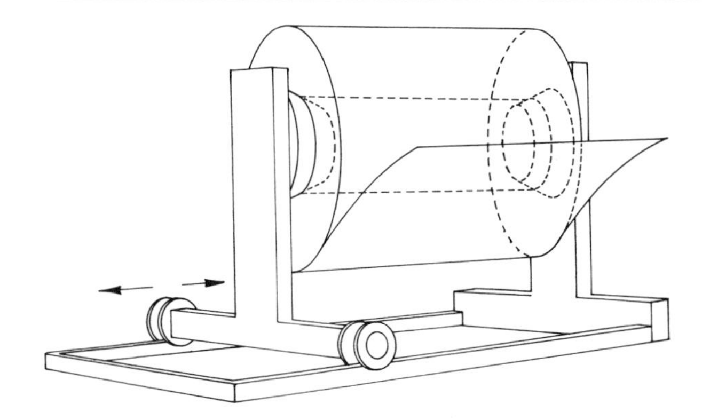

Figure 4D-6. Cone-type uncoiler

Cone-type uncoilers are good when the user is not particularly bothered by the damage of a few inside wraps. The pressure that must be applied to maintain lift and the capability of uncoiling the coil will usually damage a few of the inside wraps to the extent that they would normally not be usable without substantial waste.

More often than not, the cone-type uncoiler is nonpowered, using a drag brake of some type for resistance and requiring the machine that is using the material to pull the material off the uncoiler.

In a number of cases, the cone-type uncoiler is movable on both ends to a substantial degree in order that the coil can be more easily centered for purposes of usage. This means that lateral movement of the two halves of the uncoiler are normally independently controlled and powered.

A coil car which will run longitudinally in and out between the two halves of the uncoiler is quite practical, especially since in this situation, both ends of the coil need to be supported by the two cone units. Users of secondary steel coils, or users in the basic steel industry, such as steel mills or service warehouses, are prime users of cone-type uncoiling equipment. However, this type of uncoiler is rapidly fading from use in most segments of the industry due to the unavoidable damage resulting from the side pressure applied by the cones to support the coil.

Light-gage material is especially difficult to manipulate with cone-type uncoilers due to the increased amount of damage involved, and usually 22- or 24-gage material will be the lightest material for which such an uncoiler is used. It is much more practical when used with heavier-gage coil material.

END-SUPPORT EXPANDING-MANDREL UNCOILERS

The end-support expanding-mandrel uncoiler is a mixture of two types of uncoilers, and is actually a means to an end of practical economy which quite often defeats its own purpose.

The basic reason for the existence of the end-support expanding-mandrel uncoiler is to eliminate the need for the massive bearings involved in a cantilevered or open-end uncoiler. It was designed to support the open end of the mandrel with the weight on the mandrel of the particular coil in use at the moment.

The end-support expanding-mandrel uncoiler is quite satisfactory for prefinished or prepainted material or light-gage metal requiring inside expansion to prevent the various wraps of the coil from rubbing against one another, and when the customer is not desirous of expending the necessary funds for an open-end uncoiler.

For loading and unloading of the end-support expanding-mandrel uncoiler, the open end is normally hinged so that it can be removed and then replaced after the new coil is in position, prior to lowering the full weight of the coil onto the uncoiler. Loading of the coil is most often accomplished by means of a crane-and-sling combination, or by a coil car which can load the coil onto the

Figure 4D-7. End-support expanding-mandrel uncoiler

shaft with the end support out of the way. After the coil is put into position, the coil car is then brought out away from the uncoiler, or in some cases, it is even brought out through the center of the end-support section (Figure 4D-7).

These are deterrents to the purchase of an end-support expanding-mandrel uncoiler, because failure to replace the end support at the necessary moment in time will consistently result in shaft breakage.

The end-support expanding-mandrel uncoiler can be used with a dancer arm, a wheel, or a roller, as desired, depending on coil width, or it can be operated with a type of limit-switch activation.

Figure 4D-8. Cantilevered or open-end uncoiler

Quite often the drive system involved will consist of an adjustable-speed motor furnishing variability of speed by rheostat control. Starting and stopping of an uncoiler with this type of drive proves to be a problem, because the drive is not conducive to start-stop application, and the motor should be well protected from overheating and overloading. Continuous running is much more conducive to long life with this type of uncoiler. End-support expanding-mandrel uncoilers are used strictly for economy, and their use should be carefully researched prior to purchase.

Definitive instructions on the consequences of improper operation must of course be fully explained to all machine operators.

CANTILEVERED OR OPEN-END UNCOILERS

The cantilevered or open-end uncoiler is probably the type most used when coils are to be used completely, because loading methods are extremely simplified, especially for small uncoilers (Figure 4D-8).

An open-end uncoiler usually has expanding capability, as far as the mandrel is concerned, and this expansion can be accomplished manually, pneumatically, or hydraulically, as befits a particular need or application.

Expansion capabilities usually encompass the normal inside diameters of coils which can be best used for that particular application. If the coil is a narrow one, and is within reasonable limits as far as material thickness is concerned, its ID normally will range between 14 and 21 in., whereas wide coils usually have ID's from 16 to 21 in.

The heavier the gauge or thickness of the material, the more difficult it is to roll into coil form. Coils in the 14-gage and heavier classifications quite often are recoiled on 24 to 26-in. inside diameters.

For lighter or medium-gage materials, the mills normally use 20 or 24-in. ID's, plus or minus 1/2 in., and in the case of aluminum, especially in light and medium gages, 16 to 18-in. ID's are used.

For coils of massive proportions, such as master coils that weigh from 30,000 to 60,000 lb each, mills often will request larger-ID rolls to eliminate the problems of turning the arc.

A keeper arm is also used to contain a coil on the mandrel after the expansion process has been completed, and replaceable pads are used to accommodate coils with ID's larger than the normal expansion range, much the same as a shim is used in various instances. Though it is not at all recommended, some customers have even been known to use pieces of two-by-four and/or four-by-four lumber as pads or shims for this purpose.

A number of different drive systems are available for uncoiling purposes, and in quite a few cases the application necessitates a starting and stopping procedure. An open-end uncoiler that does not create problems when starting and stopping would certainly be of more value if the start-stop procedure is absolutely necessary. Substitution of a hydraulic drive which will maintain a constant speed is a much better solution to the aforementioned need, and this is rapidly becoming the answer in a number of cases for this type of application.

Normally the dancer arm or extended limit switch controls the function of on-off operation in order to maintain a consistent nonvarying speed. Since the coil must be started from a stationary position and must quickly arrive at the speed of the equipment which is using the metal, such as a punch press, cut-to-length line, etc., this dancer arm normally rides on the metal by means of a rubber wheel or roll, or, in cases where marring is not a problem, merely on an extended rod. This arm furnishes the capability of immediate on-off switching for start-stop control.

The normal uncoiling procedure in using an open-end uncoiler is from the top, thus creating a type of slack loop. Some uncoilers of late vintage are even capable of operating equally well from the bottom, thus allowing both light-gage and rather heavy-gage materials to be more easily uncoiled.

For instance, it is quite difficult to furnish a slack loop in 14-gage or heavier material, since it is rather difficult to flex. Uncoiling from the bottom eliminates this particular need in a particular type of material. Light-gage metals may also have a tendency to show break marks when uncoiled from the top, and thus an uncoiler that will accommodate both top and bottom uncoiling is more versatile.

Most open-end expanding-mandrel uncoilers are physically attached either to the floor or to some type of undergirding. This prevents both lateral and longitudinal movement and thus helps maintain proper alignment. When necessary, however, an open-end expanding-mandrel uncoiler can be placed on wheels, and these wheels roll laterally on rails in order to better center the coil. This can be a manual operation, or it can be mechanized, if desired, either electrically or hydraulically. Especially when using rather large coils (10,000 lb or heavier), more and more companies are purchasing coil cars for simplification of loading and unloading coil after coil, as explained in the next section.

There are many different sizes of cantilevered uncoilers, ranging from those designed for 500-lb coils to those capable of handling master steel coils weighing 60,000 lb. The needs and applications of any of these uncoilers quite often are determined by the available coil-loading capabilities.

COIL CARS

A coil car is a piece of equipment that normally travels horizontally on a pair of rails. More often than not, because of the need for substantial support, these rails are manufactured from railroad rails or similar items. The coil car is a means by which an expanding-mandrel, or "expandrel," uncoiler can be loaded with a coil without the use of a crane or fork-truck. In other words, if the crane or fork-truck places the coil on top of the coil car, the coil car can facilitate the loading of the coil as desired at any time in the future, whether the crane or fork-truck is available or not.

Most coil cars are capable of variable

Figure 4D-9. Standard coil car

horizontal movement. This movement is normally restricted by the original manufacturer to connector lines — hydraulic, electrical, or pneumatic — which are the means of connection between the coil car and the expandrel. Operational movement of the coil car in all directions is usually capable of being controlled from the expandrel which accompanies it. Exceptions to this rule are self-contained coil cars, which are completely integrated in themselves and are not physically connected to the uncoiler for power transmission.

The top of the coil car is normally a v-trough, in order that the coil might rest firmly on top of the car. The top of the coil car also normally has the capability of being raised or lowered for vertical positioning of the coil when loading it onto the expandrel, as well as for unloading coils which have not been completely used (Figure 4D-9).

Coil cars may operate on rails inset or embedded into the floor, or on rails placed directly on the floor. Coil cars also may be completely inset into the floor in a pit-type application, thus of course allowing the mandrel of the uncoiler to be placed considerably lower than usual. Although in extremely rare cases the coil car may travel a rather long distance from crane area to uncoiler, rail lengths of from 6 to 8 ft more than the maximum width of the coil are fairly standard.

Figure 4D-10. Coil car with integrated power unit

The coil car is not usually an integrated item unto itself, but rather is dependent on the expandrel to which it is attached, specifically for purposes of movement, because normally one hydraulic system or electric drive is used for both the expandrel and the coil.

There are some cases in which the coil car is built as a separate unit. Such coil cars require larger amounts of floor space, because the integrated power unit must of course be placed within the confines of the coil car itself (Figure 4D-10).

It is very important that a coil car be purchased to accommodate the maximum coil weight it will be handling at any time in the future, and even though this seems to be merely a common-sense procedure, it is quite often overlooked in preplanning of coil-handling facilities.

4E: Blast Cleaning of Structural Steel

The use of centrifugal wheel blast cleaning in structural fabricating plants has expanded rapidly in the past 10 years. Over 200 fabricating plants now have this type of equipment. Several reasons account for the increased use of blast cleaning.

A growing number of specifications call for blast cleaned steel. It has been demonstrated that coatings last considerably longer when applied over scale-free, etched surfaces. Mill scale is a brittle, nonuniform material which mechanically adheres to the steel. The scale expands and contracts at a different rate than the steel and therefore cracks, peels, and exposes the steel to oxidation. If a coating has been applied over the scale, the coating ruptures and oxidation, or rusting, begins. Removal of the scale exposes a predictable surface for subsequent coating.

New, fast-drying primers, such as inorganic zinc, require scale-free surfaces for proper application.

Rising labor costs have made wire brushing and air blasting extremely expensive. Wire brushing burnishes rather than removes tight mill scale. Air blasting is slow, dusty, uneven except when closely supervised, and very expensive.

Blast cleaning, by either air blasting or centrifugal wheel blasting, removes dry surface contaminants by impacting an abrasive material against the surface to be cleaned. Thousands of pellets are thrown at each square foot of the steel being cleaned.

DESCRIPTION OF MACHINE ELEMENTS

Centrifugal wheel blasting does not utilize compressed air. Instead, an electric-motor-driven, bladed wheel (Figure 4E-1) hurls the abrasive by centrifugal force at the steel to be cleaned. The abrasive (generally steel shot or grit) is thrown at a velocity of 14,000 fpm. Centrifugal wheels are powered by motors generally ranging from 15 to 60 hp. Each wheel throws the abrasive in a long,

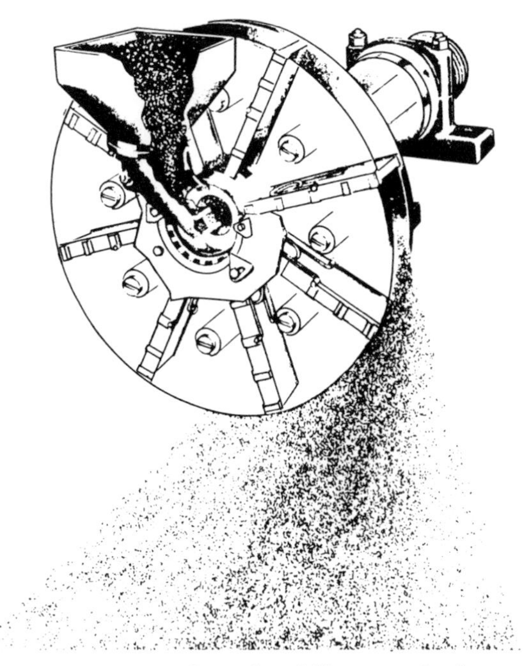

Figure 4E-1. Centrifugal blasting wheel

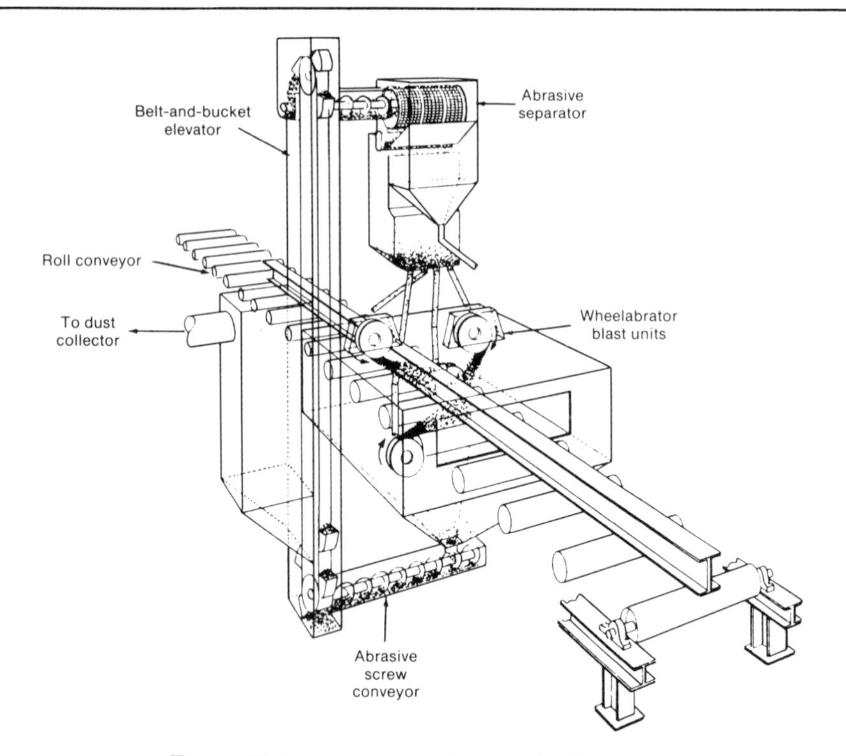

Figure 4E-2. Typical centrifugal wheel blast cleaning machine

Figure 4E-3. Arrangement of blasting wheel in a cabinet

narrow pattern about 3 in. wide and 36 in. long.

Every centrifugal wheel blasting machine consists of the following basic elements (Figure 4E-2):

1. Centrifugal wheel units
2. Cabinet or housing
3. Abrasive recirculating system
4. Work-handling system
5. Dust collector.

A number of centrifugal wheel units are arranged around the cabinet so that all surfaces of the steel being cleaned will be exposed to the abrasive blasts (Figure 4E-3). Machines that clean shapes and plate before fabrication generally are equipped with four wheel units, although machines cleaning wide plate may have six or eight wheel units. Machines cleaning after fabrication are generally equipped with eight wheels. In some special cases, machines cleaning completed fabrications may have as many as 40 wheel units.

The purpose of the housing, or cabinet, is to contain the abrasive. Cabinet size is dictated by the maximum size of the work to be cleaned. Cast steel liners are placed on the walls opposite the wheel units to absorb the "overblast" from the wheels. All openings are designed to be as abrasive-tight as possible. Generally, multiple layers of rubber curtains are placed at the entrance and exit of the cabinet to confine the abrasive.

The abrasive-recirculating system consists of a gravity hopper below the machine, a belt-and-bucket elevator to convey abrasive to the separator at the top of the machine, and the separator itself (Figure 4E-4). As abrasive, mill scale, and rust flow through the separator, they are spread into a thin curtain which falls past an airwash. A dust-collector fan pulls air through the separator, creating the airwash. Scale, fine abrasive, and dust are deflected by the airwash and are removed from the machine. The heavier, usable abrasive falls into a storage hopper and is later fed to the wheels. The work-handling system is matched to the type of work being cleaned and the preference of fabricating-plant personnel. Shapes and plate being cleaned before fabrication are often passed through the machine on a horizontal roll conveyor (Figure 4E-5). Often the conveyor extends from the storage yard to the blast machine.

Fabricated pieces often do not lend themselves to transport on a roll conveyor. Gantry cranes (Figure 4E-6), monorails, and work cars mounted on rails are used to convey fabricated pieces. These types of machines can be used for cleaning a wide variety of work (Figures 4E-7 and 4E-8). Shop cranes are usually unsuitable because they do not have adequate speed control at low speeds (2 to 15 fpm) and because it is inconvenient and uneconomical to tie up a shop crane for protracted periods at the blast machine. Dust created by the blasting process is collected by a fabric-bag-type dust collector. The dust-collector fan keeps the blast machine under negative pressure as it pulls air through the machine and to the collector. Steel leaving the blast machine is dust free and may be coated without additional preparation.

All necessary machine controls are lo-

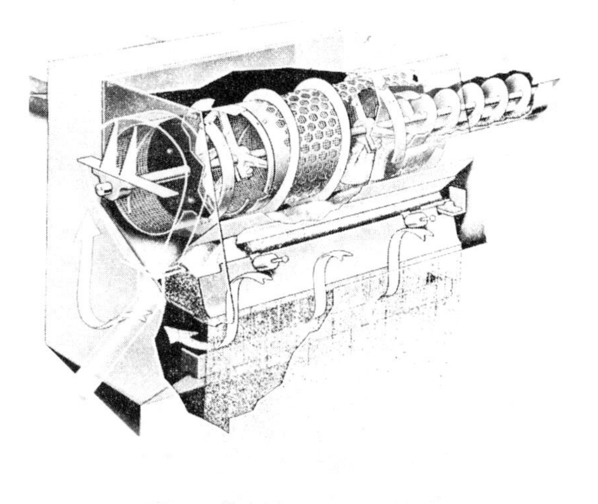

Figure 4E-4. Typical airwash-type separator

Figure 4E-5. Entrance and exit ends of roll-conveyor-type machine (note difference in cleanness of beams)

Figure 4E-6. Vertical blast cleaning machine with gantry crane

Figure 4E-7. Vertical blast cleaning machine cleaning a truss

Figure 4E-8. Vertical blast cleaning machine cleaning a rack of small shapes

cated at a central console, so that one operator can control all functions of the machine. Several refinements are available to make the machine as automatic and foolproof as possible. Sensing devices (wands or electric eyes) at the entrance of the machine automatically turn on the abrasive blast as work enters the machine and turn off the abrasive blast as the cleaned work leaves the machine. An automatic abrasive-adding unit senses the level of abrasive in the storage hopper and adds precise amounts of abrasive as needed.

ABRASIVES

The abrasives used are steel shot and steel grit. Shot is of spherical shape, whereas grit is of random shape, mostly angular. Steel shot (Figure 4E-9) is the more widely used material. It is highly refined tempered martensitic steel, heat treated to give maximum toughness and long fatigue life. Average hardness of steel shot is 45 to 51 HRC. The shot is repeatedly impacted against the steel being cleaned before it fractures. Shot is available in a number of sizes. The sizes most often used to clean structural steel are S-230, S-280, and S-330. The size designation denotes the nominal size of the pellets in thousandths of an inch — i.e., S-230 contains a high proportion of pellets 0.023 in. in diameter.

Lighter-gage plate and shapes may be cleaned with the smaller sizes of shot, but thick sections usually require larger, heavier

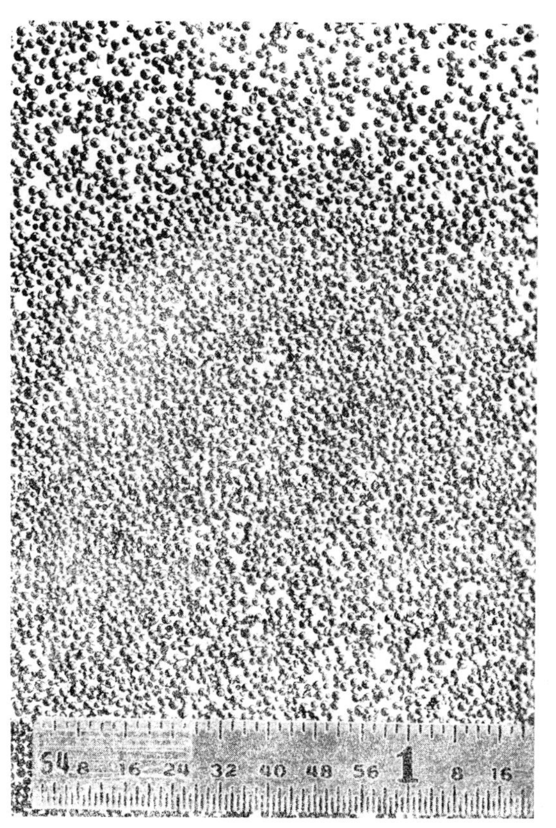

Figure 4E-9. Steel shot (S-230)

pellets to crack the thicker scale. The number of pellets per pound of abrasive increases tremendously as the size decreases. S-390 shot contains approximately 93,000 pellets per pound. S-230 contains 420,000 pellets per pound. Because cleaning speed and quality depend on coverage (or number of impacts in a given area) as well as on pellet mass, it is best to use the smallest shot that is capable of removing the toughest contaminant.

As shot impacts against the steel, it deforms the surface of the steel, producing an anchor pattern or surface profile. The anchor pattern is a nondirectional, matte finish which provides an excellent mechanical bond for primers. The anchor pattern is composed of a series of nonuniform "peaks" with adjacent "valleys." Defining an anchor pattern consists of measuring the peak-to-valley depth and counting the number of peaks per linear inch. The peak-to-valley depth increases, and the peak count decreases, with increasing shot size.

Some coatings require an anchor pattern with sharper peaks and valleys than can be

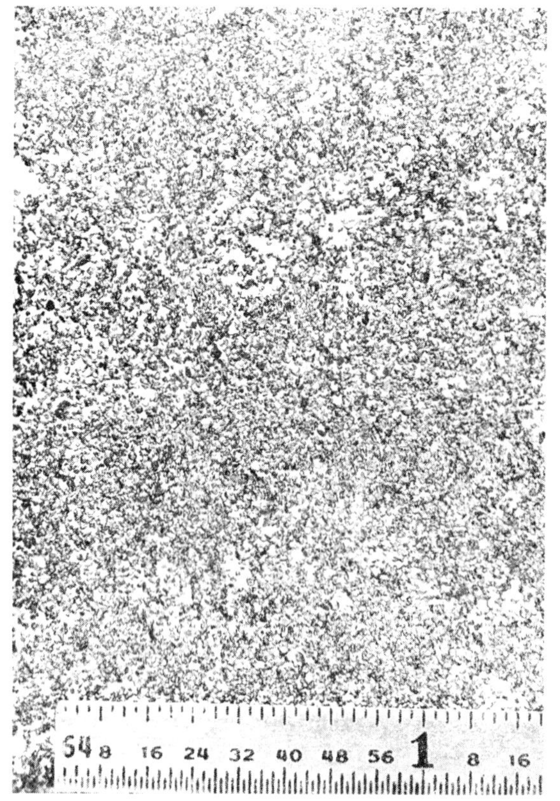

Figure 4E-10. Steel grit (G-50)

produced by shot. Steel grit, being angular, will produce the necessary anchor patterns (Figure 4E-10). Steel grit is made in three hardness ranges: soft (45 to 51 HRC), medium (56 to 60 HRC), and hard (62 to 66 HRC).

The breakdown rate, or tendency of the abrasive to fracture, increases with hardness. The harder grits therefore retain angularity through frequent fracturing, but the cost of using harder grits is higher. Not only are the harder grits consumed more rapidly than shot or soft grits, but machine maintenance also increases due to increased abrasion of machine components. Because of increased operating costs, harder grits should be used only when specified by the coatings manufacturer.

CLEANING COSTS

Cleaning costs can be expressed in terms of cost per square foot cleaned or cost per ton cleaned. Several variable factors are involved in determining cleaning costs, and therefore costs can vary widely depending on the circumstances of a particular situation. Usually, direct costs range from $0.50 to $3.00 per ton cleaned.

Total horsepower of wheel units for a given machine governs the rate at which steel can be cleaned to a particular specification. A machine with eight 30 hp wheels will clean steel at approximately twice the rate of a machine with eight 15 hp wheels.

The type of material-handling system used has a definite effect on cleaning costs. If loading and unloading of the machine is awkward and time consuming, cleaning costs will be higher than with an efficient system. One example of an efficient system is a cleaning and coating line operated by a metal building manufacturer. Structural building components weighing up to 50 lb per linear foot are attached to an overhead monorail conveyor. The monorail carries the steel at 19 fpm through the blast machine and paint-spray booth. Average cleaning cost is $0.01 per lineal foot. The type of abrasive used affects the cleaning cost as described above.

The cleanness of the steel increases as the speed through the machine decreases. Steel cleaned to "white metal" specifications will have to go through the blast machine at a much lower rate than steel cleaned to a "commercial blast cleaned" specification.

Alloy steels, such as A242 and HY80, must be cleaned at a lower rate to achieve comparable degrees of cleanness. Alloy steels have a very tenacious mill scale which is difficult to remove. Deeply rust-pitted steel is also more difficult to clean and requires lower cleaning speeds.

INCREASE IN COATING LIFE

All types of coatings are being applied over blast cleaned steel. One prominent fabricator retained an independent testing laboratory to evaluate the performance of their standard primer when applied over wire brushed and shot blasted surfaces. The primer relies on a combination of both lead and zinc chromates for rust-inhibitive properties.

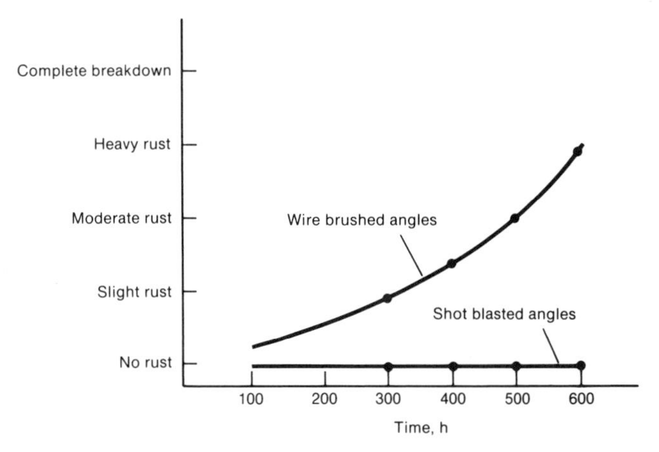

Figure 4E-11. Life of coatings applied over wire brushed and shot blasted steel angles

One set of steel angles was wire brushed while another set was blast cleaned, with an S-330 operating mix, to a commercial blast cleaned surface. The angles were then brush painted under controlled conditions. Dry-film thickness measurements were made on the shot blasted samples using a GE film-thickness gage. Due to the presence of rust and mill scale on the wire brushed samples, it was not possible to obtain significant thickness data with the gage. At the conclusion of the test, the wire brushed samples were sectioned, mounted, polished, and measured optically. The prepared samples were placed in an Atlas Type DMC twin-arc weatherometer and exposed to the deteriorating effects of twin 17-amp filtered carbon-arc lights and intermittent water spray.

Visual examinations of the samples were made after 300, 400, 500, and 600 h of exposure. At the conclusion of 600 h, the shot blasted and primed samples exhibited no rust. The wire brushed and primed samples were heavily rusted (Figure 4E-11). Significantly, paint thickness on the shot blasted surface was uniform, deviating only 0.4 mil from the average, whereas paint thickness on the wire brushed surface deviated as much as 3.0 mils from the average (Table 4E-1).

Table 4E-1. Thicknesses of Coatings Applied Over Wire Brushed and Shot Blasted Steel Angles

	Coating thicknesses, mils, for:							
	Wire brushed angle No.				Shot blasted angle No.			
	A1	A2	B1	B2	A3	A4	B3	B4
	3	5	2	1	1.4	1.5	1.2	1.5
	4	1	3	3	1.6	1.4	1.0	1.5
	3	3	2	2	1.4	1.6	1.7	1.3
	3	4	3	4	1.3	1.1	1.5	1.2
	4	3	1	1	1.4	1.2	1.2	1.7
	4	4	4	3	1.1	1.2	1.6	1.8
	5	2	5	2	1.4	1.0	1.2	1.7
	1	3	7	5	0.9	1.0	1.0	1.7
	6	2	6	4	1.2	1.5	1.2	1.5
	5	3	7	2	1.3	1.2	1.3	1.2
Average	3.8	3.0	4.0	2.7	1.3	1.3	1.3	1.5
Maximum deviation from average	2.2	2.0	3.0	2.3	0.4	0.3	0.4	0.3

Section 5
Nontraditional Machining

5A: Electrical Discharge Machining

Among the machining methods that have been conceived during the past 36 years is electrical discharge machining (EDM). This process is one of several nontraditional machining concepts. Others are electrochemical machining (ECM) and laser cutting.

There are two basic machines for electrical discharge machining: ram-type machines (Figure 5A-1) and wirecutting machines (Figure 5A-2).

The ram-type machine made its introduction in machining of dies, especially those made of hardened die steel and tungsten carbide.

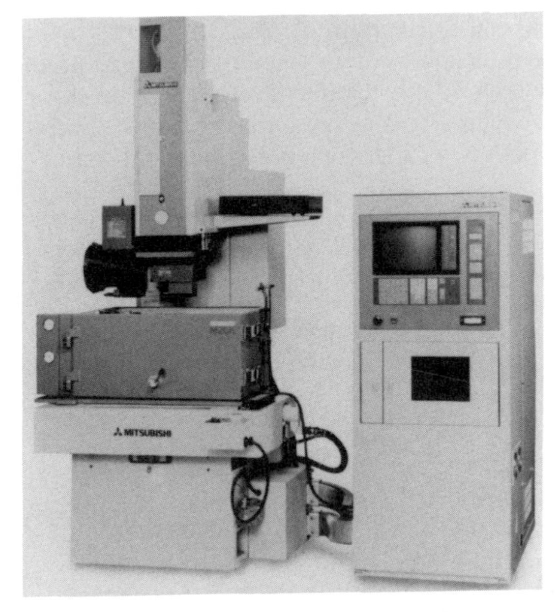

Figure 5A-1. Ram-type EDM machine

Fundamentals of Electrical Discharge Erosion

EDM is a thermoelectric process. Heat from the spark melts the metal and bulk boiling of superheated metal ejects about 1 to 10% of the molten metal. As a result, the hardness or tensile strength of the metal does not have any affect on machining by EDM. This is why EDM has always been successfully applied in machining of high-strength or hardened steels.

The thermal properties of the metal determine its machinability by EDM. Copper, tungsten, and graphite are always applied as electrode materials. This can be explained by the low wear of these materials due to their excellent thermal properties.

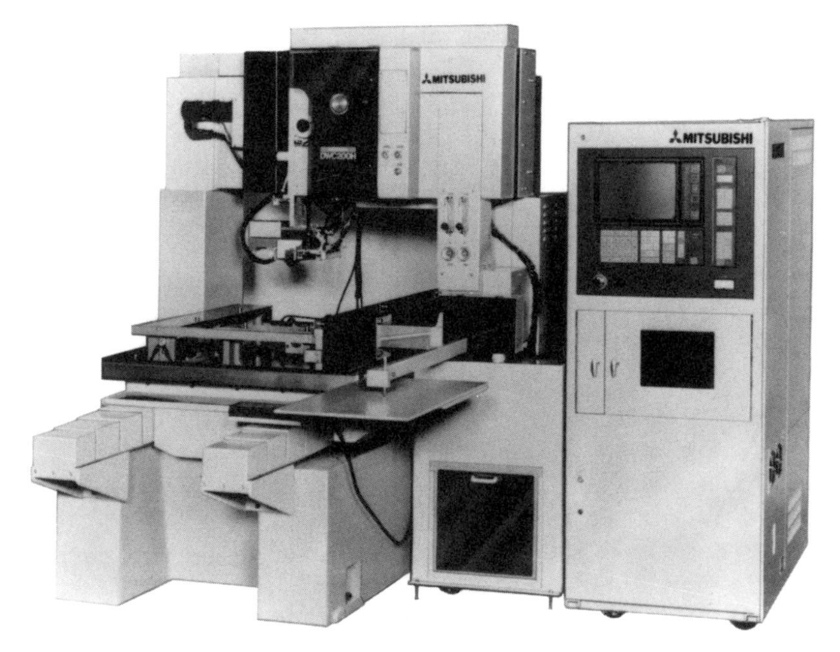

Figure 5A-2. Wirecutting EDM machine

Because 90% of the material in the molten pool of metal is not ejected at the end of the pulse, the workpiece is covered with a recast layer 1 to 30 μm thick. Sometimes this layer is desirable because its hardness and roughness hold lubricating oils. However, in some applications the recast layer has to be removed in order to avoid surface fatigue.

Setting of an EDM Machine

The dependent parameters in the EDM process are:

1. Metal-removal rate, V_W (mm^3/min)
2. Relative electrode wear, Θ (%):

$$\Theta = \frac{V_E}{V_W} \cdot 100$$

   where V_E is the electrode-removal rate.
3. Surface finish, R_t (peak to valley, μm)
4. Thickness of recast layer
5. Gap between electrode and workpiece
6. Radii of corners and edges.

These parameters are dependent on the following independent parameters:

1. Electrode material
2. Electrode polarity (+ or −)
3. Pulse current, $\bar{i}_f$ (A)
4. Pulse duration or on time, t_i (μs)
5. Pulse interval time or off time, t_o (μs)
6. Average voltage or working voltage, U (V)
7. Average current or working current, I (A)
8. Working current density, I_d (A/cm^2)
9. Open-gap voltage, U_o (V)
10. Type of dielectric
11. Flushing mode.

Dielectric-Fluid Unit

Dielectric fluid is pumped through a filter from a container (100 to 200 l) to the electrode or workpiece. Pressurized flushing through the electrode results in a conical hole in the work because sparks occur also at the sides of the electrode due to the excess of debris in the side gap. When vacuum flushing through the electrode or through the workpiece is used, a straight hole is obtained.

PRINCIPLES OF ELECTRICAL DISCHARGE MACHINING

An electrode, which serves as a cutting tool, and a workpiece are placed face to face with very little clearance (several to several tens of μm) in a dielectric fluid (kerosene-base oil specially made for electrical discharge machining is generally used). Current pulses are continually supplied to the clearance from a pulse power supply (approximately 60 to 300 V) to provide transient arc discharge (discharge retention time: 0.1 μs to 8 ms) at a high frequency so as to remove workpiece metal with a very dense energy provided by the discharge. Figure 5A-3 shows the five-step process from spark discharge, through metal machining, to the original cool state.

It sounds as if it takes a long time for steps (1) to (5), but actually the entire process is completed in 1/50 second even for rough machining, which is the slowest among the machining processes. This discharge between the entire surface of the workpiece and that of the electrode is repeated several tens to several hundred thousands of times for machining. If, therefore, electrodes are made only of such easily machinable materials as copper and graphite, workpieces can be machined highly accurately to shapes corresponding to the electrode shapes, regardless of workpiece hardness and machinability.

Surface Roughness, Electrode Wear, and Machining Speed

Because electrical discharge machining is performed by accumulation of single discharges, the amount machined, machining speed, clearance, and surface roughness increase as the single-discharge energy increases. (See Figure 5A-4). The amount of single-discharge energy depends on the single-discharge peak current (power setting) and time (pulse width). Single-discharge energy is large if the power setting and pulse width are large, and small if both settings are

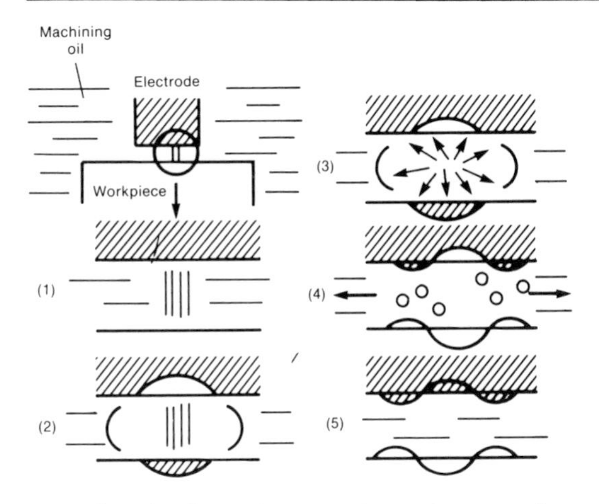

1. When the electrode approaches within several microns of the workpiece, a spark is generated at a point of the shortest distance and immediately becomes a fine arc column, or a flow of electrons at high current density, hitting the workpiece at one point. The electron flow generates heat at this point, to a temperature high enough to melt tungsten, which has a high melting point. At the same time, ions, which are generated by collision of the electron current with the dielectric fluid, heat the electrode.
2. The heat vaporizes the dielectric fluid around the flow.
3. The pressure applied to the melted workpiece and electrode is small compared with the entire surface of the workpiece and electrode, but is very large per unit area.
4. Molten metal from the workpiece is blown away into the dielectric fluid as small round blocks. The remaining portions on the edge form protrusions on both the workpiece and electrode. The protrusions become discharge points for later discharge.
5. Cool dielectric fluid flows into the hollow where molten metal has been blown off, removing the residual heat from the hollow.

Figure 5A-3. Process through which metal is machined by spark discharge and oil pressure

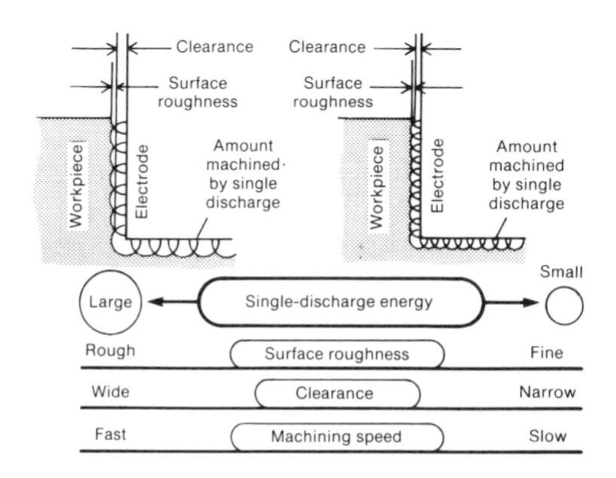

Figure 5A-4. Relations between single-discharge energy and machining characteristics

small. To obtain the desired surface roughness, the power setting and pulse width must be selected from the combinations that have the same single-discharge energy:

1. Large power setting and small pulse width
2. Small power setting and large pulse width
3. Power setting and pulse width intermediate between 1 and 2 above.

Figure 5A-5 shows methods of selecting power setting and pulse width for the same surface roughness and the relationships among methods, machining speed, and electrode consumption.

PRINCIPLES AND FEATURES OF ORBITAL-MOVEMENT EDM

This machining method provides better performance than that obtainable with conventional electrical discharge machining because orbital-movement machining controls the X- and Y-axes in addition to the Z-axis.

Improvement in Machining Shape Accuracy. Orbital-movement machining prevents loss of accuracy in the side surface direction (uneven clearance, taper, intermediate protrusion, or uneven surface) so that the entire surface can be machined evenly at any desired clearance. (See Figure 5A-6)

Less Electrode Wear. Because orbital-movement machining finishes a shape

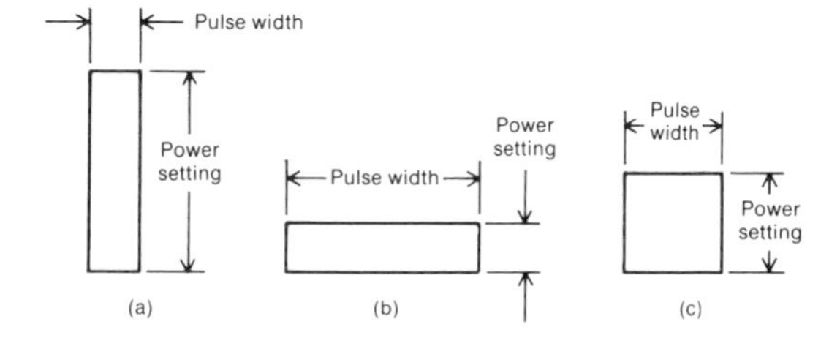

(a) Fast machining speed, more electrode consumption. (b) Slow machining speed, less electrode consumption. (c) Machining speed and electrode consumption both intermediate between (a) and (b) above.

Figure 5A-5. Method of selecting power setting and pulse width

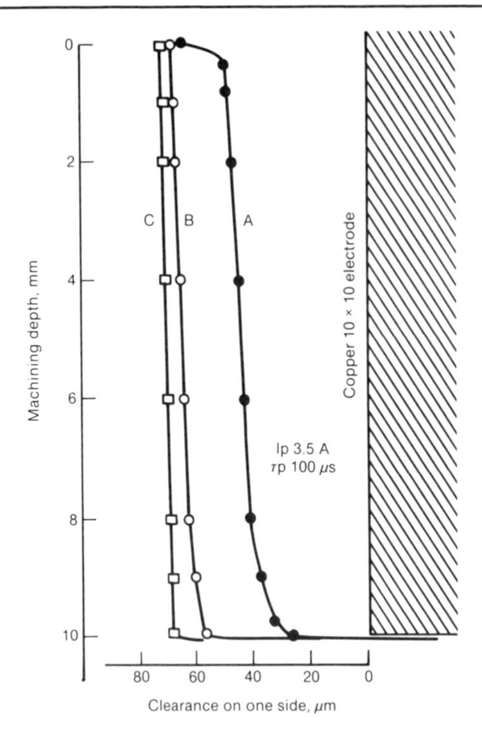

A: Conventional machining (two electrodes). B: Orbital-movement machining (one electrode), R=50 μm. C: Orbital-movement machining (two electrodes), R=50 μm.

Figure 5A-6. Improved side-surface accuracy in orbital-movement machining

by machining in the side-surface direction, the electrode wear is not concentrated on the edge, as in conventional machining methods, and consequently the effect of electrode wear is much smaller than in conventional machining methods. (See Figure 5A-7). The result is a finer surface, decreased spark hardening layer, and increased machining speed.

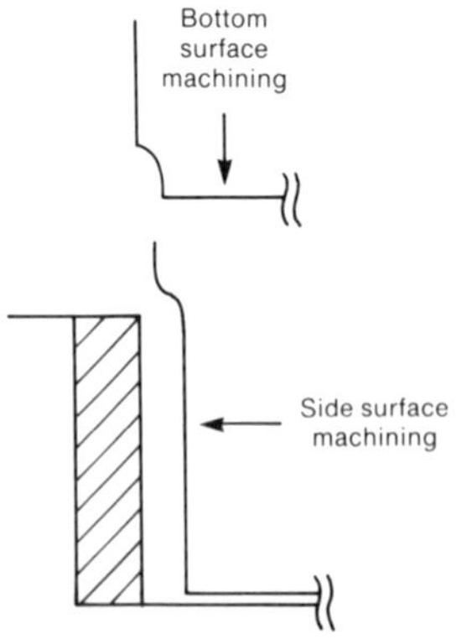

Figure 5A-7. Relationship between machining direction and shape of electrode wear

Improvement in Finish Machining Speed. Discharge area is wide, and sufficient machining current can be supplied (area effect) because machining is performed in both the bottom and side-surface directions at the same time. Orbital-movement also helps to remove machining chips more quickly, preventing generation of arcs. (See Figure 5A-8).

Reduction of Electrode Manufacturing Man-Hours. Because any values can be selected for machining dimensions by setting the orbital movement amount, two or more electrodes of different sizes for roughing and finishing are not required. An electrode manufacturing process using only one type of electrode suffices. (See Figure 5A-9).

Taper Machining With Straight Electrode. Taper machining can also be performed with a straight electrode, as an orbital-movement machining application.

Examples of taper-machining capabilities include machining of different top and bottom shapes or top and bottom radii, machining punch and die simultaneously, machining sharp edges and corner radii, and angle alteration.

WIRE ELECTRODE DISCHARGE MACHINING

Principles

Wire electrode discharge machining brings a thin wire electrode (made of brass, tungsten, or copper wire) to a workpiece and impresses a voltage between the wire electrode and workpiece, while flooding deionized water of high insulation resistance around the electrode to discharge electricity through the deionized water as a medium.

This discharge forms an arc column at the shortest distance between the wire electrode and workpiece surfaces and heats both sides locally, melting them with thermal energy as shown in Figure 5A-10(a). At the same time, the insulation water around the wire electrode is heated to vapor state and expanded rapidly to create a local explo-

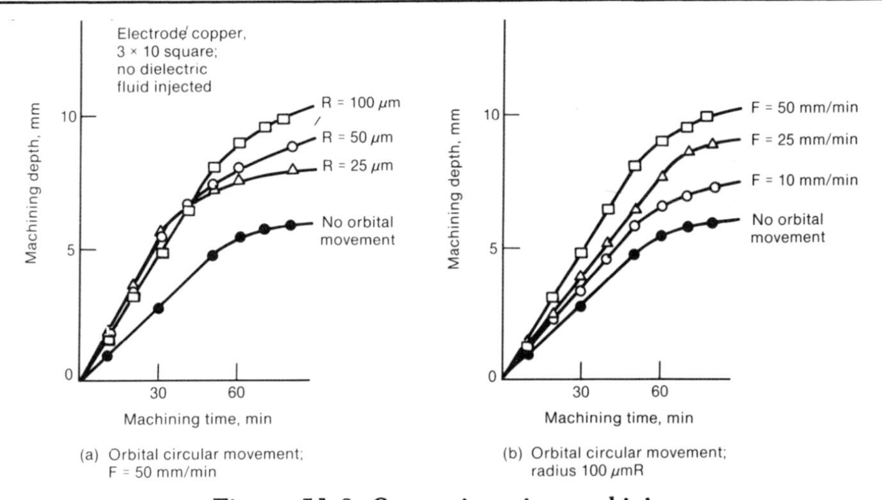

Figure 5A-8. Comparison in machining speed between orbital-movement machining and non-orbital machining

sion. The explosion pressure turns molten metal into fine metal particles, which are carried away in the water. The workpiece and wire electrode surfaces are then cooled to leave a dent as shown in Figure 5A-10(b). Because the discharge and metal removal are performed repeatedly at a very high frequency, the dent on the workpiece surface grows gradually to form a groove along the wire electrode shape as the wire electrode is fed at increasing speed.

Machining Speed

Wire electrode discharge machining defines machining speed as sectional area of machining per unit time.

$$\text{Machining speed (mm}^2\text{/min)} = \text{Machining feed speed (mm/min)} \times \text{Workpiece thickness (mm)}$$

Workpiece Thickness and Machining Speed. The machining speed of a wire electrode discharge machine is almost proportional to the machining current between the poles, as in general-purpose die-sinking electrical discharge machines. In other words, a higher peak current and short-

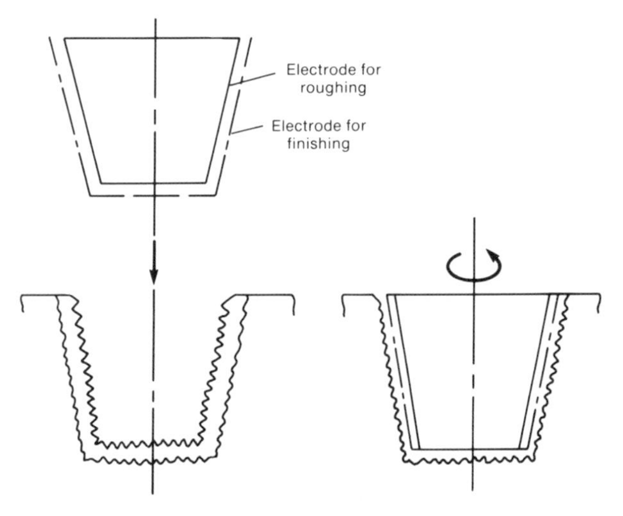

Figure 5A-9. Comparison of electrodes for conventional and orbital-movement machining processes

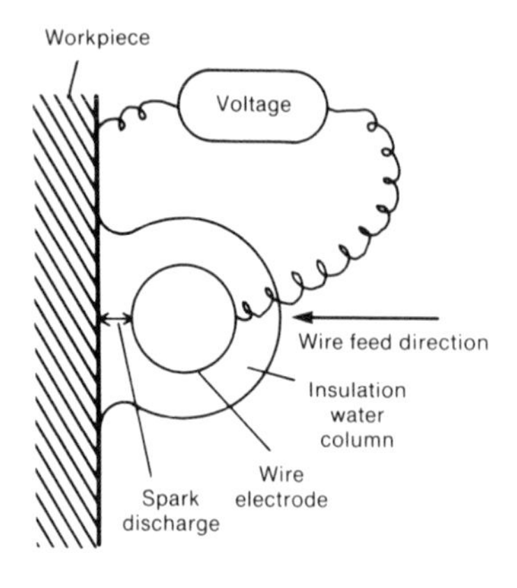

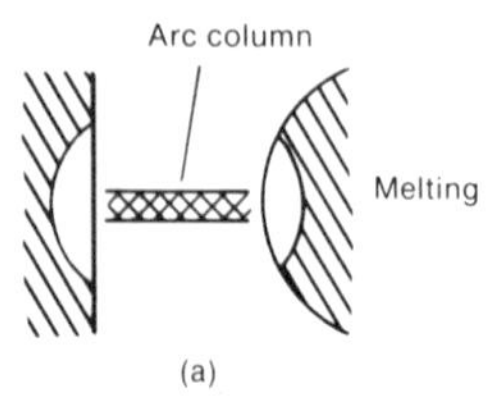

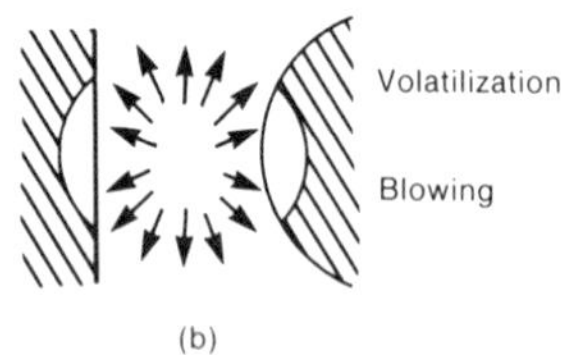

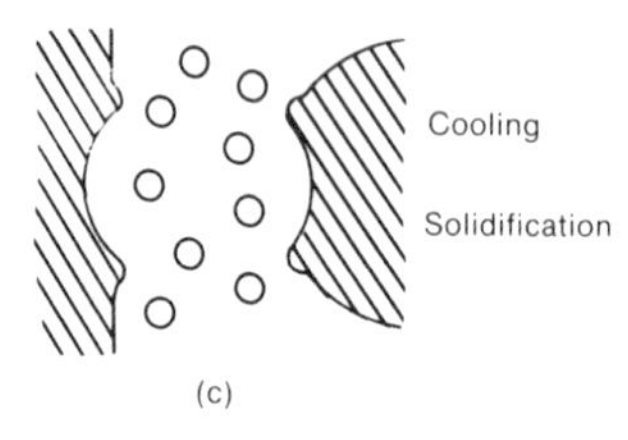

Figure 5A-10. Schematic showing principles of wire electrode discharge machining. See text.

er rest time increase the machining speed.

A greater workpiece thickness makes the machining surface area larger in the advance direction, and the area effect works effectively to improve the machining speed.

Because the peak current value is selected by machining setting, peak current is higher and thus machining speed is faster if the machining setting is larger. A larger machining setting, however, causes greater surface roughness. Figure 5A-11 shows the relation between machining speed and workpiece thickness.

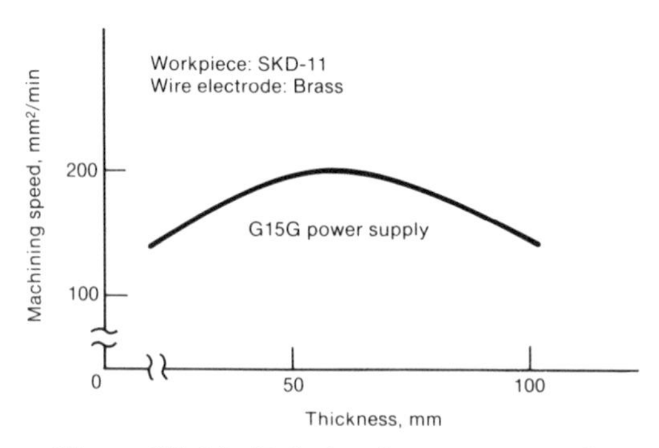

Figure 5A-11. Relation between wire electrode machining speed and workpiece thickness

Machining Accuracy

Machining accuracy in wire electrode discharge machining includes uniformity, straightness, and shape accuracy of machining groove width.

Machining Groove Width and Machining Feed Speed. Dispersion in machining groove width is all equal to that in the final dimensional accuracy because half of the width is considered as an offset value and left in the inside or outside of a size specified in the drawing to obtain the desired dimensions when a workpiece is machined.

Figure 5A-12 shows the relation between machining groove width and machining feed speed. Optimum feed occurs when the electrical condition of the machining power

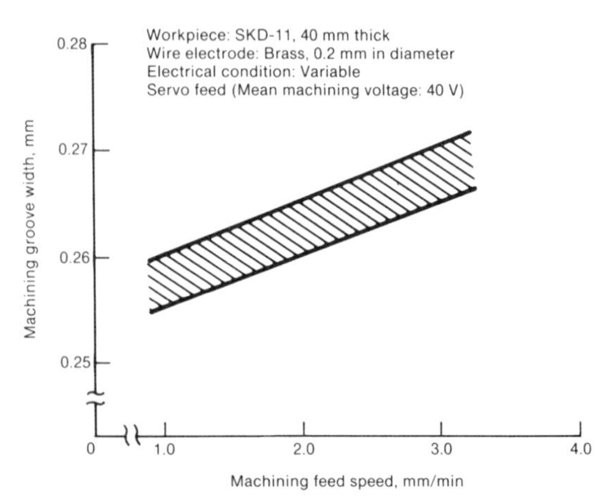

Figure 5A-12. Relation between machining groove width and machining feed speed in wire electrode discharge machining

supply is altered greatly using a mean machining voltage constant-control system. As shown in the figure, optimum feed creates an overall change in the machining groove width of approximately 15 μm even if the electrical condition is changed to significantly alter the machining feed speed.

Figure 5A-13 shows the relation between machining groove width and mean machining voltage in a constant-feed/speed system. The relationships at two machining feed speeds, F 1.4 and 2.4 mm/min, are shown with the electrical condition of machining power supply altered for each speed. As illustrated in the figure, the machining groove width varies 32 to 35 μm with the changes in the electrical condition at either speed in the constant-feed/speed system. In other words, changes in mean power greatly affect changes in groove width.

Consequently, the optimum speed system enables changes in electrical condition and machining feed speed with less effect on machining groove width than the constant-feed/speed system.

Second Cut

Second cut is a finishing process similar to that of a die-sinking electrical discharge machine. It is performed as follows: the primary machining (first cut) is made on a workpiece leaving a finish allowance; the electrical condition for the finish allowance is switched to finish condition, decreasing the offset amount gradually so as to perform two or more surface removal passes for finishing. Originally, wire electrode discharge machining was low in machining speed, but recent improvements in power-supply units allow a considerably higher speed if surface roughness of 25 to 27 μm Rmax is acceptable. As in general-purpose electrical discharge machining, however, increased speed causes greater surface roughness and corner rounding that results in lower dimensional accuracy. It is the second-cut process that solves these problems; this process provides high accuracy and good surface finish. Figure 5A-14 shows the relation of the number of sec-

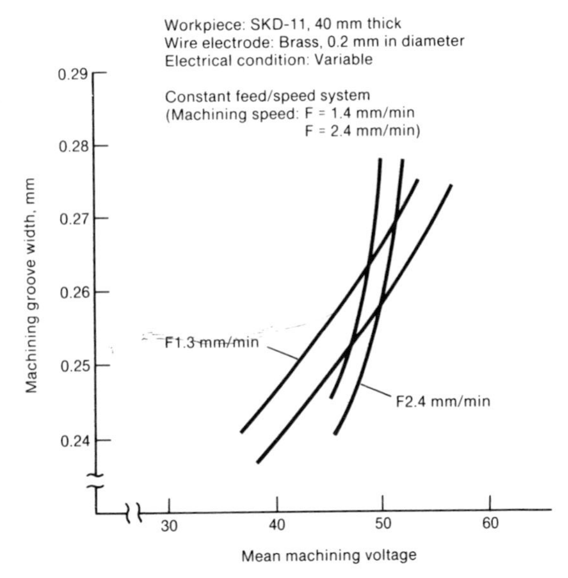

Figure 5A-13. Relation between machining groove width and mean machining voltage in a constant-feed/speed wire electrode discharge machining system

ond-cut passes to machining accuracy and surface roughness. The figure indicates that the required times for one-pass and two-pass machining have a ratio of approximately 1 to 0.3 (the latter being approximately 3 times shorter than the former). Two-pass machining also gives higher machining accuracy, if, for example, a 40-mm thick SDK-11 workpiece is to be finished at a surface roughness of approximately 10 μm Rmax.

Examples of Applications

Three examples of applications of wire electrode discharge machining, together with the advantages realized from use of the process, are:

- **Plastic molding dies,** for which a major reduction in machining time can be achieved with the integral machining of sirocco fans.
- **Progressive dies for stepping-motor cases,** for which the process provides major cost reductions with significantly shortened delivery times.
- **Dies for printed-circuit boards,** for which the automatic wire-feed option eliminates burrs — an advantage in unattended operation.

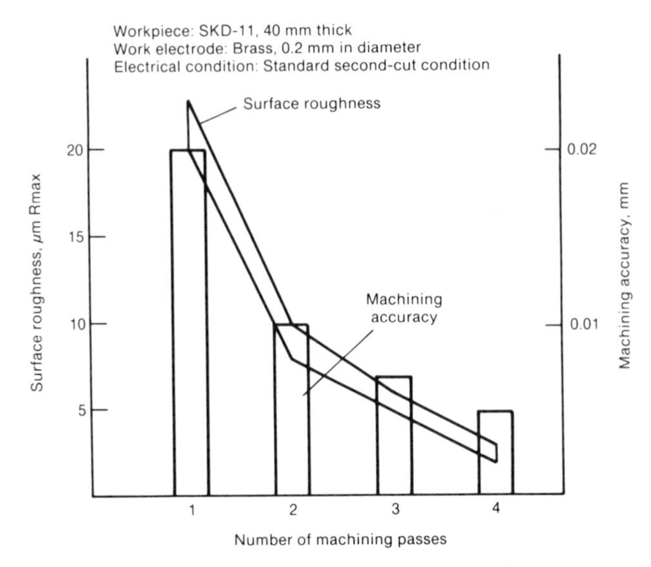

Figure 5A-14. Relation of number of second-cut passes to machining accuracy and surface roughness in wire electrode discharge machining

EFFECT OF RECAST LAYER ON SURFACE INTEGRITY AND FATIGUE LIFE OF 18% NICKEL MARAGING STEEL

Elecrical discharge machining (EDM) is known to have a large detrimental effect on fatigue strength. The nature of the EDM process, which entails removal of metal by melting and vaporization of the metal surface, inherently induces very large residual stresses which are tensile in nature and hence detrimental to fatigue resistance. The rapid cooling of the surface layer by the dielectric also results in surface and subsurface cracks and microcracks. Such cracks also tend to propagate due to the stress-concentration effect, and, under cyclic loading, fatique failure occurs prematurely.

Such deterioration in fatigue strength due to EDM has necessitated subsequent surface-layer-removal operations for heavily stressed components, or those subjected to cyclic loads. Most users would remove the recast layer and hope that this would restore the fatigue strength.

Figure 5A-15 shows the nomenclature used to define the various layers on the surface of a component machined by EDM. The recast layer (RL) is a layer of metal that was in the molten state and that solidified on the surface. The heat-affected zone (HAZ) is a layer, below the recast layer, which has been subjected to elevated temperatures sufficient to alter its mechanical and metallurgical properties. Both layers together are known as the altered metal zone (AMZ).

The Importance of Surface-Integrity Investigations

The introduction of superalloys in the aerospace industry has led to the use of very critical components in highly stressed environments. The effects that the degradation of surface integrity produced were much more noticeable and alarming. The introduction of nonconventional machining methods also introduced new processes whose effects were little understood.

One of the main concerns of aerospace manufacturers is the fatigue life of components machined by EDM. This concern has necessitated a better understanding of the effects of EDM on fatigue and of how to eliminate its detrimental effects either by past processing and layer removal or by control of the EDM process itself.

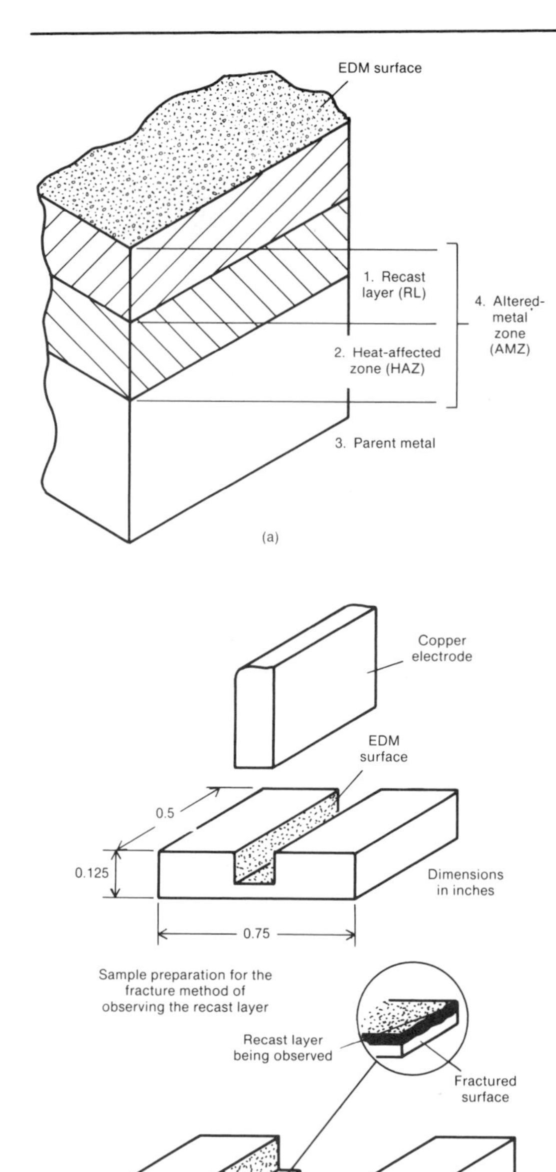

Figure 5A-15. (a) Surface layers on a component machined by EDM. (b) Fractured samples showing the fractured edges and the recast layer.

The Director of Manufacturing Research states that the following would be important contributions:

1. To determine what basic characteristics of EDM'd surfaces are influential in determining fatigue life of the machined workpiece
2. To determine how the basic parameters of EDM influence these critical surface characteristics.

It is clear that it is important to understand how material-processing methods affect the surface integrity of machined components and, in turn, how this effect on surface integrity influences the mechanical properties of the component.

The Effects of Machining Methods on Surface Integrity

There are two aspects of surface quality which must be controlled to ensure high reliability: surface roughness and surface integrity. Surface roughness is a measure of the topography or geometry of the outer surface layer of a component. Surface integrity is a term applied fairly recently to cover the metallurgical and mechanical alterations in the surface layer of a machined component.

When a surface is machined either by conventional means, such as milling or grinding, or by nonconventional methods, such as EDM, it is possible to introduce a wide variety of surface-layer alterations. Some of these alterations and their causes will be briefly discussed.

Untempered Martensite. This quite commonly occurs in chip-removal and grinding operations in machining of hardened steels. This layer results mainly from excessive overheating and sudden quenching of the surface layer. The untempered martensite layer is usually hard (above 55 HRC), and its depth depends on the temperatures reached. It is also very brittle and has a tendency to crack. In EDM the recast layer often consists of untempered martensite due to rapid cooling of the molten metal by the dielectric fluid.

Overtempered Martensite. This layer usually lies below the untempered martensite layer. It is formed by the high temperature gradient, but at a lower rate of cooling than the surface layer. It is usually softer and less brittle than the untempered martensite region. In EDM, the untempered martensite forms part or all of the heat-affected zone (HAZ) which lies below the recast layer.

Austenite Resolutioning. Maraging steels can produce a softer layer when abusively machined. At high temperatures, the austenite goes back into solution and the aging is lost. This results in a layer that is softer than the aged material.

Cracks. Micro- and macrocracks can be found in excessively abusively machined high-strength steels. The cracks will usually be quite small and detectable only by metallurgical observations. In EDM, the recast layer usually contains a large number of micro- and macrocracks, and these cracks are one of the main causes of the reduction in fatigue life. Figures 5A-16, 17, and 18 show photomicrographs of such cracks.

The Recast Layer in EDM. In EDM, the surface is subjected to intense heat which melts and vaporizes the metal. Some of that metal is redeposited on the surface to form the recast layer. This layer is porous and brittle and contains cracks which may extend into the parent metal. This layer is highly detrimental to fatigue life.

Residual Stresses. Residual stresses can be introduced by machining processes

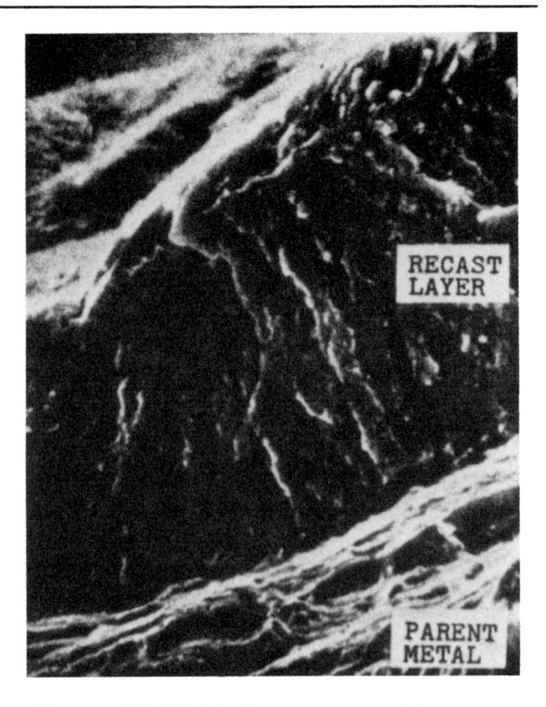

Figure 5A-17. Surface microcrack in an EDM'd workpiece of 18% Ni steel. (3000×)

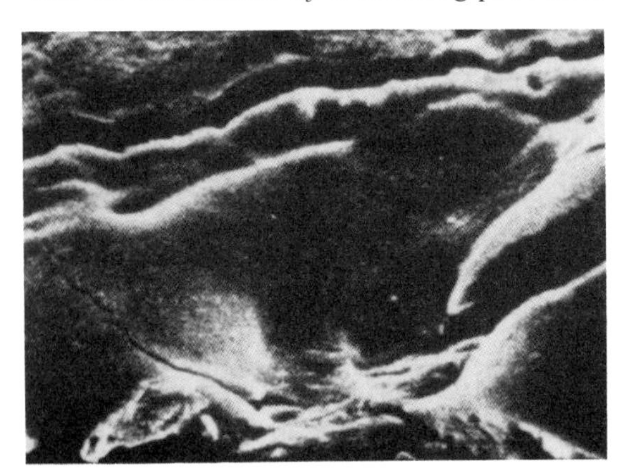

Figure 5A-16. EDM'd surface of 18% Ni maraging steel, showing cracks and discontinuities in the recast layer. (1500×)

Figure 5A-18. Fractured edge of an EDM'd sample of 18% Ni maraging steel. Current, 1 A; pulse width, 250 μs. (3000×)

such as grinding and turning. Abusive grinding produces tensile residual stresses which are detrimental to fatigue life whereas gentle grinding may introduce slightly compressive residual stresses which are beneficial to fatigue life. Any process which produces large values of tensile residual stresses is essentially detrimental to fatigue life. EDM, due to the intense heat produced on the surface, produces very high tensile residual stresses and is thus detrimental to fatigue life.

THE EFFECT OF EDM ON SURFACE INTEGRITY AND SURFACE FINISH, WITH EMPHASIS ON FATIGUE STRENGTH

The surface finishes of components machined by EDM depend on many factors, such as current settings, pulse widths, type of power supply, frequency of discharges, etc.

The surface consists of craters caused by the spark discharges (Figure 5A-19). These craters will vary in diameter and depth depending on the machine settings.

Achievement of very low values of surface finish were almost impossible before the development of new and more sophisticated power-supply systems.

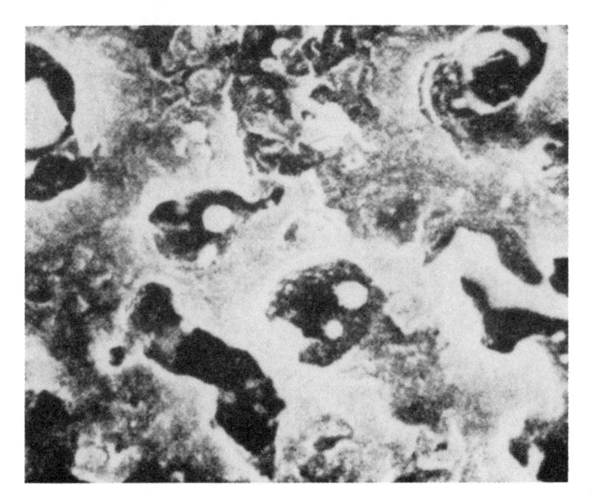

Figure 5A-19. EDM'd surface of 18% Ni maraging steel, showing craters and resolidified metal globules. (3000×)

It must be noted that topographical measurements with stylus instruments can be expected to convey only very limited information on the true nature of the surface produced by EDM. The standard stylus, or any stylus, for that matter, would be incapable of detecting the finer details in the topographical structure, such as cracks, or individual steep-sloped craters or spattered metal globules on the surface. These stylus measurements must therefore be regarded as providing limited information, even topographically in the case of EDM surfaces. Only a general indication of average amplitude is obtained, and its value is limited. Nevertheless, surface-roughness measurements are frequently made and quoted, and do provide an indication, albeit oversimplified, of the final state of the surface.

Quite evidently, the true nature of EDM'd surfaces is complex, and quantification of such a variety of factors in any single, easily expressible form can be misleading and should be treated with care.

The Effect of EDM on Surface Layers

The intense heat produced by the electrical discharges of EDM causes the metal on the surface to melt and vaporize. Some of that metal is redeposited on the surface and solidifies, forming a layer of metal that cools down very quickly from liquid to solid due to the cooling action of the dielectric fluid.

The thickness of that layer, known as the recast layer, will depend on cutting conditions, power-supply type, and material properties. The thickness of this layer may be as high as 0.004 in. or as low as 0.0001 in. or less. The recast layer is generally porous and cracked, and in many cases the cracks extend to the workpiece surface. Photomicrographs of recast layers are shown in Figures 5A-20 through 5A-25.

Below the recast layer, another metallurgical change takes place. The metal is not molten and resolidified, but has undergone changes in its properties pursuant to the temperature gradient it was subjected to. Usually, a layer of untempered and over-

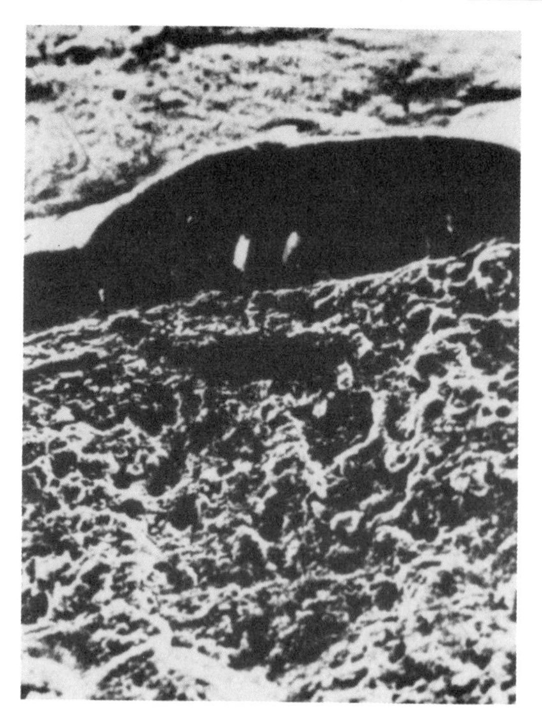

Figure 5A-20. Fractured edge of an EDM'd sample of 18% Ni maraging steel. Current, 1 A; pulse width, 100 μs. (1500×)

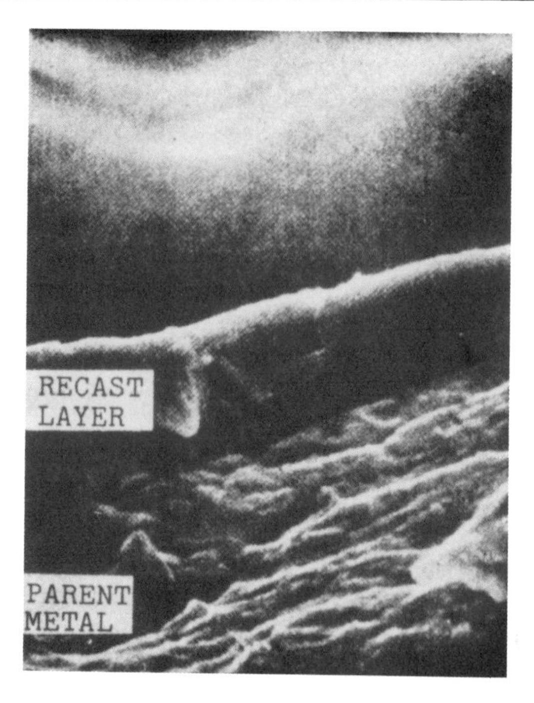

Figure 5A-22. Recast layer of EDM'd Udimet 700 before etching. Current, 6 A; pulse width, 250 μs. (3000×)

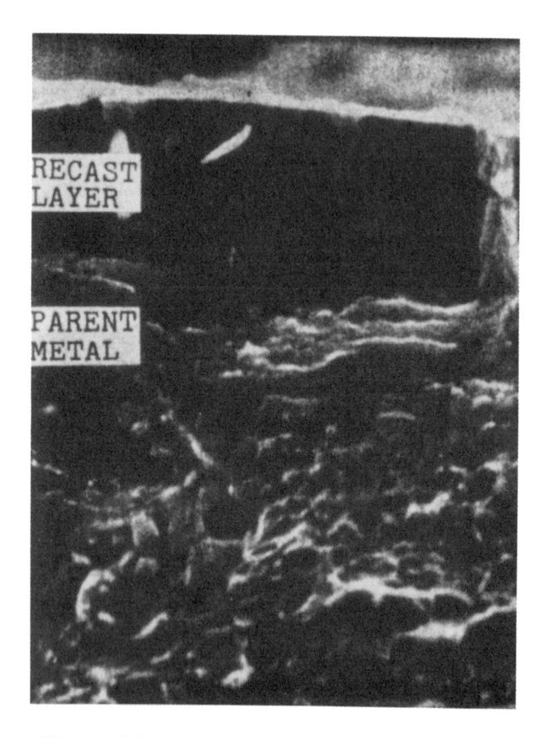

Figure 5A-21. Fractured edge of an EDM'd sample of 18% Ni maraging steel. Current, 0.5 A; pulse width, 50 μs. (1500×)

Figure 5A-23. Recast layer of EDM'd Udimet 700 after etching, showing etched γ' in the parent metal. (5000×)

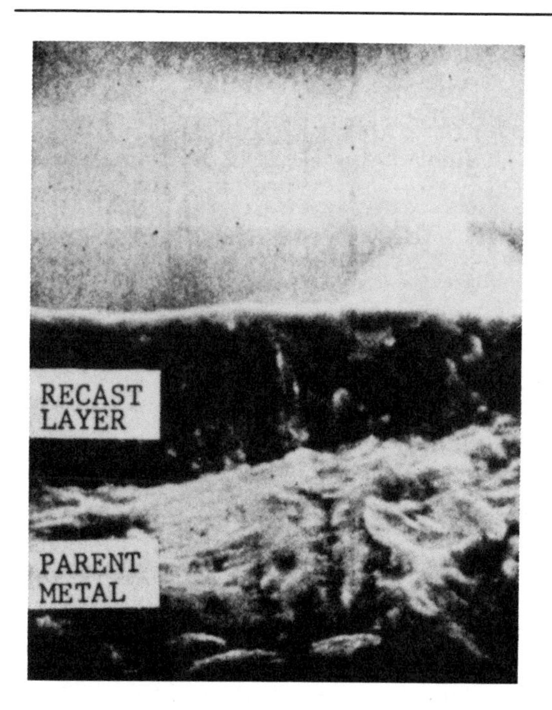

Figure 5A-24. Fractured edge of an EDM'd sample of 18% Ni maraging steel. Current, 5 A; pulse width, 100 μs. (1500×)

Figure 5A-25. Fractured edge of an EDM'd sample of 18% Ni maraging steel. Current, 1 A; pulse width, 100 μs. (1500×)

tempered martensite will be formed immediately below the melting boundary. This layer is known as the heat-affected zone (HAZ). The recast layer and the heat-affected zone together are known as the altered metal zone (AMZ).

The formation of these layers when using EDM generally lowers the fatigue strength. Use of very gentle cutting conditions does not eliminate the recast layer and HAZ, and it has been found that the fatigue strength is still greatly reduced compared with a milled or ground surface. Most aerospace manufacturers require that the EDM'd surface be removed by subsequent operations before a part is put to use in a highly stressed environment or subjected to cyclic loads.

The Effects of EDM on Surface Cracks

The surface layers of an EDM'd component usually contain micro- and macrocracks which may extend into the parent metal. These cracks form due to the extremely large residual stresses on the surface and due to the intense temperatures combined with rapid quenching by the dielectric fluid.

It is a well-known fact that surface cracks are a prime cause of fatigue failure and the main cause of the detrimental effect of EDM on fatigue strength. It is almost impossible to eliminate completely the cracks created on the EDM'd surface. Fine cutting conditions reduce, but do not eliminate, these cracks. This explains the fact that reheat treatment, although it eliminates the untempered and overtempered martensite layers, does not significantly improve fatigue life. The micro- and macrocracks formed in those layers are irreversible, and heat treatment will not eliminate them.

EXAMPLES OF APPLICATIONS OF EDM

Horizontal Machining

Product: A glass mold (three-piece)
Electrode: Copper
Workpiece: SUS 420J1
Functions: Lateral machining
Comment: All three sections of the work-

piece were produced using lateral machining.

Product: A rib mold
Electrode: Copper
Workpiece: SKD11
Functions: Lateral machining
Comment: For producing ribs, lateral machining offers both greater speed and reduced electrode wear, with stability identical to Z-axis machining.

C-Axis Machining

Application: Parts processing
Electrode: Copper
Workpiece: S55C
Functions: C-axis machining
Comment: The C-axis servo offers the same stability as the X-, Y-, and Z-axis servos.

Product: An internal helical gear
Electrode: Graphite
Workpiece: SKD61
Functions: Simultaneous Z- and C-axis machining
Comment: The minimum C-axis feed increment of 0.001 degrees ensured stable machining of this helical gear, which was 200 mm (7.88 in.) in diameter.

Copy Machining

Application: Thread cutting of an ultrahard alloy
Electrode: Copper-tungsten
Workpiece: Tungsten carbide
Functions: Simultaneous X- and Y-axis machining
Comment: The electrode was moved in a circle around the workpiece to transfer the thread pattern.

Automatic Electrode Changing

Product: A connector mold (core)
Electrode: Copper (total, 18)
Workpiece: STAVAX
Functions: Automatic tool changing (ATC), electrode-height compensation, automatic positioning compensation (APC), and mirror finishing.
Comment: The entire mold was machined and mirror-finished by a K series EDM. The over-all machining error was within $\pm 5\ \mu m$ (0.0002 in.).

3-Axis Simultaneous Machining

Product: A mold for cams used in ballpoint pens
Electrode: Copper (simple rod)
Workpiece: SKD11
Functions: Simultaneous X-, Y-, and Z-axis machining
Comment: A simple rod was moved in a helical path.

Ultralow-Electrode-Wear Machining

Product: An electronic component mold
Electrode: Copper
Workpiece: NAK80
Functions: Ultralow-wear machining
Comment: The slope control circuit prevented wear on the edges of the ribs of this mold. Machining up to a depth of 5 mm (0.197 in.) can be completed by a single electrode.

Contour Machining

Product: A washer-faced punch
Electrode: Copper (simple rod)
Workpiece: SKD11
Functions: Offset machining and contour machining by a simple rod.
Comment: The program was developed on a MEDIAPT unit used for wire-cut EDM applications.

2- and 3-Dimensional Compound Machining

Products: Punches and dies
Electrode: Copper, simple rod 21 mm (0.827 in.) in diameter
Workpiece: SKD11
Functions: Offset machining, superfine-finish finishing
Comment: The punches were produced by contour machining, and the upper parts were finished in a second pass.

5B: Use of Nontraditional Machining Processes in Producing Forging Dies and in Final Finish Machining of Near-Net-Shape Forgings

Nontraditional machining processes are finding more and more uses in today's manufacturing operations. Some offer particular advantages with difficult-to-machine materials and difficult-to-machine shapes. Some leave surface finishes that are not only smooth, but also resistant to thermal and physical fatigue. Several of these processes offer particular benefits to the manufacturers of forging dies. Such processes include:

1. Electrical discharge machining (EDM) — the most "traditional" of the nontraditional processes
2. Orbital grinding (or "abrading") to manufacture and redress graphite electrodes for EDM
3. Ultrasonic electrode-forming (or "abrading")
4. Electrochemical machining (ECM)
5. Abrasive flow machining, including both extrusion honing and orbital abrasive flow finishing.

ORBITAL GRINDING AND ULTRASONIC ABRADING TECHNIQUES OF GRAPHITE EDM ELECTRODE-FORMING

Electrical discharge machining has become the preferred machining technique for the production of molds, dies, and complex three-dimensional workpieces made of difficult-to-machine materials. The efficiency and accuracy of EDM as a full-form machining technique have been continually improving since its inception. However, to achieve the maximum efficiency from the EDM process and to make it economically justifiable on more machinable materials, the cost and inconvenience of forming the electrodes consumed as tools in EDM must be drastically improved. At present, the EDM process generally affords the user the simple privilege of conventionally machining the electrode material instead of the workpiece material.

Although EDM electrodes can be made of many different types of metals, graphite electrodes have emerged as the best EDM cutting tools. This superiority stems from graphite's excellent electrical and thermal properties, which together bring about good EDM metal-removal rates and relatively low electrode wear.

The importance of electrode wear is evidenced in the fact that in many countries the EDM process is frequently referred to as "spark erosion." It is inherent in the process that this "erosion" occurs not only on the workpiece, but on the electrode as well (although at a lower rate). If EDM parameters are chosen to keep electrode wear in the minimal range (referred to in the industry as "no-wear" machining conditions), the ratio of wear between the workpiece and the electrode can still be roughly 100 to 1 — that is, machining 1 in. (25 mm) from the workpiece removes 0.010 in. (0.25 mm) from the electrode, and seldom uniformly. Consequently, an electrode can be used for only a few workpieces (and sometimes not even one completely) before it is worn out of tolerance.

Because electrode wear is one of the major limitations of this process, EDM machining parameters are set to these "no-wear" settings to conserve the electrode as much as possible. However, although such machine settings can prolong electrode life, they prolong EDM machining time as well.

Conventional machining of an electrode is pretty much the same as machining of any other workpiece except, of course, that the electrode material is probably more machinable. In general, however, conventional machining of new electrodes (or remachining of used electrodes to re-establish dimensions reduced by wear after some use) involves significant time and cost. The time required, in relation to the time needed for conventional machining of the workpiece itself, depends primarily on the machinability of the workpiece material.

If accurate forming of electrodes were made drastically easier, electrode wear would be less critical and EDM machining parameters would be less constrained, permitting higher cutting speeds with no sacrifice in accuracy or finish.

The GrafEx Division of Extrude Hone Corporation has developed two systems — OrbiTEX™ and SoneX™ — specifically for machining of graphite electrodes. Both of these methods use low-amplitude grinding motions to form the total electrode shape and are a full order of magnitude faster and more convenient than conventional machining.

The OrbiTEX™ system is a method of *orbital abrasion* in which two horizontal platens — one holding an abrasive forming tool and the other a graphite electrode block — are oscillated in a precise orbital path with respect to each other. While orbiting, the two platens are brought together, causing the full three-dimensional shape of the abrasive forming tool to be ground into the graphite electrode material. This cutting action occurs simultaneously over the full surface of the abrasive forming tool, producing surface finishes on the electrode that are much finer than those produced by profile milling. The flow of a flushing fluid through the machining gap, combined with frequent vertical reciprocation between the platens, washes away the abraded graphite as fine particles.

The amplitudes used by an orbital abrader range from 0.010 in. (0.25 mm) to 0.250 in. (6.25 mm), with 0.060 to 0.100 in. (1.5 to 2.5 mm) being the typical amplitude range at speeds of 1000 to 1200 oscillations per minute.

Orbital abrasion produces an electrode form that is "orbitally undersized" by the amount of the orbit amplitude. If the abrasive forming tool is the same size as the form to be EDM'd, the resulting electrode is suitable for orbital (sometimes called "planetary") EDM. In orbital EDM, rather than following the conventional EDM technique of simply "sinking" the electrode straight into the metal workpiece, the electrode is instead orbited while machining the workpiece (just as in orbital abrading, the electrode is typically orbited into the abrasive forming tool). When an orbitally abraded forming tool is used, the EDM orbit is normally set to be the same as the abrading orbit was when the electrode was formed, minus the amount per side needed for the EDM "overcut" (i.e., the spark gap). Orbital EDM is rapidly gaining in popularity because it can offer such dramatic benefits in cutting speed, surface quality, and precision when compared with the conventional EDM technique.

When an orbitally abraded electrode is used in conventional, straight-sinking EDM, an "on-size" electrode must be produced by using an abrasive forming tool that is "oversize" by the amount of the orbit amplitude to be used in abrading the electrode.

The actual forming tool for the orbital abrader, when producing electrodes for use in orbital EDM, can simply be an EDM'd form itself, because a fairly rough EDM surface can easily abrade graphite. Alternatively, an abrasive forming tool may be formed by a technique of casting from a reverse model of the form to be EDM'd (Figure 5B-1).

The second electrode-forming technique, SoneX™, also uses a low-amplitude oscillating motion, but as a vertical vibration and not as a horizontal orbital oscillation. The amplitude of tool motion is much lower — typically 0.001 in. (0.025 mm); and the frequency is very much higher — typically around 20,000 cycles per *second*; hence it is called *ultrasonic abrading*. In this technique, a nonabrasive forming tool attached to a special toolholder/sonic transducer combination (often called a "sonotrode") is vibrated longitudinally at the resonant frequency of the sonotrode/forming tool. At the same time, an abrasive-filled fluid is flushed through the gap between the forming tool and the workpiece. Sometimes it is desirable to attach the graphite electrode blank to the vibrating sonotrode and then feed it into the stationary forming tool. In

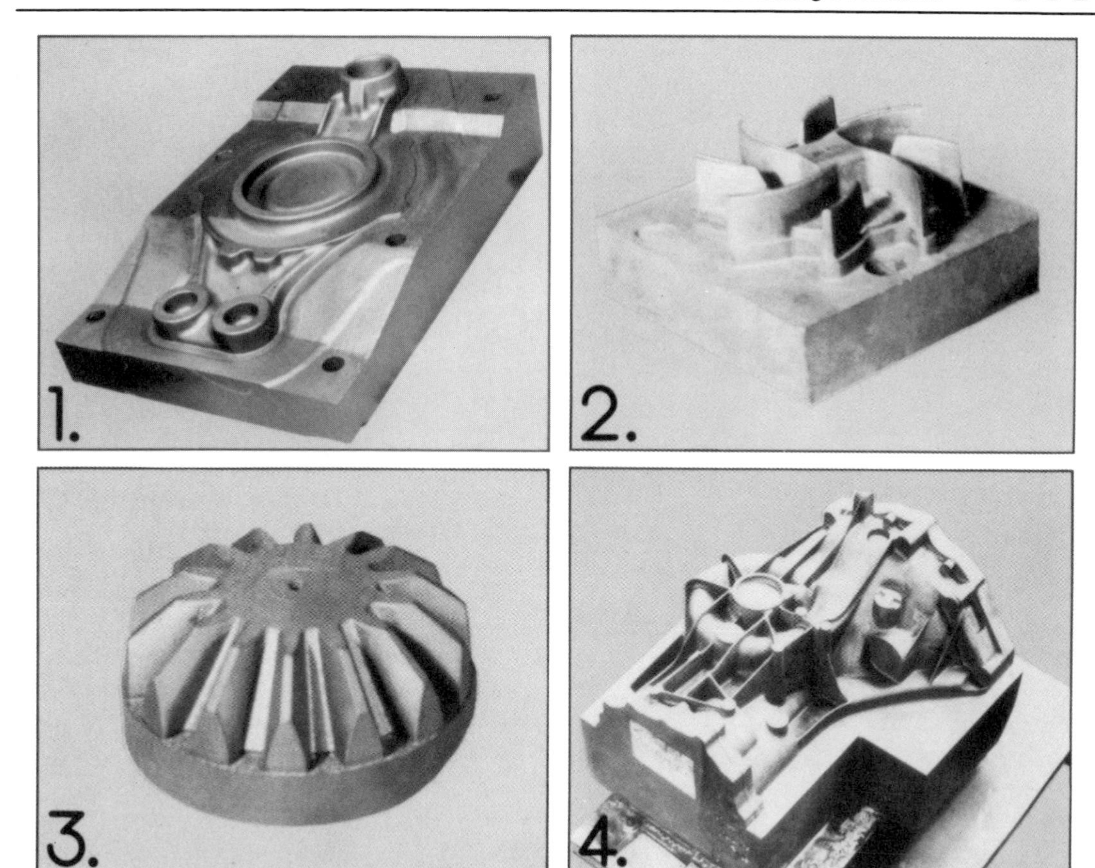

(1) Brade-spider electrode: machining time, 30 to 45 min; re-dressing time, approximately 15 min. (2) Impeller electrode: machining time, 25 to 30 min; re-dressing time, 15 to 20 min. (3) Gear electrode: machining time, 15 to 20 min; re-dressing time, 7 to 10 min. (4) Die cast transmission-housing-mold electrode: machining time, 2 to 4 h; re-dressing time, 45 to 60 min.

Figure 5B-1. Examples of graphite electrodes machined by the OrbiTEX™ method

either case, a gentle machining action is produced as the sonotrode vibrates the fine abrasive particles flowing throughout the machining gap and propels them against the graphite electrode workpiece material. These shaped vibrations, resonant with the forming tool, cause a shape that is the exact reverse of that of the forming tool to become abraded into the electrode.

SoneX™ ultrasonic electrode-forming relies on a vertical motion — the constant low-amplitude, high-frequency acoustic vibration produced by the sonotrode — to perform the required machining. The forming-tool replica produced by ultrasonic grinding will have an "overcut" (controlled by the abrasive-particle size) that usually can be made to coincide with the expected EDM "overcut."

What is particularly dramatic about this technique is its ability to produce intricately detailed electrodes with sharp internal corners and excellent surface finishes. Even very precise or subtle configurations can be produced. Very low machining forces permit manufacture of fragile electrode designs which are preserved by the gentle ultrasonic abrading action — an action in which literally thousands of tiny vibrating abrasive particles serve as the cutting tools. SoneX™ ultrasonic electrode-forming can produce electrodes that once were simply too intri-

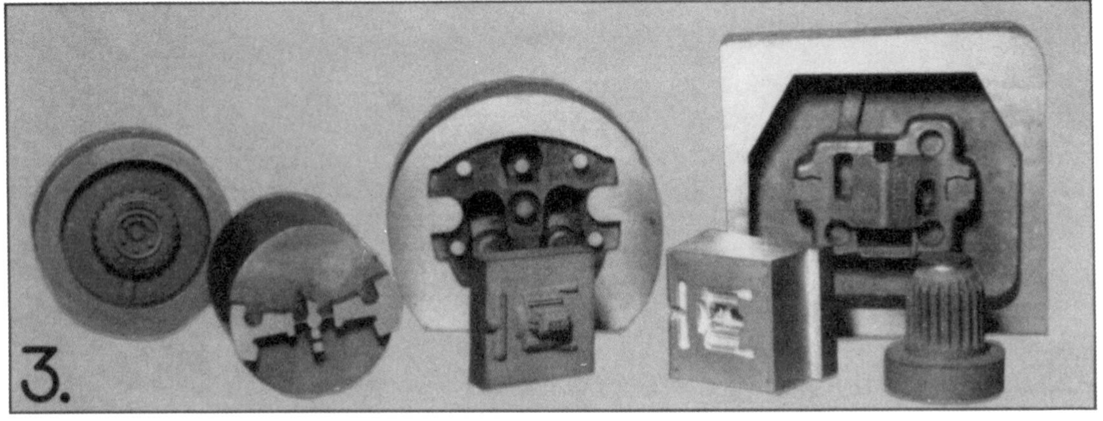

(1) EDM'd cavity (left) and electrode (right) for a medallion: machining time, 2.5 h; re-dressing time, 40 min. (2) Toy-engine-block electrode: machining time, 10 min; re-dressing time, 5 min. (3) Assorted electrodes illustrating the wide range of SoneX™ electrode-forming capabilities.

Figure 5B-2. Examples of graphite electrodes machined by the SoneX™ method

cate to be machined, or that, if producible at all, could be formed only by laborious hand engraving (Figure 5B-2).

These two new electrode-forming techniques are based on a similar theory of low-amplitude oscillitory abrading action — although they differ somewhat in application, with each approach offering its own strengths. OrbiTEX™ orbital abrading uses a low-amplitude horizontal orbiting motion to machine an "orbitally undersized" reverse impression of the forming tool into the workpiece. This system is suitable even for very large electrodes. One consideration to keep in mind, however, is that, due to the orbital motion, some details (particularly internal corner radii that are smaller than the orbit amplitude) may be lost.

On the other hand, ultrasonic abrading uses an even lower-amplitude and higher-frequency longitudinal vibration to accomplish a similar task. Ultrasonic abrading is often better suited to the forming of small or delicate electrodes. The ultrasonic abrader forming tool imparts its shaped vibrations to the abrasive grains in the fluid flowing throughout the machining gap. These tiny vibrating particles, not the forming tool, perform the actual machining. The SoneX™ system is limited to electrodes under 4 in. (100 mm) in diameter and has some other limitations in producing certain shapes and depths.

Despite their different mechanisms (and the particular benefits associated with each), these two machining techniques are strikingly similar in their over-all effects on the EDM process: the extremely efficient production of graphite electrodes with accurate dimensions and excellent surface finishes allows the use of faster EDM machining parameters (such as in negative polarity) that have previously been dismissed because of their increased tool wear. Both techniques can use the original EDM'd part itself as the forming tool or, alternatively, can use an epoxy forming tool cast from a model of a production part. And because of their improved machining speed and accuracy, both the OrbiTEX™ and SoneX™ systems can machine and re-dress worn electrodes with little of the expense and time normally spent in conventional machining of electrodes. The final result is that these improvements in electrode-forming provide for a more economical EDM process, particularly when multiple electrodes are required.

ELECTROCHEMICAL MACHINING

Another technique that improves the speed and quality of machining of forging dies is electrochemical machining, or ECM. In this process a low-voltage, high-amperage direct current electrochemically "dissolves" the die form into the workpiece — regardless of workpiece hardness — in an electrolyte made of a salt-and-water solution. Compared with EDM, ECM produces surface finishes that are generally much finer and contain no "recast layer." There is also little (in theory, no) electrode wear.

On the other hand, ECM equipment is expensive, partly because of the special materials that must counteract the corrosive nature of the electrolyte.

Another limitation is the expense of generating a tool (cathode) that will accurately form a cavity to close dimensional tolerances. An ECM cathode will not produce a form that is its exact geometric opposite, because the overcut across the machining gap will vary due to several factors, including:

1. Variations in the flow speed of the electrolyte throughout the gap, particularly at "dead" spots or eddy currents
2. Changing electrical conductivity of the electrolyte due to temperature increases (the current flow in the gap heats the electrolyte by electrical resistance) and to changing hydroxide concentrations (the metal removed as the electrolyte flows across the gap becomes — almost instantly — a metal hydroxide that changes the electrolyte conductivity as well).

Because of these factors, tool development for ECM work becomes increasingly expensive as tolerances become closer. However, for high-volume forging-die applications with moderate tolerances, this expense may be justified.

AN EDM/ECM/ORBITAL ABRADING COMBINATION

Another approach is to combine ECM with both EDM and orbital graphite grinding. For example, an orbitally abraded graphite electrode — essentially a duplicate of that used for EDM — would be used to produce a die cavity by ECM to within 0.040 in. (1.0 mm) final size. Nonuniform material removal in ECM is accepted so long as the form produced is within an envelope of 0.040 to 0.060 in. (1.0 to 1.5 mm) from final size. This final envelope is orbitally EDM'd with a duplicate (also orbitally abraded) graphite electrode. This brings the cavity to within 0.001 to 0.002 in. (0.025 to 0.05 mm) of final size. Finally, a second ECM operation — this time very brief (1 to 3 min machining time) — is used to remove the EDM recast layer and to polish the surface.

ORBITAL ABRASIVE FLOW POLISHING

Another technique for polishing EDM surfaces combines orbital abrading with Extrude Hone's special abrasive flow machining medium.

To perform this operation, first an epoxy "mandrel" is formed by duplicating an orbitally undersized electrode by casting. This mandrel is, in turn, oscillated within the workpiece cavity to uniformly displace a special abrasive flow "putty" that has been placed in the cavity. The complete and uniform displacement of this conformable medium — over even the more intricate die-surface configurations — produces a uniform stock-removal and polishing action over the entire cavity surface. Specially modified OrbiTEX™ graphite abraders are used to perform this orbital polishing technique.

For example, orbital polishing has improved the finishes of typical cavities from an original finish of 125 μin. (3.18 μm) R_a to less than 25 μin. (0.64 μm). In this case, 0.001 in. (0.025 mm) was removed from the surface while uniformly polishing it to within 0.0005 in. (0.0013 mm).

ABRASIVE FLOW FINISHING

Abrasive flow finishing, or extrusion honing, is performed by flowing a special abrasive-filled puttylike material back and forth through the passages (or over the surfaces) to be polished. This special abrasive medium will significantly abrade only the most restrictive portion of the flow passage. In cases where the desired area is not restrictive enough to produce the necessary abrasion, "restrictors" in the tooling can be used to direct and restrict the flow passage.

The surface finish produced by abrasive flow finishing has a lay that is parallel to the direction of medium flow; therefore, in polishing of dies or similar through-passages, a lay can be produced parallel with the metal flow that occurs when these parts are used.

In terms of surface roughness, abrasive flow finishing can polish surfaces of between 30 and 300 μin. (0.76 and 7.6 μm) R_a to one-tenth their original finish — and sometimes better. The dimensional change resulting from this abrasion typically falls between 20 and 50% more than the original R_t (total roughness). So, for instance, to bring an EDM'd 100-μin. (2.54-μm) R_a finish down to 10 μin. (0.25 μm), about 0.001 in. (0.025 mm) — or 1000 μin. (25.4 μm) — would be removed from the surface. In most cases, this can be held uniform throughout a passage to within ±20% or to within ±0.0002 in. (0.005 mm).

Abrasive flow finishing improves the thermal and mechanical fatigue strength of thermally machined (EDM'd) components, parts which typically experience high stress during use.

FINAL FINISH MACHINING OF NEAR-NET-SHAPE FORGINGS

Extrude Hone's abrasive flow machining process is also used to perform final finish machining operations on near-net-shape components. The uniformity and automatic consistency of the abrasion produced by this process make it well suited for maintaining the tolerances of even highly critical forged components such as turbine blades.

The ECM process is capable of more aggressive and more specific final machining. An example of ECM finish machining is the stem area of the artificial hip prosthesis. The final form is machined to a tolerance of 0.002 in. (0.05 mm), removing about 0.040 in. (1.0 mm) from the surface in a machining time of 1 min. Due to the high chromium content, the surface finish produced is 8 μin. (0.2 μm) R_a.

SUMMARY

The benefits offered by nontraditional machining techniques in the manufacture of both forging dies and forged near-net-shape products are continually expanding. Although EDM is now the "traditional" approach to machining of forging dies, the combination of EDM with orbital and/or ultrasonic electrode-abrading techniques to quickly produce *and* re-dress electrodes has dramatically enhanced the efficiency and quality of EDM die-machining operations.

Extrusion honing produces finishes that

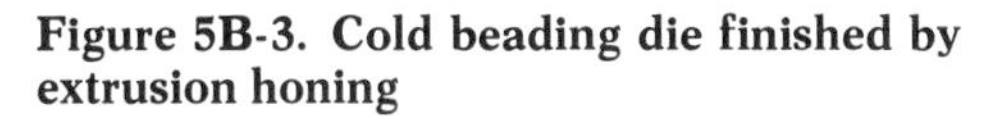

Original EDM finish of 90 to 100 μin. (2.3 to 2.5 μm) has been polished to less than 10 μin. (0.25 μm). Processing time, 10 min.

Figure 5B-3. Cold beading die finished by extrusion honing

Figure 5B-4. Cold beading die after finishing

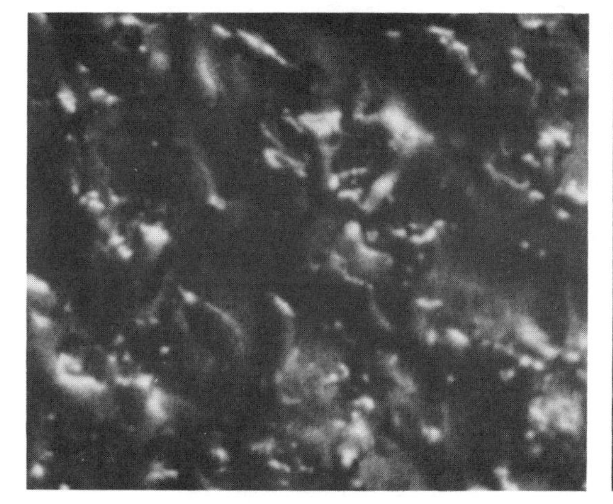

Original surface finish, 75 μin. (1.9 μm) CLA. Surface finish after 4 min of extrusion honing, 6 to 7 μin. (0.15 to 0.18 μm). Magnification, 500×.

Figure 5B-5. Photomicrographs of EDM'd die bearing surface before (left) and after (right) extrusion honing

are one-tenth of the original EDM finishes with much greater speed and uniformity than those of hand polishing.

Orbital abrasive flow polishing offers the potential to provide these benefits in finishing of blind cavities, virtually eliminating hand polishing of dies.

ECM — alone or in combination with EDM — can machine dies much more quickly than EDM alone can. In addition, the surface finish produced by ECM is both smoother and more fatigue-resistant.

As the forging industry faces new challenges, these nontraditional processes offer new tools and new strategies to meet these challenges. Typical examples of the capabilities of these processes are illustrated in Figures 5B-3 to 5B-13.

Figure 5B-6. Ceramic, tool steel, and carbide dies for extrusion, drawing, forging, cold beading, and compacting, showing the wide variety of dies that are polished by extrusion honing

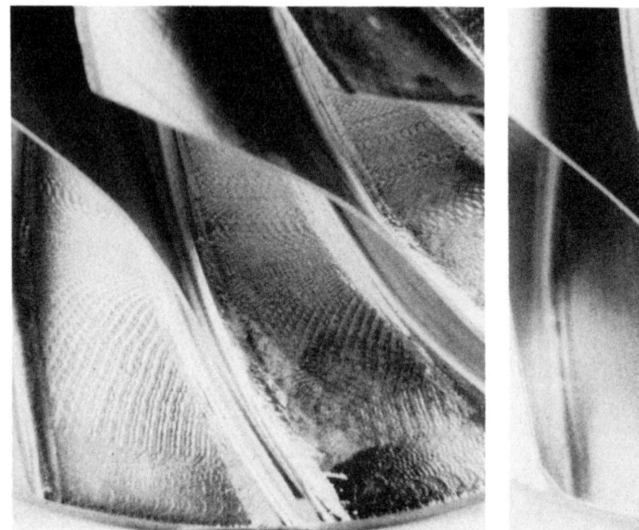

Shape within profile envelope is maintained. Hours of hand finishing are eliminated.

Figure 5B-7. Milled surfaces of an impeller before (left) and after (right) extrusion honing to improve surface finish

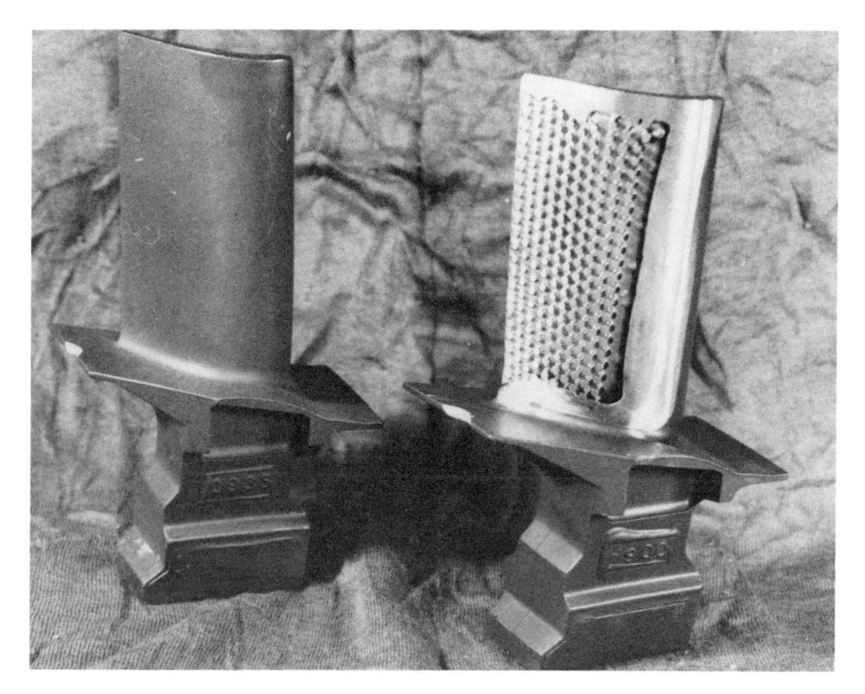

Figure 5B-8. Hollow investment cast airfoil extrusion honed to smooth complex internal cast passages and eliminate boundary-layer turbulence

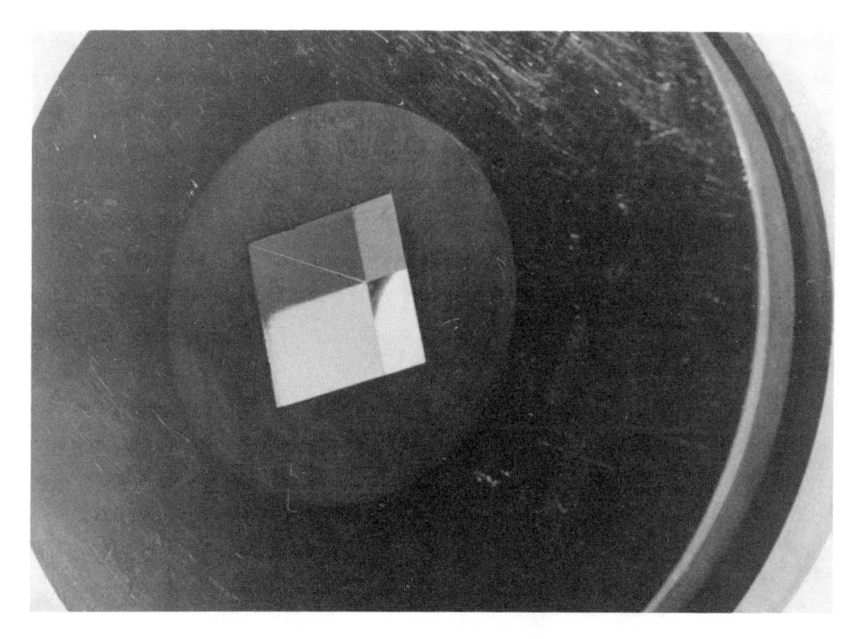

Polishing time is less than 20 min, compared with 20 h for hand polishing. Final finish is improved from 40 to 50 μin. (1.0 to 1.3 μm) to 4 to 5 μin. (0.10 to 0.13 μm). Uniform stock removal is within 0.0002 in. (0.005 mm).

Figure 5B-9. Carbide compacting die polished with a diamond-laden abrasive

This mold will be machined to final geometry by EDM and then polished by ECM.

Figure 5B-10. Die cast transmission mold machined by ECM to within 0.04 in. (1.0 mm) of final dimension

Figure 5B-11. Graphite electrode used for machining of die cast transmission mold by ECM

This machine is capable of processing a wide range of applications. It can be easily tooled for high-production applications or for exceptionally large components.

Figure 5B-12. Extrude Hone Spectrum Series abrasive flow machining unit

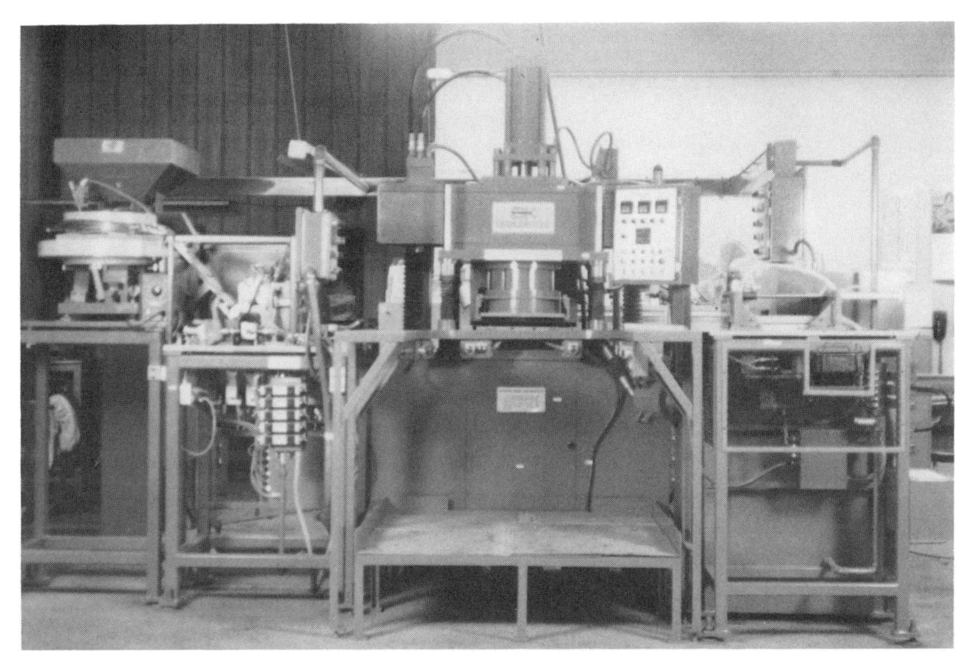

This system, used for deburring of thousands of small components in the automotive industry, includes an automatic fixture-loading station, an air/vacuum media removal and return unit, and an automated part-unloading station.

Figure 5B-13. Extrude Hone high-volume production abrasive flow machining unit

Section 6
Definitions;
Design Considerations

6A: Definitions of Terms

annealing. Softening or strain relieving of the material by application of heat above the critical temperature for the correct time interval, then cooling slowly enough to avoid hardening.

beads. 1. Narrow ridges in a part for reinforcement. 2. Narrow ridges along the edge of a part, and corresponding ridges in the die, to improve holding action in pressworking.

beading. Forming of a circumferential bead or rib in a tubular part.

bed. The portion of a hydraulic-press structure that usually rests on the foundation, forms the support for the remaining parts of the press, and is subjected to the pressing load. Also called base or bottom platen.

bend allowance. The amount of sheet metal required for making a bend around a given radius. Bend allowance (BA) is calculated by:

$$BA = D(0.01743R + 0.0078T)$$

where D is number of degrees in bend, R is inside radius of the bend, and T is thickness of metal.

bend radius (sheet metal). The radius around which the metal is formed (i.e., the inside radius of a bend).

bending and forming. The processes of bending, flanging, folding, twisting, offsetting, or otherwise shaping a portion of a blank or a whole blank, usually without materially changing the thickness of the metal.

bending brake or press. A form of open-frame, single-action press comparatively wide between the housings, with bed designed for holding long, narrow forming edges or dies. It is used for bending and forming of strip and plate.

blank. The piece of sheet material, usually flat, produced in cutting dies, and usually to be subjected to further press operations.

blank development. Because of stretching and ironing present in drawing and forming operations, it is usually necessary to arrive at the blank size and shape by trial and error. The drawing and forming dies, therefore, are made first, and then, when the blank size and shape are finally determined, the blanking die is made.

blankholder. (a) That part of a drawing or forming upper die which holds the blank by pressure against a similar surface in the mating part of the die to prevent wrinkling. This is the same as a hold-down. The pressure may be obtained through mechanical means, springs, or fluid cushions. (b) The outer slide of a double-action press usually operated by toggles or cams. Accurately, this should be called the blankholder slide.

blanking. The process of cutting out a flat piece of the size and shape necessary to produce the desired finished part.

blanking die. A die used for shearing or cutting blanks usually from flat sheets or strips. The single blanking die used for producing one blank at each stroke of the press is the simplest of all dies, consisting essentially of punch, die block, and stripper.

bolster plate. The plate attached to the top of the bed of a press, having drilled holes or T-slots for attaching the lower die or die shoe.

brakes. Friction brakes, set with compression springs, provided for stopping or holding the slide movement on a press. Brake capacity must be sufficient to stop the motion of the slide quickly and must be capable of holding the slide and its attachments at any point in its travel.

buckling. A bulge, bend, kink, or other wavy condition caused by compressive stresses.

burr. A rough ridge, edge, protuberance, or area, such as that left on metal after cutting, drilling, punching, etc.; in stamping it occurs in cutting dies because of the clearance between punch and die.

burr side. This term generally refers to the side or face of a blank or other stamping which comes in direct contact with the punch in a blanking operation, and the side or face of a blank or other stamping which comes in direct contact with the die in a piercing or perforating operation.

canning (oil canning). Distortion of a flat or nearly flat metal surface which can be deflected by finger pressure, but which will return to its original position when the pressure is removed.

capacity of a press. The related capacity of a press is the pressure, in tons, which the slide will safely exert at the bottom of the stroke in doing work within the range of the press.

clearance. (a) In punching and shearing dies, the

gap between the die and the punch. (b) In forming and drawing dies, the difference between this gap and metal thickness.

clutch. The coupling mechanism used on a mechanical power press to couple the flywheel to the crankshaft, either directly or through a gear train. A *full revolution clutch* is a type of clutch that, when tripped, cannot be disengaged until the crankshaft has completed a full revolution and the press slide a full stroke. A *part revolution clutch* is a type of clutch that can be disengaged at any point before the crankshaft has completed a full revolution and the press slide a full stroke.

coining. A squeezing operation, usually performed cold in a closed die, in which the metal is forced to flow so as to fill the shape and profile of the dies. This term is also loosely applied to other very severe and localized cold forming operations. It is usually done in a knuckle joint or coining press.

cold rolled sheet. Steel sheet processed to produce a surface superior to a hot rolled, pickled finish. In the present basic treatment, hot rolled sheets or coils are pickled, cold reduced in thickness, and usually annealed, temper or skin rolled, and roller leveled.

combination die. A press tool for doing two or more operations at each stroke of the press, one performed inside of the other. These operations most frequently are blanking and drawing.

compound die (inverted blanking punch and die). In a compound die, the die is mounted on the punch holder and the punch is fastened to the die shoe. Due to the type of construction, there is no need for a hole in the die shoe for the blanks to fall through.

connections. Connecting members that convey motion and force from a rotating member to a slide or lever on a press. Also called pitmans, connecting links (or rods or straps), or eccentric straps.

crank press. A mechanical press the slide of which is actuated by a crankshaft.

crimping. A rolling operation used to set down, or close in, a seam, usually a circular or peripheral seam. For example, the flange of a flanged can end is crimped (tightened against the ends of the can body), and the open end of a cartridge is crimped against a bullet.

crown. The top portion of a hydraulic-press structure. Cylinders and other working parts may be mounted thereon to form the crown assembly. Also called head or top platen.

cupping. The first operation in deep drawing.

curling. Forming of an edge of circular cross section along a sheet or along the end of a shell or tube, either to the inside or outside — e.g., the pinholes in sheet-metal hinges and the curled edges on cans, pots, and pans.

curling dies. Curling dies are of two general types: (a) dies used when the curling is done in the die block (female part), such as curling of sheet-metal hinges and the edge plate preparatory to seaming; and (b) dies that are ordinarily attached to the punch holder, such as those used in wiring of pots and pans.

cushion. A mechanism mounted in or under the bed of a hydraulic press to provide a controlled resistance against the work as it is displaced by the slide. The return motion may also be used for ejecting the work.

cushion, hydraulic. A *cushion* in which a hydraulic cylinder provides the resistance. Pressure in the cylinder is developed by the main ram movement. The cushion is returned to its normal position hydraulically.

cushion, hydropneumatic. A *cushion* in which a hydraulic cylinder provides the resistance. Pressure in the cylinder is developed by the main ram movement. The cushion is returned to its normal position by gas pressure acting on the hydraulic fluid in the reservoir.

cushion, pneumatic. A *cushion* in which an air cylinder provides the resistance. Air from a storage or surge tank is directly connected to the cushion cylinders.

daylight. The maximum clear distance between the pressing surfaces of a hydraulic press when the surfaces are in their usable open positions. Where a bolster is supplied, it is considered to be the pressing surface.

delayed return. A controlling device for a *cushion* to prevent its return until the main ram has returned a predetermined distance.

developed blank. A blank which yields a finished part without trimming, or with a minimum amount of trimming.

die. A complete press tool used for forming or blanking of metal. The word "die" usually designates the female member, while the male member is known as the "punch" of the die.

die clearance. For shearing or cutting operations (blanking or piercing), the distance between the punch and the die when the punch is entering the die centrally or normally. For round work, the clearance is half the difference between the diameter of the die and the diameter of the punch.

die set. A toolholder held in alignment by guide posts and bushings and consisting of a lower shoe, an upper shoe or punch holder, and the guide posts and bushings.

die shoe. A plate or block on which a die holder is mounted. A die shoe functions primarily as a base for the complete die assembly and, when used, is bolted or clamped to the bolster plate or the face of the slide.

double-action die. In the usual double-action drawing die, the blankholder (or hold-down) is fastened to the blankholder slide (outer slide), the punch is fastened to the plunger (inner slide), and the die is fastened to the bolster (bed). The double-action drawing operation is essentially as follows: The blankholder moves first and securely holds the outer edges of the blank against the mating part of the die at which time the punch starts the draw. At the completion of the draw, the punch starts back first, followed by the blankholder which has remained stationary, holding the blank during the entire draw.

double-action hydraulic press. One equipped with two moving slide members, both located on the same side of the work. Three designs are in general use:

1. **Slide within a slide.** The outer, or blankholder, slide is guided on housings or strain rods. The inner, or punch, slide is guided in the outer slide.
2. **Slide below a slide.** The lower slide is the blankholder slide. It is provided with a center hole to allow the punch to come through. Both slides are guided on housings or strain rods.
3. **Blankholder above punch slide.** Both slides are guided on housings or strain rods. Pins connect the blankholder ring (which is below the main, or punch, slide) to the blankholder slide above the main slide.

A double-action press may have a *cushion* built into the bed.

double-crank press. A crank press in which the slide is driven by two *connections*.

drawability. (a) A quantitative measure of the maximum possible reduction in a drawing process. (b) Reduction in diameter from the blank to a deep drawn shell of maximum depth.

drawing. Any of a variety of forming operations, such as deep drawing of sheet blanks, redrawing of tubular parts, and drawing of rod, wire, and tube. The usual drawing process in reference to the working of sheet metal in a press is a method for producing a cuplike form from a sheet-metal disk by holding it firmly between blankholding surfaces to prevent formation of wrinkles while punch travel produces the required shape.

drawing die. A type of die designed to produce nonflat parts such as boxes, pans, etc. Whenever practical, the die should be designed and built to finish the part in one stroke of the press, but if the part is deep in proportion to its diameter, redrawing operations are necessary.

draw ring. Ring-shape die part, either the die ring itself or a separate steel ring, over the inner edge of which the punch draws the part.

ductility. That property or characteristic of a material which permits plastic working in tension without rupture, as in drawing or stretching, and which is measured in a general way by the elongation and reduction in area of a test piece when pulled in a testing machine.

eccentric gear. A main press-drive gear with an eccentric as an integral part. The unit rotates about a common shaft with the eccentric, transmitting the gear's rotary motion into vertical slide motion through a connecting strap.

eccentric shaft. A crank with a crankpin of such size that it contains or surrounds the shaft.

ejector. A mechanism for removing work or material from between dies.

elastic limit. The maximum stress to which a material or body can be subjected and still return to its original shape and dimensions.

embossing. A process for producing raised or sunken designs in sheet material by means of male and female dies, theoretically with no change in metal thickness. Some common examples are lettering, ornamental picturing, and formation of ribs for stiffening. The heaviest embossing is usually done in knuckle-joint presses. Heavy embossing and coining are similar operations.

fixture. A tool or device for holding and accurately positioning a piece or part on a machine tool or other processing machine.

flame hardening. A heat treating method for surface hardening steel of the proper specifications in which an oxyacetylene flame heats the surface to a temperature at which subsequent cooling, usually with water or air, will give the required surface hardness.

flaring. (a) Forming of a flange on a tubular part which deviates from a plane. (b) Forming of a flange on a head.

flattening. Removal of the irregularities in a metal surface by a variety of methods, such as rolling and/or roller leveling of sheet and strip.

flywheel. A heavy rotating wheel, attached to a shaft, whose principal purpose is to store kinetic energy during the nonworking portion of a press cycle and to release energy during the working portion of a press cycle.

flywheel press. A mechanical power press in which the flywheel is mounted directly on the main crank or eccentric shaft without gearing.

foot control. A foot-operated control mechanism designed to be used with a clutch or clutch/brake control system on a press.

foot pedal (treadle). A foot-operated lever designed to operate the mechanical linkage that trips a full-revolution clutch on a press.

forming. Any change in the shape of a metal piece which does not intentionally reduce the metal thickness.

front and back plate press. A type of hydraulic press in which the crown and bed are positioned and retained by steel plate strain members, front and back. Single rectangular openings are cut in both plates to provide entry to the work area.

galling. Roughing or scratching of two contacting surfaces of metal caused by friction. Pieces of like metals rubbed together tend to gall more easily than pieces of unlike metals.

gap press. A type of hydraulic press in which the crown and bed are the upper and lower arms, respectively, of a C-shaped frame. The work area is accessible from three sides.

gate or movable-barrier device. A movable barrier arranged to enclose the point of operation before a press stroke can be started.

geared press. A press whose main crank or eccentric is connected to the driving source by one or more sets of gears.

guard. A barrier that prevents entry of a press operator's hands or fingers into the point of operation.

hems and seams. In forming and fastening sheet-metal parts together, proper use of hems and seams is very important. Single hems, double hems, wired-edge seams, grooved seams, and double seams are some examples.

hot rolled sheet. Steel sheet reduced to required thickness at a temperature above the point of scaling and therefore carrying hot mill oxide. The sheet may be flattened by cold rolling without appreciable reduction in thickness or by roller leveling, or both. Depending on the requirements, hot rolled sheet can be pickled to remove hot mill oxide and is so produced when specified.

housing press. A type of hydraulic press in which the crown and bed are separated by uprights or spacers extending from front to back through which strain rods pass.

hydraulic press. A machine in which fluid pressure is used to actuate and control the motion and force of its ram or rams.

inclinable press. A small or medium-size crank press which may be inclined (tilted backward) to facilitate ejection of finished parts by gravity. Such presses are usually of the open-back, gap-frame type with right-to-left crankshaft. They are built in a maximum size of about 200 tons. They may be, and very often are, used in the upright or vertical position, being readily adjustable, usually by a hand mechanism, to any desired inclination up to the usual maximum of 45°. The inclinable press is the most widely used type of press, being particularly adapted for blanking, piercing, forming, and shallow drawing operations on a multitude of small and medium-size parts. This type of press frequently is equipped with mechanical feeds for rapid automatic production.

ironing. The press operation of reducing the wall thickness of a shell while retaining the original thickness of the bottom, and reducing the inside diameter by only a small amount. It is the opposite of a true drawing operation, in which reducing the diameter is the primary consideration and wall thickness may even be increased slightly.

jog. An intermittent motion imparted to a press slide by momentary operation of the drive motor, after the clutch is engaged with the flywheel at rest.

joggle. (a) An offset (usually with parallel surfaces) in the surface of a sheet or part. (b) The process or act of offsetting the surface of a plate or part.

knockout, hydraulic. On a hydraulic press, a mechanism actuated by a hydraulic cylinder for the purpose of ejecting work from the tools.

knockout, mechanical. A mechanism actuated by a hydraulic-press slide on its return stroke arranged to eject work from the tools.

knockout, pneumatic. On a hydraulic press, a

mechanism actuated by a pneumatic cylinder for the purpose of ejecting work from the tools.

knuckle-joint press. A press in which the slide is directly actuated by a single toggle (or knuckle) joint, which is closed and opened by means of a connection and a crank.

lancing. Slitting and forming of a pocket-shape opening in a part.

liftout. A mechanism (also known as a knockout) that is either: (a) a crossbar below the top of the bed of a press and directly connected to the slide by rods through the top of the bed; (b) a bell crank or lever on a rocker shaft mounted in the bed and driven from the crankshaft by either a cam or an eccentric; or (c) springs or air cushions mounted directly under the bolster or to the bed of a press.

loop press. A type of hydraulic press in which the crown and bed are separated by straight side members which are an integral part of an oval-shaped continuous band.

mechanical press. A press whose slide or ram is operated by crank, eccentric, cam, toggle links, or other mechanical means, as contrasted to fluid or other means.

mill extras. Charges above the base price for sheet such as gage extra, width extra, length extra, pickling and oiling extra, etc.

multiple-slide (or plunger) press. Presses with multiple slides are built in two general types: (a) Those in which individual slides are built into the main slide in order to provide an individual adjustment for each set of tools. (b) Those in which individual slides have individual connections and cranks or eccentrics on the main shaft. This arrangement provides means for variations in length of stroke and timing as well as adjustment. Multiple-slide presses are usually equipped with transfer feeds or double roll feeds for automatic, progressive-die production of stampings.

notching. An unbalanced shearing or blanking operation in which cutting is done around only three sides (usually) of a punch.

open-back inclinable (OBI) press. A gap-frame press with an opening at the back between the two side members of the frame, and arranged to be inclinable to facilitate part removal by gravity. See also *inclinable press.*

open-back stationary (OBS) press. A gap-frame press with an opening at the back between the two side members of the frame, and arranged to be upright or permanently inclined.

open rod press. A type of hydraulic press in which the crown and bed are positioned and retained by strain rods.

pad holder. Container for a rubber pad on a hydraulic press.

perforating. Piercing of many holes, usually identical and arranged in a regular pattern, in a sheet, blank, or previously formed part. The holes are usually round, but may be of any shape. This operation is also called multiple piercing. See *piercing.*

piercing. The general term for cutting (shearing or punching) of openings such as holes and slots in sheet material, plate, or parts. This is practically the same operation as blanking; the difference is that the slug or piece produced by piercing is scrap, whereas the blank produced by blanking is the useful part. In the two cases the burr is opposite. See *burr side.*

pinch point. Any point other than the point of operation at which it is possible for a part of a press operator's body to be caught between the moving parts of a press or auxiliary equipment, between moving and stationary parts of a press or auxiliary equipment, or between the material and a moving part or parts of the press or auxiliary equipment.

plate. See *sheet.*

point of operation. The area of a press where material is actually positioned and work is performed during any process such as shearing, punching, forming, or assembling.

presence-sensing device. A device designed, constructed, and arranged to create a sensing field or area and to deactivate the clutch control of a press when an operator's hand or any other body part is within such field or area.

press component or mechanical power press component. The basic press (component) of the mechanical power press machine; that portion devoid of the tooling component, the safeguarding component(s), and auxiliary feeding components. The press component is usually a multifunctional piece of equipment that is adaptable to a limitless combination of circumstances, depending on the other components incorporated with it.

progressive dies. A series of two or more dies arranged in line for performing two or more operations on a part, one operation (single or compound) being performed in each die (at each station). Work in the form of strip is usually fed to progressive dies automatically by a roll feed; blanks or parts are fed by a transfer feed.

pull-out device. A mechanism attached to an operator's hands and connected to the upper die or slide of a press that is intended, when properly adjusted, to withdraw the operator's hands as the dies close, if the operator's hands are inadvertently within the point of operation.

punch. (a) *n.* The male part of a die, as distinguished from the female part called the die. The punch is usually the upper member of the complete die and thus is mounted on the slide (except in the inverted die). See *compound die.* (b) *v.* To pierce a hole.

redrawing. Second and subsequent deep drawing operations in which the cuplike shells are deepened and reduced in cross-sectional dimensions (sometimes in wall thickness). Redrawing is done in both single-action dies and double-action dies.

restriking. A sizing or light coining operation in which compressive strains are introduced in the stamping to counteract or offset tensile strains set up in previous operations. For example, restriking is used to counteract springback in a bedding operation.

roll feed. A mechanism for feeding strip or sheet stock to a press or other machine. The stock passes between two revolving rolls mounted one above the other, which feed it under the dies a predetermined length at each stroke of the press. Two common types of drive are the oscillating lever type and the rack-and-pinion type. The single roll feed may be used to either push or pull the stock to or from the press. The double roll feed is commonly used with wider presses (left to right) or in other cases where a single roll feed is impractical.

run. The quantity produced in one setup.

shear. (a) *n.* Any machine or tool for cutting metal or other material by the closing motion of two sharp, closely adjoining edges (e.g., squaring shear and circular shear). (b) *v.* To cut by shearing dies or blades, as in a squaring shear.

sheet. Any material or piece of uniform thickness and of considerable length and breadth as compared with its thickness is called either sheet or plate. In reference to metal, such pieces under 1/4 in. thick are called sheets, and those 1/4 in. thick and over are called plates.

shut height. The clear distance between the pressing surfaces of a hydraulic press when the slide is in its closed position.

side housings. Uprights or spacers extending from front to back used to separate the crown and the bed on a hydraulic press.

side plate press. A type of hydraulic press in which the crown and bed are separated by steel plate strain members, one on each side. The side plates are fastened to the crown and head.

single-action hydraulic press. One equipped with one moving slide member. A single-action press may have a *cushion* built into the bed.

single stroke. One complete stroke of a press slide, usually initiated from a full open (or up) position, followed by closing (or down), and then a return to the full open position.

single-stroke two-hand control. Two-hand controls for single-stroke press operations, when furnished, must conform to the following requirements: (1) *Two-Hand Trip.* Each hand control must be protected against unintentional operation and arranged by design, construction, and/or separation so that the concurrent use of both hands is required to trip the press. (2) *Holding Time.* The control system must be designed to permit an adjustment which will require concurrent pressure from both hands during the die-closing portion of the stroke. (3) *Antirepeat.* The control system must incorporate an antirepeat feature. (4) *Interrupted Stroke.* The control system must be designed to require release of all operator's hand controls before an interrupted stroke can be resumed. The fourth requirement pertains only to single-stroke two-hand controls manufactured and installed after publication of the standard establishing these requirements.

slide. The main reciprocating member on a press. Also called ram, plunger, or platen.

solid press. A type of hydraulic press in which the crown and bed are separated by side members which are an integral part of the crown and bed.

spinning. The formation of sheet metal or tubing blanks into circular-shape parts by means of a lathe, forming blocks or chucks, and hand tools or other forming tools which serve to press and shape the revolving metal about the revolving form. Many kitchen utensils, lampshades, and chandelier parts are examples of spinning. Besides forming, spinning operations include trimming, curling, beading, and bulging.

springback. The extent to which metal naturally tends to return to its original shape or position after undergoing a forming operation. This is compensated for by "overbending" or by a secondary operation of restriking (squeezing).

stamping. Pressworking of sheet metal to change its shape either with or without accompanying shearing or punching. Stampings may be divided into two general classifications. The first includes those stampings made by forming the material to the shape desired from a flat piece without cutting the metal. The operations used in this type of work are *bending and forming, drawing,* and *coining.* The second classification includes stampings that are made by shearing the metal, either to change the outline of the edges or to cut holes in the interior of the piece. The operations used in this type of work are *blanking* and *trimming.*

strain-rod nuts. Elements holding the bed or the crown to strain rods on a hydraulic press.

strain rods. The tension members joining the bed and the crown of a hydraulic press. Also called tie rods or columns.

stripper. A mechanism or die part for removing parts or material from a punch.

sweep device. A single or double arm (rod) attached to the upper die or slide of a press and intended to move the operator's hands to a safe position as the dies close, if the operator's hands are inadvertently within the point of operation.

trimming. A secondary cutting or shearing operation on previously formed, drawn, or forged parts in which the surplus metal of irregular outline or edge is sheared off to form the desired shape and size.

triple-action hydraulic press. A double-action press with a third independent movable slide built into the bed. A *cushion* is not considered a third action, but a third action may be construed to also act as a cushion.

two-hand control device. A two-hand trip that further requires concurrent pressure from both hands of the operator during a substantial part of the die-closing portion of a press stroke.

two-hand trip. A means of clutch actuation requiring the concurrent use of both hands of an operator to trip a press. Requirements for a two-hand trip are as follows: (1) The individual operator's hand controls must be protected against unintentional operation. (2) The individual operator's hand controls must be arranged, by design and construction or by separation, or both, to require the use of both hands to trip the press. (3) A control arrangement must be used that requires concurrent operation of the individual operator's hand controls.

wear plates. Replaceable elements used to face wearing surfaces on a hydraulic press.

6B: Design Considerations for Stamping

PUNCHED HOLES

In designing stampings with punched holes, it is well to take into account the fact that only about one-half the thickness of the metal is sheared cleanly to the size of the punch. The rest is torn out by the pressure exerted on the sheared slug. This produces a rough hole tapering in diameter to the size of the punch plus about 10% of the metal thickness (see Figure 6B-1). This is particularly important when the sides of the hole act as a bearing surface in a hinging application.

In specifying punched holes, a size should be specified for which a standard-size punch is available. This saves the time and expense involved in making special punches. Standard-size punches are available in almost unlimited numbers.

EXTRUDED HOLES

An extruded hole usually is formed in one operation by a punch which cleanly punches a smaller hole and then follows through to flange the sides. However, in some instances, a hole flange is too wide to be formed in one operation. This necessitates drawing metal in from outside of the hole by first drawing an embossment much larger than the hole and then, by successive steps, forming the flange and finally punching out the hole. This

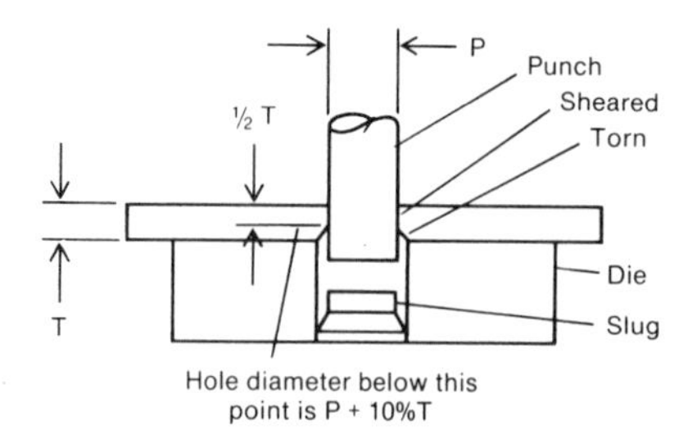

Figure 6B-1

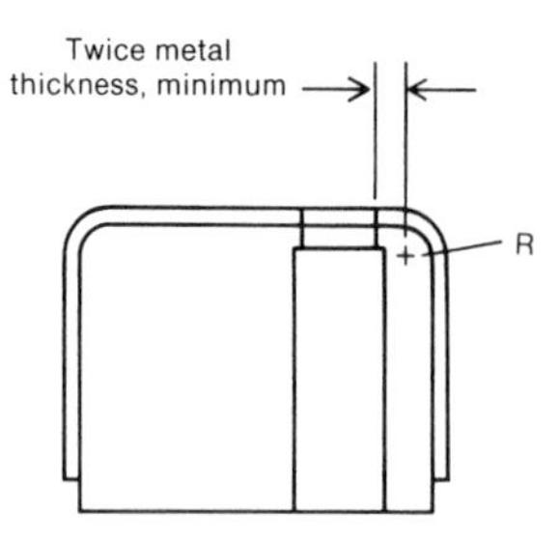

Figure 6B-2

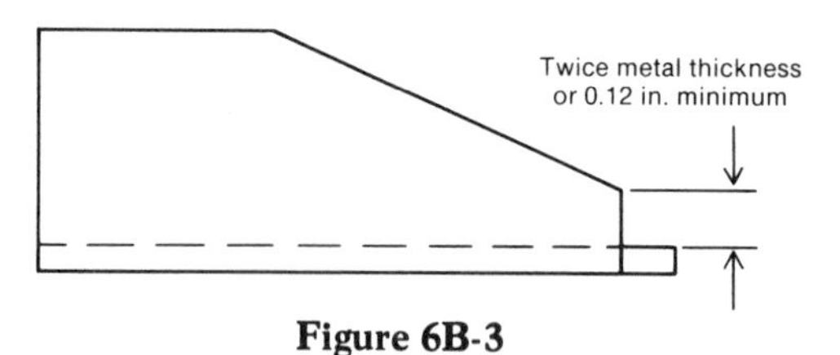

Figure 6B-3

Twice metal thickness
or 0.12 in. minimum

Figure 6B-4

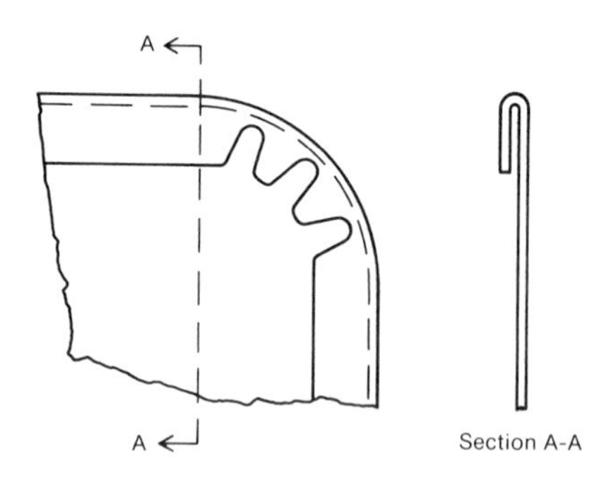

Figure 6B-5

method increases costs, and its use should be limited. One way of accomplishing this is to keep the specified flange width at an absolute minimum at all times.

PIERCED HOLES

A pierced hole is made by a sharp-pointed punch which follows through to flange the sides, forming a hole with torn or irregular edges. The distance from the edge of a hole to the tangent point of a radius should be at least twice the metal thickness (Figure 6B-2). The distance between holes or between a hole and the edge of a part should be sufficient to prevent tearing of the metal and excessive die wear.

FLANGES

Tapered flanges should not be allowed to taper to the face of the metal, but should be cut off so that the thickness of the narrowest section is at least twice the metal thickness or at least 0.12 in. measured from the inside of the metal (Figure 6B-3). Outside flanges and flanges around openings should preferably have a height of at least twice the metal thickness or a minimum of 0.12 in. on thin stampings (Figure 6B-4). A hemmed edge should be notched at the corners in order to eliminate gathering of metal in the flanging operation (Figure 6B-5). Whenever design permits, allowance should be made for distortion and variation in tab shape and hole size to permit piercing and notching of the blank (Figure 6B-6).

In order to facilitate the trimming operation, the minimum flange width should be twice the metal thickness (Figures 6B-4 and 6B-7). The permissible flange condition when a sharp edge is not objectionable is shown in Figure 6B-8. The absence of a flange requires expensive trimming if dimension X in Figure 6B-9 is maintained and the edge is held even, because the trimming must be done on a horizontal plane.

In designing a box-shape part, if the strength of the part is not involved, forming it is less expensive than drawing (Figure 6B-10). A corner cut at 45 degrees in the blank, as shown in Figure 6B-10(a), permits folding flanges and often requires no further

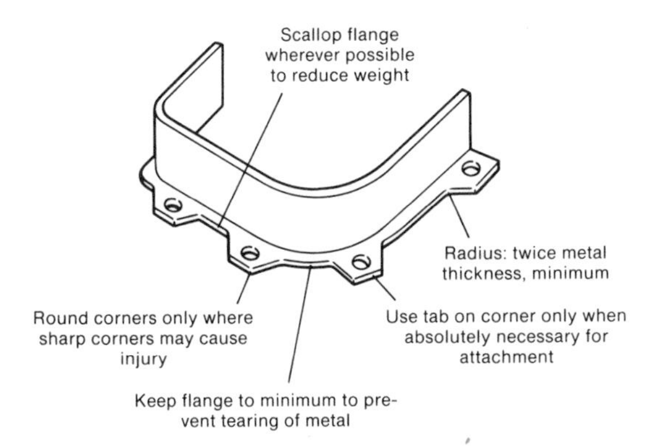

Figure 6B-6

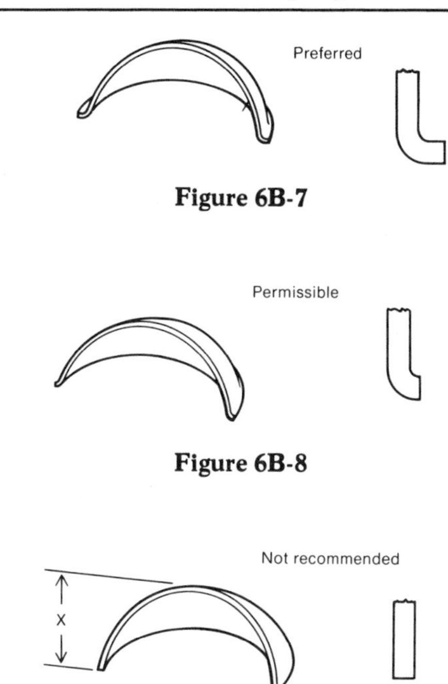

Figure 6B-7

Figure 6B-8

Figure 6B-9

trim. Notching corners in the blank, as shown in Figure 6B-10(b), has the same advantages as the 45-degree corners but often is more desirable for appearance. A continuous flange, as shown in Figure 6B-10(c), requires a draw operation.

When flanges that extend over only a portion of a highly stressed part are necessary, a notch or circular hole should be used in order to eliminate tearing of the metal (Figure 6B-11). A circular hole (type A relief) is used when the maximum possible flange height is necessary. A notch (type B relief) is used when the flange can be shorter than the maximum possible height.

Inside radii on stampings at bends should not be less than the thickness of the metal, if possible (Figure 6B-12). Larger radii facilitate production. A radius which is four times the metal thickness or greater at the bottom of a drawn stamping is preferable, because a small radius requires extra operations and extra dies. In using small radii on the punch of a die, the metal is liable to fracture between punch and die (Figure 6B-13), whereas if the punch has a large radius the pinching action can be avoided. When a flanged piece is bent in the shape of an offset, both the inside and outside radii should be as large as possible (Figure 6B-14).

Depth of Draw

The depth of draw should be kept as shallow as practicable in order to reduce the number of die operations and to minimize

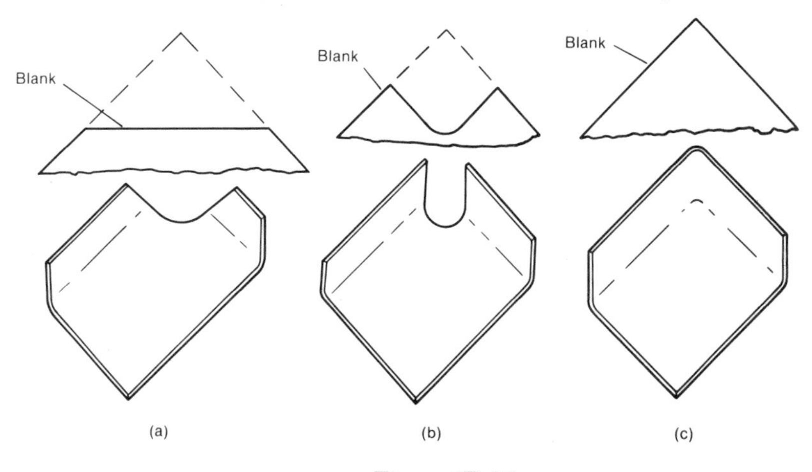

Figure 6B-10

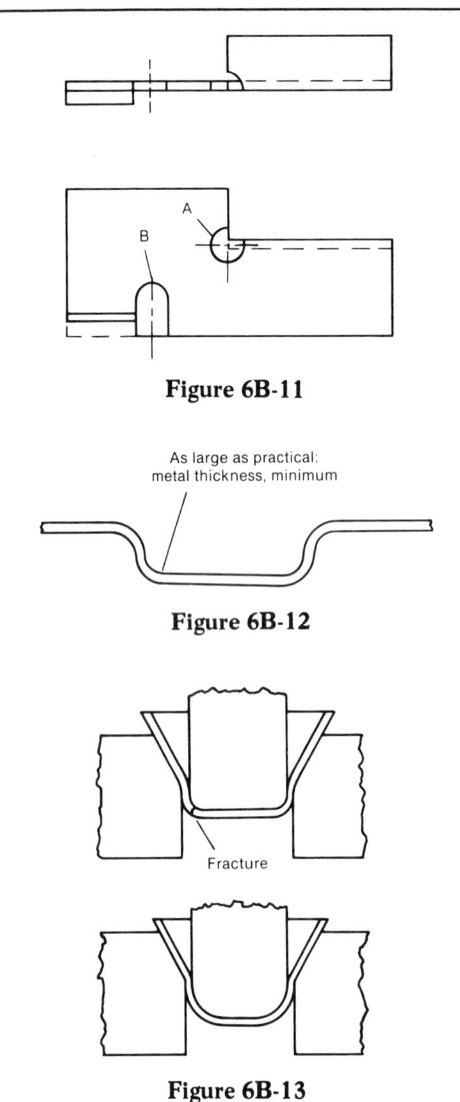

Figure 6B-11

Figure 6B-12

Figure 6B-13

cost. As the depth increases beyond that attainable with a single operation, the cost increases.

NESTING OF BLANKS

Development of blanks for stamped parts, particularly smaller ones, must be studied with consideration of the arrangement on the sheet to minimize scrap (Figure 6B-15). It is sometimes more economical to make a part in two pieces, to reduce scrap. Parts should also be studied for corners or flanges projecting at the top, bottom, or sides of the sheet that could be cut off without sacrificing strength or welding surface.

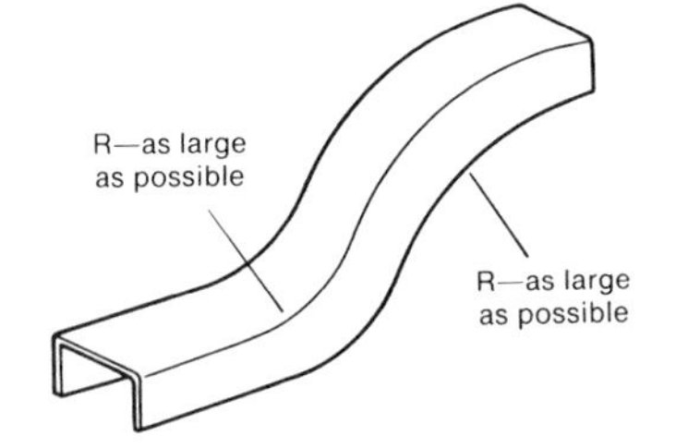

Figure 6B-14

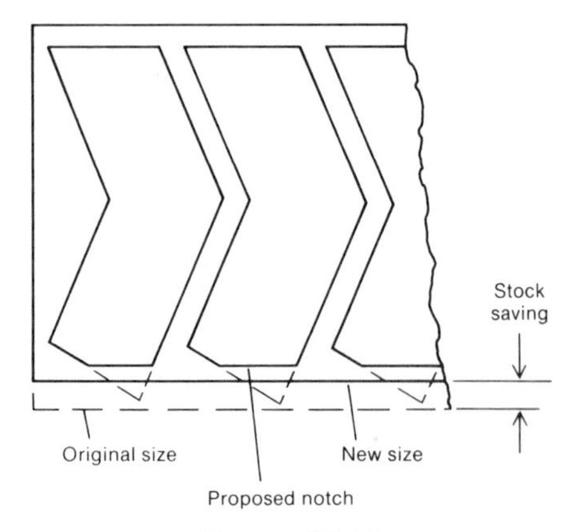

Figure 6B-15

Parts should be designed so that straight edges can be maintained on the flat blanks of formed parts wherever possible, in the interest of economy and ease of manufacture. This permits the blank to be sheared from flat stock at less cost because of the elimination of expensive blanking dies (Figure 6B-16).

NOTCHES

Notches fall into two groups: notches that are part of the design and that are provided for clearance, attachment, locating, etc.; and notches that are added to flanges to facilitate forming of the part, such as at a corner when the flange is to be formed by a flanging operation.

Notches in highly stressed parts should never be specified with a sharp "V" at the vertex because this sharp "V" might provide the starting point for a tear. The radius specified in this case should be as large as possible or a minimum of twice the metal thickness. This type of notch is usually added in the blank. Therefore, allowance should be made

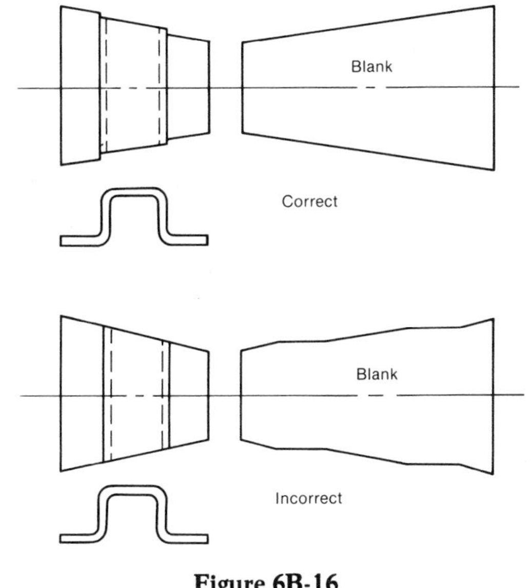

Figure 6B-16

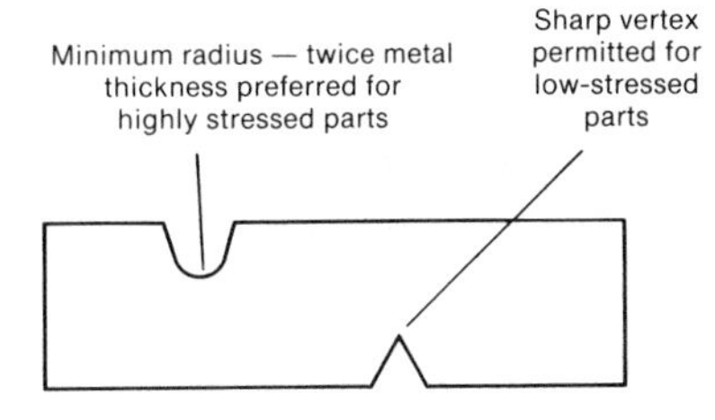

Figure 6B-17

in design for distortion that might be caused in the forming operation (Figure 6B-17).

A sharp vertex in a notch is allowed on less highly stressed parts when it will aid in lowering die costs. However, if the shape and design of the part permit, allowance for distortion should be made so that the notch can be added in the blank (Figure 6B-17).

SLOTS

Slots of regular shape designed to compensate for inaccuracies in manufacture or to provide for adjustment are dimensioned for size by over-all length and width dimensions, and for location by dimensions to their centerlines (Figure 6B-18). Slots in sheet metal made by standard punches are specified as shown.

A slot which is intended to perform a mechanical function and whose size and location are subject to gage inspection is treated

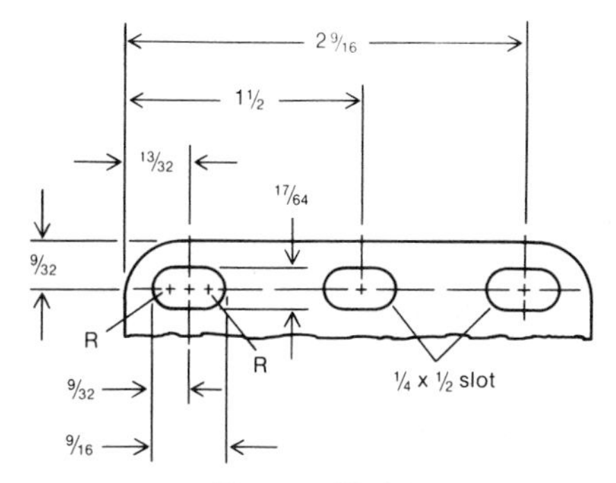

Figure 6B-18

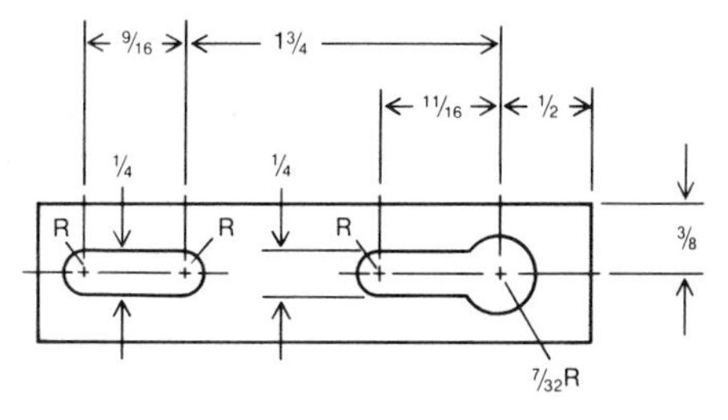

Figure 6B-19

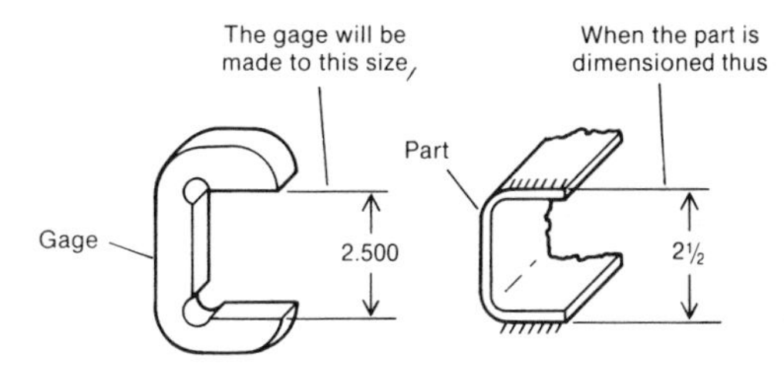

Figure 6B-20

as two partial holes separated by space and dimensioned as such (Figure 6B-19).

GAGE INSPECTION

A proven and effective way of ensuring interchangeability of parts from multiple sources and of avoiding rejection of usable material is to specify on the drawings exactly the extent to which the parts made therefrom are subject to inspection by gages (Figure 6B-20). Where this method of dimensioning is used, parts that have portions which match with mating parts should have those portions dimensioned to gage sizes, and indication should be made that they are to be checked by gages to ensure interchangeability. Nominal dimensions should be specified for these mating portions of parts along with a statement that they must conform to nominally sized gages within specified tolerances. Nominal dimensions are used because they supply gage makers with the proper gage sizes.

Matching form contours of mating parts are signaled thusly: / / / / / / / / / / /, and the extent of the signaled area is dimensioned. Signaling should be shown in as few views as possible, preferably in the view which best covers its full extent. It may be shown at any convenient angle to the surface and spaced about the same as crosshatching, extending 1/8 to 1/4 in. from the surface line depending on the size of the part.

In the channel section shown in Figure 6B-20, the marking of the signaled surfaces (those marked / / / / /) indicates that they match with mating parts and should be checked by a gage. The 2½-in. dimension which connects them is a nominal dimension (having no tolerance), because it denotes the size of the gage, which is made to exact size.

The tolerance shown in Figure 6B-21 indicates that the part is acceptable if it is produced in a size range which extends from *maximum size* (completely filling the gage opening) to *minimum size* (loose in the gage to the specified tolerance). A feeler of the tolerance thickness (shown in Figure 6B-21), may be used to determine the minimum size.

Form-Contour Tolerances

Where it is considered inadvisable to specify a tolerance on each dimension of a contour or shape due to possible objectionable accumulation of these tolerances, as

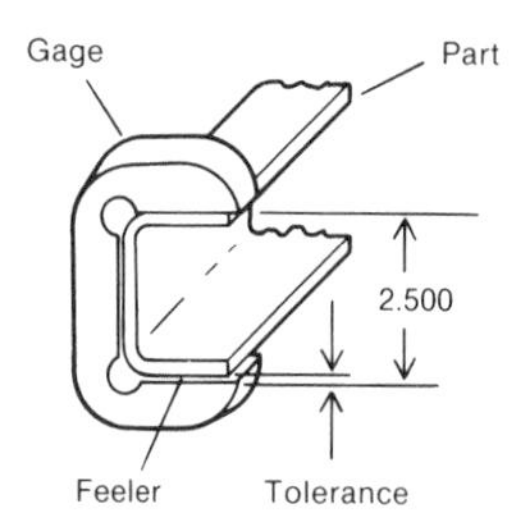

Figure 6B-21

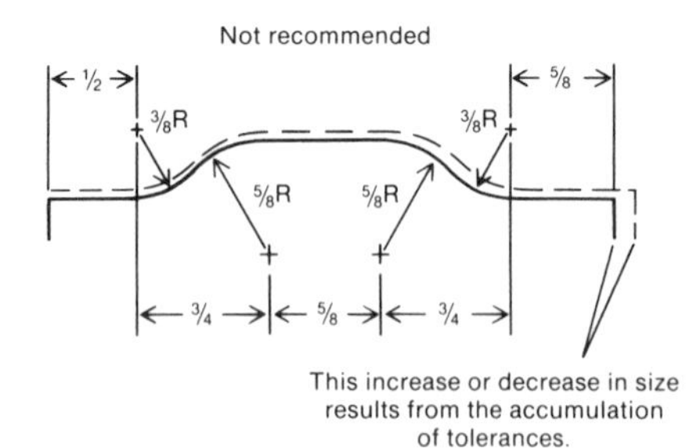

Figure 6B-22

shown in Figure 6B-22, a note should be used to express them, as illustrated in Figure 6B-23.

Drilled-Hole Tolerances

Tolerances for drilled holes should be as wide as possible. This will often permit a variety of drill sizes to be used and will cover oversize holes made by incorrectly ground drills. Express the hole size in limits or tolerances wherever possible and avoid specifying the drill number or letter on the drawing.

Drilled Holes Without Specified Tolerances

In the case of holes not requiring close tolerances (bolt clearance holes, lightening holes, etc.), the general note "unless otherwise specified" above the title block will govern.

Punched-Hole Tolerances

Tolerances for punched holes should be as wide as possible, except those for solid rivets which should be +0.015, −0.000 in., and those for draw rivets, which should be ±0.005 in.

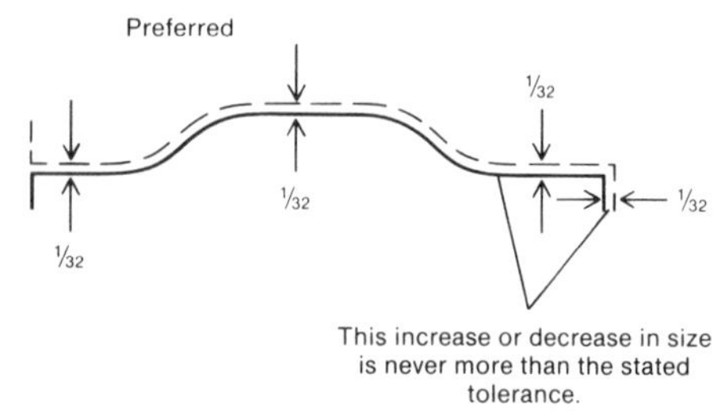

Figure 6B-23

Reamed-Hole Tolerances

The maximum limit on reamed holes is the nominal size plus 0.0005 in., wherever possible. This is the size of hole which a new reamer will produce. The minimum size should be determined by considering the fit desired and proportioning the tolerances between the hole and the mating part so that maximum economy will result. This frequently means that the hole will carry a larger tolerance than the mating part, because it is more economical to hold an external diameter to closer tolerances than an internal one. The 0.001-in. tolerance on the shaft usually does not increase the cost of the grinding operation and permits the use of a new reamer until it has worn to 0.003 in. undersize.

PRESS CAPABILITIES

Press Size and Type. The stroke, the slide and bed areas, and the shut height or die space should be sufficiently large to easily accommodate any dies to be used.

Crankshaft. The crankshaft may be called the heart of the press. It is usually this member on which the strengths of the frame, connection screws, tie rods, and other vital parts are based. The design of the crankshaft, including the bearing arrangement, has much to do with the tonnage that can be exerted safely at the bottom of the stroke. This should be considered carefully when designing other parts of the press.

Stroke Necessary. For blanking or shallow forming work, standard press strokes usually are sufficient. For drawing work, the stroke generally should be slightly more than twice the depth of the formed article to ensure easy removal of the work from the press. The sizes of both the article being made and the dies determine the stroke required.

Die Space. The distance from the bed to the slide with the stroke down and adjustment up, on some presses, is sufficient for the ordinary run of work. The die space can be increased when so ordered by shortening the connection a limited amount, or by lengthening the pattern of the frame. One of these methods must be employed to maintain the standard die space with an increased stroke. When the pressure occurs at quite a distance above the bottom of the stroke or when the length of the stroke required is comparatively long, as in drawing presses, the torsional load and gearing must be considered.

On some single-crank presses with very long strokes, on double-crank presses of great width or with long strokes, and on all large single- and double-crank presses, "twin gearing" (use of a gear on each end of the crankshaft) is employed. This arrangement increases the gearing strength and the torsional capacity of the crankshaft and, in wide double-crank presses, reduces the torsional deflection between the cranks. Under these conditions, the load at the bottom of the stroke will still be limited by these additional considerations.

Additional Pressure Required. Additional pressure to offset the pressure of die cushions and other attachments, such as those used for drawing work, should be included in specifying the tonnage required of the press.

Method of Feeding. The direction of feeding and the size of the sheet, blank, or article also may determine the size and type of press needed.

Index

E

Q

R